T0348747

Advances in
DEVELOPMENTAL
BIOLOGY

VOLUME **18**

Cardiovascular
Development

Advances in
DEVELOPMENTAL BIOLOGY

VOLUME **18**

Cardiovascular Development

Edited by

ROLF BODMER
Burnham Institute for Medical Research, La Jolla

AMSTERDAM • BOSTON • HEIDELBERG • LONDON
NEW YORK • OXFORD • PARIS • SAN DIEGO
SAN FRANCISCO • SINGAPORE • SYDNEY • TOKYO
Academic Press is an imprint of Elsevier

Elsevier
Radarweg 29, PO Box 211, 1000 AE Amsterdam, The Netherlands
Linacre House, Jordan Hill, Oxford OX2 8DP, UK

First edition 2008

Library of Congress Cataloging-in-Publication Data
A catalog record for this book is available from the Library of Congress

British Library Cataloguing in Publication Data
A catalogue record for this book is available from the British Library

ISBN: 978-0-444-53014-1
ISSN: 1574-3349

For information on all Elsevier publications
visit our website at books.elsevier.com

Pinted in the United States of America
Transferred to Digital Printing, 2015

Utpal Banerjee
Department of Molecular, Cell and Developmental Biology, Department of Biological Chemistry, Molecular Biology Institute, University of California, Los Angeles, California.

D. Woodrow Benson
Department of Pediatrics, Division of Cardiology, MLC 7042, University of Cincinnati, Cincinnati Children's Hospital Medical Center, Cincinnati, Ohio.

Brian L. Black
Cardiovascular Research Institute and Department of Biochemistry and Biophysics, University of California San Francisco, San Francisco, California.

Rolf Bodmer
Center for Neurosciences and Aging, The Burnham Institute for Medical Research, La Jolla, California.

Benoit G. Bruneau
Department of Pediatrics and Biomedical Sciences Program, Gladstone Institute of Cardiovascular Disease, University of California, San Francisco, California.

Vincent M. Christoffels
Heart Failure Research Center, Academic Medical Center, University of Amsterdam, Amsterdam, The Netherlands.

Sylvia M. Evans
Skaggs School of Pharmacy and Pharmaceutical Sciences, University of California, San Diego, La Jolla, California.

Cory J. Evans
Department of Molecular, Cell and Developmental Biology, University of California, Los Angeles, California.

Ann C. Foley
Burnham Institute for Medical Research, La Jolla, California.

Rosa M. Guzzo
Burnham Institute for Medical Research, La Jolla, California.

Volker Hartenstein,
Department of Molecular, Cell and Developmental Biology, Molecular Biology Institute, University of California, Los Angeles, California.

Willem M.H. Hoogaars
Heart Failure Research Center, Academic Medical Center, University of Amsterdam, Amsterdam, The Netherlands.

Yessenia M. Ibarra
Burnham Institute for Medical Research, La Jolla, California.

Xingqun Liang
Skaggs School of Pharmacy and Pharmaceutical Sciences, University of California, San Diego, La Jolla, California.

Jiandong Liu*
Center for Neuroscienes and Aging, The Burnham Institute for Medical Research, La Jolla, California.
Graduate Program in Molecular, Cellular and Developmental Biology, University of Michigan, Ann Arbor, Michigan.

Lolitika Mandal
Department of Molecular, Cell and Developmental Biology, University of California, Los Angeles, California.

Jörg Männer
Department of Anatomy and Embryology, Georg-August-University of Göttingen, Göttingen 37075, Germany.

Julian A. Martinez-Agosto
Department of Pediatrics, Mattel Children's Hospital at UCLA, University of California, Los Angeles, California.

Mark Mercola†
Burnham Institute for Medical Research, La Jolla, California.

Antoon F.M. Moorman
Heart Failure Research Center, Academic Medical Center, University of Amsterdam, Amsterdam, The Netherlands.

Gregory E. Morley
The Leon H. Charney Division of Cardiology, New York University School of Medicine, New York, New York.

---

* Present Address: Department of Biochemistry and Biophysics, University of California, San Francisco, San Francisco, California
† Present Address: Weill Cornell Medical College, New York, New York

Judith M. Neugebauer
Department of Neurobiology and Anatomy, University of Utah, Salt Lake City, Utah.

Li Qian[‡]
Center for Neurosciences and Aging, The Burnham Institute for Medical Research, La Jolla, California.
Graduate Program in Molecular, Cellular and Developmental Biology, University of Michigan, Ann Arbor, Michigan.

Vijaya Ramachandran
Department of Pediatrics, Division of Cardiology, MLC 7042, University of Cincinnati, Cincinnati Children's Hospital Medical Center, Cincinnati, Ohio.

Anabel Rojas
Cardiovascular Research Institute and Department of Biochemistry and Biophysics, University of California San Francisco, San Francisco, California.

Pilar Ruiz-Lozano
Burnham Institute for Medical Research, La Jolla, California.

Eva Samal
Gladstone Institute of Cardiovascular Disease, Departments of Pediatrics and Biochemistry and Biophysics, University of California, San Francisco, California.

Elaine L. Shelton
Division of Molecular Cardiovascular Biology, Cincinnati Children's Hospital Medical Center, Cincinnati, Ohio.

Sergey A. Sinenko
Department of Molecular, Cell and Developmental Biology, University of California, Los Angeles, California.

Deepak Srivastava
Gladstone Institute of Cardiovascular Disease, Departments of Pediatrics and Biochemistry and Biophysics, University of California, San Francisco, California.

Dina Myers Stroud
The Leon H. Charney Division of Cardiology, New York University School of Medicine, New York, New York.

Yunfu Sun
Skaggs School of Pharmacy and Pharmaceutical Sciences, University of California, San Diego, La Jolla, California.

‡ Present Address: Gladstone Institute of Cardiovascular Disease, San Francisco, California

Jeffrey A. Towbin
Pediatric Cardiology, Texas Children's Hospital, Baylor College of Medicine, Houston, Texas.

Josette Ungos
Laboratory of Molecular Genetics, NICHD, NIH, Bethesda, Maryland.

Brant M. Weinstein
Laboratory of Molecular Genetics, NICHD, NIH, Bethesda, Maryland.

H. Joseph Yost
Department of Neurobiology and Anatomy, University of Utah, Salt Lake City, Utah.

Katherine E. Yutzey
Division of Molecular Cardiovascular Biology, Cincinnati Children's Hospital Medical Center, Cincinnati, Ohio.

Shan-Shan Zhang
Department of Pediatrics and Biomedical Sciences Program, Gladstone Institute of Cardiovascular Disease, University of California, San Francisco, California.

CONTENTS

In this volume of *Advances in Developmental Biology* we cover a wide spectrum of current excitements and in-depth reviews of topics in cardio-vascular research, ranging from hematopoiesis in *Drosophila* to congenital heart disease in humans. A particular effort has been made to explore fundamental mechanisms of how the heart becomes an exquisitely well-functioning organ in a variety of organisms. The insights on conserved mechanisms in forming a heart have made comparative approaches of studying heart development increasingly informative. It is thus expected that many aspects of human heart disease and development will be addressed in a variety of model systems, including mice, zebrafish, and flies, with increasing popularity.

The chapters of this volume reflect the rapid advances and new mechanisms involved in the heart development field in recent years. For the simple heart of *Drosophila*, the model system in which cardiac specification was first pioneered at the molecular level, a reasonably com-prehensive understanding of cardiac induction has now been achieved (Chapter 1 by Qian et al. in this book), and understanding hematopoiesis and vasculogenesis are not far behind revealing interesting parallels to vertebrates (Chapter 11 by Evans et al. and Chapter 12 by Ungos and Weinstein in this book). In the more complex vertebrate heart, the multi-tude of relevant transcription factor interactions is beginning to be unra-veled (Chapter 3 by Shelton and Yutzey, Chapter 4 by Rojas and Black, Chapter 6 by Sun et al., Chapter 10 by Zhang and Bruneau, and Chapter 14 by Ramachandran and Benson in this book), as well as the signaling path-ways (Chapter 5 by Guzzo et al. and Chapter 8 by Neugebauer and Yost in this book). Exciting new regulatory mechanisms in cardiac development have also been discovered, such as that of microRNAs (Chapter 7 by Sama and Srivastava in this book) and inductive influences from the epicardium (Chapter 13 by Manner and Ruiz-Lozano in this book). Major new advances have been achieved in elucidating the development of the con-duction system (Chapter 2 by Hoogaars et al., Chapter 9 by Stroud and Morley, and Chapter 10 by Zhang and Bruneau in this book). Insights from these studies have been instrumental in significantly furthering our under-standing human congenital heart disease (Chapter 14 by Ramachandran and Benson and Chapter 15 by Towbin in this book).

Despite these advances, we still do not quite understand how the heart is actually but together, although its embryology (Chapter 2 by

Hoogaars et al.) and contributing progenitors and their potential have been studied in some detail (Chapter 5 by Guzzo et al. and Chapter 6 by Evans in this book). Thus, an important outstanding problems in cardiac development is the control of morphogenesis, which requires regional specification of cell identities, cell interaction, and migration, as well as the three-dimensional orchestration of cardiac morphology. This process, in addition, is likely modulated by the contractile properties and actual beating of the developing heart. Thus, heart development studies will continue to be exciting.

ROLF BODMER
Burnham Institute for Medical Research

CHAPTER 1

Heart Development in *Drosophila*

Li Qian,[*,†,1] **Jiandong Liu**[*,†,2] and **Rolf Bodmer**[*]

Contents

[*] Center for Neuroscienes and Aging, The Burnham Institute for Medical Research, La Jolla, California
[†] Graduate Program in Molecular, Cellular and Developmental Biology, University of Michigan, Ann Arbor, Michigan
[1] Present Address: Gladstone Institute of Cardiovascular Disease, San Francisco, California
[2] Present Address: Department of Biochemistry and Biophysics, University of California, San Francisco, San Francisco, California

Advances in Developmental Biology, Volume 18
ISSN 1574-3349, DOI: 10.1016/S1574-3349(07)18001-7

Abstract   The *Drosophila* heart is a simple linear heart tube that is reminiscent
                    of the primitive tubular heart found in early vertebrate embryos. The
                    evolutionary conservation of the *Drosophila* heart and powerful
                    genetic tools for analyzing gene functions and interactions make
                    Drosophila a unique system to study the molecular mechanisms of
                    development governing cardiac specification and differentiation. The
                    use of *Drosophila* as a model system has recently been extended to
                    study genes involved in cardiac aging and function. In this chapter, we
                    summarize the current knowledge and recent findings in *Drosophila*
                    cardiac specification and differentiation, as well as touch on new
                    research on heart physiology and aspects of aging. The potential
                    of using *Drosophila* as a model system to identify genetic loci
                    contributing to polygenic cardiac disorders is to be discussed.

1. INTRODUCTION

The *Drosophila* heart, also called the dorsal vessel, is a linear tube that is reminiscent of the primitive vertebrate embryonic heart. Although the final heart structure in *Drosophila* is very different from that in vertebrates, many of the basic elements for cardiac specification and differentiation are conserved. Because of the simplicity in structure and availability of powerful genetic tools, the *Drosophila* heart has emerged as a pioneering model system for unraveling the basic genetic and molecular mechanisms of cardiac development, function, and aging. Here we summarize recent discoveries of signaling pathways and genetic networks involved in sub-dividing cardiac mesoderm and orchestrating cardiac specification and morphogenesis. In addition, we briefly introduce the newly emerging field of cardiac function and cardiac aging in *Drosophila*. The insights obtained will be of significant help in defining related events during vertebrate cardiogenesis and eventually elucidating the etiology of human cardiac disease.

2. MORPHOLOGY OF THE *DROSOPHILA* HEART

2.1. Embryology of heart formation

In the past decade, a cascade of genes involved in mesoderm determination and positioning in the blastoderm of the *Drosophila* embryo have been identified (Boulay et al., 1987; Thisse et al., 1988, 1991; Kosman et al., 1991;

Leptin et al., 1992; St Johnston and Nusslein-Volhard, 1992), among them the basic helix-loop-helix transcription factor *twist* is one of the key components (Leptin, 1991). *twist* induces pan mesodermal expression of several transcription factors, such as the MADS-box gene *dMef-2* (Nguyen et al., 1994; Bour et al., 1995; Lilly et al., 1995), the homeobox gene *tinman* (Azpiazu and Frasch, 1993; Bodmer, 1993), and zinc finger gene *zfh-1* (Su et al., 1999), all of which are important for patterning and differentiation of the mesoderm. With the onset of gastrulation, the *Drosophila* mesoderm spreads beneath the ectoderm along the ventral midline and migrates dorsally as a monolayer of cells toward the dorsal edge of the germ band (Leptin and Grunewald, 1990). As a result, different portions of the mesoderm are exposed to different contexts and signaling systems according to their positions relative to the overlying ectoderm. Thus, the primordia of the mesodermal tissues become determined, as the mesoderm is gradually subdivided along both the dorsoventral and anteroposterior (A-P) axis. The cardiac progenitors are selected at the most dorsal edge of the mesoderm on either side of the embryo. Subsequently, the presumptive heart precursors rearrange and form a dorsal row of myocardial cells, which is flanked by an adjacent row of pericardial cells.

During germ band retraction, the myocardial cells undergo a typical mesenchyme–epithelium transition (EMT), resulting in an apical-basal polarization of the myocardial cells (Zaffran et al., 1995; Fremion, 1999; Haag et al., 1999; Chartier et al., 2002; Qian et al., 2005a,b). The bilateral rows of cardiac cells migrate toward the dorsal midline where they establish close contacts. As a result, the two rows of myocardial cells fuse at the dorsal midline to form a central cavity, the lumen of the heart, pumping hemolymph throughout the body in an open circulatory system. Along the A-P axis, the narrower anterior part of the heart is called aorta while the wider posterior portion is called heart proper (Gajewski et al., 2000; Molina and Cripps, 2001; Ponzielli et al., 2002; Lo and Frasch, 2003). During metamorphosis at the pupal stage, the heart undergoes substantial remodeling to give rise to the adult heart that contains functional ostium (openings to the hemolymph) and segmental valves (Rizki, 1978). The fly heart contracts rhythmically throughout the life of the fly (Wessells and Bodmer, 2004; Wessells et al., 2004; Bodmer et al., 2005; Monier et al., 2005; Ocorr et al., 2007b).

2.2. Comparison between the fly and the vertebrate heart

Although there are obvious morphological differences in heart structure between *Drosophila* and vertebrates, significant similarities in their embryonic origin and initial tubular structure have been shown (Bodmer, 1995; Bodmer and Venkatesh, 1998; Bodmer and Frasch, 1999; Bodmer et al., 2005). The *Drosophila* heart primordium forms bilaterally at the dorsalmost edge of the mesoderm, which migrates further dorsally to meet their contralateral counterparts at the dorsal midline. As they migrate, the prospective myocardial cells undergo cell shape changes during the process of EMT and finally fuse together to form the lumen (Zaffran et al., 1995; Fremion, 1999; Haag et al., 1999; Chartier et al., 2002).

Similarly, in vertebrates once the uncommitted splanchnic mesoderm is specified to a cardiogenic fate, the heart precursors also reorganize by an EMT and migrate along the anterior intestinal portal to converge at the midline of the embryo, where they form the cardiac crescent that folds ventrally, resulting in the fusion of the cardiac primordium and formation of the linear heart tube (Fishman and Chien, 1997; Harvey, 2002; Brand, 2003; Olson, 2004). In particular, cardiac specification in both systems involves inductive signals across germ layers, although in vertebrates the cardiac-inducing activity is mainly of endodermal origin while in fly the cardiac inductive signals emanate from the dorsal ectoderm (Frasch, 1995; Wu et al., 1995; Cripps and Olson, 2002; Lockwood and Bodmer, 2002).

The similarity also extends to the conservation of key factors that determine the initial formation of the heart as well as the genetic and molecular mechanisms that orchestrate cardiac specification and morphogenesis, discussed below. Recent studies on cardiac function and aging suggest that late events regulating cardiac performance, which have previously been thought to be divergent from that of the vertebrate, may also share common pathways and regulatory networks.

3. GENETIC CONTROL OF CARDIAC INDUCTION

3.1. Inductive signals for cardiac mesoderm formation: *decapentaplegic* and *wingless*

During gastrulation in *Drosophila*, the presumptive mesoderm invaginates and forms a monolayer of cells in close contact with the ectoderm. As the migrating cells reach the dorsal edge of the ectoderm, the ventral portion of the mesoderm can be distinguished from the dorsal portion, which harbors the primordia for the cardiac and visceral mesoderm. Unlike the endodermal origin of cardiac-inducing activity in vertebrates (Harvey, 2002), secreted signals emanating from the dorsal ectoderm of the fly are crucial for cardiac mesoderm formation, including the signaling molecules encoded by *decapentaplegic* (*dpp*) and *wingless* (*wg*).

dpp encodes a BMP-like TGF-*b* family member and is required for the subdivision of the mesoderm along its dorsoventral axis (Frasch, 1995). In this process, *dpp* signaling acts to maintain expression of one of its key targets, the homeobox transcription factor encoded by *tinman*, within the dorsal mesoderm (Frasch, 1995). Both Dpp and Tinman are essential for establishing dorsal mesoderm fates (Frasch, 1995; Azpiazu et al., 1996). During gastrulation and mesoderm migration, *dpp* is expressed in a broadband of cells within the dorsal ectoderm and signals across germ layers to pattern the underlying dorsal mesoderm. Mutant embryos that lack the activity of Dpp do not form cardiac or visceral mesoderm progenitors. In these mutants, *tinman* expression is eliminated in the dorsal mesoderm. *tinman* is likely to be a direct target of *dpp* signaling since molecular analyses have identified several specific *dpp*-responsive Smad-binding sites within the *tinman* enhancer (Xu et al., 1998; Yin and Frasch, 1998; Raftery and Sutherland, 1999).

Like Dpp, the secreted Wg protein, a homologue of the vertebrate Wnts, is also released to the mesoderm from the overlying ectoderm, but in a pattern of segmentally repeated transverse stripes. Loss-of-*wg*-function results in the absence of heart formation, indicating that the Wg signal is also required, in addition to Dpp, for cardiac specification (Wu et al., 1995; Park et al., 1996). In contrast to vertebrates, where noncanonical Wnt signaling promotes cardiogenesis (Pandur et al., 2002), in the fly the canonical Wg pathway specifies heart (Wu et al., 1995; Azpiazu et al., 1996; Park et al., 1996; Riechmann et al., 1997). Whether noncanonical Wg is also involved in *Drosophila* cardiogenesis is still under investigation. Many of the factors that transduce the canonical Wg signal, including Pangolin (Pan/dTCF/LEF-1) and its mesodermal target encoded by *sloppy-paired*, also function in cardiac induction (Park et al., 1996, 1998; Lee and Frasch, 2000).

Since *dpp* and *wg* are both required for cardiac mesoderm induction, it has been proposed that the cardiac mesoderm is positioned in the areas of the mesoderm that receive both signals (Lockwood and Bodmer, 2002). During stages for cardiac mesoderm induction, *dpp* and *wg* are oriented in an orthogonal pattern of expression to one another. Genetic studies demonstrated that cardiac progenitors are induced precisely within the mesodermal region where the *wg* domains intersect with the dorsal *dpp* domains. Moreover, misexpression of both *wg* and *dpp* is able to induce ectopic expression of the cardiac markers like *tinman* and *even-skipped* (*eve*), suggesting that the convergence of *wg* and *dpp* signaling is not only necessary but also sufficient for inducing cardiac competence (Lockwood and Bodmer, 2002).

3.2. Transcriptional regulation of cardiac induction: *tinman*, *pannier*, and *dorsocross*

In addition to inductive signals from the ectoderm, activation of a series of transcription factors within the mesoderm is another central event for cardiac primordium specification. *tinman* is the first regulatory gene in any species known to be expressed in the precardiac mesoderm and to function in specifying cardiac precursors (Azpiazu and Frasch, 1993; Bodmer, 1993; Bodmer and Frasch, 1999). During gastrulation, *tinman* is initially expressed throughout the entire mesoderm. Subsequently, *tinman* expression becomes restricted to the dorsal mesoderm. Abrogation of *tinman* function results in a complete loss of dorsal mesodermal derivatives, including cardiac, visceral muscle cells, and a subset of somatic muscle cells. This suggests that *tinman* is required for all three of these tissues (Azpiazu and Frasch, 1993; Bodmer, 1993). Finally, *tinman* expression becomes further restricted to the myocardial and pericardial progenitor cells of the heart, and it is also thought to function in differentiation, function, and aging of the heart (see below).

Genetic experiments suggest that although *tinman* is required, it is not sufficient to promote cardiogenesis. One explanation is that Tinman may only function in conjunction with other factors in providing heart-forming competence in mesoderm. The fly GATA protein encoded by *pannier (pnr)* is one of these postulated transcription factors and is also required for cardiac progenitors specification (Gajewski et al., 1999; Klinedinst and Bodmer, 2003). *pannier* is coexpressed with *tinman* in a narrow dorsal domain during cardiac mesoderm induction. Overexpression of both together gives rise to robust, ectopic expression of heart markers, suggesting these two genes act synergistically in cardiac specification (Klinedinst and Bodmer, 2003), which is consistent with *in vitro* studies of cardiac gene expression in vertebrate cell culture (Durocher et al., 1997; Lee et al., 1998; Sepulveda et al., 1998). In addition to being a target of *tinman*, *pannier* is also a target of *wg* and *dpp*. In turn, *pannier* is required for maintaining the late phase of *wg* (S. Klinedinst and R.B., unpublished data), *dpp* expression in the dorsal ectoderm, and *tinman* expression in the dorsal mesoderm. It has been proposed that at early stages in heart development, *tinman*, *wg*, and *dpp* initiate *pannier* expression in the heart-forming region, and *pannier*, *wg*, and *dpp* maintain *tinman* expression in the cardiac mesoderm. Later, *tinman* and *pannier* maintain each other in the heart progenitors, and ectodermal *pannier* maintains *wg* and *dpp* in the overlying dorsal ectoderm (Klinedinst and Bodmer, 2003).

A set of T-box genes, *dorsocross1(doc1)*, *dorsocross2(doc2)*, and *dorsocross 3 (doc3)*, has been identified as key mediators of Dpp and Wg signaling during cardiac mesoderm induction (Reim et al., 2003; Hamaguchi et al., 2004; Reim *et al.*, 2005). These three genes are arranged as a cluster within the genome, are expressed in identical patterns, and function redundantly with each other. *doc* genes initiate their expression in the presumptive amnioserosa at blastoderm stage, dependent on *dpp* expression, and then in the dorsal ectoderm and dorsal mesoderm. Subsequently, the *doc* genes are expressed in segmental patterns in the mesoderm at intersects of Wg and Dpp signals (Reim et al., 2003; Hamaguchi et al., 2004). *doc* genes act together with *tinman* to activate *pannier* expression, then they interact with *tinman* and *pannier* to promote cardiogenesis (Reim and Frasch, 2005). As another early determinant for cardiac mesoderm formation, *doc* is required for specification of all myocardial and pericardial cells except the *even-skipped*-expressing pericardial cells (EPCs). In *doc* mesodermal loss-of-function mutants, heart formation is severely reduced (Reim and Frasch, 2005). Taken together, we conclude that the homeobox gene *tinman*, GATA factor *pannier*, and T-box factors *dorsocross* are major players in early cardiac mesoderm specification. The interaction and cross-regulation between them are of significant importance for cardiogenesis, which have been suggested to function in a homologous fashion in vertebrates (Garg et al., 2003).

4. CARDIAC CELL SPECIFICATION AND DIFFERENTIATION

4.1. A-P positioning of cardiac progenitor cells

4.1.1. Hedgehog and Ras signaling

Similar to the heart in vertebrates at the linear tube stage, the *Drosophila* "dorsal vessel" is also built with an A-P polarity. The fly's heart tube, in addition, is constructed in a metamerically (segmentally) repeated pattern of cellular identities along the A-P axis within each segment (Fig. 1). This stereotyped arrangement of cell types first manifests itself at stages for cardiac progenitor specification (late stage 10 to early stage 11). Distinct groups of cardiac progenitors can be distinguished within the *tinman*-positive cardiogenic region, which are positioned at discrete locations along the A-P axis in each segment (Fig. 2). Among them are those expressing the homeobox genes *ladybird early* (*lbe*), *eve*, and the COUP transcription factor *seven-up* (*svp*), each marking a subpopulation of cardiac progenitors (Frasch et al., 1987; Jagla et al., 1997; Su et al., 1999; Ward and Skeath, 2000; Lo and Frasch, 2001; Han et al., 2002; Liu et al., 2006). Studies have shown that *hedgehog* (*hh*) has a direct role in generating the repeated pattern of these identity genes' expression independently of its requirement for maintaining *wg* expression in the mesoderm (Azpiazu et al., 1996; Park et al., 1996; Ponzielli et al., 2002; Liu et al., 2006).

hh is required for the formation of two clusters of cardiac progenitors that are precisely positioned relative to the segmentally repeated ectodermal *hh* stripes. The *eve*-expressing clusters are immediately anterior and the *svp*-expressing cells are immediately posterior to the *hh* stripes, whereas the Lbe clusters are further removed from the *hh* stripes (Fig. 2; Ponzielli et al., 2002; Liu et al., 2006). In the absence of *hh* activity, neither Eve- nor Svp-positive cells are formed, but *lbe* expression is expanded to encompass most if not all of the entire cardiogenic region. The formation of *eve* cells also relies on local activation of Ras signaling in the dorsal mesoderm (Carmena et al., 1998; Halfon et al., 2000). *hh* signaling apparently determines the level of Ras pathway activation by transcriptionally regulating *rhomboid* (*rho*) expression (Liu et al., 2006), which codes for a protease that is required for processing the EGF receptor ligand Spitz (Spi). Consistent with this idea is the observation that mesodermal

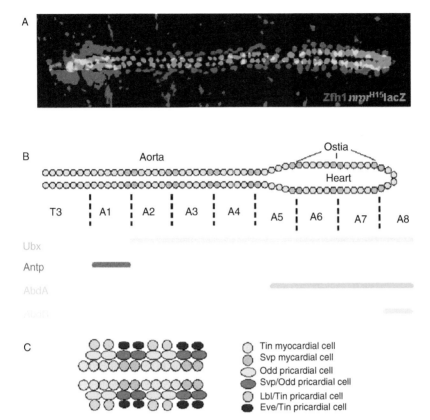

FIGURE 1   Schematic representation of the *Drosophila* heart. (A) Confocal picture showing a stage 16 *Drosophila* heart stained for T-box transcription factor Neuromancer (*nmr*[H15]lacZ enhancer trap labels myocardial cell nuclei) and homeobox protein Zfh-1 (antibody labels pericardial cell nuclei). (B) Myocardium of the heart tube is composed of *tinman*-expressing (Tin, light blue) and *seven-up*-expressing cells (Svp, pink). Ostia are also marked by Wingless. Segmental register of the homeotic selector genes Ultra-bithorax (Ubx), Antennapedia (Antp), Abdominal-A (AbdA), and Abdominal-B (AbdB) is indicated [compare specific segment allocations by Ponzielli et al. (2002) and Perrin et al. (2004)]. (C) Schematic diagram of two segments along the mature embryonic heart shows myocardial and pericardial cell types as defined by combinations of transcription factors expression [adapted from Han and Bodmer (2003)].

overexpression of *ras* can partially restore *eve* expression in *hh* mutant (*hh*-off, *wg*-on). In addition, *hh* seems to also function independently of Ras in repressing *lbe* expression, via inhibition of the Ci repressor. Thus, the metameric pattern and relative positioning along the A-P axis of *eve*- versus *lbe*-expressing cells in the cardiac mesoderm are generated by ectodermal Hh signaling, which seems to function both upstream of and in parallel to Ras signaling (Liu et al., 2006).

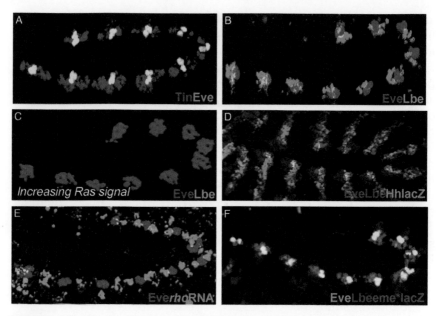

FIGURE 2 Cardiac specification and differentiation. (A) Eve labels a subset of Tinman-positive cardiac progenitors in late stage 11 embryos. (B) In each segment, Eve-expressing progenitors are located just posterior adjacent to the Lbe-expressing cardiac progenitors and immediately anterior to the ectodermal *hh* stripes (D). (C) Ras signaling promotes Eve cells formation at the expense of Lbe cells. (E) *rho*-expressing cells alternate with Eve cells at the dorsal mesodermal edge. (F) Triple labeling showing colocalization of mutated EME reporter with Eve and Lbe at stage 12.

In each hemisegment, the *lbe*-expressing progenitors develop into two Lbe-positive myocardial cells and two Lbe-positive pericardial cells. The *eve*-expressing progenitors differentiate into two Eve pericardial cells (EPCs) and two dorsal muscles (DA1 and DO2). The *svp*-expressing progenitors give rise to two Svp myocardial cells and two Svp pericardial cells [Fig. 1C; for detailed lineages, see Ward and Skeath (2000), Alvarez et al. (2003), and Han and Bodmer (2003)]. As the *Drosophila* heart develops, the myocardial cells are placed dorsally to the pericardial cells to form a continuous row of cells at the dorsal margin of the mesoderm, while maintaining the stereotyped cardiac cell arrangement within each segment (Fig. 1C).

4.1.2. Hox genes

In addition to the segmental A-P polarity of the heart tube, a broad subdivision of the dorsal vessel into three major regions is also observed (Fig. 1A and B): (1) the anterior aorta or "outflow track," a possible archetype of the vertebrate outflow system, forms at the anterior end of the heart tube (T3–A1) and requires a group of *lbe*-expressing ectodermal cells in close contact with the outflow track (Zikova et al., 2003); (2) the (posterior) "aorta" forms in segments A2–A4 and is narrower than the more posterior portion; and (3) the "heart" proper forms posteriorly (A5–A8) and contains specialized myocardial cells forming hemolymph inlet–outlet valves called "ostia" (Rizki, 1978). The later two regions are separated by an internal valve (Molina and Cripps, 2001; Ponzielli et al., 2002; Lo and Frasch, 2003; Zikova et al., 2003). The exact segmental origin of the heart segments had not been entirely clear, and a reevaluation suggests that the anterior aorta (anterior to *svp* or *Ultrabithorax* (*Ubx*) expression in Fig. 1B) derives from T1–T3 segments, which shifts the register by one segment (A1–A7; Perrin et al., 2004).

Hox genes are the principal determinants that control this A-P subdivision of the heart tube (Lo et al., 2002; Lovato et al., 2002; Ponzielli et al., 2002; Lo and Frasch, 2003; Perrin et al., 2004). In the embryonic heart, specific *hox* genes are expressed in distinct regional domains along the A-P axis of the heart (Fig. 1B). For example, *antennapedia* (*antp*) expression is confined to the boundary of the posterior aorta and the prospective outflow track; *ubx* is primarily expressed in the posterior aorta; *abdominal A* (*abdA*) is restricted to the (posterior) heart; and *abdominal B* (*abdB*) is only present in the four most posterior myocardial cells (Lo and Frasch, 2003). Importantly, the functions of these *hox* genes are required for discriminating between these cardiac regions. For instance, in *abdA* mutants, the (posterior) heart is transformed to assume an aorta fate, as expected according to the "posterior prevalence" rule of the Hox code (McGinnis and Krumlauf, 1992). At the molecular level, the specific posterior cardiac marker gene expression (e.g., *wg*, *troponinC-akin1*) is abolished, while aorta markers (e.g., Ubx) are upregulated in the posterior heart tube. Conversely, uniform expression of *abdA* in the mesoderm or throughout the whole heart tube results in transformation of the aorta into "posterior" heart, which expresses posterior markers, is wider, and forms posterior-specific "ostium" structures (Lo et al., 2002; Lovato et al., 2002; Perrin et al., 2004). These and other data suggest that the *hox* genes function autonomously within the linear heart tube to specify regional cardiac identities.

4.2. Cardiac homeobox factors

As mentioned above, cardiac progenitors that express the homeobox genes *eve*, *lbe*, and *svp* are the first groups of cells to be distinguished within the *tinman*-expressing cardiac mesoderm in a segmentally repeated pattern. These factors are thought to serve as identity genes as they are essential for the progenitor cells to maintain their normal gene expression and identities during subsequent differentiation (Frasch et al., 1987; Jagla et al., 1997, 2002; Su et al., 1999; Ward and Skeath, 2000; Lo and Frasch, 2001; Han et al., 2002; Han and Bodmer, 2003; Liu et al., 2006).

Among the identity genes, the regulation of mesodermal *eve* expression has been studied in detail. Eve progenitors are the first cell type present in cardiac mesoderm. It has been demonstrated that a small but necessary and sufficient enhancer in *eve*'s 3' regulatory region depends on a combinatorial code of patterned gene activity, including that of *twist, tinman, wg, dpp*, and *ras/MAPK*. (Carmena et al., 1998; Halfon et al., 2000; Knirr and Frasch, 2001; Han et al., 2002) Elimination of *eve* specifically in the mesoderm by a genomic rescue construct lacking the mesodermal *eve* enhancer ("*eve* meso minus") indicates that *eve* is indeed required for the specification of the Eve progenitors that give rise to the EPCs and DA1/DO2 body wall muscles, as well as indirectly for the proper functioning and performance of the adult heart myocardium (Fujioka et al., 2005).

lbe is expressed in cells anteriorly adjacent to the *eve* cells, and the antagonistic interaction between them results in their mutually exclusive expression domains (Jagla et al., 1997, 2002; Han et al., 2002; Liu et al., 2006). In *lbe* mutants, *eve* expression is expanded, while overexpression of *lbe* suppresses formation of the Eve cluster. Similarly, in "*eve* meso minus" embryos, the *lbe* clusters are expanded (Fujioka et al., 2005), and pan-mesodermal expression of *eve* strongly inhibits *lbe* expression. The repressive effect of Lbe on *eve* expression is likely through binding directly to the *eve* mesodermal enhancer (eme) as a repressor since mutating the Lbe-binding sites in eme causes a derepression (of a reporter construct) in the *lbe* expression domain. Remarkably, this mutant enhancer is no longer sensitive to repression by pan-mesodermal *lbe* overexpression *in vivo* (Han et al., 2002).

svp is expressed posteriorly to *eve* (Lo and Frasch, 2001) Initially, Svp progenitor cells coexpress *tinman*. As the heart develops further, *tinman* expression is gradually lost in the differentiating *svp* cells, which is achieved by a repressive action of Svp on *tinman* expression. Within each segment, *svp* is eventually expressed in only two of six pairs of myocardial cells and some pericardial cells that exclude *tinman* expression (Lo and Frasch, 2001). The two *svp*-expressing myocardial cells are morphologically and functionally distinct and form ostia in the posterior "heart" region. Elimination of *svp* expression in the late embryonic heart by over-expressing *tinman* compromises ostia formation (Lo et al., 2002; Lo and Frasch, 2003; Monier et al., 2005; Zaffran et al., 2006).

odd-skipped (*odd*) is another transcription factor expressed in cardiac mesoderm in a pattern partially overlapping with *svp* but later in development than *eve, lbe*, and *svp*. Although the function of *odd* in the heart is not yet known, lineage-tracing experiments indicate that *odd*-expressing precursors eventually develop into pericardial cells (Ward and Skeath, 2000).

4.3. Cardiac T-box factors

In *Drosophila*, two sets of T-box genes have been shown to be involved in heart development. As mentioned above, the activity of the *doc* triplet of genes is critical for early cardiac mesoderm induction. After cardiac specification, the *doc* genes are reexpressed in a subset of myocardial cells that coexpress *svp*. The major role of late embryonic *doc* function is to suppress *tinman* and thereby restrict its expression to four of six myocardial cells [and vice versa—*doc* is repressed by Tinman in myocardial cells (Reim and Frasch, 2005; Zaffran et al., 2006)]. In cardiac mesoderm-only *tinman* mutants, *doc* is expanded to all myocardial cells and results in much of the heart tube being transformed into ostium-like structures, suggesting that *doc* genes are likely to be required for "ostia" identity (Zaffran et al., 2006).

Another group of T-box genes is the Tbx20 homologues in *Drosophila*—the *neuromancer1* (*nmr1*, *H15*) and *neuromancer2* (*nmr2*, *midline*) gene pair (Miskolczi-McCallum et al., 2005; Qian et al., 2005a; Reim et al., 2005). Loss- and gain-of-function studies show that *nmr* is required for correct specification of cardiac progenitors, especially for activation of *tinman* expression in myocardial cells (Qian et al., 2005a; Reim et al., 2005). In addition, a GATA factor encoded by *pannier* is essential for its cardiac expression, and acts synergistically with *tinman* in promoting *nmr* expression. Moreover, reducing *nmr* activity in the absence of *pannier* aggravates the deficit in cardiac specification, suggesting a genetic interaction between *nmr* and *pannier* (Qian et al., 2005a) Interestingly, loss of *nmr* function eventually results in defective cardiac morphogenesis (Section 5), as judged by the misaligned cardiac cells at the dorsal midline (Miskolczi-McCallum et al., 2005; Qian et al., 2005a; Reim and Frasch, 2005; Reim et al., 2005). Investigation into the underlying mechanism reveals that *nmr* controls acquisition of the appropriate cell polarity in the heart tube, thus influencing the alignment of the heart cells (Qian et al., 2005a). Therefore, we conclude Tbx20/*nmr* genes play multiple roles, ranging from cardiac specification, morphogenesis to cardiac function (L.Q. and R.B., unpublished data).

4.4. Myocardin and dHand

Drosophila myocardin-related transcription factor (DMRTF), a SAP family chromatin-remodeling protein, and *Drosophila* Hand (dHand), a basic loop-helix-loop protein, have been studied in the fly (Kolsch and Paululat, 2002; Han et al., 2004; Han and Olson, 2005; Han et al., 2006; Sellin et al., 2006). Both of them have vertebrate orthologues (myocardin and Hand) that are indispensable for heart development (Srivastava, 1999; Wang et al., 2001). Even though homologous recombination-mediated

DMRTF knockout mutants do not exhibit developmental cardiac defects (possibly due to compensation by maternally inherited DMRTF), over-expression of a dominant-negative form of DMRTF in the mesoderm severely affects mesodermal cell migration, resulting in a defective heart with gaps or misaligned cells. *In vitro* biochemical studies indicate that DMRTF is likely a transcriptional coactivator of *Drosophila* serum response factor (DSRF), which is also highly conserved (Han et al., 2004). These findings suggest an ancient protein partnership involved in cardiac cell migration.

dHand is expressed in three major cell types that comprise the circulatory system: the myocardium, the pericardial cells, and hematopoetic lymph gland cells. (Kolsch and Paululat, 2002; Han and Olson, 2005; Han et al., 2006; Sellin et al., 2006). Correspondingly, *dHand* expression is activated by *tinman* and *pannier* in cardiac cells, and *serpent* (encoding another GATA factor) in hematopoetic cells. *dHand* mutants have a hypoplastic myocardium and have defective pericardial and lymph gland cells, which can be rescued by cardiac expression of apoptotic inhibitor p35 (Han et al., 2006). Thus, *dHand* seems to function at the nexus of genetic pathways regulating both cardiogenesis and hematopoiesis.

4.5. microRNAs

microRNAs (miRNAs) are 21–22 nucleotide small noncoding RNAs that posttranscriptionally target other mRNA transcripts for degradation or translational inhibition. They do so by binding to 3' UTRs of mRNAs (reviewed in Ambros, 2001; Bartel, 2004). Hundreds of miRNAs have thus far been identified in many species, including flies and humans, but their exact roles in biological processes are largely unexplored and only a few targets have been characterized (Ambros, 2003). The studies of miRNAs in *Drosophila* heart development were greatly enhanced by work on *dmiR-1* (Kwon et al., 2005; Sokol and Ambros, 2005), the single orthologue of vertebrate *miR-1-1* and *miR-1-2* that target the transcripts for cardiac-enriched basic helix-loop-helix transcription factor Hand2 and play a crucial role in mouse cardiogenesis (Zhao et al., 2005). *dmiR-1* is highly expressed in the presumptive mesoderm and all muscular derivatives, including the heart. This expression is regulated by Twist and Dmef2 (Sokol and Ambros, 2005). Genetic evidence indicates that *dmiR-1* is involved in maintaining muscle gene expression and cardiac progenitor cell differentiation, probably through targeting the Notch ligand *delta* mRNA for translational inhibition (Kwon et al., 2005).

In the fly heart, asymmetric cell division is a well-established mechanism for specifying alternative cardiac cell fates. For example, the membrane-associated protein encoded by *numb* segregates into one but not the other of the two daughter cells cell division, and this process is essential for the distinction of its cell fate from that of its sibling devoid of Numb. During these asymmetric cell divisions, the transmembrane proteins Notch and Sanpodo are also essential for lineage decisions and terminal cell differentiation (Park et al., 1998; Ward and Skeath, 2000; Jagla et al., 2002; Alvarez et al., 2003; Han and Bodmer, 2003). In *dmiR-1* mutants, Eve and Dmef2 cell populations are expanded and *tinman* is ectopically expressed, reminiscent of mutants lacking Notch signaling. In support, *in vivo* overexpression of *dmiR-1* is sufficient to inhibit expression of Delta (Kwon et al., 2005). This function of *dmiR-1* indicates a novel mechanism of regulating Notch signaling in cardiac lineage decisions.

5. CONTROL OF CARDIAC CELL POLARITY AND MORPHOGENESIS

5.1. Epithelial polarity of cardiac cells

After the cardiac progenitors are specified, the myocardial cells on either side of the embryo line up in a single row, flanked by pericardial cells. These two rows of cardiac cells migrate toward the dorsal midline where they establish close contact with their contralateral counterparts. During this process, the bilateral myocardial cells undergo a typical EMT, resulting in an apical-basal polarization of the myocardial cells. As a result, two bilateral rows of myocardial cells fuse at the dorsal midline to form the heart tube enclosing a lumen (Zaffran et al., 1995; Fremion, 1999; Haag et al., 1999; Chartier et al., 2002; Qian et al., 2005b).

On contact between the bilateral myocardial rows, the cells change cell shape during EMT and acquire their appropriate polarity characteristics (Fig. 3). This transition and the coordination of correct cell–cell contacts in the forming heart tube are largely unexplored. It is also not known if there is a causal relationship between cardiac cell polarity and patterning or cellular alignment of the forming heart tube at the dorsal midline. However, several recent papers describe the discovery of new factors and potential mechanisms involved in *Drosophila* cardiac morphogenesis (see below). Various markers of cardiac cell polarity have also been discovered, which will help elucidate the relationship between cell polarity and morphogenesis of the heart.

We can divide the development of the embryonic heart into two phases: (1) cardiac specification and differentiation, which is defined by the combined activities of inductive signals and transcription factors that specify and correctly position the cardiac mesoderm (as described in Sections 3 and 4); and (2) cardiac morphogenesis, which is concerned with the cellular events that carry out the program of cell migration, cell polarization, assembly into a linear tube, and terminal differentiation of

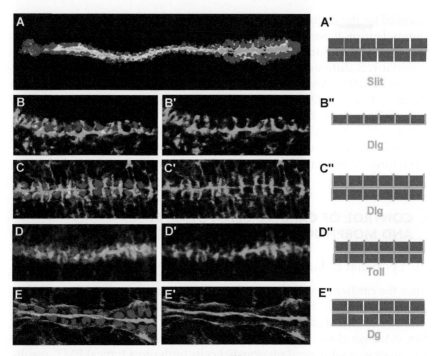

FIGURE 3   Cardiac cell polarity markers. *nmr^{H15}*lacZ-labeled myocardial cell nuclei are in red and cell polarity markers in green. (A, A′) After dorsal closure, Slit is highly accumulated at the dorsal midline where bilateral myocardial cell rows contact each other. (B, B′, B″) Before closure, Dlg is distributed at basal and lateral sides of myocardial cells. (C, C′, C″) As the two bilateral rows of myocardial cells come in contact, Dlg is redistributed to the apical(dorsal)-lateral sides of the myocardial cells. (D, D′, D″) Toll shows a similar pattern of subcellular localization as Dlg after dorsal closure. (E, E′, E″) Dg labels the apical and basal sides of the myocardial cells but is excluded from the lateral sides.

previously specified cell types (see below; Zaffran et al., 1995; Fremion, 1999; Haag et al., 1999; Chartier et al., 2002; Qian et al., 2005a,b; Wang et al., 2005).

5.2. Molecules essential for cardiac morphogenesis

The types of genes that have been identified so far are different between the two phases of heart development, phase 1 genes are mainly involved in signaling and transcriptional regulation, whereas most characterized phase 2 genes are involved in morphogenesis code for cell adhesion and extracellular matrix (ECM) molecules, membrane-associated glycoproteins, and intracellular cell membrane-associated proteins. Not surprisingly, abnormalities of phase 2 gene mutants have been found to affect cardiac morphogenesis and not any of the phase 1 specification events. For example, *robo* or *slit* mutants cause myocardial patterning defects, but specification of cell identities does not seem to be significantly altered, nor are the numbers of myocardial or pericardial cells (Qian et al., 2005b; MacMullin and Jacobs, 2006). Interestingly, T-box20/*nmr*-encoded transcription factors are an exception, since they seem to be involved in both specification and morphogenesis of the heart (Qian et al., 2005a). It is important to note that in contrast to *nmr* mutants, which affect cardiac myocyte polarity before and after assembly of the heart tube, *robo* or *slit* mutant myocardial cells are correctly polarized initially but do not properly undergo repolarization after the bilateral rows of heart cells reach the dorsal midline (Fig. 3; Qian et al., 2005b).

Five different types of gene products essential for cardiac morphogenesis can be distinguished at this point: transcription factors [Nmr (Qian et al., 2005a)], cell adhesion molecules [faint sausage, laminin A, E-cadherin (Yarnitzky and Volk, 1995; Haag et al., 1999)], ECM molecules [Pericardin, Dystroglycan (Chartier et al., 2002; Deng et al., 2003)], guidance molecules [Slit, Robo (Qian et al., 2005b; MacMullin and Jacobs, 2006)], and cell polarity genes [Fig. 3; PDZ protein discs large, α-Spectrin, Armadillo/β-catenin, Toll, and others (Fremion, 1999; Chartier et al., 2002; Qian et al., 2005a; Wang et al., 2005)]. Multiple mechanistic involvements may exist for some gene products. For instance, *dystroglycan* may also be considered a cell polarity gene because of its expression pattern and its function in follicle cell polarization (Deng et al., 2003). It is intriguing to look at the ways that these molecules genetically interact with each other to orchestrate the morphogenetic process of heart tube formation. For example, the fact that *nmr* influences Slit localization, but not vise versa, suggests that these genes form a genetic hierarchy controlling cardiac morphogenesis (Qian et al., 2005b).

A screen for mutants that alter the form of the heart tube late in embryogenesis identified the metabolic mevalonate pathway to be crucial for the association between myocardial and pericardial cells during cardiac morphogenesis (Yi et al., 2006). Thus, the cooperation between ECM, signal transduction and epithelial polarization, and adhesion molecules eventually leads to the higher ordered structure (and function) of the heart.

6. MYOCYTE REPROGRAMMING DURING METAMORPHOSIS

After the larval heart is formed at the end of embryogenesis, the cardiac tube increases in size during the larval stages and then undergoes extensive morphological and functional changes during metamorphosis to form the adult organ by a process called myocyte reprogramming or remodeling (Curtis et al., 1999; Molina and Cripps, 2001; Zikova et al.,

2003; Monier et al., 2005). The contractile myocardium of the adult heart originates from the larval posterior aorta, while the terminal chamber of the adult heart is derived from the anterior larval "heart" portion of the cardiac tube in the same segment (Figs. 1A,B and 4A,B). The posterior two segments of the larval heart are eliminated by histolysis during metamorphosis. During this process, cardiac myocyte morphology is modified in several aspects: (1) the adult heart is wider than its larval predecessor, except the terminal chamber, which is thinner than its larval counterpart in segment 5; (2) a new muscle layer not expressing *tinman* and of unknown origin associates ventrally with the heart tube (Molina and Cripps, 2001); (3) myofibril content and orientation changes; for example, in the terminal chamber, the myofibers are reoriented from transverse in the larvae to longitudinal in the adult; (4) in contrast to larvae, the adult heart is heavily innervated both anteriorly (glutaminergic nerve endings on the conical chambers in A1 and A2) and posteriorly (peptidergic nerve endings in A4 and A5) (Dulcis and Levine, 2003, 2005; Monier et al., 2005); and (5) *svp*-expressing myocytes of the posterior larval aorta (A1–A4) develop into functional ostia in the adult by increasing in size and remodeling their shape, which enables them to regulate the entry of hemolymph during relaxation of the cardiac cycle. *In vivo* time-lapse analysis and cell-tracing experiments indicate that the adult heart arises primarily by the remodeling of larval myocytes in a continuous and progressive process apparently without cell proliferation of the *tinman*- and *svp*-expressing larval cardiac myocytes (Monier et al., 2005). The remodeled cardiac tube in the adult remains an open circulatory system and has the ability to contract rhythmically and efficiently throughout the fly's life (reviewed in Bodmer et al., 2005).

This cardiac remodeling process is accompanied by transcriptional reprogramming, mainly by an anterior shift of expression of *wg*, *Ih* (encodes the *Drosophila* homologue of the vertebrate hyperpolarization-activated cyclic nucleotide-gate channel, HCN), and *NdaeI* (encodes a Na^+ driven anion exchanger involved in cellular ionic homeostasis). Moreover, reprogramming is genetically tightly controlled, mainly by *hox* genes *ubx* and *abdA* (Monier et al., 2005). Loss- and gain-of-function studies suggest that *ubx* repression in *tinman*-expressing myocytes from segment A1–A4 is required to drive the remodeling of the posterior aorta into adult heart, and AbdA activity is cell autonomously involved in remodeling of segment A5 myocytes to form the terminal chamber. The cardiac remodeling occurs during metamorphosis at a time that is coincident with the peak release of ecdysone. Indeed, ecdysone signaling was found to provide temporally regulated input signal driving cardiac remodeling (Monier et al., 2005): Introduction of either a dominant-negative form of ectysone receptor (EcR) or RNAi-mediated knockdown of EcR autonomously inhibited metamorphosis of the heart. The cardiac

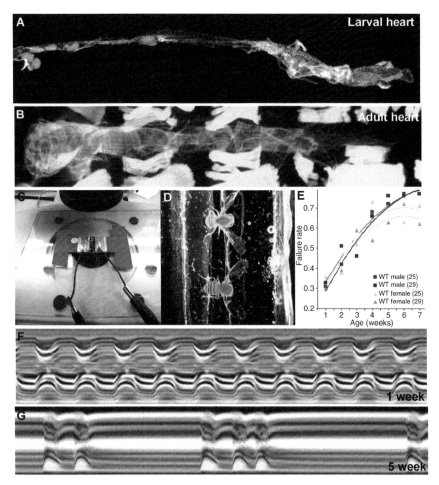

FIGURE 4 Myocyte reprogramming and cardiac function. (A,B) Larval (A) and adult (B) cardiac tubes, stained with phalloidin for polymerized actin (F-actin), are shown at the same magnification (pictures kindly provided by B. Monier and L. Perrin). (C,D) Electrical pacing setup (Wessells and Bodmer, 2004). (C) Microscope slide with conductive aluminum foil electrodes separated a 1-cm gap for mounting flies to be paced. The two foil strips are separated by a space of ∼1 cm and connected to a square wave stimulator. (D) Flies mounted for electrical pacing: conductive jelly is spread along the edges of the foil strips, leaving a fly-sized gap in between. Flies are then arrayed in a line with their heads touching the jelly on one side and the tips of their abdomens touching the jelly on the other side. (E) Rate of electrical pacing-induced "heart failure" as a function of age after a 30-s train of external stimuli in outbred wild-type flies [adapted from Wessells et al. (2004)]. (F,G) M-mode from movies taken at the level of the third abdominal segment of a 1-week-old (F) and a 5-week-old fly (G) with surgically exposed hearts (pictures kindly provided by K. Ocorr). Notice the pronounced arrhythmias at old age (G) compared to younger flies (F) (Ocorr et al., 2007b).

remodeling effect of ectysone signaling is achieved in part through regulation of Ubx expression and in part through modifying AbdA activity. Taken together, cardiac myocyte reprogramming is controlled by the ecdysone-mediated metamorphosis program.

7. CARDIAC FUNCTION AND CARDIAC AGING

The power of *Drosophila* genetics makes this fly not only an excellent model for studying heart development but recently also for investigating heart function and cardiac aging. Not much is understood on cardiac function compared to the fruitful field of early development. However, a number of assays have been developed to evaluate the different aspects of heart function, including heart rate, contractility, rhythmicity, ejection volume, electrical activity, and response to pacing stress.

Extracellular electrical recordings have been used to monitor a fly's electrocardiogram (ECG) (Johnson et al., 2000, 2001; Ocorr et al., 2007b). Intracellular electrical recordings have also been successful using suction and floating electrodes (Papaefthmiou and Theophilidis, 2001; Lalevee et al., 2006). To trace the movement of the heart in intact animals, a photodiode transistor-based assay has been developed (Dowse et al., 1995; Johnson et al., 1998; Nichols et al., 1999; Zornik et al., 1999; Mispelon et al., 2003). A more sensitive way for tracking the heart wall movements is achieved by an edge-tracing system, also possible in intact animals (Wessells and Bodmer, 2004; Wessells et al., 2004). Similarly, optical coherence tomography (OCT), a noncontact-type ultrasound, has been successfully adapted to flies and served as a tool to identify dilated cardiomyopathy phenotypes in adult flies (Wolf et al., 2006). To measure cardiac performance under duress, a stress assay has been developed, in which the heart failure rate (cardiac arrest or fibrillation) is calculated in response to external electrical pacing (Paternostro et al., 2001; Wessells and Bodmer, 2004; Wessells et al., 2004). Moreover, high-speed movies recorded from semiintact adult fly heart preparations allow detailed image-based analysis (e.g., of M-mode traces) and characterization of contractility and cardiac output, as well as a variety of heart rhythm parameters (Ocorr et al., 2007b). Cardiac muscle tension of the adult heart using carbon fibers, akin to single mammalian cardiac myocyte measurements (Yasuda et al., 2005), has also been achieved (Ocorr et al., 2007b).

Using these methods, it was shown that *Drosophila* mutants for the KCNQ potassium channel, which in humans is responsible for the late phase of cardiac repolarization and causing arrhythmias when malfunctioning, also exhibit significant arrhythmias and prolonged contractions in flies (Ocorr et al., 2007b). Manipulation of another (two-pore) potassium channel, Ork1, has a profound influence on the heartbeat frequency of the fly's larval heart, suggestive of a novel regulator of heart rate (Lalevee et al., 2006). Furthermore, ATP-dependent potassium channel function of the SUR (ABC) family of proteins, dSUR, protects the heart from hypoxia and stress-induced failure (Nasonkin et al., 1999; Akasaka et al., 2006). Interestingly, similar mutations in genes encoding sarcomeric contractile proteins, such as in *troponin I* and *sarcoglycan*, which cause dilated cardiomyopathies in humans, also cause a dilated phenotypes in flies (Wolf et al., 2006). A number of other genes, including *SERCA* and calcium channel *cacophony*, have also been shown to influence heart rate (Johnson et al., 1997, 2002; Dulcis and Levine, 2005; Dulcis et al., 2005; Ray and Dowse, 2005; reviewed in Bodmer et al., 2005; Jennings et al., 2006; Sanyal et al., 2006).

As in humans, cardiac performance declines with age in *Drosophila* (Paternostro et al., 2001; Wessells and Bodmer, 2004; Wessells et al., 2004; Ocorr et al., 2007b). As for heart development, basic aspects of the signaling pathways and hormone-regulatory networks that control or accompany the process of "cardiac aging" might be conserved. Senescence of cardiac function manifests itself in progressively decreasing heart rate, increasing arrhythmias and increasing rate in pacing-induced heart failure with age (Fig. 4C–G). This process is significantly influenced by heart-autonomous insulin-IGF receptor (InR) and TOR signaling, which suggests that InR-TOR signaling can directly modulate the aging process of individual organs in addition to its well-known function in regulating overall life span (Wessells et al., 2004; Luong et al., 2006). For example, heart-specific expression of InR causes accelerated aging of heart function, whereas dPTEN or dFOXO maintains a well-performing heart until "old fly age." These studies provide the first evidence that functional aging can be modulated in an organ-autonomous fashion, and *Drosophila* is thus a first genetic model system of organ senescence.

8. *DROSOPHILA* MODEL FOR HUMAN CARDIAC DISEASE

Cardiac disease is one of the leading causes of death in the world, thus elucidating the etiology of cardiac disease is of significant importance to the public. There are several categories of cardiac disease, congenital malformations and other cardiomyopathies, conduction defects, hypertension, and so on, that involve a wide spectrum of genes encoding transcription factors, cytoskeletal proteins, ion channel-associated proteins, for example. Despite steady progress, the majority of heart disease genes remain to be identified. More importantly, the polygenic traits involving interaction between various genetic loci in this disease is largely unexplored.

Because *Drosophila* is a highly tractable genetic model, it has a number of advantages for the analysis of human cardiac disease genes (reviewed in Bier and Bodmer, 2004; Bier, 2005; Ocorr et al., 2007a). First, access to the complete sequence of the *Drosophila melanogaster* genome allows cross-genome analysis to identify homologous genes in fly, which are associated with human disease. One of the databases for exploring human disease genes in flies is "Homophila" (http://homophila.sdsc.edu) (Reiter et al., 2001; Chien et al., 2002). To date, many homologous genes that cause a wide spectrum of human disease ranging from neurodegenerative disease, cancer, developmental defects, metabolic disorders to cardiaovascular disease have been identified in flies (Feany and Bender, 2000; Muqit and Feany, 2002; Rulifson et al., 2002; Bier, 2005). Second, "Drosophilists" have developed an ever-increasing repertoire of techniques and tools that make a wide variety of genetic and molecular analyses achievable (Adams and Sekelsky, 2002; St Johnston, 2002). The legendary UAS/Gal4 system (with the Gal80-ts modification) allows manipulation of gene expression in a highly precise spatial and temporal manner (Brand and Perrimon, 1993; McGuire et al., 2004). Usage of various transposon elements and collection of deficiency kits greatly facilitate the process of mutant characterization. By taking advantage of these tools and rapid life cycle, large-scale genetic screens are an obvious strength of *Drosophila*, including identification of modifiers of cardiac disease genes such as *tinman/NKX2-5* (L.Q., J.L., and R.B., unpublished data). Importantly, an increasing volume of literature suggests that cardiac development and function are controlled by an evolutionarily highly conserved genetic network between *Drosophila* and vertebrate (Bodmer, 1995; Bodmer and Venkatesh, 1998; Cripps and Olson, 2002; Harvey, 2002; Zaffran and Frasch, 2002; Brand, 2003; Bodmer et al., 2005). One challenge in human cardiac disease is to track the complex traits caused by polygenic variants and to define the genetic interaction between them. *Drosophila*, undoubtedly, is the only simple genetic model with a pumping, myogenic heart where such challenge can be addressed.

9. CONCLUSIONS AND PERSPECTIVES

Since the discovery of first cardiogenic gene—*tinman*, the past decade has witnessed tremendous progress in genetic and molecular studies of heart development and function in *Drosophila*, which has led to significant advances in our understanding of specific genes and pathways in vertebrate cardiogenesis. The evolutionary conservation of the fly heart and powerful genetic tools for analyzing gene function and interaction make the fly a distinctive model for heart development. A number of "cardiac" screens have already been conducted (Bidet et al., 2003; Kim et al., 2004;

Wessells and Bodmer, 2004; Wolf et al., 2006; Yi et al., 2006), therefore characterization of candidate gene function in heart physiology will be an important focus in the coming years. Understanding how these new genes are integrated into known pathways or coupled with other cardiac regulatory genes remains another major challenge. In addition to the classical transcriptional regulation, it will also be possible to investigate other layers of regulation, for instance posttranscriptional regulation (i.e., RNAi, miRNA), alterations in chromatin conformation, and epigenetic mechanism (i.e., DNA methylation, histone acetylation).

In the postgenomic era where the genome sequence of several model organisms is available, one critical advantage provided by *Drosophila* is the ability to perform functional analysis of candidate genes involved in multigenetic human (heart) diseases. It becomes obvious that the factors identified so far are, for the most part, the major components in genetic networks controlling early heart development, and mutations in them usually cause embryonic lethality. A majority of these factors are also expressed at multiple developmental times and may act not only during heart development but also in the adult and aging organism. Moreover, subtle heart defects can be caused by haploinsufficiency, hypomorphic, or dominant-negative mutations. Modifier screens or cardiac-specific RNAi screens may help explore these possibilities and identify important new gene functions.

In vertebrates, accumulating evidence supports the existence of cardiac stem cell and cardiomyocyte regeneration. Although in the fly it is not known yet whether the heart also has stem cells that contribute to regeneration, the knowledge gained on the different aspects of normal cardiogenesis and cardiac physiology in *Drosophila* will nevertheless offer valuable guidance in future cardiac research. Ultimately, the simple but powerful fruit fly will help us elucidate the etiology of human cardiac disease and design efficient therapies for congenital cardiac disease as well as age-related cardiac disorders.

ACKNOWLEDGMENTS

We apologize that we have not done justice and given appropriate consideration to all interesting aspects of research in "*Drosophila* heart development," which is mainly due to space considerations. We thank Laurent Perrin and Karen Ocorr for critical reading of the chapter and for giving us advice and suggestions. This work was supported by the American Heart Association (L.Q. and J.L.) and by NHLBI of the National Institutes of Health (R.B.).

REFERENCES

Adams, M.D., Sekelsky, J.J. 2002. From sequence to phenotype: Reverse genetics in *Drosophila melanogaster*. Nat. Rev. Genet. 3, 189–198.

Akasaka, T., Klinedinst, S., Ocorr, K., Bustamante, E.L., Kim, S.K., Bodmer, R. 2006. The ATP-sensitive potassium (KATP) channel-encoded dSUR gene is required for *Drosophila* heart function and is regulated by tinman. Proc. Natl. Acad. Sci. USA 32, 11999–12004.

Alvarez, A.D., Shi, W., Wilson, B.A., Skeath, J.B. 2003. Pannier and pointedP2 act sequentially to regulate *Drosophila* heart development. Development 13, 3015–3026.

Ambros, V. 2001. microRNAs: Tiny regulators with great potential. Cell 7, 823–826.

Ambros, V. 2003. MicroRNA pathways in flies and worms: Growth, death, fat, stress, and timing. Cell 6, 673–676.

Azpiazu, N., Frasch, M. 1993. Tinman and bagpipe: Two homeo box genes that determine cell fates in the dorsal mesoderm of *Drosophila*. Genes Dev. 7(B), 1325–1340.

Azpiazu, N., Lawrence, P.A., Vincent, J.P., Frasch, M. 1996. Segmentation and specification of the *Drosophila* mesoderm. Genes Dev. 24, 3183–3194.

Bartel, D.P. 2004. MicroRNAs: Genomics, biogenesis, mechanism, and function. Cell 2, 281–297.

Bidet, Y., Jagla, T., Da Ponte, J.P., Dastugue, B., Jagla, K. 2003. Modifiers of muscle and heart cell fate specification identified by gain-of-function screen in *Drosophila*. Mech. Dev. 9, 991–1007.

Bier, E. 2005. *Drosophila*, the golden bug, emerges as a tool for human genetics. Nat. Rev. Genet. 1, 9–23.

Bier, E., Bodmer, R. 2004. *Drosophila*, an emerging model for cardiac disease. Gene 342, 1–11.

Bodmer, R. 1993. The gene tinman is required for specification of the heart and visceral muscles in *Drosophila*. Development 3, 1301–1306.

Bodmer, R. 1995. Heart development in *Drosophila* and its relationship to vertebrate systems. Trends Cardiovasc. Med. 5, 21–27.

Bodmer, R., Venkatesh, T.V. 1998. Heart development in *Drosophila* and vertebrates: Conservation of molecular mechanisms. Dev. Genet. 3, 181–186.

Bodmer, R., Frasch, M. 1999. Genetic determination of *Drosophila* heart development. In: *Heart Development* (N. Rosenthal, R. Harvey, Eds.), San Diego, London, New York: Academic Press, pp. 65–90.

Bodmer, R., Wessells, R.J., Johnson, E.C., Dowse, H. 2005. Heart development and function. In: *Comprehensive Molecular Insect Science (vol. 2)* (L.I. Gilbert, K. Iatrou, S. Gill, Eds.), Elsevier, vols. 1–7, pp. 199–250. Amsterdam, The Netherlands.

Boulay, J.L., Dennefeld, C., d Alberga, A. 1987. The *Drosophila* developmental gene snail encodes a protein with nucleic acid binding fingers. Nature 6146, 395–398.

Bour, B.A., O'Brien, M.A., Lockwood, W.L., Goldstein, E.S., Bodmer, R., Taghert, P.H., Abmayr, S.M., Nguyen, H.T. 1995. *Drosophila* MEF2, a transcription factor that is essential for myogenesis. Genes Dev. 6, 730–741.

Brand, A.H., Perrimon, N. 1993. Targeted gene expression as a means of altering cell fates and generating dominant phenotypes. Development 2, 401–415.

Brand, T. 2003. Heart development: Molecular insights into cardiac specification and early morphogenesis. Dev. Biol. 1, 1–19.

Carmena, A., Gisselbrecht, S., Harrison, J., Jimenez, F., Michelson, A.M. 1998. Combinatorial signaling codes for the progressive determination of cell fates in the *Drosophila* embryonic mesoderm. Genes Dev. 24, 3910–3922.

Chartier, A., Zaffran, S., Astier, M., Semeriva, M., Gratecos, D. 2002. Pericardin, a *Drosophila* type IV collagen-like protein is involved in the morphogenesis and maintenance of the heart epithelium during dorsal ectoderm closure. Development 13, 3241–3253.

Chien, S., Reiter, L.T., Bier, E., Gribskov, M. 2002. Homophila: Human disease gene cognates in *Drosophila*. Nucleic Acids Res. 1, 149–151.

Cripps, R.M., Olson, E.N. 2002. Control of cardiac development by an evolutionarily conserved transcriptional network. Dev. Biol. 1, 14–28.

Curtis, N.J., Ringo, J.M., Dowse, H.B. 1999. Morphology of the pupal heart, adult heart, and associated tissues in the fruit fly, *Drosophila melanogaster*. J. Morphol. 3, 225–235.

Deng, W.M., Schneider, M., Frock, R., Castillejo-Lopez, C., Gaman, E.A., Baumgartner, S., Ruohola-Baker, H. 2003. Dystroglycan is required for polarizing the epithelial cells and the oocyte in *Drosophila*. Development 1, 173–184.

Dowse, H., Ringo, J., Power, J., Johnson, E., Kinney, K., White, L. 1995. A congenital heart defect in *Drosophila* caused by an action-potential mutation. J. Neurogenet. 3, 153–168.

Dulcis, D., Levine, R.B. 2003. Innervation of the heart of the adult fruit fly, *Drosophila melanogaster*. J. Comp. Neurol. 4, 560–578.

Dulcis, D., Levine, R.B. 2005. Glutamatergic innervation of the heart initiates retrograde contractions in adult *Drosophila melanogaster*. J. Neurosci. 2, 271–280.

Dulcis, D., Levine, R.B., Ewer, J. 2005. Role of the neuropeptide CCAP in *Drosophila* cardiac function. J. Neurobiol. 3, 259–274.

Durocher, D., Charron, F., Warren, R., Schwartz, R., Nemer, M. 1997. The cardiac transcription factors Nkx2-5 and GATA-4 are mutual cofactors. EMBO J. 18, 5687–5696.

Feany, M.B., Bender, W.W. 2000. A *Drosophila* model of Parkinson's disease. Nature 404, 394–398.

Fishman, M.C., Chien, K.R. 1997. Fashioning the vertebrate heart: Earliest embryonic decisions. Development 11, 2099–2117.

Frasch, M. 1995. Induction of visceral and cardiac mesoderm by ectodermal Dpp in the early *Drosophila* embryo. Nature 6521, 464–467.

Frasch, M., Hoey, T., Rushlow, C., Doyle, H., Levine, M. 1987. Characterization and localization of the even-skipped protein of *Drosophila*. EMBO J. 3, 749–759.

Fremion, F. 1999. The heterotrimeric protein Go is required for the formation of heart epithelium in *Drosophila*. J. Cell Biol. 145, 1063–1076.

Fujioka, M., Wessells, R.J., Han, Z., Liu, J., Fitzgerald, K., Yusibova, G.L., Zamora, M., Ruiz-Lozano, P., Bodmer, R., Jaynes, J.B. 2005. Embryonic even skipped-dependent muscle and heart cell fates are required for normal adult activity, heart function, and lifespan. Circ. Res. 11, 1108–1114.

Gajewski, K., Choi, C.Y., Kim, Y., Schulz, R.A. 2000. Genetically distinct cardial cells within the *Drosophila* heart. Genesis 1, 36–43.

Gajewski, K., Fossett, N., Molkentin, J.D., Schulz, R.A. 1999. The zinc finger proteins Pannier and GATA4 function as cardiogenic factors in *Drosophila*. Development 24, 5679–5688.

Garg, V., Kathiriya, I.S., Barnes, R., Schluterman, M.K., King, I.N., Butler, C.A., Rothrock, C.R., Eapen, R.S., Hirayama-Yamada, K., Joo, K., Matsuoka, R., Cohen, J.C., et al. 2003. GATA4 mutations cause human congenital heart defects and reveal an interaction with TBX5. Nature 424, 443–447.

Haag, T.A., Haag, N.P., Lekven, A.C., Hartenstein, V. 1999. The role of cell adhesion molecules in *Drosophila* heart morphogenesis: Faint sausage, shotgun/DE-cadherin, and laminin A are required for discrete stages in heart development. Dev. Biol. 1, 56–69.

Halfon, M.S., Carmena, A., Gisselbrecht, S., Sackerson, C.M., Jimenez, F., Baylies, M.K., Michelson, A.M. 2000. Ras pathway specificity is determined by the integration of multiple signal-activated and tissue-restricted transcription factors. Cell 1, 63–74.

Hamaguchi, T., Yabe, S., Uchiyama, H., Murakami, R. 2004. *Drosophila* Tbx6-related gene, Dorsocross, mediates high levels of Dpp and Scw signal required for the development of amnioserosa and wing disc primordium. Dev. Biol. 2, 355–368.

Han, Z., Bodmer, R. 2003. Myogenic cells fates are antagonized by Notch only in asymmetric lineages of the *Drosophila* heart, with or without cell division. Development 13, 3039–3051.

Han, Z., Olson, E.N. 2005. Hand is a direct target of Tinman and GATA factors during *Drosophila* cardiogenesis and hematopoiesis. Development 15, 3525–3536.

Han, Z., Fujioka, M., Su, M., Liu, M., Jaynes, J.B., Bodmer, R. 2002. Transcriptional integration of competence modulated by mutual repression generates cell-type specificity within the cardiogenic mesoderm. Dev. Biol. 2, 225–240.

Han, Z., Li, X., Wu, J., Olson, E.N. 2004. A myocardin-related transcription factor regulates activity of serum response factor in *Drosophila*. Proc. Natl. Acad. Sci. USA 34, 12567–12572.

Han, Z., Yi, P., Li, X., Olson, E.N. 2006. Hand, an evolutionarily conserved bHLH transcription factor required for *Drosophila* cardiogenesis and hematopoiesis. Development 6, 1175–1182.

Harvey, R.P. 2002. Patterning the vertebrate heart. Nat. Rev. Genet. 7, 544–556.

Jagla, K., Jagla, T., Heitzler, P., Dretzen, G., Bellard, F., Bellard, M. 1997. ladybird, a tandem of homeobox genes that maintain late wingless expression in terminal and dorsal epidermis of the *Drosophila* embryo. Development 1, 91–100.

Jagla, T., Bidet, Y., Da Ponte, J.P., Dastugue, B., Jagla, K. 2002. Cross-repressive interactions of identity genes are essential for proper specification of cardiac and muscular fates in *Drosophila*. Development 4, 1037–1047.

Jennings, N.S., Smethurst, P.A., Graham, M.N., Knight, C., O'Connor, M.N., Joutsi-Korhonen, L., Stafford, P., Stephens, J., Gamer, S.F., Harmer, I.J., Farndale, R.W., Watkins, N.A., Ouwehand, W.H. 2006. Production of calmodulin-tagged proteins in *Drosophila* Schneider S2 cells: A novel system for antigen production and phage antibody isolation. J. Immunol. Methods 316, 75–83.

Johnson, E., Ringo, J., Dowse, H. 1997. Modulation of *Drosophila* heartbeat by neurotransmitters. J. Comp. Physiol. [B] 2, 89–97.

Johnson, E., Ringo, J., Bray, N., Dowse, H. 1998. Genetic and pharmacological identification of ion channels central to the *Drosophila* cardiac pacemaker. J. Neurogenet. 1, 1–24.

Johnson, E., Ringo, J., Dowse, H. 2000. Native and heterologous neuropeptides are cardioactive in *Drosophila melanogaster*. J. Insect Physiol. 8, 1229–1236.

Johnson, E., Ringo, J., Dowse, H. 2001. Dynamin, encoded by shibire, is central to cardiac function. J. Exp. Zool. 2, 81–89.

Johnson, E., Sherry, T., Ringo, J., Dowse, H. 2002. Modulation of the cardiac pacemaker of *Drosophila*: Cellular mechanisms. J. Comp. Physiol. [B] 3, 227–236.

Kim, Y.O., Park, S.J., Balaban, R.S., Nirenberg, M., Kim, Y. 2004. A functional genomic screen for cardiogenic genes using RNA interference in developing *Drosophila* embryos. Proc. Natl. Acad. Sci. USA 1, 159–164.

Klinedinst, S.L., Bodmer, R. 2003. Gata factor Pannier is required to establish competence for heart progenitor formation. Development 13, 3027–3038.

Knirr, S., Frasch, M. 2001. Molecular integration of inductive and mesoderm-intrinsic inputs governs even-skipped enhancer activity in a subset of pericardial and dorsal muscle progenitors. Dev. Biol. 1, 13–26.

Kolsch, V., Paululat, A. 2002. The highly conserved cardiogenic bHLH factor Hand is specifically expressed in circular visceral muscle progenitor cells and in all cell types of the dorsal vessel during *Drosophila* embryogenesis. Dev. Genes Evol. 10, 473–485.

Kosman, D., Ip, Y.T., Levine, M., Arora, K. 1991. Establishment of the mesoderm-neuroectoderm boundary in the *Drosophila* embryo. Science 5028, 118–122.

Kwon, C., Han, Z., Olson, E.N., Srivastava, D. 2005. MicroRNA1 influences cardiac differentiation in *Drosophila* and regulates Notch signaling. Proc. Natl. Acad. Sci. USA 52, 18986–18991.

Lalevee, N., Monier, B., Senatore, S., Perrin, L., Semeriva, M. 2006. Control of cardiac rhythm by ORK1, a *Drosophila* two-pore domain potassium channel. Curr. Biol. 15, 1502–1508.

Lee, H.H., Frasch, M. 2000. Wingless effects mesoderm patterning and ectoderm segmentation events via induction of its downstream target sloppy paired. Development 24, 5497–5508.

Lee, Y., Shioi, T., Kasahara, H., Jobe, S.M., Wiese, R.J., Markham, B.E., Izumo, S. 1998. The cardiac tissue-restricted homeobox protein Csx/Nkx2.5 physically associates with the zinc finger protein GATA4 and cooperatively activates atrial natriuretic factor gene expression. Mol. Cell Biol. 6, 3120–3129.

Leptin, M. 1991. Twist and snail as positive and negative regulators during *Drosophila* mesoderm development. Genes Dev. 9, 1568–1576.

Leptin, M., Grunewald, B. 1990. Cell shape changes during gastrulation in *Drosophila*. Development 1, 73–84.

Leptin, M., Casal, J., Grunewald, B., Reuter, R. 1992. Mechanisms of early *Drosophila* mesoderm formation. Dev. Suppl. 23–31.

Lilly, B., Zhao, B., Ranganayakulu, G., Paterson, B.M., Schulz, R.A., Olson, E.N. 1995. Requirement of MADS domain transcription factor D-MEF2 for muscle formation in *Drosophila*. Science 5198, 688–693.

Liu, J., Qian, L., Wessells, R.J., Bidet, Y., Jagla, K., Bodmer, R. 2006. Hedgehog and RAS pathways cooperate in the anterior-posterior specification and positioning of cardiac progenitor cells. Dev. Biol. 2, 373–385.

Lo, P.C., Frasch, M. 2001. A role for the COUP-TF-related gene seven-up in the diversification of cardioblast identities in the dorsal vessel of *Drosophila*. Mech. Dev. 1–2, 49–60.

Lo, P.C., Frasch, M. 2003. Establishing A-P polarity in the embryonic heart tube: A conserved function of Hox genes in *Drosophila* and vertebrates?Trends Cardiovasc. Med. 5, 182–187.

Lo, P.C., Skeath, J.B., Gajewski, K., Schulz, R.A., Frasch, M. 2002. Homeotic genes autonomously specify the anteroposterior subdivision of the *Drosophila* dorsal vessel into aorta and heart. Dev. Biol. 2, 307–319.

Lockwood, W.K., Bodmer, R. 2002. The patterns of wingless, decapentaplegic, and tinman position the *Drosophila* heart. Mech. Dev. 1–2, 13–26.

Lovato, T.L., Nguyen, T.P., Molina, M.R., Cripps, R.M. 2002. The Hox gene abdominal-A specifies heart cell fate in the *Drosophila* dorsal vessel. Development 21, 5019–5027.

Luong, N., Davies, C.R., Wessells, R.J., Graham, S.M., King, M.T., Veech, R., Bodmer, R., Oldham, S.M. 2006. Activated FOXO-mediated insulin resistance is blocked by reduction of TOR activity. Cell Metab. 2, 133–142.

MacMullin, A., Jacobs, J.R. 2006. Slit coordinates cardiac morphogenesis in *Drosophila*. Dev. Biol. 1, 154–164.

McGinnis, W., Krumlauf, R. 1992. Homeobox genes and axial patterning. Cell 2, 283–302.

McGuire, S.E., Mao, Z., Davis, R.L. 2004. Spatiotemporal gene expression targeting with the TARGET and gene-switch systems in *Drosophila*. Sci. STKE 220, p 16.

Miskolczi-McCallum, C.M., Scavetta, R.J., Svendsen, P.C., Soanes, K.H., Brook, W.J. 2005. The *Drosophila melanogaster* T-box genes midline and H15 are conserved regulators of heart development. Dev. Biol. 2, 459–472.

Mispelon, M., Thakur, K., Chinn, L., Owen, R., Nichols, R. 2003. A nonpeptide provides insight into mechanisms that regulate *Drosophila melanogaster* heart contractions. Peptides 10, 1599–1605.

Molina, M.R., Cripps, R.M. 2001. Ostia, the inflow tracts of the *Drosophila* heart, develop from a genetically distinct subset of cardial cells. Mech. Dev. 1, 51–59.

Monier, B., Astier, M., Semeriva, M., Perrin, L. 2005. Steroid-dependent modification of Hox function drives myocyte reprogramming in the *Drosophila* heart. Development 23, 5283–5293.

Muqit, M.M., Feany, M.B. 2002. Modelling neurodegenerative diseases in *Drosophila*: A fruitful approach? Nat. Rev. Neurosci. 3, 237–243.

Nasonkin, I., Alikasifoglu, A., Ambrose, C., Cahill, P., Cheng, M., Sarniak, A., Egan, M., Thomas, P.M. 1999. A novel sulfonylurea receptor family member expressed in the embryonic *Drosophila* dorsal vessel and tracheal system. J. Biol. Chem. 41, 29420–29425.

Nguyen, H.T., Bodmer, R., Abmayr, S.M., McDermott, J.C., Spoerel, N.A. 1994. D-mef2: A *Drosophila* mesoderm-specific MADS box-containing gene with a biphasic expression profile during embryogenesis. Proc. Natl. Acad. Sci. USA 16, 7520–7524.

Nichols, R., McCormick, J., Cohen, M., Howe, E., Jean, C., Paisley, K., Rosario, C. 1999. Differential processing of neuropeptides influences *Drosophila* heart rate. J. Neurogenet. 1–2, 89–104.

Ocorr, K., Akasaka, T., Bodmer, R. 2007a. Age related cardiac disease model of *Drosophila*. Mech. Ageing Dev. 128(1), 112–116.

Ocorr, K., Reeves, N.L., Wessells, R.J., Fink, M., Chen, H.S., Akasaka, T., Yasuda, S., Metzger, J.M., Giles, W., Posakony, J.W., Bodmer, R. 2007b. KCNQ potassium channel mutations cause cardiac arrhythmias in *Drosophila* that mimic the effects of aging. Proc. Natl. Acad. Sci. USA 104, 3943–3948.

Olson, E.N. 2004. A decade of discoveries in cardiac biology. Nat. Med. 5, 467–474.

Pandur, P., Lasche, M., Eisenberg, L.M., Kuhl, M. 2002. Wnt-11 activation of a non-canonical Wnt signalling pathway is required for cardiogenesis. Nature 6898, 636–641.

Papaefthmiou, C., Theophilidis, G. 2001. An *in vitro* method for recording the electrical activity of the isolated heart of the adult *Drosophila melanogaster*. *In Vitro* Cell. Dev. Biol. Anim. 7, 445–449.

Park, M., Wu, X., Golden, K., Axelrod, J.D., Bodmer, R. 1996. The wingless signaling pathway is directly involved in *Drosophila* heart development. Dev. Biol. 1, 104–116.

Park, M., Yaich, L.E., Bodmer, R. 1998. Mesodermal cell fate decisions in *Drosophila* are under the control of the lineage genes numb, Notch, and sanpodo. Mech. Dev. 1–2, 117–126.

Paternostro, G., Vignola, C., Bartsch, D.U., Omens, J.H., McCulloch, A.D., Reed, J.C. 2001. Age-associated cardiac dysfunction in *Drosophila melanogaster*. Circ. Res. 10, 1053–1058.

Perrin, L., Monier, B., Ponzielli, R., Astier, M., Semeriva, M. 2004. *Drosophila* cardiac tube organogenesis requires multiple phases of Hox activity. Dev. Biol. 2, 419–431.

Ponzielli, R., Astier, M., Chartier, A., Gallet, A., Therond, P., Semeriva, M. 2002. Heart tube patterning in *Drosophila* requires integration of axial and segmental information provided by the Bithorax Complex genes and hedgehog signaling. Development 19, 4509–4521.

Qian, L., Liu, J., Bodmer, R. 2005a. Neuromancer Tbx20-related genes (H15/midline) promote cell fate specification and morphogenesis of the *Drosophila* heart. Dev. Biol. 2, 509–524.

Qian, L., Liu, J., Bodmer, R. 2005b. Slit and Robo control cardiac cell polarity and morphogenesis. Curr. Biol. 24, 2271–2278.

Raftery, L.A., Sutherland, D.J. 1999. TGF-beta family signal transduction in *Drosophila* development: From Mad to Smads. Dev. Biol. 2, 251–268.

Ray, V.M., Dowse, H.B. 2005. Mutations in and deletions of the Ca^{2+} channel-encoding gene cacophony, which affect courtship song in *Drosophila*, have novel effects on heartbeating. J. Neurogenet. 1, 39–56.

Reim, I., Lee, H.H., Frasch, M. 2003. The T-box-encoding Dorsocross genes function in amnioserosa development and the patterning of the dorsolateral germ band downstream of Dpp. Development 14, 3187–3204.

Reim, I., Mohler, J.P., Frasch, M. 2005. Tbx20-related genes, mid and H15, are required for tinman expression, proper patterning, and normal differentiation of cardioblasts in *Drosophila*. Mech. Dev. 9, 1056–1069.

Reiter, L.T., Potocki, L., Chien, S., Gribskov, M., Bier, E. 2001. A systematic analysis of human disease-associated gene sequences in *Drosophila* melanogaster. Genome Res. 11, 1114–1125.

Riechmann, V., Irion, U., Wilson, R., Grosskortenhaus, R., Leptin, M. 1997. Control of cell fates and segmentation in the *Drosophila* mesoderm. Development 15, 2915–2922.

Rizki, T.M., Rizki, R.M. 1978. Larval adipose tissue of homoeotic bithorax mutants of *Drosophila*. Dev. Biol. 65, 476–482.

Rulifson, E.J., Kim, S.K., Nusse, R. 2002. Ablation of insulin-producing neurons in flies: Growth and diabetic phenotypes. Science 5570, 1118–1120.

Sanyal, S., Jennings, T., Dowse, H., Ramaswami, M. 2006. Conditional mutations in SERCA, the Sarco-endoplasmic reticulum Ca(2+)-ATPase, alter heart rate and rhythmicity in *Drosophila*. J. Comp. Physiol. [B] 3, 253–263.

Sellin, J., Albrecht, S., Kolsch, V., Paululat, A. 2006. Dynamics of heart differentiation, visualized utilizing heart enhancer elements of the *Drosophila melanogaster* bHLH transcription factor Hand. Gene Expr. Patterns 4, 360–375.

Sepulveda, J.L., Belaguli, N., Nigam, V., Chen, C.Y., Nemer, M., Schwartz, R.J. 1998. GATA-4 and Nkx-2.5 coactivate Nkx-2 DNA binding targets: Role for regulating early cardiac gene expression. Mol. Cell. Biol. 6, 3405–3415.

Sokol, N.S., Ambros, V. 2005. Mesodermally expressed *Drosophila* microRNA-1 is regulated by Twist and is required in muscles during larval growth. Genes Dev. 19, 2343–2354.

Srivastava, D. 1999. HAND proteins: Molecular mediators of cardiac development and congenital heart disease. Trends Cardiovasc. Med. 9(1–2), 11–18.

St Johnston, D. 2002. The art and design of genetic screens: *Drosophila melanogaster*. Nat. Rev. Genet. 3, 176–188.

St Johnston, D., Nusslein-Volhard, C. 1992. The origin of pattern and polarity in the *Drosophila* embryo. Cell 2, 201–219.

Su, M.T., Fujioka, M., Goto, T., Bodmer, R. 1999. The *Drosophila* homeobox genes zfh-1 and even-skipped are required for cardiac-specific differentiation of a numb-dependent lineage decision. Development 14, 3241–3251.

Thisse, B., Stoetzel, C., Gorostiza-Thisse, C., Perrin-Schmitt, F. 1988. Sequence of the twist gene and nuclear localization of its protein in endomesodermal cells of early *Drosophila* embryos. EMBO J. 7, 2175–2183.

Thisse, C., Perrin-Schmitt, F., Stoetzel, C., Thisse, B. 1991. Sequence-specific transactivation of the *Drosophila* twist gene by the dorsal gene product. Cell 7, 1191–1201.

Wang, D., Chang, P.S., Wang, Z., Sutherland, L., Richardson, J.A., Small, E., Krieg, P.A., Olson, E.N. 2001. Activation of cardiac gene expression by myocardin, a transcriptional cofactor for serum response factor. Cell 7, 851–862.

Wang, J., Tao, Y., Reim, I., Gajewski, K., Frasch, M., Schulz, R.A. 2005. Expression, regulation, and requirement of the toll transmembrane protein during dorsal vessel formation in *Drosophila melanogaster*. Mol. Cell. Biol. 10, 4200–4210.

Ward, E.J., Skeath, J.B. 2000. Characterization of a novel subset of cardiac cells and their progenitors in the *Drosophila* embryo. Development 22, 4959–4969.

Wessells, R.J., Bodmer, R. 2004. Screening assays for heart function mutants in *Drosophila*. Biotechniques 37, 2–7.

Wessells, R.J., Fitzgerald, E., Cypser, J.R., Tatar, M., Bodmer, R. 2004. Insulin regulation of heart function in aging fruit flies. Nat. Genet. 12, 1275–1281.

Wolf, M.J., Amrein, H., Izatt, J.A., Choma, M.A., Reedy, M.C., Rockman, H.A. 2006. *Drosophila* as a model for the identification of genes causing adult human heart disease. Proc. Natl. Acad. Sci. USA 5, 1394–1399.

Wu, X., Golden, K., Bodmer, R. 1995. Heart development in *Drosophila* requires the segment polarity gene wingless. Dev. Biol. 2, 619–628.

Xu, X., Yin, Z., Hudson, J.B., Ferguson, E.L., Frasch, M. 1998. Smad proteins act in combination with synergistic and antagonistic regulators to target Dpp responses to the *Drosophila* mesoderm. Genes Dev. 15, 2354–2370.

Yarnitzky, T., Volk, T. 1995. Laminin is required for heart, somatic muscles, and gut development in the *Drosophila* embryo. Dev. Biol. 2, 609–618.

Yasuda, S., Townsend, D., Michele, D.E., Favre, E.G., Day, S.M., Metzger, J.M. 2005. Dystrophic heart failure blocked by membrane sealant poloxamer. Nature 7053, 1025–1029.

Yi, P., Han, Z., Li, X., Olson, E.N. 2006. The mevalonate pathway controls heart formation in *Drosophila* by isoprenylation of Ggamma1. Science 5791, 1301–1303.

Yin, Z., Frasch, M. 1998. Regulation and function of tinman during dorsal mesoderm induction and heart specification in *Drosophila*. Dev. Genet. 3, 187–200.

Zaffran, S., Frasch, M. 2002. Early signals in cardiac development. Circ. Res. 6, 457–469.

Zaffran, S., Astier, M., Gratecos, D., Guillen, A., Semeriva, M. 1995. Cellular interactions during heart morphogenesis in the *Drosophila* embryo. Biol. Cell 1–2, 13–24.

Zaffran, S., Reim, I., Qian, L., Lo, P.C., Bodmer, R., Frasch, M. 2006. Cardioblast-intrinsic Tinman activity controls proper diversification and differentiation of myocardial cells in *Drosophila*. Development 20, 4073–4083.

Zhao, Y., Samal, E., Srivastava, D. 2005. Serum response factor regulates a muscle-specific microRNA that targets Hand2 during cardiogenesis. Nature 7048, 214–220.

Zikova, M., Da Ponte, J.P., Dastugue, B., Jagla, K. 2003. Patterning of the cardiac outflow region in *Drosophila*. Proc. Natl. Acad. Sci. USA 21, 12189–12194.

Zornik, E., Paisley, K., Nichols, R. 1999. Neural transmitters and a peptide modulate *Drosophila* heart rate. Peptides 1, 45–51.

CHAPTER 2

Morphogenesis of the Vertebrate Heart

Willem M.H. Hoogaars, Vincent M. Christoffels, and **Antoon F.M. Moorman**

Contents

Heart Failure Research Center, Academic Medical Center, University of Amsterdam, Amsterdam, The Netherlands

Advances in Developmental Biology, Volume 18
ISSN 1574-3349, DOI: 10.1016/S1574-3349(07)18002-9

Abstract   The adult four-chambered heart of higher vertebrates functions as a sophisticated pump, driving a pulmonary and a systemic circulation that have been separated during evolution. During cardiac development, assemblage of all the different components that make this structure functional is achieved by complex morphogenetic processes that we are just beginning to appreciate. Starting as two cardiac progenitor fields residing in the mesodermal layer of the embryonic disc, the heart begins to shape when these heart fields fuse and fold to form the heart tube. Subsequently, myocardial components are added to both poles of this tube and formation of the functional components of the adult heart, such as the chambers and the cardiac conduction system, is initiated. The molecular and cellular processes that control these morphogenetic processes are emerging and involve multiple gene programs controlled by conserved transcriptional regulators, such as T-box factors, Nkx2–5, and GATA4. In this chapter, we will highlight the morphological changes that the developing heart undergoes before the mature four-chambered heart emerges. Furthermore, we will take a closer look at recent progress that has been made in deciphering the molecular pathways underlying these processes.

1. INTRODUCTION

The adult heart is a vital and complex organ which supports the circulation of the blood flow, which is necessary for distribution and exchange of nutrients, oxygen, and waste products. During vertebrate evolution, the heart underwent remarkable adaptations to facilitate two separate circulations in the heart of higher vertebrates such as birds and mammals. This resulted in the four-chambered heart as we know it, pumping blood through the body via the systemic circulation and providing oxygenation of the blood via the pulmonary circulation. During heart development, all the components required for a functional heart are shaped, beginning with two heart fields, which fuse and fold to form a linear heart tube. This tube elongates and bends to the right after which the ventricles and atria are formed at discrete sites. Septation of the chambers subsequently takes place to provide the barrier necessary for the physical separation of the pulmonary and systemic flows of blood. The cardiac conduction system provides the components necessary for the generation of the electrical stimulus as well as the means to propagate this impulse to the myocardial cells of the chambers, which results in the heart beat and the proper sequential contraction of the chambers. The correct development and alignment of these different components of the adult heart is highly regulated by conserved gene programs which determine cell fate by regulating important cellular processes such as differentiation, proliferation, and migration. Identification of several key players and their role in these different processes during cardiogenesis provided much insight into the morphogenetic mechanisms underlying heart development and will be discussed in this chapter. We will focus on molecular and morphogenetic aspects of important processes that shape the heart, such as early heart development, patterning, chamber development, development of the arterial and venous poles, and formation of the conduction system.

2. EARLY HEART DEVELOPMENT

2.1. The heart-forming fields

The heart is of mesodermal origin and is the first organ which is formed during embryogenesis by a sequence of complex morphogenetic processes. During gastrulation, mesodermal cardiac progenitor cells migrate from the primitive streak to the cranial side of the embryonic disc, forming two heart fields, which join at the midline to form a crescent-shaped cardiac region known as the cardiac crescent (Fig. 1A). This region is referred to as the primary heart-forming field which subsequently folds to form the tubular heart. This primary heart tube consists of an inner layer, the endocardium and an outer layer, the myocardium (Fig. 1B). The earliest cardiac fate of mesodermal progenitor cells can be demonstrated by the expression of *Mesp1* and *Mesp2*, two basic helix-loop-helix (bHLH) transcription factors (Kitajima et al., 2000; Saga et al., 2000). Lineage analyses show that cells expressing these transcription factors contribute to all the cardiac lineages, that is, the myocardium, the endocardium, the epicardium, and also the mesenchyme of the cushions (Saga et al., 2000). At a later stage when the heart fields fuse to form the crescent, transcription factors GATA4 and Nkx2–5 label the cardiac lineage and expression of these factors is induced by the expression of the soluble proteins of the bone morphogenetic protein (BMP) and fibroblast growth factor (FGF) family (Brand, 2003). These pathways are conserved from fly to mouse and play an important role in cardiogenesis, although in higher vertebrates these transcription factors are not essential for the formation of the heart tube, suggesting redundancy and involvement of other factors yet to be identified (Brand, 2003).

Classic labeling experiments in chicken suggested that myocardium is subsequently added or recruited from mesoderm to both the venous and arterial pole of the primary heart tube (Stalsberg and de Haan, 1969). More studies, based on dye labeling experiments in chicken embryos, substantiated the presence of an additional source of cardiac progenitor cells, which contributes to the development of the arterial pole of the tube and which resides outside the conventional primary heart-forming field in the pharyngeal mesoderm (Mjaatvedt et al., 2001; Waldo et al., 2001). This was also observed in mouse, where the expression of the *lacZ* gene,

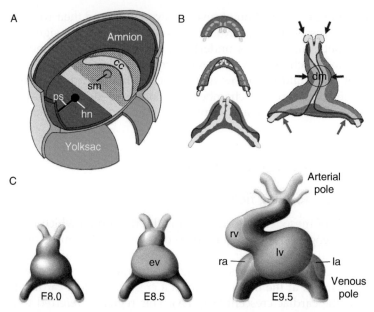

FIGURE 1   Early stages of heart development. (A) In the flat human embryonic disc, the two bilateral progenitor fields at the cranial side of the embryonic disc have joined in the midline to form the cardiac crescent (cc), or primary heart-forming field. (B) Dorsal view of the cardiac crescent, which folds to form the heart tube consisting of an inner layer, the endocardium (yellow), and an outer layer, the myocardium (gray). Note that the red line depicts the peripheral region of the crescent, which becomes the ventral–caudal region subsequent to folding, and which gives rise to the systemic venous system recruited from *Tbx18*-positive mesoderm (red arrows). The blue line depicts the central boundary of the crescent, which becomes positioned dorsally subsequent to folding, and which gives rise to the myocardial descendants of the *Isl1*-positive second heart field (blue arrows). (C) Stages of heart development after formation of the primary heart tube. At embryonic day (E) 8.0 in mouse, the heart tube consists of primary myocardium, which displays a primitive phenotype. At E8.5, the first sign of chamber formation is observed at the ventral side of the tube, where the embryonic ventricle is formed. Subsequently at E9.5, the heart tube begins to loop and the atria are formed at the outer curvatures of the heart tube. dm, dorsal mesocardium; ev, embryonic ventricle; hn, Hensen's node; la, left atrium; lv, left ventricle; ps, primitive streak; ra, right atrium; rv, right ventricle; sm, stomatopharyngeal membrane.

integrated in the *Fgf10* locus, and DiI labeling of pharyngeal mesoderm revealed that myocardial cells at the arterial pole of the tube are recruited from pharyngeal mesoderm (Kelly et al., 2001). Lineage analysis of cells expressing the LIM-homeodomain gene *Isl1* revealed a cardiogenic precursor field, that contributes to the arterial pole, right ventricle, and also to the atria (Cai et al., 2003). *Isl1*-knockout mice die early in development and lack outflow tract, right ventricle, and part of the atria, suggesting that *Isl1*-positive cells residing in the pharyngeal mesoderm and mediastinal mesoderm of the dorsal mesocardium (Fig. 1B) are not only crucial for the contribution of cells to the arterial pole, but also to the venous pole. In addition, this study and other experiments identified the pathways important for the regulation of the cardiac progenitors in the second heart-forming field. These pathways involve transcription factors such as GATA factors, Nkx2–5, Foxh1, Mef2c and soluble factors BMPs and FGFs, which act downstream of and/or in conjunction with Isl1 (Dodou et al., 2004; von Both et al., 2004; Buckingham et al., 2005; Yang et al., 2006). Since none of these genes specifically marks either the primary or the second heart-forming field, it is unclear whether the second heart-forming field represents a distinct cardiac lineage or whether it roots from the same lineage as myocardial cells of the primary heart-forming field (Abu-Issa et al., 2004; Moorman et al., 2007). Nonetheless, we employ the term second heart-forming field to refer to the cell population that is recruited to both poles of the tube.

2.2. Patterning and looping of the tube

While myocardium is added at both poles, the heart tube undergoes a looping process, which results in the parallel arrangement of the future chamber compartments. Starting as a nearly symmetric structure, the tube becomes asymmetrically arranged as a result of the rightward looping of the tube (Fig. 1C). Graded expression of different genes along the anteroposterior, dorsoventral, and left–right axis is crucial for establishing cellular identity during looping of the heart tube and formation of the chambers. Over the time, there has been dispute about the point in time when cells of the cardiogenic region become arranged in a pattern corresponding to their future location. Classic labeling experiments in chicken, in which cells in the primitive streak already appeared to have the positional identity corresponding with their future location, led to the segmental model of heart development (Fig. 2A) (Rosenquist, 1970; Garcia-Martinez and Schoenwolf, 1993). This model assumes that the heart-forming regions and the heart tube can be divided into segments along the anteroposterior axis, which are predisposed to contribute to the different compartments of the mature heart. However, other experiments in both chicken and mouse, using labeling techniques as well as genetic tools to label cells in the early heart-forming region, demonstrated that these cells are not yet definitively organized and that the first evidence of anteroposterior patterning is not established until later stages in the cardiac crescent (Stalsberg and de Haan, 1969; Redkar et al., 2001; Meilhac et al., 2003). When the heart tube is formed, polarity along the anteroposterior axis can also be observed by measuring electrical activity, with cells displaying the highest beat rate always located at the inflow region, or posterior part, of the heart. This results in unidirectional contractions, which generate a flow of blood before a conduction system is present.

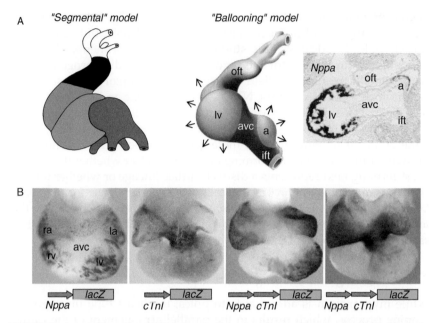

FIGURE 2 Local formation of the chamber is a morphogenetic process which depends on the local repression of gene programs in the primary myocardium. (A) Two models of heart development. On the left, the segmental model and on the right the ballooning model. The section on the right shows expression of *Nppa* in myocardium of the forming chambers in the mouse heart at E9.5. (B) Local repression of the regulatory sequences of *Nppa* in the atrioventricular canal. Repressive elements present in a 0.7-kb regulatory fragment of *Nppa* are responsible for local repression in the atrioventricular canal. This promoter fragment drives expression of the *lacZ* reporter gene (blue staining) in a pattern largely mimicking the endogenous *Nppa* pattern in the chamber myocardium, but not in the atrioventricular canal. When the *Nppa* promoter is cloned upstream of a promoter fragment of a gene, *cTnI*, which drives expression of the reporter gene in the atrioventricular canal, the expression of *lacZ* is repressed in this region. Mutation of a T-box binding element or an Nkx binding element, depicted as a red asterisk, abolishes this repression, thus showing that T-box factor and an Nkx factor (Nkx2–5) are responsible for the repression of *Nppa* promoter activity in the primary myocardium of the atrioventricular canal. a, atrium; avc, atrioventricular canal; ift, inflow tract; la, left atrium; lv, left ventricle; oft, outflow tract; ra, right atrium; rv, right ventricle.

Retinoic acid signaling is essential for anteroposterior patterning of the primary heart tube. Deficiency of retinoic acid leads to expansion of the ventricular, or anterior, compartment and excess of retinoic acid results in expansion of atrial, or posterior, structures (Yutzey et al., 1994; Hochgreb et al., 2003). Retinaldehyde-specific dehydrogenase 2 (RALDH2), which is responsible for retinoic acid synthesis, was found to be the key component of retinoic acid signaling during embryogenesis and heart development (Niederreither et al., 2001). Several genes are sensitive for retinoic acid signaling including the transcription factor Tbx5, which is expressed in the cardiac crescent and later prevails in the posterior part of the developing heart, with high expression in the sinoatrial region (Bruneau et al., 1999). RALDH2 deficiency results in reduced levels of *Tbx5* expression in the sinoatrial region and excess of retinoic acid in the heart tube leads to high levels of *Tbx5* in the right ventricle and outflow tract, suggesting that *Tbx5* acts downstream of retinoic acid signaling in the heart (Liberatore et al., 2000; Niederreither et al., 2001). Accordingly, hearts of *Tbx5*-deficient mice are severely malformed showing hypoplasia of the posterior structures of the tube (Bruneau et al., 2001).

With rightward looping of the heart tube, the heart becomes an asymmetric structure, which is governed by a left–right patterning pathway seemingly independent of pathways determining normal left–right identity of the different components of the body and the future heart. The molecular pathways involved in the regulation of looping of the heart tube are till date largely unknown; however, the transforming growth factor-β (TGF-β) superfamily has been implied to play an important role in this process. Block or expansion of Bmp-signaling in the heart tube of *Xenopus* and zebrafish and cardiac-specific deletion of *ALK5*, a type 1 TGF-β receptor, in mice leads to a block of looping (Chen et al., 1997; Charng et al., 1998; Breckenridge et al., 2001). Furthermore, mice lacking transcription factors dHand/Hand2, Nkx2–5, and Mef2c are embryonically lethal before the looping of the heart tube is completed, suggesting these factors may also be involved in the looping of the heart tube. It is not known, however, whether this observed arrest of looping morphogenesis is secondary to the myocardial defects seen in these mice (Lyons et al., 1995; Lin et al., 1997; Srivastava et al., 1997).

The homeobox transcription factor Pitx2 was found to be an important regulator of the molecular program governing the asymmetric morphogenesis of several organs including the heart. Mice lacking the Pitx2c isoform, which is responsible for left–right morphogenesis, display a range of cardiac morphogenetic defects due to laterality defects, including right isomerism (Kitamura et al., 1999). It is unlikely, however, that Pitx2c controls the looping process itself since these mice do not have any looping defects, suggesting a role for this transcription factor in establishment of left–right identity at later stages. Interestingly, Pitx2 has also been linked to the second heart-forming field. *Pitx2* is asymmetrically expressed in the left part of the second heart-forming field and was found to be regulated by the T-box transcription factor Tbx1 and Nkx2–5 in this region (Nowotschin et al., 2006). Conditional inactivation of *Pitx2* in the second heart-forming field results in severe outflow tract defects (Ai et al., 2006; Nowotschin et al., 2006). Just before looping commences, after anteroposterior patterning has been determined at E8–E8.5 in mouse, dorsoventral patterning is observed by the expression of several genes at the ventral side of the tube (Moorman and Christoffels, 2003). Genes like *Nppa/Anf*,

Smpx/Chisel, *Msg1/Cited1*, and *Hand1/eHand* are expressed in the ventral region of the heart tube where the embryonic ventricle is formed, and mark the initiation of chamber development (Moorman and Christoffels, 2003; Christoffels et al., 2004a). Chamber formation is an event that is tightly regulated by conserved transcription factors, such as T-box factors, which ensure local initiation and restriction of a chamber-specific gene program, as will be discussed in the next chapter.

3. CHAMBER DEVELOPMENT

3.1. Primary versus working myocardium

The classic view presented by the segmental model of heart development is that all the components of the adult heart are already represented in the primary heart tube in an anteroposterior fashion (Fig. 2A). However, the discovery of the second heart-forming field led to the realization that after the linear heart tube has been formed, large parts of the developing heart are added from the adjacent mesenchyme. Only the progenitors of the future left ventricle reside in the linear heart tube. Recent experiments in chicken show that a mesenchymal precursor pool at the venous pole of the heart tube displays the highest proliferation rate and that the primary myocardium of the heart tube itself exhibits low, if any, proliferative activity (Soufan et al., 2006). This is consistent with the idea that the initial elongation of the heart tube mainly occurs by recruitment of myocardial precursors from a cardiac precursor pool and not by proliferation of primary myocardium. Furthermore, a local increase in proliferation is observed at the outer curvature when the heart tube starts to loop, which coincides with the local initiation of chamber formation (Soufan et al., 2006). Myocardium of the outer curvature at this stage obtains a distinct molecular and structural phenotype, which is marked by the initiation of a distinct gene program, whereas the myocardium of the outflow tract, the atrioventricular canal (AVC), the inflow tract, and also the inner curvatures retain the more primitive myocardial phenotype of the primary heart tube. Myocardium of the developing and mature chambers is termed working myocardium since these cells exhibit all the features necessary for force production, including high conduction velocity of the electrical impulse, good cell-to-cell coupling, and well-developed sarcomeric structures (Table 1). This model of heart development has been dubbed by the ballooning model of chamber formation (Moorman and Christoffels, 2003), since working myocardium expands, or balloons, at localized regions of the primary heart tube (Fig. 2B).

Until recently, it was not known whether the gene program responsible for this localized expansion of the chambers and the working myocardial phenotype is regionally activated and not active in the parts that do not form

TABLE 1 Phenotype of myocardial cells

Phenotype	"Primary" myocardium	"Nodal" myocardium	His– Purkinje	"Working" myocardium
Automaticity	High *Hcn4*	High *Hcn4*	High *Hcn4*	Low
Conduction velocity	Low	Low	High *Cx40*	High *Cx40/43*
Contractility	Low	Low	Low	High
SR activity	Low	Low	Low	High
Proliferation	Low	Low	Low	High

Characteristic features of different myocardial components of the developing and adult heart. Nodal myocardium of the sinuatrial node and the atrioventricular node shares molecular and morphological features with the primary myocardium of the heart tube, such as high automaticity, low conduction velocity, low contractility and poorly developed sarcomeres and sarcoplasmic reticular structure (SR-activity). High automaticity in both primary and nodal myocardium coincides with the expression of the pacemaker channel *Hcn4*. Cells of the working myocardium of the chambers develop a phenotype which makes these cells specialized for contraction and conduction of the electrical impulse, with a high abundance of fast conducting gap junction proteins *Cx40* and *Cx43*. They also have well-developed sarcomeres and sarcoplasmic reticular structure and display a high proliferation rate during development. Cells of the His-purkinje system display an intermediate phenotype, partly retaining the high automaticity and primitive phenotype of the primary myocardium but also displaying high conduction velocity necessary for the fast propagation of the electrical impulse to the working myocardium.

working myocardium, or whether the chamber-specific gene program is locally repressed. Furthermore, the transcriptional regulators governing the chamber-specific gene program were unknown. Expression and regulation of natriuretic precursor peptide type A (*Nppa*) has been extensively studied in this perspective, because it marks the myocardium of the forming chambers (Fig. 2B). Transcriptional activity of this gene in working myocardium of the chambers is regulated by conserved transcription factors including Nkx2–5, GATA4, and Tbx5 (Bruneau et al., 2001; Hiroi et al., 2001; Garg et al., 2003; Small and Krieg, 2003). Several studies, using knockout mouse models, show that these transcription factors are important for chamber development, with homozygous knock out of these genes causing hypoplastic development of the chambers or even a lack of chamber formation. In addition, mutations in *Nkx2–5, GATA4,* and *Tbx5* in human resulting in haploinsufficiency are a frequent cause of heart defects, including septal defects (Basson et al., 1997; Schott et al., 1998; Bruneau et al., 1999; Li et al., 2000; Garg et al., 2003; Pashmforoush et al., 2004).

However, the transcription factors mentioned earlier are also present in the primary myocardium where *Nppa* expression is absent, indicating that other factors are involved in the regulation of *Nppa* in the parts of the tube that do not form working myocardium. Studies of transgenic mice with *lacZ* controlled by 0.7-kb *Nppa* regulatory sequences, demonstrated that these sequences drive expression of β-galactosidase in a pattern mimicking that of endogenous *Nppa* in the working myocardium (Fig. 2B).

Repressive elements present in the *Nppa* promoter overrule the expression of β-galactosidase in the AVC driven by *cTnI* regulatory sequences. This repression of *lacZ* expression was mediated by the transcription factors Nkx2–5 and Tbx2 and/or Tbx3, since mutation of either a NK-binding element or a T-box binding element in the 0.7-kb *Nppa* regulatory sequences resulted in expression of the transgene in the AVC and outflow tract (Habets et al., 2002, 2003) (Fig. 2B). These experiments suggested a repressive transcriptional control of the chamber-gene program in regions of the heart tube that do not obtain the chamber phenotype.

3.2. Regionalized control of chamber-specific gene expression

The T-box factors, Tbx2 and Tbx3, are expressed in the developing heart and act as transcriptional repressors of gene activity displaying high homology in both structure and function (He et al., 1999; Lingbeek et al., 2002; Paxton et al., 2002; Christoffels et al., 2004b; Hoogaars et al., 2004). During heart development, *Tbx2* and *Tbx3* show an overlap in the expression (Fig. 3A), suggesting a possible redundancy in function in early development. In mouse, *Tbx2* is expressed in the inflow tract, AVC, and outflow tract of the developing heart from about E8.5 onward (Fig. 3A). At later stages, *Tbx2* expression gradually fades in the heart (Christoffels et al., 2004b). Expression of *Tbx3* is observed in the heart throughout development and adult stages, initially overlapping with *Tbx2* in the inflow region and AVC (Fig. 3A). At later stages, expression of *Tbx3* becomes restricted to the proximal components of the conduction system (Hoogaars et al., 2004). Apart from myocardial expression, both *Tbx2* and *Tbx3* are expressed in the cushion mesenchyme of the AVC and outflow tract (Fig. 3A and not shown). Both factors are also expressed in the second heart-forming field. The expression patterns of *Tbx2* and *Tbx3* in the developing heart are conserved in chicken and human, suggesting the role of these factors is evolutionary preserved in vertebrates (Christoffels et al., 2004b; Hoogaars et al., 2004).

Recent experiments in which the functional role of Tbx2 during heart development was examined, revealed that Tbx2 can repress the development of the identity of the working myocardium. Overexpression of *Tbx2* in the heart tube using the βMHC-promoter resulted in lack of chamber formation and repression of *Nppa* and the gap junction gene *Cx40* (Fig. 3B) (Christoffels et al., 2004b). Conversely, *Tbx2* homozygous knockout mice die early in development of severe cardiovascular defects and lack an atrioventricular boundary as can be demonstrated by the expression of the chamber-specific genes *Nppa*, *Cx40*, *Smpx*, and *Cited1* (Harrelson et al., 2004). These data suggest that Tbx2 is important for the repression of the chamber-specific gene programs in the primary myocardium, resulting in the preservation of the primitive slow conducting phenotype. More recent studies from our laboratory show that Tbx3 has a comparable role during cardiogenesis (unpublished observations).

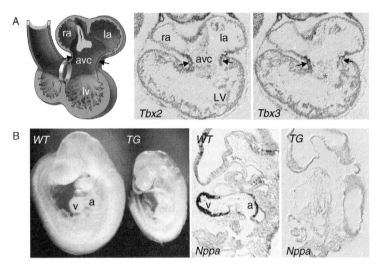

FIGURE 3 Expression of T-box factors Tbx2 and Tbx3 in the primary myocardium of the developing heart represses chamber development. (A) Section of E9.5 heart, showing expression of Tbx2 and Tbx3 in the atrioventricular canal (black arrows) and cushion mesenchyme (red asterisk). (B) Overexpression of Tbx2 using the βMHC-promoter results in block of chamber development and expression of chamber-specific genes. βMHC-Tbx2 transgenic mouse embryos of E9.5 (TG) have a linear heart tube, while wild-type littermates (WT) show chamber formation. Sections on the right show expression of Nppa in wild-type embryos, which is absent in βMHC-Tbx2 embryos. a, atrium; avc, atrioventricular canal; ev, embryonic ventricle; la, left atrium; lv, left ventricle; ra, right atrium.

The primary myocardium, which expresses *Tbx2* and *Tbx3*, continues to display a low proliferation rate as opposed to the fast expanding and proliferating working myocardium. This makes it conceivable that another task for these transcriptional regulators could be the suppression of the proliferation rate of myocardial cells. It is well established that these transcription factors are involved in cell cycle regulation. In addition, expression of *Tbx2* was found to be tightly regulated during cell cycle progression, thus making it likely that Tbx2 plays a role in cell cycle regulation (Bilican and Goding, 2006). Both Tbx2 and Tbx3 are amplified in a number of cancers and both factors are able to control the expression of several cell cycle regulators resulting in immortalization of cells (Jacobs et al., 2000; Rowley et al., 2004; Vance et al., 2005). In cell culture, fibroblasts undergo a process called senescence, which results in cell cycle arrest. This process is activated by accumulation of the tumor suppressor protein p53, which is protected against degradation by the negative cell cycle regulator p19[ARF]. Additionally, p53 activates transcription of *Cdkn1a* which encodes p21, a protein that prevents cells to enter the S-phase or DNA replication phase of the cell cycle resulting in cell cycle arrest. Tbx2 and Tbx3 are capable of repressing the transcriptional activity of the *Cdkn2a* locus, which encodes p19^{ARF}. This results in the degradation of p53 and the subsequent immortalization of fibroblast cells in culture (Brummelkamp et al., 2002; Lingbeek et al., 2002). Tbx2 can also invoke senescence bypass through repression of p21 (Prince et al., 2004; Vance et al., 2005). Accordingly, one would expect the myocardial components of the heart expressing these factors to display a high proliferation rate. This, however, is not the case. Moreover, the observations that Tbx2 and Tbx3 induce senescence bypass are derived from experiments which use cell culture systems. There is no proof, as yet, that Tbx2 and/or Tbx3 are involved in the regulation of p53-dependent pathways in cardiac muscle cells and other tissues (Davenport et al., 2003; Harrelson et al., 2004; Jerome-Majewska et al., 2005). Taken together, these data suggest that other pathways should be involved in the regulation of the cell cycle in the myocardium, or that these transcription factors play an opposite role in cell cycle regulation in myocardium. In support of this latter notion, is the recent observation that Tbx2 may be involved in the repression of proliferation in primary myocardium, ensuring low proliferation rate by the repression of *Nmyc1*, a gene important for the expansion of chamber myocardium (Cai et al., 2005). Mice deficient of T-box factor Tbx20 show a lack of chamber development and ectopic *Tbx2* expression in the heart, which is accompanied with the repression of *Nmyc1* (Cai et al., 2005). Moreover, direct binding of Tbx2 to a T-box binding element present in the promoter of this gene resulted in the repression of transcriptional activity of *Nmyc1*. Cyclin A2 was also downregulated in these mice, although it was not established whether this was a direct cause of Tbx2 misexpression in the heart (Cai et al., 2005).

These data show that Tbx2 is important for the retention of the primary myocardial phenotype in slowly proliferating regions of the heart tube. The primary myocardium of the inflow tract, the AVC, and the outflow tract is subsequently involved in two important processes during heart development. First, primary myocardium of the outflow tract and the AVC induces epithelium-to-mesenchyme transformation of endocardial cells, resulting in the formation of cushion mesenchyme. This cushion mesenchyme is important for the development of the septa which divides the systemic and pulmonary circulation, thereby participating in the correct alignment of the chambers. Second, as will be discussed later on, the nodal components of the conduction system originate from the primary myocardium.

3.3. Modes of cardiac transcriptional regulation

It is imperative to appreciate that the regulation of downstream target genes by T-box factors is complex since transcriptional activators of *Nppa*, such as Tbx5, GATA4, and Nkx2-5 are present in both primary and working myocardium. The transcriptional regulation of chamber-specific genes therefore depends on the physical and genetic interactions of their transcriptional regulators, such as T-box factors, present in the different components of the developing heart. All T-box factors share a conserved region called the T-box which is highly preserved between all T-box family members. The T-box domain accounts for the DNA-binding ability of these transcription factors to conserved DNA-binding sites called T-box binding element (Bollag et al., 1994; Plageman and Yutzey, 2005). The consequence of this preserved T-box domain is that different T-box factors can target the same genes, depending on their spatial distribution and their specific DNA-binding affinity for the specific T-box binding element. In addition, some T-box factors can target specific genes through binding to variant T-box binding sites (Lingbeek et al., 2002). T-box factors can act as repressor or activator of transcription, depending on the presence of additional activation or repression domains (Naiche et al., 2005; Plageman and Yutzey, 2005). Therefore, the expression of T-box target genes is dependent on the presence of the different T-box factors in their expression domain, and may be fine-tuned through competition between different T-box factors. *In vitro*, these interactions can be demonstrated with reporter-gene assays using the promoter region of *Nppa* or *Cx40*, which contain multiple conserved T-box binding elements (Bruneau et al., 2001; Hiroi et al., 2001). Activity of the *Nppa* promoter is synergistically activated by Tbx5 and Nkx2–5. Cotransfection with either Tbx2 or Tbx3 resulted in dose-dependent abrogation of this synergistic activation, thus showing these factors can compete with Tbx5 to regulate *Nppa* gene-transcription (Christoffels et al., 2004b; Hoogaars et al., 2004). Regulation of *Nppa* is dependent on the DNA-binding capacity of these proteins, since isoforms of Tbx2 or Tbx3 with a mutation in the T-box domain, can not bind to the promoter and are not able to exert this repression.

These reporter-gene assays also show that physical interactions between T-box factors and other transcription factors, such as Nkx2.5 and Gata4, can result in synergistic regulation of target genes. Mutations specifically affecting these interactions are known to cause congenital heart disease in human (Garg et al., 2003). Recent experiments identified other protein partners of T-box factors cooperating to regulate specific target genes, although the exact function of these protein complexes in heart development is, as yet, not fully understood (Murakami et al., 2005; Koshiba-Takeuchi et al., 2006). In this respect, it is intriguing to see that a protein involved in the deacetylation of genes, HDAC1, interacts with Tbx2 to repress the *Cdkn1a* promoter activity in melanoma cells (Vance et al., 2005).

Recent lines of evidence also pointed out that T-box factors may regulate expression of other T-box factors. As mentioned earlier, analysis of knockout mice of the T-box factor Tbx20, which is expressed in primary as well as working myocardium in early stages of heart development, revealed ectopic expression of *Tbx2* (Cai et al., 2005; Singh et al., 2005; Stennard et al., 2005). In addition, one of these studies demonstrated that Tbx20 is capable of directly suppressing the transcriptional activity of the

Tbx2 promoter, and that Tbx20 binds to a T-box binding element present in the *Tbx2* promoter sequence (Cai et al., 2005). Analysis of *Tbx5*-deficient mice, using micro-array analysis, revealed downregulation of *Tbx3*, placing Tbx5 upstream of Tbx3, although it is not known, as yet, whether this regulation is direct or indirect (Mori et al., 2006).

In summary, T-box factors are important transcriptional regulators of chamber-specific genes and accordingly control the formation and restriction of the working myocardial phenotype in the developing heart. The transcriptional regulation of T-box target genes is highly dependent on protein–protein interactions with other protein partners as well as the availability of the different T-box factors.

3.4. Role of Bmp signaling in chamber formation and septation

BMPs are soluble proteins that play an important role in many developmental processes through paracrine and autocrine signaling pathways resulting in Smad- and TAK1-mediated transcriptional regulation of downstream target genes. During vertebrate heart development, several members of this family are expressed in the heart where they are involved in different processes. Expression of *Bmp10* is restricted to working myocardium of the developing chambers, where it is required for normal proliferation and maturation of myocardial cells. *Bmp10*-deficient mice display severe cardiac defects including hypoplastic ventricles and absence of trabecular myocardium as a result of proliferation defect (Chen et al., 2004). In addition, a conditional knockout mouse model for *Nkx2–5* in the ventricles shows hypertrabeculation as a result of *Bmp10* overexpression, suggesting Nkx2–5 controls Bmp10-mediated proliferation in trabecular myocardium (Pashmforoush et al., 2004).

Bmp2 and *Bmp4* are expressed early in the cardiac crescent stage where they induce expression of *Nkx2–5* and *Gata4* and are important for cardiomyocyte differentiation (Brand, 2003). At later stages, at the time chamber development is initiated, these factors become restricted to AVC and outflow tract myocardium. BMPs have been implied to regulate the expression of T-box factors in the heart and in other tissues like the limbs (Suzuki et al., 2004). Experiments in chicken embryos showed that expression of *Tbx2*, and to a lesser extent *Tbx3*, can be induced by beads soaked in Bmp2 protein (Yamada et al., 2000). Cardiac-specific deletion of *Bmp2* in mice resulted in a lack of atrioventricular restriction as demonstrated by the expansion of chamber markers *Nppa*, *Cx40*, and *Chisel/Smpx*. Furthermore, these mice lack *Tbx2* expression, suggesting a role for Bmp2 in the restriction of the chamber phenotype through regulation of *Tbx2* expression (Ma et al., 2005; Rivera-Feliciano and Tabin, 2006). Conditional knockout of *BmpR1* in the *Isl1*-positive second heart-forming field resulted in the ablation of *Tbx3* expression in the AVC myocardium. Regulatory sequences of the *Tbx3* gene contain Smad sites capable of inducing transcriptional activity of the

Tbx3 promoter *in vitro* through Smad1 and 4 (Yang et al., 2006). Regulation of *Tbx2* and *Tbx3* mediated by BMP signaling seems to be specific since neither *Tbx5* or *Tbx20* are affected in these models.

Apart from a role in repression of chamber identity, both Bmp2 and Bmp4 are important for epithelium-to-mesenchyme transition in the outflow tract and the AVC. During this process, signals from the myocardium to the endocardium result in the invasion of epithelial cells into the cardiac jelly, thereby forming cushion mesenchyme, which is essential for the development of septa and valves of the heart. *Bmp4*-deficient mice and conditional myocardial-specific inactivation of *Bmp2* in mice result in defects of atrioventricular cushion formation and septation defects (Jiao et al. 2003; Ma et al. 2005). Furthermore, inactivation of the type 1A-Bmp receptor (Bmpr1a), a major receptor for Bmp2 and Bmp4 signaling, in the endocardium disrupted endocardial cushion formation (Ma et al. 2005). Since *Tbx2* and *Tbx3* are expressed in cushion mesenchyme and the myocardium flanking the cushions, these T-box factors might be involved in epithelium-to-mesenchyme transition and in the development of the cushion mesenchyme flanking the primary myocardium (Fig. 3A). Together, these data place BMPs high in the transcriptional hierarchy that dictate morphogenesis during heart development, where they play diverse roles in early development as well as later in development during chamber development and septation by regulating the expression of transcription factors like T-box factors and Nkx2–5.

4. DEVELOPMENT OF THE ARTERIAL AND VENOUS POLES OF THE HEART

4.1. Introduction

The arterial and venous poles of the heart serve as the outflow and the inflow tract, respectively, of the blood flow in the heart. In the formed adult heart of higher vertebrates, the outflow tract is segregated and comprises the aorta, which provides the systemic circulation, and the pulmonary artery, through which the pulmonary circulation is guided. The inflow tract consists of the inferior and superior caval veins, which make up the systemic venous return, draining the blood from the systemic circulation into the right atrium, and the pulmonary veins, which make up the pulmonary venous return that drains the oxygenated blood from the pulmonary circulation into the left atrium.

During heart development, the outflow tract and the inflow tract form by recruitment of mesenchymal cells to the myocardial lineage, as will be discussed later. In addition, cells migrating from the neural crest make an important contribution to the septation process of the outflow tract, which is crucial for the segregation of the outflow tract that subsequently gives rise to the aorta and pulmonary trunk (Kirby et al., 1983). Early in heart development, elongation of the primary heart tube prior to the appearance of the chambers occurs mainly by recruitment of mesodermal cells at both poles (Fig. 1B). At this stage, the tube consists of progenitors of the future left ventricle only, while the myocardium of the outflow tract, right ventricle, and of large parts of the atria is recruited from the *Isl1*-positive second heart-forming field. Whereas the chambers obtain the phenotype of working myocardium, the rest of the developing heart including the outflow tract and the systemic venous return component of the inflow tract retains the phenotype of the primary myocardium, which is marked by the absence of chamber markers such as Nppa and Cx40.

4.2. Development of the arterial pole

The molecular mechanisms underlying the development of the outflow tract have recently received much attention owing to the discovery of the second heart-forming field, which was found to be the main contributor to the formation of this structure (Buckingham et al., 2005). Tbx1 is a key regulator in outflow tract development, possibly by regulating proliferation in the second heart-forming field, thereby effecting recruitment of mesodermal cells to the arterial pole (Baldini, 2005; Zhang et al., 2006). FGF signaling has also been identified to play a major role in outflow tract development. *Fgf8* expression in the second heart-forming field is required for cell survival and proliferation in this region and normal outflow tract development, as demonstrated by studies that conditionally deleted *Fgf8* in the second heart-forming field (Ilagan et al., 2006; Park et al., 2006). Several studies show that Tbx1 acts upstream of Fgf8 (Vitelli et al., 2002; Hu et al., 2004; Vitelli et al., 2006), although Fgf8 also seems to function independent of Tbx1 (Vitelli et al., 2006). In addition, other transcription factors that play important roles in the cell population of the second heart-forming field are also necessary for the normal formation and elongation of the outflow tract, including Pitx2c, Isl1, Mef2c, and FoxH1 (Buckingham et al., 2005).

4.3. Development of the venous pole

The inflow tract of the heart tube can be morphologically distinguished as an initially symmetric component consisting of two conduits draining into the heart at the venous pole at E8 in mouse. At this stage, the inflow tract consists entirely of the systemic venous return since the pulmonary venous return has not yet formed. In lower vertebrates, like fish, the compartment in which the systemic venous return drains into the heart is called sinus venosus. This structure is not observed as such in the adult heart of higher vertebrates. In addition, recent experiments based on morphological and molecular observations show that in mammals, no sinus venosus can be distinguished upstream from the atrium at the venous pole at any given time point in development (Soufan et al., 2004).

Thus, the convergence of the systemic venous return should be considered as an integral part of the forming atrium and never as a separate entity like the sinus venosus in lower vertebrates.

Approximately two days later in development (E10–11 in mouse), the systemic venous return; at this stage, the common cardinal veins become asymmetrical arranged, protrude into the pericardial cavity from the pericardial mesenchyme, and become surrounded by primary, that is, *Cx40* and *Nppa* negative, myocardium. These myocardial structures are termed sinus horns and represent the precursors of the adult systemic venous return. In human, the right sinus horn will become the dominant structure of the systemic venous return, forming the myocardial component of the inferior and superior caval veins, which is incorporated into the right atrium to form the smooth-walled part of the right atrium called the sinus venarum. The left sinus horn diminishes in size and ends up as the coronary sinus, forming the confluence of the cardiac veins, which drain into the heart at the border of the right side of the AVC. In contrast to the human condition, in mouse both the left and the right sinus horns persist as the left and right superior caval veins.

The data from our laboratory show that the myocardium, which surrounds the initial inflow tract of the heart tube and which expresses the homeodomain transcription factor Nkx2-5, becomes incorporated into the developing atria, including the venous valves (Soufan et al., 2004; Christoffels et al., 2006). The myocardial precursors of the sinus horns are recruited at a later stage from pericardial mesenchyme that expresses the T-box factor Tbx18 (Christoffels et al., 2006). In addition, lineage analysis experiments of *Nppa*-expressing cells and *Nkx2-5*-expressing cells show that these "atrial" myocardial lineages do not contribute to the development of the sinus horns (Christoffels et al., 2006). Interestingly, the *Tbx18*-positive pericardial mesenchyme is *Isl1*-negative, suggesting this recruitment does not involve precursors of the second heart-forming field (Fig. 1B). The importance of Tbx18 in the development of the systemic venous return was demonstrated in *Tbx18*-deficient mice, which have a defect in sinus horn development, and in which myocardial differentiation was impaired and the common cardinal veins remained embedded in the pericardial mesenchyme (Christoffels et al., 2006). The pulmonary veins, which constitute the pulmonary venous return draining into the left atrium, form later in development from the mediastinal mesenchyme (Christoffels et al., 2006).

5. DEVELOPMENT OF THE CARDIAC CONDUCTION SYSTEM

5.1. Definition of the mature cardiac conduction system

The cardiac conduction system is a crucial functional component of the vertebrate heart. It consists of cells of myogenic origin that are responsible for the generation and the propagation of the electrical impulse to the

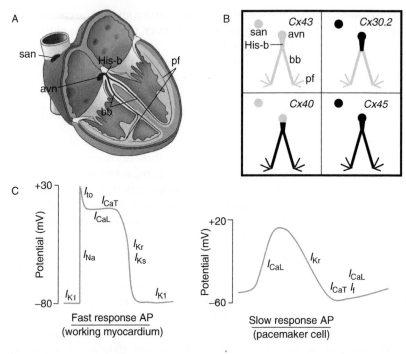

Current	Channel (gene)	Function	Loss of function
I_f (in)	Hcn4 (*Hcn4*)	Diastolic depolarization	san dysfunction/bradycardia (M+H)
I_{CaL} (in)	Cav1.2 (*Cacna1c*)	Upstroke, diastolic depolarization	Decreased contraction and heart rate (Z+M) Timothy syndrome; san dysfunction/bradycardia (H)
I_{CaL} (in)	Cav1.3 (*Cacna1d*)	Upstroke, diastolic depolarization	Bradycardia, AV-block, arrythmia (M)
I_{CaT} (in)	Cav3.1 (*Cacna1g*)	Diastolic depolarization	san dysfunction, AV-block (M)
I_{Kr} (out)	ERG1 (*Kcnh2*)	AP duration, repolarization	Bradycardia (M), long QT2 syndrome (H)
I_{St} (in)	?	Diastolic depolarization?	?

FIGURE 4 Molecular determinants of the adult conduction system. (A) The components of the cardiac conduction system of the adult heart. avn, atrioventricular node; bb, bundle branches; His-b, His bundle; pf, Purkinje fibers; san, sinuatrial node; scv, superior caval vein. (B) Scheme showing the expression of gap junction proteins in the adult cardiac conduction system. *Cx43* is not expressed in the conduction system of adult mouse heart. *Cx40* is expressed in the fast conducting His–Purkinje system. *Cx45* is expressed in all the conduction system components and *Cx30.2* is expressed in the nodal

cardiomyocytes of the chambers. Based on this definition, the cardiac conduction system of the adult heart can be subdivided into the nodal components, which display the ability of impulse generation, and the His–Purkinje system, which displays high conduction velocities, necessary for impulse propagation. The term central conduction system (CCS) has been introduced in the past without exact definition, but is used further on to refer to the proximal components of the conduction system, that is, the sinuatrial node, the atrioventricular node, the His bundle, and the interventricular ring.

The nodal components comprise the sinuatrial node, located at the border of the right atrium and the superior caval vein, and the atrioventricular node, located at the atrioventricular junction (Fig. 4A). The sinuatrial node acts as the dominant pacemaker of the heart and is crucial for the initiation of the heart beat. The atrioventricular node also has pacemaker characteristics and, in addition, acts as a relay point, delaying the electrical impulse at the atrioventricular junction thereby ensuring proper sequential contraction of the atrial and ventricular chambers. The His–Purkinje system includes the His bundle, or atrioventricular bundle, which propagates the electrical impulse to the components of the ventricular conduction system. These consist of the bundle branches on both sides of the interventricular septum, and Purkinje fibers embedded in the ventricular myocardium (Fig. 4A). This configuration of the fast conducting components of the conduction system results in the apex-to-base excitation of the ventricles. The myocardial cells of the His–Purkinje system display high conduction velocities and good cell-to-cell coupling, which makes these cells exquisitely suited for fast conduction. In contrast to what the term conduction system may lead one to suspect, the nodal components of the conduction system are in fact not part of the fast conducting network. Instead, these pacemaker cells are poorly coupled and exhibit low conduction velocities. In addition, nodal cells have few myofilaments, low abundance of mitochondria, and poorly developed sarcoplasmic reticulum (Virágh and Challice, 1977, 1980; Boyett et al., 2000).

Apart from conductive properties and the morphological phenotype, another important feature of myocardial cells of the conduction system, especially the nodes, is pacemaker activity or automaticity, which is defined by the ability to spontaneously generate an electrical impulse necessary to trigger contraction of the working myocardial cells of the

---

components plus the His bundle. (C) Typical action potential of working myocardial cell and a pacemaker cell with the major contributing ion currents. (D) Table showing the major currents responsible for the pacemaker phenotype, the corresponding proteins/ genes, their role in action potential, and the phenotype of channel-deficiency in zebrafish (Z), mouse (M), and/or human (H). AP, action potential; AV-block, atrioventricular block; in, inward current; out, outward current; san, sinuatrial node.

chambers. This can be demonstrated by recording the electrical action potential of myocardial cells. Working myocardial cells display a so-called fast-response action potential, which is characterized by a negative maximum diastolic depolarization and a fast upstroke of the resting membrane potential followed by partial repolarization, the plateau phase, and a subsequent repolarization back to the resting membrane potential (Fig. 4C). Conversely, pacemaker cells display a so-called low-response action potential, characterized by a relatively positive maximum diastolic depolarization of the membrane, a slow upstroke velocity, a lack of partial repolarization and a relatively long action potential duration (Fig. 4C). Automaticity of pacemaker cells is the result of slow diastolic depolarization of the membrane potential until it reaches the threshold for the next action potential (Fig. 4C). The pacemaker cells of the nodes display high automaticity and accordingly prominent diastolic depolarization, resulting in a high pacemaker rate. Myocardial cells of the His–Purkinje system also display diastolic depolarization and can therefore function as escape pacemakers (Callewaert et al., 1984; Schram et al., 2002).

Although the precise molecular landscape that accounts for the hallmarks of the different components of the conduction system has not yet been established, the electrical and conductive characteristics of the conduction system correlates with the unique composition of gap junction proteins and ion channel proteins identified in these cells.

5.2. Gap junctions and impulse propagation

At the molecular level, several genes are linked to the functional properties of myocardial conduction cells. Gap junction channels located at the intercellular junction, or intercalated disc, have been shown to be crucial for the propagation of the impulse. Both morphological and immunological studies have demonstrated a high abundance of these membrane structures in the working myocardium and the His–Purkinje system. Gap junction channels are composed of 12 connexin proteins that allow the propagation of the electrical impulse by passage of ions. Depending on the availability of different connexin isoforms, gap junction channels can be composed of one connexin isoform (homotypic/homomeric channel) or multiple connexin isoforms (heterotypic/heteromeric channel), resulting in gap junctions with unique conductance properties (Moreno, 2004).

Cx40 is the dominant connexin isoform in the fast conducting components of the conduction system of the adult heart and is also expressed in the atria, whereas *Cx43* is exclusively expressed in the working myocardium of the ventricles and the atria (van Kempen et al., 1996; Miquerol et al., 2004) (Fig. 4B; Table 1). Analysis of *Cx40*-deficient mice revealed the necessity of Cx40 in the electrical coupling of myocardial cells of the conduction system and atrial myocardium, since these mice display atrioventricular block, bundle branch block, and altered atrial conduction (Kirchhoff et al., 1998; Simon et al., 1998; Bagwe et al., 2005). In contrast, cells of the nodes have a low abundance of these fast conducting gap junctions (Boyett et al., 2000; Dobrzynski et al., 2003). This accounts for the poor electrical coupling and low conduction velocities observed in the nodes. In addition, two other gap junction proteins, Cx30.2 and Cx45, are expressed in the conduction system. Expression of *Cx45* is found in the entire conduction system while *Cx30.2* is expressed in the nodes and the His bundle where it overlaps with *Cx45* (Fig. 4B) (Coppen et al., 1999a,b; Kreuzberg et al., 2005). Gap junction channels composed of these proteins have much lower conductance properties compared with Cx40 and Cx43 channels (Kreuzberg et al., 2005). This could contribute to the electrical coupling and the, albeit slow, propagation of impulse from the sinuatrial node and the atrioventricular node to working myocardium of the chambers (Kreuzberg et al., 2005). Confusingly, *Cx30.2*-deficient mice show accelerated conduction velocity in the atrioventricular node (Kreuzberg et al., 2006). Cx45 could account for this acceleration, since it is the only isoform coexpressed with Cx30.2 in the atrioventricular node centre. In addition, Cx45 homotypic channels exhibit higher conductance properties compared with Cx45/Cx30.2 heterotypic/heteromeric channels and Cx30.2 homotypic channels (Kreuzberg et al., 2005, 2006). Thus, the expression of different connexin isoforms in the conduction system provides a molecular base for the conductive properties of the conduction system.

5.3. Ion channels and pacemaker phenotype

Different voltage-dependent ion currents contribute to the characteristic electrical phenotype of pacemaker cells, including the hyperpolarization-activated cation or funny current (I_h or I_f), L-type Ca^{2+} current (I_{CaL}), T-type Ca^{2+} current (I_{CaT}), delayed rectifier current (I_K), and sustained inward current (I_{st}) (Fig. 4C and D). Furthermore, pacemaker activity is regulated by the autonomic nervous system, which controls the heart rate, and is highly dependent on the concentration of intracellular cAMP and the activity of cAMP-dependent protein kinase (DiFrancesco, 1993; Baruscotti et al., 2005). Sympathetic stimulation, modulated via β-adrenergic receptors, induces inward currents active during the diastolic depolarization, including I_f, I_{CaL}, and I_{st}, while deactivating the outward I_K current, which results in acceleration of the heart rate. Conversely, parasympathetic stimulation via acetylcholine lowers the heart rate by inhibiting cAMP-dependent modulation of ion channels and by activating the acetylcholine-sensitive K^+ current, I_{KAch}, encoded by *Kir3.1* (Schram et al., 2002).

Automaticity of conduction system cells is predominantly regulated by I_f (DiFrancesco, 1993). I_f is activated upon hyperpolarization of the membrane and provides an inward sodium current that contributes to the spontaneous depolarization of the membrane potential. Regulation of the heartbeat by expression of I_f is conserved in all vertebrate species, and inhibition of this current results in bradycardia and sinuatrial node dysfunction in fish, mouse, and human (Warren et al., 2001; Stieber et al., 2003; Milanesi et al., 2006). The molecular component of I_f is the hyperpolarization-activated cyclic nucleotide-gated family of proteins (*HCN*), which is encoded by four genes in mammals, *Hcn1–4* (Ludwig et al., 1998; Moosmang et al., 2001; Robinson and Siegelbaum, 2003). Hcn4 is prevalent in the developing and mature components of the conduction system and is the dominant isoform in the sinuatrial node and atrioventricular node (Moosmang et al., 2001; Dobrzynski et al., 2003; Efimov et al., 2004). Accordingly, deficiency of this channel in mice causes almost complete loss of I_f and slower contraction of the embryonic heart, resulting in prenatal death between E9.5 and E11.5 (Moosmang et al., 2001; Stieber et al., 2003). Mutations resulting in nonfunctional HCN4 in humans cause bradycardia and sick sinus syndrome, showing the importance of this pacemaker channel in the regulation of the heart rate (Schulze-Bahr et al., 2003; Milanesi et al., 2006). Although expressed in all conduction system components, HCN transcripts are much more abundant in the sinuatrial node compared with the Purkinje fibers, providing a possible molecular explanation for the high pacing rate of the sinuatrial node compared with the His–Purkinje system (Shi et al., 1999).

In addition to I_f, calcium-dependent inward currents have also been shown to play an important role in the regulation of pacemaker activity. L-type calcium currents (I_{CaL}) provide an inward Ca^{2+} current and are important for the spontaneous diastolic depolarization and upstroke of the pacemaker action potential (Verheijck et al., 1999; Schram et al., 2002). L-type calcium channels are composed of several pore-forming subunits, and the main subunit expressed in the heart is the α1C ion-conducting subunit $Ca_v1.2$, encoded by *Cacna1c*, which is enriched in the sinuatrial node (Bohn et al., 2000). Loss of function of $Ca_v1.2$ in humans, as seen in Timothy syndrome, results in conduction system defects like bradycardia, AV block, and arrythmia (Splawski et al., 2004, 2005). $Ca_v1.3$, the α1D ion-conducting subunit of the voltage-gated L-type calcium channel, is also expressed in the SAN, and deficiency of $Ca_v1.3$ in mice results in SAN dysfunction and bradycardia (Platzer et al., 2000; Zhang et al., 2002; Mangoni et al., 2003). The role of T-type calcium currents in pacemaker activity was demonstrated in mice deficient of the protein $Ca_v3.1/\alpha_{1g}$, which is a pore-forming subunit of the T-type calcium channel that is mainly expressed in the conduction system (Bohn et al., 2000). These mice lack I_{CaT} and display bradycardia and impaired AV conduction, suggesting a role for I_{CaT} in conduction and regulation of the heart rate (Mangoni et al., 2006).

The outward delayed rectifier current I_K is divided into a slow component (I_{Ks}) and a fast component (I_{Kr}), and plays an important role in repolarization of the action potential. I_{Kr} is also involved in the regulation of pacemaker activity in mouse. This was demonstrated in mice deficient of ERG1, encoding the α-subunit of I_{Kr}, which display sinus bradycardia (Lees-Miller et al., 2003). Expression of *minK*, which encodes the β-subunit of I_{Ks}, is found in the developing conduction system components, but no report exists on its role in pacemaking or conduction (Kondo et al., 2003).

In addition, several currents are not, or at a much lower level, expressed in pacemaker cells compared with working myocardial cells. Absence or low amount of the inwardly rectifying K^+current (I_{K1}) also contributes to the diastolic depolarization characteristic to pacemaker cells (Fig. 4C). I_{K1} is important for maintaining the resting membrane potential in working myocardial cells and loss of this current in these cells, encoded by *Kir2.1/2.2*, results in spontaneous activity (Miake et al., 2002). The transient outward current (I_{to}), encoded by $K_v4.2/4.3$, is important for early phase-1 repolarization in working myocardial cells, which pacemaker cells lack (Schram et al., 2002).

Thus, it can be concluded that the electrical properties of the different components of the cardiac conduction system and the working myocardium of the chambers appears to be dependent on the composition of the different channel proteins present in these cells. Although the physiological contribution of these genes to the function of the cardiac conduction system is emerging, the molecular pathways regulating their distribution and transcriptional activity are less well defined. In addition, the spatial distribution and expression of these genes during heart development is, to date, still inadequately established. Elucidating the expression and regulation of these genes during heart development will therefore provide much needed information about the function as well as the origin of the conduction system.

5.4. Early delineation of the conduction system

The development of the cardiac conduction system and its origin have been heavily debated since its discovery and initial morphologic description. It is now well established that the origin of these cells is myocardial, although the electrical phenotype of these cells resembles in some aspects that of neuronal cells which initially led to the opposing notion of an extracardiac neurogenic origin for these cells. This view was supported by the discovery that neuronal markers, such as the neuroligament NF-M and HNK1 in rabbit, are expressed in the developing conduction system (Gorza et al., 1988; Nakamura et al., 1994; Vitadello et al., 1996). Neural crest cells, which originate from the neural folds, were thought to be the primary neurogenic source of the conduction system (Stoller and Epstein, 2005). However, myocardial cells of the heart tube display coordinate contraction before neural crest cells reach the developing heart. In addition, lineage experiments at the end of the last century demonstrated that virally tagged neural crest do not end up in the conduction system and furthermore ascertained that myocardial cells are the source of the cardiac conduction system (Cheng et al., 1999). Although discussion remains because of conflicting results, which trace labeled neural crest cells to the proximal conduction system (Gorza et al., 1988; Poelmann and Gittenberger-de Groot, 1999; Nakamura et al., 2006), these observations make it unlikely that neural crest cells make an important contribution to the formation and development of the conduction system. Conditional knockout of the transcription factor HF-1b in neural crest cells results in a reduction of neural crest-derived cardiac neuron markers, suggesting neural crest cells may play a role in the parasympathetic innervation of the conduction system (St. Amand et al., 2006).

The discussion of (Pennisi et al., 2002; Moorman and Christoffels, 2003; Christoffels et al., 2004a) was focused on the mechanism underlying the specification and differentiation of the cardiac conduction system. One hypothesis is based on the assumption that these cells are recruited from myocardial precursors and subsequently obtain a specialized conduction phenotype and slow proliferation rate. Although this recruitment mechanism has been proposed to pertain to both the nodal and the fast conducting components of the conduction system, this concept is mainly based on the formation of the His–Purkinje system (Pennisi et al., 2002; Gourdie et al., 2003). Cheng et al. (1999) showed with their lineage experiments that clones of myocardial cells virally tagged in the embryonic heart tube give rise to both conduction and working myocardial cells. Several studies showed that formation of the fast conducting components of the conduction system can be induced from working myocardial cells by paracrine signals from the endocardium, thus confirming the "recruitment" theory for these cells. In chicken, expression of the secreted factor *endothelin 1* induces expression of several genes which mark the ventricular conduction system including the chicken ortholog of *Cx40*, *Cx42* (Gourdie et al., 1998). Another paracrine factor secreted by endocardium, *neuregulin-1*, induces formation of conduction system cells. A transgenic mouse line, dubbed *cardiac conduction system (CCS)-lacZ*, with the *lacZ* gene integrated in an as yet undefined locus, displays pronounced expression of the transgene in all the developing and mature components of the conduction system. Expression of the *lacZ* reporter gene is induced by neuregulin in cultured hearts of *CCS-lacZ* embryos (Rentschler et al., 2002). Induction of the nodal pacemaker phenotype by endothelin has also been suggested (Gassanov et al., 2004). In addition, molecular and physiological observations of the developing zebrafish heart suggested that the slow conducting myocardium of the atrioventricular region develops by induction of paracrine factors such as neuregulin and notch derived from the endocardium (Milan et al., 2006). However, there is no direct proof that these paracrine signals contribute to the formation of the nodes in higher vertebrates, since conduction system defects in endothelin-deficient mice have not been reported. Neuregulin-deficient mice die too early to allow analysis of the effect of neuregulin deficiency on conduction system development (Kurihara et al., 1995; Meyer and Birchmeier, 1995).

In higher vertebrates, it is thought that the nodes develop specifically from the slow proliferating primary myocardium of the heart tube and retain their primitive phenotype (Moorman and Christoffels, 2003; Christoffels et al., 2004a) (Fig. 5). This notion correlates with the ballooning model of chamber formation mentioned earlier and is based on morphological, physiological, and molecular observations. Several morphological studies on the ultrastructure of adult and developing nodal conduction cells refer to these cells as being less differentiated than the neighboring working myocardium of the chambers, with a low abundance of gap junctions, myofibrils and mitochondria, and a low proliferation rate (Virágh and Challice, 1977, 1980; Masson-Pevet et al., 1979; Bleeker et al., 1980; Erokhina and Rumyantsev, 1986) (Table 1). In accordance with these observations, molecular data show that the nodes express Tbx3, which analogous to Tbx2, may suppress the working myocardial phenotype (Hoogaars et al., 2004). This "nodal" phenotype is highly reminiscent of that of the primary myocardium of the heart tube, which is retained by the outflow tract, the AVC, and the inflow tract. In the primary heart tube, the highest pacemaker activity is located at the venous pole.

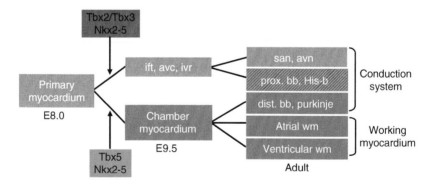

FIGURE 5 Scheme for chamber development and formation of the cardiac conduction system. A transcriptional program, including T-box factor Tbx5 and Nkx2–5 (green box) induces the formation of the chambers at the outer curvature of the heart tube in mouse. Repression of this chamber program by T-box factors Tbx2 and Tbx3 and Nkx2–5 retains in part the primary phenotype of the heart tube at regions that become the nodal components of the conduction system. avc, atrioventricular canal; avn, atrioventricular node; dist. bb, distal bundle branches; His-b, His bundle; ift, inflow tract; ivr, interventricular ring; prox. bb, proximal bundle branches; san, sinuatrial node; wm, working myocardium.

The mechanism that ensures the high pacemaker rate at the venous pole is not yet known, but will undoubtedly involve transcription factors that regulate activity of important pacemaker genes like *Hcn4*. Observations of the electrical activity of chicken embryos revealed that subsequent to the initiation of chamber development, corresponding with E9.5 in mouse, an adult-like electrocardiogram can already be observed (Fig. 6A) (Paff et al., 1968; Moorman et al., 2005). This suggests that the AVC and inflow tract have the same function as the nodes in the adult heart. Thus, chamber development and the retention of primary myocardium results in an adult-like electrical configuration of the developing heart (Fig. 6A). This electrical configuration coincides with the initiation of *Cx40* and *Cx43* expression. Furthermore, a dramatic reduction of I_f with

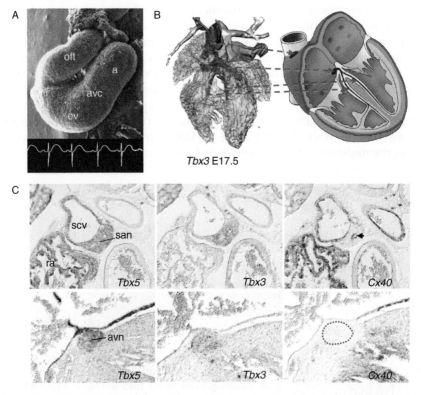

FIGURE 6 Development of the cardiac conduction system. (A) An adult-like electrocardiogram can already be recorded from an embryonic chicken heart at looping stage (HH18) when chamber formation is initiated. (B) Expression of Tbx3 at E17.5 is localized in all the central conduction system components. (C) Sections showing coexpression of *Tbx3* and *Tbx5* in the sinuatrial node and the atrioventricular node at E17.5 in mouse. Note that Tbx5 is also expressed in working myocardium of the atrium. Tbx3 expression is complementary to expression of fast conducting gap-junction protein Cx40.

corresponding downregulation of *HCN* transcripts is seen in mouse ventricular cells between E9.5 and E18.5, thus showing a shift from spontaneous to nonspontaneous myocardial phenotype in working myocardium during development (Yasui et al., 2001).

Data obtained from several transgenic models suggest that the nodes are derived from the primary myocardium of the AVC and venous pole. *CCS-lacZ* mice and other transgenic models expressing *lacZ* in the developing conduction system, the *Gata-6* enhancer-*lacZ*, and *minK-lacZ* show expression of the transgene in the primary myocardium of the AVC and/or the venous pole prior to the formation of the nodes (Kupershmidt et al., 1999; Davis et al., 2001; Rentschler et al., 2001). However, the disadvantage of these transgenic models is that they are either restricted to certain parts of the conduction system or show expression in myocardium that does not contribute to the conduction system. Recent data from our laboratory show that atrial working myocardium flanking the sinuatrial node does not contribute to the formation of this structure, suggesting that sinuatrial node myocardium escapes chamber formation and differentiates from primary myocardium.

5.5. Factors involved in the formation of the conduction system

Although we gained some insight into the origin of the cardiac conduction system, the transcriptional program responsible for the development of the nodes or the fast conducting components and the molecular pathways involved remain clouded. A number of transcription factors have been identified to participate in the development and maturation of the conduction system. The above-mentioned zinc finger protein HF-1b has been implied in the development of the atrioventricular and ventricular conduction system, based on data obtained from *HF-1b* knockout mice. These mice display conduction defects such as atrioventricular block and ventricular arrhythmias and show deregulation of *Cx40, Cx43,* and *minK* expression in the ventricular conduction system (Nguyen-Tran et al., 2000). The transcription factor Nkx2–5 plays an important role in the development of the atrioventricular conduction system and ventricular conduction system. This is visible in patients haploinsufficient for *Nkx2–5,* who have cardiac defects including atrioventricular block and bundle branch block (Schott et al., 1998; Benson et al., 1999). Analysis of heterozygous *Nkx2–5* mutant mice revealed hypoplastic atrioventricular node and His bundle and hypocellularity of the ventricular conduction system, suggesting this transcription factor is essential in the development and differentiation of these components of the conduction system (Jay et al., 2004). Interestingly, no records exist of sinuatrial node dysfunction in *Nkx2–5*-haploinsufficient humans or mice. In addition, recent experiments in our laboratory show that *Nkx2–5* is not expressed in the sinuatrial node primordium during early heart development. Thus, Nkx2–5 does not seem to play a significant role in the development of this nodal structure as opposed to the atrioventricular node. Expression of the transcriptional cofactor Hop is dependent on Nkx2–5 in the heart and is enriched in the conduction system of the adult heart (Chen et al., 2002; Shin et al., 2002). Adult *Hop*-knockout mice show conduction defects in the fast conducting components of the cardiac conduction system without any structural changes. Furthermore, a decrease of *Cx40* expression was observed in these mice, suggesting an important function for Hop in the conduction system is to regulate *Cx40* (Ismat et al., 2005). Transcription factor Tbx5 has also been implicated in conduction system development. *Tbx5* is expressed in the sinuatrial node and atrioventricular node throughout development (Fig. 6C). Heterozygous mutations in *TBX5* in humans cause conduction system defects including bradycardia, atrioventricular block, atrial fibrillation, and sinuatrial node dysfunction. Furthermore, analysis of mice displaying haploinsufficiency of *Tbx5* shows that this factor is important for maturation and development of the atrioventricular node and bundle branches (Moskowitz et al., 2004). Tbx3 is an interesting candidate for the regulation of the "nodal" phenotype since this T-box factor is expressed in the primary myocardium of the AVC and the inflow tract, which at later stages develop into the CCS components (Fig. 6B and C) (Hoogaars et al., 2004). As mentioned, Tbx3 acts as repressor of chamber-specific genes and genes important for the function of the fast conducting components of the conduction system like *Cx40*. Heart defects, such as septum defects, have been reported in patients with ulnar-mammary syndrome, who exhibit haploinsufficiency of *TBX3* (Meneghini et al., 2006). However, in these patients, no reports of conduction system defects exist, which suggest that low amounts of Tbx3 are enough to exert its potential function in the conduction system.

Future experiments are aimed at finding the molecular pathways in which these transcription factors are involved, which will enhance our understanding of the development of the cardiac conduction system.

6. CONCLUSIONS AND FUTURE DIRECTIONS

The development of the fully functional four-chambered heart depends upon a complex network of proteins which genetically and physically interact to regulate different gene programs. Genes responsible for important cardiomyocyte functions, such as electrical coupling, contraction and pacemaker activity, and the molecular pathways that regulate these genes, are emerging. Conserved transcription factors such as T-box factors and Nkx2–5 function as important factors, which decide the fate of myocardial cell populations in the developing heart. Two distinct populations of myocardial cells can be observed in early cardiogenesis, primary myocardium and working myocardium, which are segregated by the expression or repression of certain genes that generate the cellular phenotype needed for their specific tasks in the adult heart. When the linear heart tube loops and chambers are shaped, the primary myocardium located at the inflow tract, AVC, inner curvatures, and outflow tract does not participate in the differentiation into the working myocardial phenotype, but instead conserves the primary "nodal" phenotype. It is important to realize that the primary myocardium is important for the formation of two important components in the developing heart, the cushion mesenchyme of the AVC and outflow tract, which contributes to valve formation and septation, and the nodal components of the conduction system. Other transcriptional regulators, such as Tbx5 and Nkx2–5, control the local formation of the working myocardium at the outer curvatures of the heart tube, resulting in the development of the atria and the ventricles. Although some insight has been gained by the identification of important cardiac transcriptional regulators, the precise molecular mechanism and the pathways involved in chamber formation and other morphogenetic processes during heart development remains enigmatic. Future experiments will be aimed at the identification of all molecular determinants, such as ion channels, gap junction proteins, and cell cycle regulators, which regulate the specific functions and phenotypes found in the different cardiac lineages. Large-scale screens, such as microarrays and chromatin-immuno precipitation, for potential downstream target genes of important cardiac transcription factors, will allow further insight into the molecular pathways involved in heart development.

REFERENCES

Abu-Issa, R., Waldo, K., Kirby, M.L. 2004. Heart fields: One, two or more? Dev. Biol. 272, 281–285.

Ai, D., Liu, W., Ma, L., Dong, F., Lu, M.F., Wang, D., Verzi, M.P., Cai, C., Gage, P.J., Evans, S., Black, B.L., Brown, N.A., et al. 2006. Pitx2 regulates cardiac left-right asymmetry by patterning second cardiac lineage-derived myocardium. Dev. Biol. 296, 437–449.

Bagwe, S., Berenfeld, O., Vaidya, D., Morley, G.E., Jalife, J. 2005. Altered right atrial excitation and propagation in connexin40 knockout mice. Circulation 112, 2245–2253.

Baldini, A. 2005. Dissecting contiguous gene defects: TBX1. Curr. Opin. Genet. Dev. 15, 279–284.

Baruscotti, M., Bucchi, A., DiFrancesco, D. 2005. Physiology and pharmacology of the cardiac pacemaker ("funny") current. Pharmacol. Ther. 107, 59–79.

Basson, C.T., Bachinsky, D.R., Lin, R.C., Levi, T., Elkins, J.A., Soults, J., Grayzel, D., Kroumpouzou, E., Traill, T.A.L.-S.J., Renault, B., Kucherlapati, R., Seidman, J.G., et al. 1997. Mutations in human TBX5 (corrected) cause limb and cardiac malformation in Holt-Oram syndrome. Nat. Genet. 15, 30–35.

Benson, D.W., Silberbach, G.M., Kavanaugh-McHugh, A., Cottrill, C., Zhang, Y., Riggs, S., Smalls, O., Johnson, M.C., Watson, M.S., Seidman, C.E., Plowden, J., Kugler, J.D. 1999. Mutations in the cardiac transcription factor NKX2.5 affect diverse cardiac developmental pathways. J. Clin. Invest. 104, 1567–1573.

Bilican, B., Goding, C.R. 2006. Cell cycle regulation of the T-box transcription factor tbx2. Exp. Cell Res. 312, 2358–2366.

Bleeker, W.K., Mackaay, A.J.C., Masson-Pevet, M., Bouman, L.N., Becker, A.E. 1980. Functional and morphological organization of the rabbit sinus node. Circ. Res. 46, 11–22.

Bohn, G., Moosmang, S., Conrad, H., Ludwig, A., Hofmann, F., Klugbauer, N. 2000. Expression of T- and L-type calcium channel mRNA in murine sinoatrial node. FEBS Lett. 481, 73–76.

Bollag, R.J., Siegfried, Z., Cebra-Thomas, J.A., Garvey, N., Davison, E.M., Silver, L.M. 1994. An ancient family of embryonically expressed mouse genes sharing a conserved protein motif with the T locus. Nat. Genet. 7, 383–389.

Boyett, M.R., Honjo, H., Kodama, I. 2000. The sinoatrial node, a heterogeneous pacemaker structure. Cardiovasc. Res. 47, 658–687.

Brand, T. 2003. Heart development: Molecular insights into cardiac specification and early morphogenesis. Dev. Biol. 258, 1–19.

Breckenridge, R.A., Mohun, T.J., Amaya, E. 2001. A role for BMP signalling in heart looping morphogenesis in *Xenopus*. Dev. Biol. 232, 191–203.

Brummelkamp, T.R., Kortlever, R.M., Lingbeek, M., Trettel, F., MacDonald, M.E., van Lohuizen, M., Bernards, R. 2002. TBX-3, the gene mutated in ulnar-mammary syndrome, is a negative regulator of p19ARF and inhibits senescence. J. Biol. Chem. 277, 6567–6572.

Bruneau, B.G., Logan, M., Davis, N., Levi, T., Tabin, C.J., Seidman, J.G., Seidman, C.E. 1999. Chamber-specific cardiac expression of Tbx5 and heart defects in Holt-Oram syndrome. Dev. Biol. 211, 100–108.

Bruneau, B.G., Nemer, G., Schmitt, J.P., Charron, F., Robitaille, L., Caron, S., Conner, D.A., Gessler, M., Nemer, M., Seidman, C.E., Seidman, J.G. 2001. A murine model of Holt-Oram syndrome defines roles of the T-box transcription factor Tbx5 in cardiogenesis and disease. Cell 106, 709–721.

Buckingham, M., Meilhac, S., Zaffran, S. 2005. Building the mammalian heart from two sources of myocardial cells. Nat. Rev. Genet. 6, 826–837.

Cai, C.L., Liang, X., Shi, Y., Chu, P.H., Pfaff, S.L., Chen, J., Evans, S. 2003. Isl1 identifies a cardiac progenitor population that proliferates prior to differentiation and contributes a majority of cells to the heart. Dev. Cell. 5, 877–889.

Cai, C.L., Zhou, W., Yang, L., Bu, L., Qyang, Y., Zhang, X., Li, X., Rosenfeld, M.G., Chen, J., Evans, S. 2005. T-box genes coordinate regional rates of proliferation and regional specification during cardiogenesis. Development 132, 2475–2487.

Callewaert, G., Carmeliet, E., Vereecke, J. 1984. Single cardiac Purkinje cells: General electrophysiology and voltage-clamp analysis of the pace-maker current. J. Physiol. 349, 643–661.

Charng, M.J., Frenkel, P.A., Lin, Q., Yamada, M., Schwartz, R.J., Olson, E.N., Overbeek, P., Schneider, M.D., Yumada, M. 1998. A constitutive mutation of ALK5 disrupts cardiac looping and morphogenesis in mice. Dev. Biol. 199, 72–79.

Chen, F., Kook, H., Milewski, R., Gitler, A.D., Lu, M.M., Li, J., Nazarian, R., Schnepp, R., Jen, K., Biben, C., Runke, G., Mackay, J.P., et al. 2002. Hop is an unusual homeobox gene that modulates cardiac development. Cell. 110, 713–723.

Chen, H., Shi, S., Acosta, L., Li, W., Lu, J., Bao, S., Chen, Z., Yang, Z., Schneider, M.D., Chien, K.R., Conway, S.J., Yoder, M.C., et al. 2004. BMP10 is essential for maintaining cardiac growth during murine cardiogenesis. Development 131, 2219–2231.

Chen, J.N., van Eeden, F.J.M., Warren, K.S., Chin, A., Nusslein-Volhard, C., Haffter, P., Fishman, M.C. 1997. Left-right pattern of cardiac BMP4 may drive asymmetry of the heart in zebrafish. Development 124, 4373–4382.

Cheng, G., Litchenberg, W.H., Cole, G.J., Mikawa, T., Thompson, R.P., Gourdie, R.G. 1999. *Development* of the cardiac conduction system involves recruitment within a multipotent cardiomyogenic lineage. Development 126, 5041–5049.

Christoffels, V.M., Burch, J.B.E., Moorman, A.F.M. 2004a. Architectural plan for the heart: Early patterning and delineation of the chambers and the nodes. Trends Cardiovasc. Med. 14, 301–307.

Christoffels, V.M., Hoogaars, W.M.H., Tessari, A., Clout, D.E.W., Moorman, A.F.M., Campione, M. 2004b. T-box transcription factor Tbx2 represses differentiation and formation of the cardiac chambers. Dev. Dyn. 229, 763–770.

Christoffels, V.M., Mommersteeg, M.T.M., Trowe, M.O., Prall, O.W.J., de Gier-de Vries, C., Soufan, A.T., Bussen, M., Schuster-Gossler, K., Harvey, R.P., Moorman, A.F.M., Kispert, A. 2006. Formation of the venous pole of the heart from an Nkx2-5-negative precursor population requires Tbx18. Circ. Res. 98, 1555–1563.

Coppen, S.R., Kodama, I., Boyett, M.R., Dobrzynski, H., Takagishi, Y., Honjo, H., Yeh, H.I., Severs, N.J. 1999a. Connexin45, a major connexin of the rabbit sinoatrial node, is co-expressed with connexin43 in a restricted zone at the nodalcrista terminalis border. J. Histochem. Cytochem. 47, 907–918.

Coppen, S.R., Severs, N.J., Gourdie, R.G. 1999b. Connexin45 (alpha 6) expression delineates an extended conduction system in the embryonic and mature rodent heart. Dev. Genet. 24, 82–90.

Davenport, T.G., Jerome-Majewska, L.A., Papaioannou, V.E. 2003. Mammary gland, limb and yolk sac defects in mice lacking Tbx3, the gene mutated in human ulnar mammary syndrome. Development 130, 2263–2273.

Davis, D.L., Edwards, A.V., Juraszek, A.L., Phelps, A., Wessels, A., Burch, J.B.E. 2001. A GATA-6 gene heart-region-specific enhancer provides a novel means to mark and probe a discrete component of the mouse cardiac conduction system. Mech. Dev. 108, 105–119.

DiFrancesco, D. 1993. Pacemaker mechanisms in cardiac tissue. Annu. Rev. Physiol. 55, 455–472.

Dobrzynski, H., Nikolski, V.P., Sambelashvili, A.T., Greener, I.D., Yamamoto, M., Boyett, M.R., Efimov, I.R. 2003. Site of origin and molecular substrate of atrioventricular junctional rhythm in the rabbit heart. Circ. Res. 93, 1102–1110.

Dodou, E., Verzi, M.P., Anderson, J.P., Xu, S.M., Black, B.L. 2004. Mef2c is a direct transcriptional target of ISL1 and GATA factors in the anterior heart field during mouse embryonic development. Development 131, 3931–3942.

Efimov, I.R., Nikolski, V.P., Rothenberg, F., Greener, I.D., Li, J., Dobrzynski, H., Boyett, M. 2004. Structure-function relationship in the AV junction. Anat. Rec. A Discov. Mol. Cell. Evol. Biol. 280, 952–965.

Erokhina, I.L., Rumyantsev, P.P. 1986. Ultrastructure of DNA-synthesizing and mitotically dividing myocytes in sinoatrial node of mouse embryonal heart. J. Mol. Cell. Cardiol. 18, 1219–1231.

Garcia-Martinez, V., Schoenwolf, G.C. 1993. Primitive streak origin of the cardiovascular system in avian embryos. Dev. Biol. 159, 706–719.

Garg, V., Kathiriya, I.S., Barnes, R., Schluterman, M.K., King, I.N., Butler, C.A., Rothrock, C.R., Eapen, R.S., Hirayama-Yamada, K., Joo, K., Matsuoka, R., Cohen, J.C., et al. 2003. GATA4 mutations cause human congenital heart defects and reveal an interaction with TBX5. Nature 424, 443–447.

Gassanov, N., Er, F., Zagidullin, N., Hoppe, U.C. 2004. Endothelin induces differentiation of ANP-EGFP expressing embryonic stem cells towards a pacemaker phenotype. FASEB J. 18, 1710–1712.

Gorza, L., Schiaffino, S., Vitadello, M. 1988. Heart conduction system: A neural crest derivative? Brain Res. 457, 360–366.

Gourdie, R.G., Wei, Y., Kim, D., Klatt, S.C., Mikawa, T. 1998. Endothelin-induced conversion of embryonic heart muscle cells into impulse-conducting purkinje fibers. Proc. Natl. Acad. Sci. USA 95, 6815–6818.

Gourdie, R.G., Harris, B.S., Bond, J., Justus, C., Hewett, K.W., O'Brien, T.X., Thompson, R.P., Sedmera, D. 2003. Development of the cardiac pacemaking and conduction system. Birth Defects Res. C Embryo Today 69, 46–57.

Habets, P.E.M.H., Moorman, A.F.M., Clout, D.E.W., van Roon, M.A., Lingbeek, M., Lohuizen, M., Christoffels, V.M. 2002. Cooperative action of Tbx2 and Nkx2.5 inhibits

ANF expression in the atrioventricular canal: Implications for cardiac chamber formation. Genes Dev. 16, 1234–1246.

Habets, P.E.M.H., Moorman, A.F.M., Christoffels, V.M. 2003. Regulatory modules in the developing heart. Cardiovasc. Res. 58, 246–263.

Harrelson, Z., Kelly, R.G., Goldin, S.N., Gibson-Brown, J.J., Bollag, R.J., Silver, L.M., Papaioannou, V.E. 2004. Tbx2 is essential for patterning the atrioventricular canal and for morphogenesis of the outflow tract during heart development. Development 131, 5041–5052.

He, M., Wen, L., Campbell, C.E., Wu, J.Y., Rao, Y. 1999. Transcription repression by *Xenopus* ET and its human ortholog TBX3, a gene involved in ulnar-mammary syndrome. Proc. Natl. Acad. Sci. USA 96, 10212–10217.

Hiroi, Y., Kudoh, S., Monzen, K., Ikeda, Y., Yazaki, Y., Nagai, R., Komuro, I. 2001. Tbx5 associates with Nkx2-5 and synergistically promotes cardiomyocyte differentiation. Nat. Genet. 28, 276–280.

Hochgreb, T., Linhares, V.L., Menezes, D.C., Sampaio, A.C., Yan, C.Y., Cardoso, W.V., Rosenthal, N., Xavier-Neto, J. 2003. A caudorostral wave of RALDH2 conveys antero-posterior information to the cardiac field. Development 130, 5363–5374.

Hoogaars, W.M.H., Tessari, A., Moorman, A.F.M., de Boer, P.A.J., Hagoort, J., Soufan, A.T., Campione, M., Christoffels, V.M. 2004. The transcriptional repressor Tbx3 delineates the developing central conduction system of the heart. Cardiovasc. Res. 62, 489–499.

Hu, T., Yamagishi, H., Maeda, J., McAnally, J., Yamagishi, C., Srivastava, D. 2004. Tbx1 regulates fibroblast growth factors in the anterior heart field through a reinforcing autoregulatory loop involving forkhead transcription factors. Development 131, 5491–5502.

Ilagan, R., bu-Issa, R., Brown, D., Yang, Y.P., Jiao, K., Schwartz, R.J., Klingensmith, J., Meyers, E.N. 2006. Fgf0008 is required for anterior heart field development. Development 133, 2435–2445.

Ismat, F.A., Zhang, M., Kook, H., Huang, B., Zhou, R., Ferrari, V.A., Epstein, J.A., Patel, V.V. 2005. Homeobox protein Hop functions in the adult cardiac conduction system. Circ. Res. 96, 898–903.

Jacobs, J.J.L., Keblusek, P., Robanus Maandag, E., Kristel, P., Lingbeek, M., Nederlof, P.M., van Welsem, T., van de Vijver, M.J., Koh, E.Y., Daley, G.Q., van Lohuizen, M. 2000. Senescence bypass screen identifies Tbx2, which represses *Cdkn2a* ($p19^{ARF}$) and is amplified in a subset of human breast cancers. Nat. Genet. 26, 291–299.

Jay, P.Y., Harris, B.S., Maguire, C.T., Buerger, A., Wakimoto, H., Tanaka, M., Kupershmidt, S., Roden, D.M., Schultheiss, T.M., O'Brien, T.X., Gourdie, R.G., Berul, C.I., et al. 2004. Nkx2-5 mutation causes anatomic hypoplasia of the cardiac conduction system. J. Clin. Invest. 113, 1130–1137.

Jerome-Majewska, L.A., Jenkins, G.P., Ernstoff, E., Zindy, F., Sherr, C.J., Papaioannou, V.E. 2005. Tbx3, the ulnar-mammary syndrome gene, and Tbx2 interact in mammary gland development through a p19^{Arf}/p53-independent pathway. Dev. Dyn. 234, 922–933.

Jiao, K., Kulessa, H., Tompkins, K., Zhou, Y., Batts, L., Baldwin, H.S., Hogan, B.L. 2003. An essential role of Bmp4 in the atrioventricular septation of the mouse heart. Genes Dev. 17, 2362–2367.

Kelly, R.G., Brown, N.A., Buckingham, M.E. 2001. The arterial pole of the mouse heart forms from Fgf10-expressing cells in pharyngeal mesoderm. Dev. Cell. 1, 435–440.

Kirby, M.L., Gale, T.F., Stewart, D.E. 1983. Neural crest cells contribute to normal aortico-pulmonary septation. Science 220, 1059–1061.

Kirchhoff, S., Nelles, E., Hagendorff, A., Kruger, O., Traub, O., Willecke, K. 1998. Reduced cardiac conduction velocity and predisposition to arrhythmias in connexin40-deficient mice. Curr. Biol. 8, 299–302.

Kitajima, S., Takagi, A., Inoue, T., Saga, Y. 2000. MesP1 and MesP2 are essential for the development of cardiac mesoderm. Development 127, 3215–3226.

Kitamura, K., Miura, H., Miyagawa-Tomita, S., Yanazawa, M., Katoh-Fukui, Y., Suzuki, R., Ohuchi, H., Suehiro, A., Motegi, Y., Nakahara, Y., Kondo, S., Yokoyama, M., et al. 1999. Mouse Pitx2 deficiency leads to anomalies of the ventral body wall, heart, extra- and periocular mesoderm and right pulmonary isomerism. Development 126, 5749–5758.

Kondo, R.P., Anderson, R.H., Kupershmidt, S., Roden, D.M., Evans, S.M. 2003. Development of the cardiac conduction system as delineated by minK-lacZ. J. Cardiovasc. Electrophysiology 14, 383–391.

Koshiba-Takeuchi, K., Takeuchi, J.K., Arruda, E.P., Kathiriya, I.S., Mo, R., Hui, C.C., Srivastava, D., Bruneau, B.G. 2006. Cooperative and antagonistic interactions between Sall4 and Tbx5 pattern the mouse limb and heart. Nat. Genet. 38, 175–183.

Kreuzberg, M.M., Sohl, G., Kim, J.S., Verselis, V.K., Willecke, K., Bukauskas, F.F. 2005. Functional properties of mouse connexin30.2 expressed in the conduction system of the heart. Circ. Res. 96, 1169–1177.

Kreuzberg, M.M., Schrickel, J.W., Ghanem, A., Kim, J.S., Degen, J., Janssen-Bienhold, U., Lewalter, T., Tiemann, K., Willecke, K. 2006. Connexin30.2 containing gap junction channels decelerate impulse propagation through the atrioventricular node. Proc. Natl. Acad. Sci. USA 103, 5959–5964.

Kupershmidt, S., Yang, T., Anderson, M.E., Wessels, A., Niswender, K.D., Magnuson, M.A., Roden, D.M. 1999. Replacement by homologous recombination of the minK gene with lacZ reveals restriction of minK expression to the mouse cardiac conduction system. Circ. Res. 84, 146–152.

Kurihara, Y., Kurihara, H., Oda, H., Maemura, K., Nagai, R., Ishikawa, T., Yazaki, Y. 1995. Aortic arch malformations and ventricular septal defect in mice deficient in endothelin-1. J. Clin. Invest. 96, 293–300.

Lees-Miller, J.P., Guo, J., Somers, J.R., Roach, D.E., Sheldon, R.S., Rancourt, D.E., Duff, H.J. 2003. Selective knockout of mouse ERG1 B potassium channel eliminates I_{Kr} in adult ventricular myocytes and elicits episodes of abrupt sinus bradycardia. Mol. Cell Biol. 23, 1856–1862.

Li, Q.Y., Newbury-Ecob, R.A., Terret, J.A., Wilson, D.I., Curtis, A.R., Yi, C.H., Gebuhr, T., Bullen, P.J., Robson, S.C., Strachan, T., Bonnet, D., Lyonnet, S., et al. 2000. Holt-Oram syndrome is caused by mutations in TBX5, a member of the brachyury (T) gene family. Nat. Genet. 15, 21–29.

Liberatore, C.M., Searcy-Schrick, R.D., Yutzey, K.E. 2000. Ventricular expression of Tbx5 inhibits normal heart chamber development. Dev. Biol. 223, 169–180.

Lin, Q., Schwarz, J., Bucana, C., Olson, E.N. 1997. Control of mouse cardiac morphogenesis and myogenesis by transcription factor MEF2C. Science 276, 1404–1407.

Lingbeek, M.E., Jacobs, J.J., van Lohuizen, M. 2002. The T-box repressors TBX2 and TBX3 specifically regulate the tumor suppressor gene p14ARF via a variant T-site in the initiator. J. Biol. Chem. 277, 26120–26127.

Ludwig, A., Zong, X., Jeglitsch, M., Hofmann, F., Biel, M. 1998. A family of hyperpolarization-activated mammalian cation channels. Nature 393, 587–591.

Lyons, I., Parsons, L.M., Hartley, L., Li, R., Andrews, J.E., Robb, L., Harvey, R.P. 1995. Myogenic and morphogenetic defects in the heart tubes of murine embryos lacking the homeobox gene Nkx2–5. Gene Dev. 9, 1654–1666.

Ma, L., Lu, M.F., Schwartz, R.J., Martin, J.F. 2005. Bmp2 is essential for cardiac cushion epithelial-mesenchymal transition and myocardial patterning. Development 132, 5601–5611.

Mangoni, M.E., Couette, B., Bourinet, E., Platzer, J., Reimer, D., Striessnig, J., Nargeot, J. 2003. Functional role of L-type $Ca_v1.3$ Ca^{2+} channels in cardiac pacemaker activity. Proc. Natl. Acad. Sci. USA 100, 5543–5548.

Mangoni, M.E., Traboulsie, A., Leoni, A.L., Couette, B., Marger, L., Le, Q.K., Kupfer, E., Cohen-Solal, A., Vilar, J., Shin, H.S., Escande, D., Charpentier, F., et al. 2006. Bradycardia and slowing of the atrioventricular conduction in mice lacking CaV3.1/{alpha}1G T-type calcium channels. Circ. Res. 98, 1422–1430.

Masson-Pevet, M., Bleeker, W.K., Gros, D. 1979. The plasmamembrane of leading pacemaker cells in the rabbit sinus node. A qualitative and quantitative ultrastructural analysis. Circ. Res. 45, 621–629.

Meilhac, S.M., Kelly, R.G., Rocancourt, D., Eloy-Trinquet, S., Nicolas, J.F., Buckingham, M.E. 2003. A retrospective clonal analysis of the myocardium reveals two phases of clonal growth in the developing mouse heart. Development 130, 3877–3889.

Meneghini, V., Odent, S., Platonova, N., Egeo, A., Merlo, G.R. 2006. Novel TBX3 mutation data in families with ulnar-mammary syndrome indicate a genotype-phenotype relationship: Mutations that do not disrupt the T-domain are associated with less severe limb defects. Eur. J. Med. Genet. 49(2), 151–158.

Meyer, D., Birchmeier, C. 1995. Multiple essential functions of neuregulin in development. Nature 378, 386–390.

Miake, J., Marban, E., Nuss, H.B. 2002. Biological pacemaker created by gene transfer. Nature 419, 132–133.

Milan, D.J., Giokas, A.C., Serluca, F.C., Peterson, R.T., MacRae, C.A. 2006. Notch1b and neuregulin are required for specification of central cardiac conduction tissue. Development 133, 1125–1132.

Milanesi, R., Baruscotti, M., Gnecchi-Ruscone, T., DiFrancesco, D. 2006. Familial sinus bradycardia associated with a mutation in the cardiac pacemaker channel. N. Engl. J. Med. 354, 151–157.

Miquerol, L., Meysen, S., Mangoni, M., Bois, P., Van Rijen, H.V., Abran, P., Jongsma, H., Nargeot, J., Gros, D. 2004. Architectural and functional asymmetry of the his-purkinje system of the murine heart. Cardiovasc. Res. 63, 77–86.

Mjaatvedt, C.H., Nakaoka, T., Moreno-Rodriguez, R., Norris, R.A., Kern, M.J., Eisenberg, C.A., Turner, D., Markwald, R.R. 2001. The outflow tract of the heart is recruited from a novel heart-forming field. Dev. Biol. 238, 97–109.

Moorman, A.F.M., Christoffels, V.M. 2003. Cardiac chamber formation: Development, genes and evolution. Physiol. Rev. 83, 1223–1267.

Moorman, A.F.M., Christoffels, V.M., Anderson, R.H. 2005. Anatomic substrates for cardiac conduction. Heart Rhythm 2, 875–886.

Moorman, A.F.M., Christoffels, V.M., Anderson, R.H., van den Hoff, M.J.B. 2007. The heart-forming fields: One or multiple? Philos. Trans. R. Soc. Lond. B, in press.

Moosmang, S., Stieber, J., Zong, X., Biel, M., Hofmann, F., Ludwig, A. 2001. Cellular expression and functional characterization of four hyperpolarization activated pacemaker channels in cardiac and neuronal tissues. Eur. J. Biochem. 268, 1646–1652.

Moreno, A.P. 2004. Biophysical properties of homomeric and heteromultimeric channels formed by cardiac connexins. Cardiovasc. Res. 62, 276–286.

Mori, A.D., Zhu, Y., Vahora, I., Nieman, B., Koshiba-Takeuchi, K., Davidson, L., Pizard, A., Seidman, C.E., Seidman, J.G., Chen, X.J., Henkelman, R.M., Bruneau, B.G., et al. 2006. Tbx5-dependent rheostatic control of cardiac gene expression and morphogenesis. Dev. Biol. 297, 566–586.

Moskowitz, I.P.G., Pizard, A., Patel, V.V., Bruneau, B.G., Kim, J.B., Kupershmidt, S., Roden, D., Berul, C.I., Seidman, C.E., Seidman, J.G. 2004. The T-Box transcription factor Tbx5 is required for the patterning and maturation of the murine cardiac conduction system. Development 131, 4107–4116.

Murakami, M., Nakagawa, M., Olson, E.N., Nakagawa, O. 2005. A WW domain protein TAZ is a critical coactivator for TBX5, a transcription factor implicated in Holt-Oram syndrome. Proc. Natl. Acad. Sci. USA 102, 18034–18039.

Naiche, L.A., Harrelson, Z., Kelly, R.G., Papaioannou, V.E. 2005. T-Box genes in vertebrate development. Annu. Rev. Genet. 39, 219–239.

Nakamura, T., Ikeda, T., Shimokawa, I., Inoue, Y., Suematsu, T., Sakai, H., Iwasaki, K., Matsuo, T. 1994. Distribution of acetylcholinesterase activity in the rat embryonic heart with reference to HNK-1 immunoreactivity in the conduction tissue. Anat. Embryol. 190, 367–373.

Nakamura, T., Colbert, M.C., Robbins, J. 2006. Neural crest cells retain multipotential characteristics in the developing valves and label the cardiac conduction system. Circ. Res. 98, 1547–1554.

Nguyen-Tran, V.T., Kubalak, S.W., Minamisawa, S., Fiset, C., Wollert, K.C., Brown, A.B., Ruiz-Lozano, P., Barrere-Lemaire, S., Kondo, R., Norman, L.W., Gourdie, R.G., Rahme, M.M., et al. 2000. A novel genetic pathway for sudden cardiac death via defects in the transition between ventricular and conduction system cell lineages. Cell 102, 671–682.

Niederreither, K., Vermot, J., Messaddeq, N., Schuhbaur, B., Chambon, P., Dolle, P. 2001. Embryonic retinoic acid synthesis is essential for heart morphogenesis in the mouse. Development 128, 1019–1031.

Nowotschin, S., Liao, J., Gage, P.J., Epstein, J.A., Campione, M., Morrow, B.E. 2006. Tbx1 affects asymmetric cardiac morphogenesis by regulating Pitx2 in the secondary heart field. Development 133, 1565–1573.

Paff, G.H., Boucek, R.J., Harrell, T.C. 1968. Observations on the development of the electrocardiogram. Anat. Rec. 160, 575–582.

Park, E.J., Ogden, L.A., Talbot, A., Evans, S., Cai, C.L., Black, B.L., Frank, D.U., Moon, A.M. 2006. Required, tissue-specific roles for Fgf0008 in outflow tract formation and remodeling. Development 133, 2419–2433.

Pashmforoush, M., Lu, J.T., Chen, H., Amand, T.S., Kondo, R., Pradervand, S., Evans, S.M., Clark, B., Feramisco, J.R., Giles, W., Ho, S.Y., Benson, D.W., et al. 2004. Nkx2–5 pathways and congenital heart disease; loss of ventricular myocyte lineage specification leads to progressive cardiomyopathy and complete heart block. Cell 117, 373–386.

Paxton, C., Zhao, H., Chin, Y., Langner, K., Reecy, J. 2002. Murine Tbx2 contains domains that activate and repress gene transcription. Gene 283, 117–124.

Pennisi, D.J., Rentschler, S., Gourdie, R.G., Fishman, G.I., Mikawa, T. 2002. Induction and patterning of the cardiac conduction system. Int. J. Dev. Biol. 46, 765–775.

Plageman, T.F., Jr., Yutzey, K.E. 2005. T-box genes and heart development: Putting the "T" in heart. Dev. Dyn. 232, 11–20.

Platzer, J., Engel, J., Schrott-Fischer, A., Stephan, K., Bova, S., Chen, H., Zheng, H., Striessnig, J. 2000. Congenital deafness and sinoatrial node dysfunction in mice lacking class D L-type Ca^{2+} channels. Cell 102, 89–97.

Poelmann, R.E., Gittenberger-de Groot, A.C. 1999. A subpopulation of apoptosis-prone cardiac neural crest cells targets to the venous pole: Multiple functions in heart development. Dev. Biol. 207, 271–286.

Prince, S., Carreira, S., Vance, K.W., Abrahams, A., Goding, C.R. 2004. Tbx2 directly represses the expression of the p21(WAF1) cyclin-dependent kinase inhibitor. Cancer Res. 64, 1669–1674.

Redkar, A., Montgomery, M., Litvin, J. 2001. Fate map of early avian cardiac progenitor cells. Development 128, 2269–2279.

Rentschler, S., Vaidya, D.M., Tamaddon, H., Degenhardt, K., Sassoon, D., Morley, G.E., Jalife, J., Fishman, G.I. 2001. Visualization and functional characterization of the developing murine cardiac conduction system. Development 128, 1785–1792.

Rentschler, S., Zander, J., Meyers, K., France, D., Levine, R., Porter, G., Rivkees, S.A., Morley, G.E., Fishman, G.I. 2002. Neuregulin-1 promotes formation of the murine cardiac conduction system. Proc. Natl. Acad. Sci. USA 99, 10464–10469.

Rivera-Feliciano, J., Tabin, C.J. 2006. Bmp2 instructs cardiac progenitors to form the heart-valve-inducing field. Dev. Biol. 295, 580–588.

Robinson, R.B., Siegelbaum, S.A. 2003. Hyperpolarization-activated cation currents: From molecules to physiological function. Annu. Rev. Physiol. 65, 453–480.

Rosenquist, G.C. 1970. Location and movements of cardiogenic cells in the chick embryo: The heart forming portion of the primitive streak. Dev. Biol. 22, 461–475.

Rowley, M., Grothey, E., Couch, F.J. 2004. The role of Tbx2 and Tbx3 in mammary development and tumorigenesis. J. Mammary Gland Biol. Neoplasia 9, 109–118.

Saga, Y., Kitajima, S., Miyagawa-Tomita, S. 2000. Mesp1 expression is the earliest sign of cardiovascular development. Trends Cardiovasc. Med. 10, 345–352.

Schott, J.-J., Benson, D.W., Basson, C.T., Pease, W., Silberbach, G.M., Moak, J.P., Maron, B.J., Seidman, C.E., Seidman, C.E. 1998. Congenital heart disease caused by mutations in the transcription factor NKX2–5. Science 281, 108–111.

Schram, G., Pourrier, M., Melnyk, P., Nattel, S. 2002. Differential distribution of cardiac ion channel expression as a basis for regional specialization in electrical function. Circ. Res. 90, 939–950.

Schulze-Bahr, E., Neu, A., Friederich, P., Kaupp, U.B., Breithardt, G., Pongs, O., Isbrandt, D. 2003. Pacemaker channel dysfunction in a patient with sinus node disease. J. Clin. Invest. 111, 1537–1545.

Shi, W., Wymore, R., Yu, H., Wu, J., Wymore, R.T., Pan, Z., Robinson, R.B., Dixon, J.E., McKinnon, D., Cohen, I.S. 1999. Distribution and prevalence of hyperpolarization-activated cation channel (HCN) mRNA expression in cardiac tissues. Circ. Res. 85, e1–e6.

Shin, C.H., Liu, Z., Passier, R., Zhang, C., Wang, D., Harris, T.M., Yamagishi, H., Richardson, J.A., Childs, G., Olson, E.N. 2002. Modulation of cardiac growth and development by HOP, an unusual homeodomain protein. Cell 110, 725–735.

Simon, A.M., Goodenough, D.A., Paul, D.L. 1998. Mice lacking connexin40 have cardiac conduction abnormalities characteristic of atrioventricular block and bundle branch block. Curr. Biol. 8, 295–298.

Singh, M.K., Christoffels, V.M., Dias, J.M., Trowe, M.O., Petry, M., Schuster-Gossler, K., Burger, A., Ericson, J., Kispert, A. 2005. *Tbx20* is essential for cardiac chamber differentiation and repression of *Tbx2*. Development 132, 2697–2707.

Small, E.M., Krieg, P.A. 2003. Transgenic analysis of the atrialnatriuretic factor (ANF) promoter: Nkx2–5 and GATA-4 binding sites are required for atrial specific expression of ANF. Dev. Biol. 261, 116–131.

Soufan, A.T., van den Hoff, M.J.B., Ruijter, J.M., de Boer, P.A.J., Hagoort, J., Webb, S., Anderson, R.H., Moorman, A.F.M. 2004. Reconstruction of the patterns of gene expression in the developing mouse heart reveals an architectural arrangement that facilitates the understanding of atrial malformations and arrhythmias. Circ. Res. 95, 1207–1215.

Soufan, A.T., van den Berg, G., Ruijter, J.M., de Boer, P.A.J., van den Hoff, M.J.B., Moorman, A.F.M. 2006. Regionalized sequence of myocardial cell growth and proliferation characterizes early chamber formation. Circ. Res. 99, 545–552.

Splawski, I., Timothy, K.W., Sharpe, L.M., Decher, N., Kumar, P., Bloise, R., Napolitano, C., Schwartz, P.J., Joseph, R.M., Condouris, K., Tager-Flusberg, H., Priori, S.G., et al. 2004. Ca$_v$1.2 calcium channel dysfunction causes a multisystem disorder including arrhythmia and autism. Cell 119, 19–31.

Splawski, I., Timothy, K.W., Decher, N., Kumar, P., Sachse, F.B., Beggs, A.H., Sanguinetti, M.C., Keating, M.T. 2005. Severe arrhythmia disorder caused by cardiac L-type calcium channel mutations. Proc. Natl. Acad. Sci. USA 102, 8089–8096.

Srivastava, D., Thomas, T., Lin, Q., Kirby, M.L., Brown, D., Olson, E.N. 1997. Regulation of cardiac mesodermal and neural crest development by the bHLH transcription factor, dHAND. Nat. Genet. 16, 154–160.

St. Amand, T.R., Lu, J.T., Zamora, M., Gu, Y., Stricker, J., Hoshijima, M., Epstein, J.A., Ross, J.J., Jr., Ruiz-Lozano, P., Chien, K.R. 2006. Distinct roles of HF-1b/Sp4 in ventricular and neural crest cells lineages affect cardiac conduction system development. Dev. Biol. 291, 208–217.

Stalsberg, H., de Haan, R.L. 1969. The precardiac areas and formation of the tubular heart in the chick embryo. Dev. Biol. 19, 128–159.

Stennard, F.A., Costa, M.W., Lai, D., Biben, C., Furtado, M.B., Solloway, M.J., McCulley, D.J., Leimena, C., Preis, J.I., Dunwoodie, S.L., Elliott, D.E., Prall, O.W., et al. 2005. Murine T-box transcription factor Tbx20 acts as a repressor during heart development, and is essential for adult heart integrity, function and adaptation. Development 132, 2451–2462.

Stieber, J., Herrmann, S., Feil, S., Loster, J., Feil, R., Biel, M., Hofmann, F., Ludwig, A. 2003. The hyperpolarization-activated channel HCN4 is required for the generation of pacemaker action potentials in the embryonic heart. Proc. Natl. Acad. Sci USA 100, 15235–15240.

Stoller, J.Z., Epstein, J.A. 2005. Cardiac neural crest. Semin. Cell Dev. Biol. 16, 704–715.

Suzuki, T., Takeuchi, J., Koshiba-Takeuchi, K., Ogura, T. 2004. Tbx genes specify posterior digit identity through Shh and BMP signaling. Dev. Cell. 6, 43–53.

van Kempen, M.J.A., Vermeulen, J.L.M., Moorman, A.F.M., Gros, D.B., Paul, D.L., Lamers, W.H. 1996. Developmental changes of connexin40 and connexin43 mRNA-distribution patterns in the rat heart. Cardiovasc. Res. 32, 886–900.

Vance, K.W., Carreira, S., Brosch, G., Goding, C.R. 2005. Tbx2 is overexpressed and plays an important role in maintaining proliferation and suppression of senescence in melanomas. Cancer Res. 65, 2260–2268.

Verheijck, E.E., van Ginneken, A.C.G., Wilders, R., Bouman, L.N. 1999. Contribution of L-type Ca^{2+} current to electrical activity in sinoatrial nodal myocytes of rabbits. Am. J. Physiol. 276, H1064–H1077.

Virágh, S.Z., Challice, C.E. 1977. The development of the conduction system in the mouse embryo heart. II. Histogenesis of the atrioventricular node and bundle. Dev. Biol. 56, 397–411.

Virágh, S.Z., Challice, C.E. 1980. The development of the conduction system in the mouse embryo heart. III. The development of sinus muscle and sinoatrial node. Dev. Biol. 80, 28–45.

Vitadello, M., Vettore, S., Lamar, E., Chien, K.R., Gorza, L. 1996. Neurofilament M mRNA is expressed in conduction system myocytes of the developing and adult rabbit heart. J. Mol. Cell. Cardiol. 28, 1833–1844.

Vitelli, F., Taddei, I., Morishima, M., Meyers, E.N., Lindsay, E.A., Baldini, A. 2002. A genetic link between Tbx1 and fibroblast growth factor signaling. Development 129, 4605–4611.

Vitelli, F., Zhang, Z., Huynh, T., Sobotka, A., Mupo, A., Baldini, A. 2006. Fgf0008 expression in the Tbx1 domain causes skeletal abnormalities and modifies the aortic arch but not the outflow tract phenotype of Tbx1 mutants. Dev. Biol. 295, 559–570.

von Both, I., Silvestri, C., Erdemir, T., Lickert, H., Walls, J.R., Henkelman, R.M., Rossant, J., Harvey, R.P., Attisano, L., Wrana, J.L. 2004. Foxh1 is essential for development of the anterior heart field. Dev. Cell. 7, 331–345.

Waldo, K.L., Kumiski, D.H., Wallis, K.T., Stadt, H.A., Hutson, M.R., Platt, D.H., Kirby, M.L. 2001. Conotruncal myocardium arises from a secondary heart field. Development 128, 3179–3188.

Warren, K.S., Baker, K., Fishman, M.C. 2001. The slow mo mutation reduces pacemaker current and heart rate in adult zebrafish. Am. J. Physiol. Heart Circ. Physiol. 281, H1711–H1719.

Yamada, M., Revelli, J.P., Eichele, G., Barron, M., Schwartz, R.J. 2000. Expression of chick Tbx-2, Tbx-3, and Tbx-5 genes during early heart development: Evidence for BMP2 induction of Tbx2. Dev. Biol. 228, 95–105.

Yang, L., Cai, C.L., Lin, L., Qyang, Y., Chung, C., Monteiro, R.M., Mummery, C.L., Fishman, G.I., Cogen, A., Evans, S. 2006. Isl1Cre reveals a common Bmp pathway in heart and limb development. Development 133, 1575–1585.

Yasui, K., Liu, W., Opthof, T., Kada, K., Lee, J., Kamiya, K., Kodama, I. 2001. If current and spontaneous activity in mouse embryonic ventricular myocytes. Circ. Res. 88, 536–542.

Yutzey, K.E., Rhee, J.T., Bader, D. 1994. Expression of the atrial-specific myosin heavy chain AMHC1 and the establishment of anteroposterior polarity in the developing chicken heart. Development 120, 871–883.

Zhang, Z., Xu, Y., Song, H., Rodriguez, J., Tuteja, D., Namkung, Y., Shin, H.S., Chiamvimonvat, N. 2002. Functional roles of $Ca_v1.3$ ($alpha_{1D}$) calcium channel in sinoatrial nodes: Insight gained using gene-targeted null mutant mice. Circ. Res. 90, 981–987.

Zhang, Z., Huynh, T., Baldini, A. 2006. Mesodermal expression of Tbx1 is necessary and sufficient for pharyngeal arch and cardiac outflow tract development. Development 133, 3587–3595.

CHAPTER 3

Heart Development and T-box Transcription Factors: Lessons from Avian Embryos

Elaine L. Shelton and **Katherine E. Yutzey**

Abstract

Members of the T-box (Tbx) family of transcription factors have diverse regulatory functions in the developing heart and are associated with several human genetic syndromes that include cardiac malformations. Studies in chicken embryos and other animal model systems have been used to define T-box gene regulatory networks important for heart formation with parallels in limb outgrowth and patterning. In the heart, T-box gene regulatory networks function in early cardiac lineage diversification and morphogenesis as well as in

Division of Molecular Cardiovascular Biology, Cincinnati Children's Hospital Medical Center, Cincinnati, Ohio

Advances in Developmental Biology, Volume 18
ISSN 1574-3349, DOI: 10.1016/S1574-3349(07)18003-0

valvuloseptal development and conduction system maturation. These regulatory networks include inductive interactions with developmentally important signaling pathways and regulation of downstream target gene expression in complexes with other cardiac transcription factors. An additional feature of cardiac T-box gene networks is complex regulatory relationships among multiple T-box genes expressed in a given cell type. The focus of this chapter is T-box family members and the regulatory networks that control diverse aspects of heart development with particular emphasis on experimental embryological approaches in the chicken model system.

1. INTRODUCTION

The vertebrate T-box (Tbx) gene family includes ~20 different members with diverse roles in embryonic development and associations with human genetic disease. The original T-box gene, *T* or *brachyury*, was cloned in 1990 and its DNA-binding domain was termed the T-box (Herrmann et al., 1990). Additional members of T-box gene family have been identified in organisms ranging from hydra to humans and extensive research by many investigators has provided insights into T-box protein structure, transcriptional activity, genetic targets, developmental regulatory functions, and associated disease mechanisms (reviewed in Papaioannou, 2001). Studies performed in *Drosophila, Xenopus*, zebrafish, avians, and mice have taken advantage of the strengths of each system to examine T-box gene regulation of developmental and disease processes (reviewed in Naiche et al., 2005). T-box genes are involved in the development and patterning of many organ systems and embryonic structures including the heart, limb, eye, central axis, and face. In addition, T-box genes are subject to regulation by or induce the expression of developmentally important signaling molecules, such as retinoic acid (RA), bone morphogenetic proteins (BMPs), fibroblast growth factors (FGFs), and Wnts, in different organ systems (reviewed in Naiche et al., 2005). T-box proteins can act as transcriptional activators or repressors with a variety of cofactors to regulate expression of genes involved in cell lineage determination, differentiation, and maturation (reviewed in Tada and Smith, 2001). Overall, T-box genes are integrated into regulatory networks that control patterning, growth, and maturation of many cell types and tissues in the developing embryo.

Multiple T-box genes are expressed in the developing heart where they regulate cardiac lineage determination, chamber formation, valvuloseptal development, and conduction system maturation (Fig. 1) (reviewed in Plageman and Yutzey, 2005). T-box genes also are important for the development of cells derived from neural crest or proepicardium that migrate into the heart. The necessity of T-box genes in cardiac development is

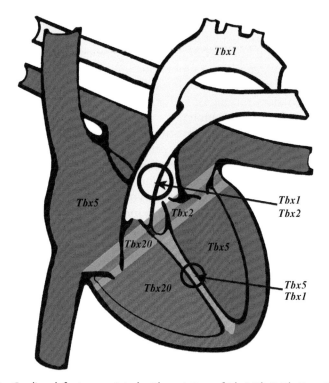

FIGURE 1   Cardiac defects associated with mutation of *Tbx1, Tbx2, Tbx5,* or *Tbx20.* Homozygous loss of *Tbx1* in mice or deletion of the DiGeorge criticial region of human chromosome 22 that includes *Tbx1* is associated with OFT malformations and aortic arch anomalies (yellow) as well as ventricular septal defects (orange). Likewise, homozygous loss of *Tbx2* leads to OFT malformations (yellow) as well as an expansion of chamber myocardium in the AV canal region (green). Cardiac malformations associated with human Holt–Oram syndrome patients or observed in *Tbx5* heterozygous mutant mice include hypoplastic left ventricle and ASDs (red) in addition to ventricular septal defects and conduction system anomalies (orange). *Tbx20* is expressed predominantly in the right ventricle and valve precursors, and loss of Tbx20 function inhibits myocardial maturation and valve remodeling (blue). Please see text for references.

further demonstrated by the association of mutations in several T-box genes with human genetic syndromes that include congenital heart malformations (reviewed in Packham and Brook, 2003). Genetic studies in mice have been used to define the requirements for individual T-box genes in the development of specific cardiac structures (reviewed in Stennard and Harvey, 2005). Experiments in avian embryos, which are more amenable to embryonic manipulation, have further defined complex regulatory relationships involving T-box genes and developmentally important signaling pathways. The cellular processes regulated by T-box genes include differentiation, proliferation, migration and maturation of cardiac myocytes, valve progenitors, neural crest, and epicardial precursors. Taken together, the regulation of heart development by T-box genes encompasses most of the major cardiac cell lineages, and T-box deficiencies often lead to defective cardiac organogenesis and embryonic lethality. A broad spectrum of experimental approaches in multiple model systems is being used to dissect the precise regulatory mechanisms that control T-box gene expression and function during heart development.

The chicken embryo has proven to be an effective animal model system for analysis of regulatory processes that control normal and abnormal development. Historically, the chicken has been used for experimental embryological approaches including fate mapping, microsurgeries, quail–chick chimeras, and embryonic cell culture (Stern, 2005). Discoveries made in the chicken system include the initial identification of neural crest, regulative properties of gastrulating embryos, mechanisms of left–right asymmetry, and signaling centers for limb patterning. In the molecular era, additional techniques including retrovirus infections, electroporation, and RNA interference (RNAi) are available for introducing targeted genetic alterations in specific embryonic structures with precise developmental timing (Nakamura et al., 2004). These approaches add to more traditional strategies of embryonic exposure to growth factors or pharmacological agents for the experimental analysis of developmental processes. Together these tools make the chicken embryo a powerful system for the elucidation of regulatory networks that control development of the heart and related systems.

The focus of this chapter is the role of T-box genes in heart development with particular attention to research conducted in the avian model system. Studies in chicken embryos have been useful for manipulation of specific T-box functions in cardiac structures and for analysis of T-box gene regulatory networks. In particular, experimental embryological approaches in avian embryos have been used to identify signaling pathways and inductive interactions that control T-box gene activity during development of the heart and limbs. Conservation of these regulatory interactions has been demonstrated in other vertebrate model systems, which underscores the importance of T-box gene regulation of the critical developmental processes required for normal heart formation.

2. T-box REGULATORY NETWORKS AND CARDIAC CELL LINEAGE DEVELOPMENT

Several T-box genes, including *Tbx2*, *Tbx3*, *Tbx5*, and *Tbx20*, are expressed in the primary heart field of developing vertebrate embryos (Fig. 2). Studies performed in avian embryos have demonstrated that expression of *Tbx2*, *Tbx3*, and *Tbx20* is induced by BMP signaling in cardiogenic mesoderm while *Tbx5* expression in the primitive heart tube is responsive to retinoid

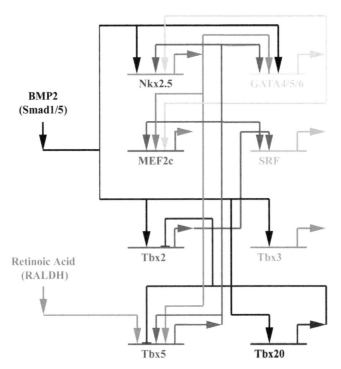

FIGURE 2   Transcriptional regulatory networks in the primary heart field include multiple T-box genes. Reported regulatory interactions of Tbx2, Tbx3, Tbx5, and Tbx20 with other transcription factors expressed in the heart are indicated by individually colored lines. Transcriptional activation is indicated by arrows, and repression is indicated by perpendicular bars. Inductive relationships with BMP and RA signaling pathways are shown as black and gray arrows, respectively. Aspects of this regulatory network are included in Cripps and Olson (2002), and the construction of transcriptional regulatory circuits is based on concepts in Davidson (2006). Please see text for additional relevant references.

signaling (Liberatore et al., 2000; Yamada et al., 2000; Niederreither et al., 2001; Plageman and Yutzey, 2004). Experiments in cell culture and genetic analyses in mice suggest that coordination of these T-box genes promotes early cardiac gene expression and differentiation, but demonstrate that none is independently required for cardiac cell specification and determination (Bruneau et al., 2001; Hiroi et al., 2001; Davenport et al., 2003; Harrelson et al., 2004; Cai et al., 2005; Singh et al., 2005; Stennard et al., 2005; Takeuchi et al., 2005). While Tbx2, Tbx3, Tbx5, and Tbx20 are not individually required for initiation of cardiomyogenic differentiation, each has been implicated in aspects of cardiac lineage maturation and heart patterning in conjunction with other transcription factors expressed in the heart.

Tbx5 is dynamically expressed in the avian embryonic heart consistent with a role in early heart tube patterning and chamber specification.

The first indication of *Tbx5* expression is in the anterior lateral heart-forming region or cardiac crescent of late gastrulation stage embryos (Bruneau et al., 1999; Liberatore et al., 2000). During heart looping, *Tbx5* expression becomes restricted to the posterior portion of the heart tube corresponding to the presumptive atria and left ventricle. Anteroposterior patterning of the heart is subject to regulation by RA signaling, and RA treatment of avian embryos produces an anterior expansion of *Tbx5* cardiac expression (Yutzey et al., 1994, 1995; Liberatore et al., 2000). Likewise, *Tbx5* expression is absent from the posterior primitive heart tube of mouse embryos that lack retinoid signaling due to the targeted mutation of the RA-metabolizing enzyme gene *raldh2* (Niederreither et al., 2001). Tbx5 is required for development of the venous segments of the heart as demonstrated by hypomorphic atrial development and loss of atrial gene expression in mice lacking Tbx5 (Bruneau et al., 2001). Similarly, ectopic expression of *Tbx5* throughout the primitive heart tube of transgenic embryos results in reduced formation of the ventricular trabeculae and loss of ventricular-specific gene expression (Liberatore et al., 2000). Taken together, these results support a role for Tbx5 in the proper development of the atria and left ventricle, but Tbx5 is not essential for cardiac specification or early heart tube differentiation (Bruneau et al., 2001).

Other T-box family members, including Tbx2 and Tbx20, also function in patterning and morphogenesis of the primitive heart. Studies in avian embryos identified BMP2 as a critical inducing factor in primary cardiac lineage specification and differentiation, and *Tbx2* and *Tbx20* are among the cardiac transcription factor genes induced by BMP signaling in the heart-forming region (Schultheiss et al., 1997; Yamada et al., 2000; Plageman and Yutzey, 2004). During heart tube formation and looping, *Tbx2*, a transcriptional repressor, is initially expressed throughout the primary myocardium of the early heart tube and then becomes progressively restricted to the primary myocardium of the outflow tract (OFT), atrioventricular (AV) canal, and atria (Yamada et al., 2000; Habets et al., 2002; Christoffels et al., 2004b). *Tbx20* is expressed more widely throughout the linear heart tube, but later in development, its expression is concentrated in the AV canal, OFT, and atria (Iio et al., 2001; Plageman and Yutzey, 2004; Yamagishi et al., 2004). Expression of both *Tbx2* and *Tbx20* is induced by BMP signaling in the AV canal of avian embryos (Yamada et al., 2000; Shelton and Yutzey, 2007). Additional genetic manipulation of BMP signaling in mice confirms that expression of *Tbx2* and *Tbx20* in the AV canal and endocardial cushions is dependent on this regulatory pathway (Ma et al., 2005; Rivera-Feliciano and Tabin, 2006). Together these analyses demonstrate an inductive interaction between BMP signaling and expression of *Tbx2* and *Tbx20* in the primary heart field and specialized regions of the OFT and AV canal.

Targeted mutagenesis and RNAi studies in mice revealed direct genetic interactions between Tbx20 and Tbx2 in the regulation of myocardial cell proliferation (Cai et al., 2005; Singh et al., 2005; Stennard et al., 2005; Takeuchi et al., 2005). Mice lacking Tbx20 demonstrate hypoplasia of the primitive myocardium and decreased chamber maturation. One of the most obvious alterations in gene expression in these embryos is increased expression of *Tbx2* throughout the cardiac crescent and primitive heart tube. Further molecular analyses indicate that Tbx20 can act directly on *Tbx2* regulatory sequences to repress gene expression (Cai et al., 2005). In addition, the reduced proliferation of the cardiomyocytes was accompanied by decreased expression of *N-myc* gene, which was attributed to relief of repression by Tbx2 (Cai et al., 2005). Together these analyses define critical roles for Tbx20 and Tbx2 in the regulation of cardiomyocyte proliferation. Tbx5 has also been implicated in proliferation of cardiomyocytes in the primitive heart tube of *Xenopus* embryos and also during later stages of ventricular morphogenesis (Hatcher et al., 2001; Goetz et al., 2006). Therefore, multiple T-box proteins may act together to regulate each other's expression as well as expression of critical cell cycle regulatory genes.

Coordination of Tbx2, Tbx5, and Tbx20 function occurs during specification and maturation of diversified regions of chamber and primitive myocardium (Fig. 3). Tbx2 is associated with repression of chamber myocardium differentiation in the AV canal, inflow tract (IFT), and OFT, while Tbx5 induces expression of chamber maturation genes such as *natriuretic precursor peptide A* (*Nppa*) and *connexin 40* (*Cx40*) (Bruneau et al., 2001; Christoffels et al., 2004b). In regions where Tbx2 and Tbx5 are coexpressed, Tbx2 competes with Tbx5 and represses expression of chamber myocardial genes including *Nppa* (Habets et al., 2002). Decreased expression of *Nppa* and other chamber-specific genes also is apparent in mice lacking Tbx20, potentially through increased repressor function of Tbx2 (Cai et al., 2005; Singh et al., 2005; Stennard et al., 2005; Takeuchi et al., 2005). Taken together, these results demonstrate that Tbx2, subject to regulation by Tbx20, maintains primitive myocardium in the AV canal, IFT, and OFT by repressing chamber maturation gene programs. In contrast, Tbx5 is expressed in the chamber myocardium and activates genes characteristic of chamber maturation. Since *Tbx2, Tbx5,* and *Tbx20* expression overlaps in several regions of the heart, most notably in the AV canal, the balance of these proteins and hypothesized cofactors is critical for normal cardiac lineage maturation and morphogenesis (Stennard and Harvey, 2005).

3. T-box REGULATION OF CARDIAC VALVULOSEPTAL AND CONDUCTION SYSTEM DEVELOPMENT

In addition to having a role in patterning the posterior venous segments of the primitive heart, Tbx5 is important for atrial and ventricular septation as is evident by the septal defects associated with Holt–Oram syndrome in humans and in animal model studies. Holt–Oram syndrome,

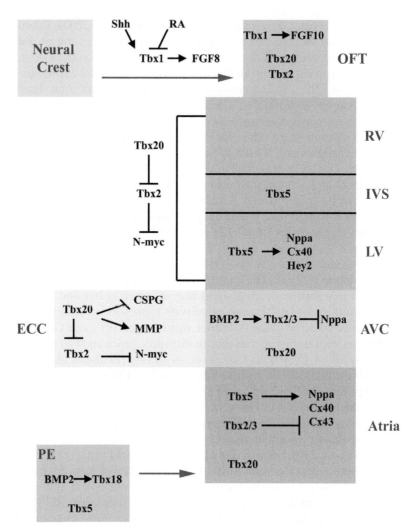

FIGURE 3 T-box gene regulatory interactions in heart organogenesis. Regions of the heart (OFT, outflow tract; RV, right ventricle; IVS, interventricular septum; LV, left ventricle; AVC, atrioventricular canal; ECC, endocardial cushions) are shown as colored blocks with relevant T-box regulatory interactions indicated. Neural crest and proepi-cardium (PE) are indicated as sources of cells that migrate into the heart. Please see text for references and further details.

caused by mutations in *TBX5*, is an autosomal dominant disorder characterized by congenital cardiac malformations, including septal defects, and forelimb anomalies (Basson et al., 1997; Li et al., 1997). *Tbx1* also has been implicated in ventricular septation through its association with DiGeorge syndrome (Baldini, 2004). During ventricular septation in chicken embryos, *Tbx5* is expressed in the left but not right ventricle, and it has been hypothesized that the limit of *Tbx5* expression determines the position of the interventricular septum (Takeuchi et al., 2003b; Plageman and Yutzey, 2004). In support of this hypothesis, misexpression of *Tbx5* throughout the developing chicken heart using *in ovo* electroporation and retroviral infection results in loss or malpositioning of the interventricular septum (Takeuchi et al., 2003b).

Tbx5 also has been implicated in atrial septation. Atrial septal defects (ASDs) are a defining feature of Holt–Oram syndrome and mice with *Tbx5* haploinsufficiency also display severe ASDs (Basson et al., 1999; Bruneau et al., 2001). In addition, human genetic studies have identified α-*myosin heavy chain* (MHC) as a Tbx5 target gene important for atrial septum development (Ching et al., 2005). Supporting studies in the chicken system demonstrated that targeted loss of atrial MHC using morpholinos *in ovo* resulted in ablation of the atrial septum (Ching et al., 2005). Additional Tbx5-responsive genes, including *photoreceptor cadherin*, are expressed during atrial septation in mouse embryos (Plageman and Yutzey, 2006). Together these studies demonstrate a role for Tbx5 in development of both the ventricular and atrial septa.

T-box genes also function in heart valve development. Of the T-box genes expressed in the heart, *Tbx20* is the most highly expressed in the endocardial cushions of the AV canal and OFT as well as in the remodeling mitral and tricuspid valves (Stennard et al., 2003; Plageman and Yutzey, 2004; Yamagishi et al., 2004; Shelton and Yutzey, 2007). Recent studies provide evidence that Tbx20 promotes proliferation and inhibits maturation of valve progenitor cells. In cultured endocardial cushion cells isolated from chicken embryos, Tbx20 gain of function was accomplished using a Tbx20-expressing adenovirus while inhibition of endogenous Tbx20 expression was achieved with transfection of Tbx20-specific siRNA (Shelton and Yutzey, 2007). In these experiments, increased Tbx20 function promotes cell proliferation, whereas loss of Tbx20 leads to decreased cell proliferation, with corresponding effects on *N-myc* gene expression. In addition, increased Tbx20 expression leads to decreased expression of chondroitin sulfate proteoglycans (CSPGs) and increased expression of matrix metalloproteinases (MMPs) consistent with the unremodeled state of the endocardial cushions (Shelton and Yutzey, 2007). Supporting evidence for Tbx20 function in valve formation is provided by inhibition of valvulogenesis in mouse embryos with low dose expression of Tbx20 siRNA (Takeuchi et al., 2005). The results of these studies indicate a role for Tbx20 in maintaining the undifferentiated proliferative state of the endocardial cushion mesenchyme prior to extracellular matrix organization and remodeling of the developing valves.

Tbx2, Tbx3, and Tbx5 all have been implicated in development of the specialized conduction system. The cardiac conduction system is composed of the sinoatrial node, AV node, common bundle of His, left and right bundle branches, and the distal Purkinje fiber network (Fishman, 2005). *Tbx2* and *Tbx3* have overlapping expression and potentially redundant functions in the developing conduction system. However, distinct compartments of the conduction system have differential T-box gene expression with *Tbx2* predominant in peripheral conduction system and *Tbx3* specific to the central conduction system (Hoogaars et al., 2004). While *Tbx3* is generally recognized as a specific marker of the central conduction system, no conduction system abnormalities were observed in *Tbx3* null mice or in human patients with ulnar–mammary syndrome, a disorder caused by mutations in *TBX3* (Bamshad et al., 1997; Davenport et al., 2003; Hoogaars et al., 2004). In contrast, mice with targeted mutagenesis of *Tbx2* are deficient for heart chamber maturation and have misexpression of genes associated with the cardiac conduction system (Harrelson et al., 2004). However, mice with reduced expression of both *Tbx2* and *Tbx3*, as a result of decreased BMP signaling in the heart, exhibit normal early conduction system specification and patterning (Yang et al., 2006). Therefore, further studies are necessary to define the contributions of Tbx2 and Tbx3 to specialized conduction system development.

Tbx5 also is expressed in the developing conduction system, and patients with Holt–Oram syndrome, as well as mice haploinsufficient for *Tbx5*, display conduction system anomalies (Bruneau et al., 1999, 2001; Moskowitz et al., 2004). Conduction system anomalies also have been reported for human patients with mutations in the Tbx5 cofactor gene *NKX2-5* (Schott et al., 1998). In transfection studies, Tbx5 and Nkx2.5 cooperatively activate the expression of *Nppa* and *Cx40* (Bruneau et al., 2001; Stennard et al., 2003; Plageman and Yutzey, 2004), both of which are expressed in the conduction system (Houweling et al., 2002; Coppen et al., 2003). Additionally, Tbx2 can interact with Nkx2.5 and directly repress *Nppa* (Habets et al., 2002). Because Tbx2 and Tbx5 share binding partners and transcriptional targets, it has been suggested that a tightly regulated balance between the repressor Tbx2 and the activator Tbx5 is necessary for proper development of the conduction system (Habets et al., 2002; Christoffels et al., 2004a). However, the mechanism of their coordinated regulation in distinct populations of specialized cardiomyocytes at specific developmental time points is still unclear.

4. T-ʙᴏx GENE REGULATION OF CELLS THAT MIGRATE INTO THE HEART

The human *Tbx1* gene is located in the DiGeorge syndrome critical region of chromosome 22 associated with congenital heart and craniofacial malformations (Packham and Brook, 2003). Studies in animal model systems have confirmed the importance of Tbx1 to the development of neural crest-derived structures, including the aortic arch arteries and cardiac OFT. *Tbx1* is expressed in pharyngeal arch mesoderm and anterior endoderm that provide critical guidance cues and inducing factors to the migrating neural crest (Garg et al., 2001; Xu et al., 2004; Roberts et al., 2005). *Tbx1* also is expressed in the secondary heart field that contributes to the cardiac OFT, supporting a role for Tbx1 in the development of cardiac structures that are not derived from neural crest (Xu et al., 2004; Nowotschin et al., 2006). In mice, null mutations in *Tbx1* lead to several developmental anomalies observed in human patients with DiGeorge syndrome, including abnormal aortic arch development, OFT anomalies, and ventricular septal defects (Jerome and Papaioannou, 2001; Lindsay et al., 2001; Merscher et al., 2001). However haploinsufficiency of Tbx1 in mice does not fully recapitulate the 22q11 deletion syndrome in patients. It has been hypothesized that additional genes present in the deleted region of chromosome 22 modulate Tbx1 function and affect pharyngeal development, thereby contributing to DiGeorge syndrome phenotypes (Moon et al., 2006).

Tbx1 is part of complex regulatory networks that govern neural crest migration and maturation in the pharyngeal arches and cardiac OFT. In chicken embryos, *Sonic hedgehog* (Shh) is expressed in the anterior pharyngeal endoderm, and implantation of Shh beads on the surface of the pharyngeal arches results in expanded *Tbx1* expression (Garg et al., 2001). Additionally, *Shh* null mice have disrupted pharyngeal arch development and decreased *Tbx1* expression (Garg et al., 2001). RA signaling also is present in the pharyngeal arches, and disruption of retinoid signaling can result in developmental defects similar to those observed in DiGeorge syndrome (Lammer et al., 1985). Studies performed in quail embryos demonstrated that antagonism of retinoid signaling resulted in abnormal and ectopic *Tbx1* expression (Roberts et al., 2005). Likewise, implantation of RA beads or RA treatment of whole chicken embryos inhibits the expression of *Tbx1*, indicating that retinoic signaling represses *Tbx1* expression during pharyngeal arch development (Roberts et al., 2005). Moreover, studies in genetically manipulated mouse embryos demonstrated that Tbx1 is required for the expression of *Fgf8* in pharyngeal endoderm and loss of *Fgf8* results in neural crest-related defects in aortic arch arteries and the OFT (Vitelli et al., 2002, 2006; Zhang et al., 2006). Together, these studies demonstrate that *Tbx1* expression is responsive to Shh and RA signaling, and that Tbx1 can regulate *Fgf8* expression required for normal development of the pharyngeal arches and cardiac OFT.

Tbx5 and Tbx18 have been implicated in the development of the proepicardium and epicardial-derived cell lineages (Hatcher et al., 2004; Schlueter et al., 2006). The proepicardium is a transient embryonic structure that gives rise to mesenchymal cells that migrate across the surface of the heart to form the epicardium and its derivatives in the coronary vasculature (Manner et al., 2001). In chicken embryos, both *Tbx5* and *Tbx18* are expressed in the proepicardium located adjacent to the right horn of the sinus venosus and also in the epicardial cell layer surrounding the heart (Hatcher et al., 2004; Schlueter et al., 2006). Manipulation of Tbx5 expression in the proepicardium with retrovirus-mediated transgenesis or Tbx5-specific morpholinos demonstrated a role for Tbx5 in cell proliferation and migration out of the proepicardium (Hatcher et al., 2004). Although it has not been specifically examined, RA signaling is present in the proepicardium (Manner et al., 2001) and may contribute to regulating *Tbx5* expression, as has been demonstrated in the primitive heart tube (Liberatore et al., 2000). The function of Tbx18 in epicardial cell lineages has not been determined, and defects in epicardial development or coronary vasculogenesis were not apparent in *Tbx18* null mice (Bussen et al., 2004). However, studies in avian embryos showed that *Tbx18* expression in the proepicardium is sensitive to precise levels of BMP signaling (Schlueter et al., 2006). The lack of epicardial defects in *Tbx18* null mice may be indicative of redundant functions for Tbx5 and Tbx18 in proepicardial development.

5. T-box COFACTORS AND TARGET GENES IN THE DEVELOPING HEART

T-box family member transcription factor function has been studied using a variety of molecular approaches and cell culture assays. T-box proteins regulate transcription through binding to T-box elements (TBEs) with the DNA consensus sequence (NGTGNNA) (Coll et al., 2002). TBEs are often clustered or are adjacent to binding sites for other cardiac transcription factors, such as Nkx2.5 or GATA4, in cardiac gene regulatory sequences (reviewed in Plageman and Yutzey, 2005). T-box proteins can be grouped into subfamilies that act as transcriptional activators or repressors depending on the particular molecular or cellular context. In addition, individual T-box proteins including Tbx20 and Tbx2 contain both activator and repressor domains (Paxton et al., 2002; Stennard et al., 2003; Plageman and Yutzey, 2004). A limited number of T-box target genes have been identified, and they act in multiple cardiac compartments to regulate a variety of cellular functions as described above. The most extensively characterized T-box target gene in the heart is *Nppa*, which is subject to positive regulation by Tbx5 and negative regulation by Tbx2 or Tbx3, as demonstrated in transfection studies and in genetically manipulated mice (Bruneau et al., 2001; Habets et al., 2002; Christoffels et al., 2004b). Tbx20 can activate or repress *Nppa* expression in transfection studies depending on the cell type (Stennard et al., 2003; Plageman and Yutzey, 2004). In addition, the loss of

Nppa expression in mice lacking Tbx20 could be due to direct activation or the strong induction of the repressor Tbx2 throughout the myocardium (Cai et al., 2005). The apparent regulation of *Nppa* expression by the precise balance of T-box factors expressed in localized regions of the heart at specific times during development may be a typical mechanism for coordination of gene expression and patterning by T-box genes during embryogenesis.

Cardiac differentiation and gene regulation are coordinately controlled by T-box proteins acting in concert with other cardiac transcription factors, such as Nkx2.5, GATA4, Mef2c, and SRF (Fig. 2). Regulatory elements involved in cardiac expression of genes such as *Nppa, Cx40, Cx43, MYH6,* and *N-myc* include TBEs located adjacent to NKX-binding elements (NKEs) and/or GATA consensus binding sites (Bruneau et al., 2001; Hiroi et al., 2001; Stennard et al., 2003; Hoogaars et al., 2004; Linhares et al., 2004; Plageman and Yutzey, 2004; Cai et al., 2005; Ching et al., 2005). Biochemical studies have revealed physical interactions between Tbx5, GATA4, and Nkx2.5 proteins through specific peptide domains, and deficiencies in these interactions may be related to congenital heart malformations observed with mutations of these genes in the human population (Hiroi et al., 2001; Garg et al., 2003). There is extensive evidence for T-box gene autoregulation or regulation of other T-box family member gene expression in the same tissue (Sun et al., 2004; Brown et al., 2005; Cai et al., 2005; Takeuchi et al., 2005). Additional T-box downstream targets include *SRF, MEF2c, GATA4,* and *Nkx2.5* genes that are dynamically expressed in the developing myocardium and are required for normal heart formation (Cripps and Olson, 2002; Heicklen-Klein and Evans, 2004; Barron et al., 2005; Takeuchi et al., 2005). The extensive interaction between these factors at levels of gene regulation and protein function is evidence for reinforcing regulatory networks that include multiple T-box genes important for cardiac lineage specification and maturation.

T-box transcription factor function is further regulated by coactivators and chromatin remodeling as well as by control of subcellular localization. Tbx5 cofactors include the WW-domain protein TAZ, the BAF60c subunit of the SWI–SNF chromatin-remodeling complex, and the histone acetyltransferase TIP60, all of which enhance Tbx5 activation of known target genes (Lickert et al., 2004; Barron et al., 2005; Murakami et al., 2005). Additional factors have been identified that affect nucleocytoplasmic shuttling of T-box proteins. The nuclear localization signals of T-box proteins are atypical and there is evidence that multiple protein domains may control nuclear import (Collavoli et al., 2003; Stoller and Epstein, 2005). Additional modifications or protein–protein interactions may contribute to the regulation of T-box protein subcellular localization. The PDZ-LIM domain protein LMP4, first identified in chicken embryos, associates with Tbx5 or

Tbx4, but not with Tbx2 or Tbx3, and sequesters Tbx5 in the cytoplasm localized to actin filaments (Krause et al., 2004). This interaction provides a mechanism for control of nuclear versus cytoplasmic localization of Tbx5 protein, thereby affecting expression of downstream target genes including *Nppa*, in subpopulations of cardiomyocytes and epicardial cells in the developing chicken heart (Camarata et al., 2006). Together these molecular modulators of Tbx protein function provide exquisite multifaceted regulation of this family of transcription factors.

6. T-BOX REGULATORY NETWORKS AND DEVELOPMENT OF OTHER ORGAN SYSTEMS

Several of the T-box genes expressed in the developing heart also are important for limb patterning and morphogenesis (Fig. 4). Early reports of T-box gene expression in avian and other vertebrate embryos noted differential expression of T-box genes in the developing limbs (Gibson-Brown et al., 1998). The observation that *Tbx5* is restricted to the forelimb bud and *Tbx4* is predominant in the hindlimb bud led to the hypothesis

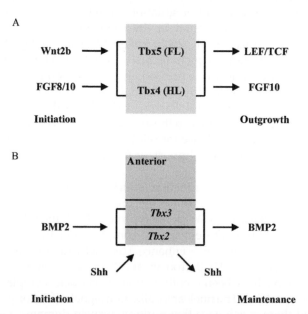

FIGURE 4   T-box gene regulatory networks and limb outgrowth and patterning. (A) Tbx5 and Tbx4 are included in regulatory networks that control forelimb (FL) and hindlimb (HL) initiation (left) and outgrowth (right). (B) Anteroposterior patterning of the forelimb and hindlimb is regulated by posterior expression of *Tbx3* and *Tbx2* in feed-forward regulatory relationships with BMP2 and Shh signaling. Please see text for references.

that differential T-box activity was responsible for limb identity. This was tested directly in chicken embryos with ectopic expression of Tbx5 or Tbx4 in the developing limb buds, using electroporation and retroviral vectors (Rodriguez-Esteban et al., 1999; Takeuchi et al., 1999). In these experiments, ectopic expression of Tbx5 in the hindlimb resulted in the outgrowth of a wing-like structure, and ectopic expression of Tbx4 in the flank induced a supernumerary hindlimb, thus supporting a role for these genes in limb specification. However, a study in mouse embryos, in which Tbx4 was ectopically expressed in *Tbx5* null forelimbs, implicates these genes in limb outgrowth but not in specification of forelimb versus hindlimb identity (Minguillon et al., 2005).

Tbx2 and *Tbx3* are expressed together in the posterior regions of both forelimbs and hindlimbs, but expression differs in chicken and mouse embryos (Naiche et al., 2005). In chicken embryos, *Tbx2* is expressed only in the most posterior digit region, whereas *Tbx3* is expressed in the posterior two digits (Suzuki et al., 2004). Ectopic expression of Tbx2 or Tbx3 in the limb bud results in posterior transformation of digits, leading to the conclusion that Tbx2 and Tbx3 together specify the most posterior fourth digit and that Tbx3 alone is sufficient to specify the third digit. In addition, ectopic expression of Tbx3 in the flank shifts the position of the limb along the anteroposterior axis. In mice, *Tbx2* and *Tbx3* are expressed at both anterior and posterior margins of the limb buds (Suzuki et al., 2004). Loss of Tbx3 results in mild defects in the development of the posterior limb margins, and duplication of digit 4 of the hindlimb was observed in *Tbx2* null embryos (Davenport et al., 2003; Harrelson et al., 2004). These studies in mice support the observations made in avian embryos that the precise dosage of Tbx2 and Tbx3 is related to posterior limb identity. Overall, the extensive research on *Tbx* genes in the developing limbs implicates Tbx2, Tbx3, Tbx4, and Tbx5 in limb outgrowth and patterning, but not in specific cell lineage determination events.

Studies in chicken embryos also have been used to identify regulatory interactions between T-box genes and FGF, BMP, SHH, and Wnt signaling pathways in the developing limb. In many cases, T-box genes have been found to be induced by the same signaling molecules that they in turn regulate. For example, ectopic expression of FGF10 in the flank induces Tbx5 expression, which can directly activate *Fgf10* gene expression (Ng et al., 2002; Agarwal et al., 2003). Likewise, BMP-infiltrated bead implantation leads to induction of *Tbx2* and *Tbx3* expression, and ectopic expression of Tbx2 or Tbx3 in the limb bud leads to increased expression of *Bmp2* (Suzuki et al., 2004). Similar feed-forward mechanisms have been observed for Tbx2 and SHH, and for Tbx5 and Wnt signaling molecules (Gibson-Brown et al., 1998; Ng et al., 2002; Agarwal et al., 2003; Takeuchi et al., 2003a; Suzuki et al., 2004). Additional studies in other embryonic structures, such as the developing eye, lung bud, and nasal placode, have revealed similar regulatory relationships between BMP and FGF signaling pathways and T-box genes (Koshiba-Takeuchi et al., 2000; Firnberg and Neubuser, 2002; Sakiyama et al., 2003; Leconte et al., 2004). The observation that T-box genes are both upstream and downstream of the same signaling molecules in a variety of organ systems may represent a common mechanism for amplification of pathways that control growth and morphogenesis during embryogenesis.

Research on the developing limbs and other embryonic structures has uncovered complex T-box regulatory networks that include signaling pathways involved in growth and morphogenesis rather than specific cell lineage determination events. Similar mechanisms may be operating in the heart, where T-box genes have been associated with cell proliferation and chamber morphogenesis. However, cardiac T-box gene targets also include genes, such as *Nppa* and *Cx40*, which are differentially expressed in specialized cell lineages. The regulation of cardiac lineage-related genes by T-box proteins may be a consequence of associations with Nkx2.5 and GATA factors acting in close proximity within the same genetic regulatory elements. Therefore, it is likely that T-box gene regulatory networks in the heart have dual functions in the control of cell proliferation and chamber morphogenesis as well as in cardiac lineage-specific gene expression, depending on the specific cellular and molecular context.

7. CONCLUSIONS AND PERSPECTIVES

The regulation and function of T-box genes in the developing heart has been an active area of research for the past several years. Work by many investigators using state-of-the-art experimental approaches in a broad spectrum of animal model systems has identified critical functions for individual T-box genes in heart tube patterning, morphogenesis, cell proliferation, valvuloseptal development, and conduction system maturation. Studies of avian embryos have been particularly useful for the identification of inductive interactions that control T-box expression and function, and for establishing regulatory parallels between heart and limb development. A limited number of direct T-box target genes expressed in the heart have been identified and their functions include cell cycle regulation and chamber lineage maturation. In many cases, these target genes are subject to regulation by multiple T-box transcription factors expressed together in specific cardiac structures. Much of our knowledge of T-box gene function in the heart is based on manipulation of individual family members. Therefore, further work is necessary to determine how multiple T-box proteins with different transcriptional functions act together to control specific localization and levels of gene expression in the developing heart.

Haploinsufficiency of Tbx5, Tbx1, and Tbx3 is associated with human genetic syndromes, and heterozygous mutations in T-box cofactor genes, GATA4 and NKX2-5, also are associated with congenital cardiac malformations (Schott et al., 1998; Garg et al., 2003). Gene targeting and experimental embryological approaches have been used extensively to examine the regulation of early heart formation and patterning by T-box genes. Less is known of how altered T-box function leads to clinically significant cardiac malformations such as atrial and ventricular septal defects, arrhythmias, or OFT anomalies. Analyses of the later events of heart formation and function will require more targeted conditional approaches or hypomorphic alleles that do not lead to embryonic lethality. Resources and analytical methods for these studies are becoming increasingly available and will likely yield clinically significant information on T-box gene function and cardiac malformations.

At present, a limited number of T-box transcription factor target genes expressed in distinct cardiac lineages and structures have been identified. These include genes involved in cell proliferation, cell–cell communication, cardiac muscle contraction, and extracellular matrix remodeling. However, these genes likely do not represent the full complement of genes regulated by T-box proteins in the developing heart, based on diverse cardiac phenotypes that arise with altered T-box gene function. In addition, the T-box genes themselves are subject to regulation by a variety of signaling pathways important for heart development, and relatively little is known of the transcriptional mechanisms that control T-box gene expression. A current challenge is to understand how the identified regulatory interactions fit together in complex networks to control morphogenesis and function of the developing heart.

ACKNOWLEDGMENTS

We thank Joy Lincoln, Santanu Chakraborty, and Heather Evans-Anderson for critical suggestions. This work was supported by an AHA Ohio Valley Affiliate predoctoral fellowship 0515153B to ELS, NIH grant HL082716 to KEY, and NIH/NHLBI SCCOR in Pediatric Heart Development and Disease P50 HL074728.

REFERENCES

Agarwal, P., Wylie, J.N., Galceran, J., Arkhitko, O., Li, C., Deng, C., Grosschedl, R., Bruneau, B.G. 2003. Tbx5 is essential for forelimb bud initiation following patterning of the limb field in the mouse embryo. Development 130, 623–633.

Baldini, A. 2004. DeGeorge syndrome: An update. Curr. Opin. Cardiol. 19, 201–204.

Bamshad, M., Lin, R.C., Law, D.J., Watkins, W.C., Krakowiak, P.A., Moore, M.E., Franceschini, P., Lala, R., Holmes, L.B., Gebuhr, T.C., Bruneau, B.G., Schinzel, A., et al. 1997. Mutations in human TBX3 alter limb, apocrine, and genital development in ulnar-mammary syndrome. Nat. Genet. 16, 311–315.

Barron, M.R., Belaguli, N.S., Zhang, S.X., Trinh, M., Iyer, D., Merlo, X., Lough, J.W., Parmacek, M.S., Bruneau, B.G., Schwartz, R.J. 2005. Serum response factor, an enriched cardiac mesoderm obligatory factor, is a downstream target for Tbx genes. J. Biol. Chem. 280, 11816–11828.

Basson, C.T., Bachinsky, D.R., Lin, R.C., Levi, T., Elkins, J.A., Soults, J., Grayzel, D., Kroumpouzou, E., Traill, T.A., Leblanc-Straceski, J., Renault, B., Kucherlapati, R., et al. 1997. Mutations in human *TBX5* cause limb and cardiac malformation in Holt-Oram syndrome. Nat. Genet. 15, 30–35.

Basson, C.T., Huang, T., Lin, R.C., Bachinsky, D.R., Weremowicz, S., Vaglio, A., Bruzzone, R., Quadrelli, R., Lerone, M., Romeo, G., Silengo, M. 1999. Different TBX5 interactions in heart and limb defined by Holt-Oram syndrome mutations. Proc. Natl. Acad. Sci. USA 96, 2919–2924.

Brown, D.D., Martz, S.N., Binder, O., Goetz, S.C., Price, B.M.J., Smith, J.C., Conlon, F.L. 2005. Tbx5 and Tbx20 act synergistically to control vertebrate heart morphogenesis. Development 132, 553–563.

Bruneau, B.G., Logan, M., Davis, N., Levi, T., Tabin, C.J., Seidman, J.G., Seidman, C.E. 1999. Chamber-specific cardiac expression of *Tbx5* and heart defects in Holt-Oram syndrome. Dev. Biol. 211, 100–108.

Bruneau, B.G., Nemer, G., Schmitt, J.P., Charron, F., Robitaille, L., Caron, S., Conner, D.A., Gessler, M., Nemer, M., Seidman, C.E., Seidman, J.G. 2001. A murine model of Holt-Oram syndrome defines roles of the T-box transcription factor Tbx5 in cardiogenesis and disease. Cell 106, 709–721.

Bussen, M., Petry, M., Schuster-Gossler, K., Leitges, M., Gossler, A., Kispert, A. 2004. The T-box transcription factor Tbx18 maintains the separation of anterior and posterior somite compartments. Genes Dev. 18, 1209–1221.

Cai, C.L., Zhou, W., Yang, L., Bu, L., Qyang, Y., Zhang, X., Li, X., Rosenfeld, M.G., Chen, J., Evans, S. 2005. T-box genes coordinate regional rates of proliferation and regional specification during cardiogenesis. Development 132, 2475–2487.

Camarata, T., Bimber, B., Kulisz, A., Chew, T.-L., Yeung, J., Simon, H.G. 2006. LMP4 regulates Tbx5 protein subcellular localization and activity. J. Cell Biol. 174, 339–348.

Ching, Y.H., Ghosh, T.K., Cross, S.J., Packham, E.A., Honeyman, L., Loughna, S., Robinson, T.E., Dearlove, A.M., Ribas, G., Bonser, A.J., Thomas, N.R., Scotter, A.J., et al. 2005. Mutation in myosin heavy chain 6 causes atrial septal defect. Nat. Genet. 37, 423–428.

Christoffels, V.M., Burch, J.B., Moorman, A.F. 2004a. Architectural plan for the heart: Early patterning and delineation of the chambers and the nodes. Trends Cardiovasc. Med. 14, 301–307.

Christoffels, V.M., Hoogaars, W.M.H., Tessari, A., Clout, D.E.W., Moorman, A.F.M., Campione, M. 2004b. T-box transcription factor Tbx2 represses differentiation and formation of the cardiac chambers. Dev. Dyn. 229, 763–770.

Coll, M., Seidman, J.G., Muller, C.W. 2002. Structure of the DNA-bound T-box domain of human TBX3, a transcription factor responsible for ulnar-mammary syndrome. Structure 10, 343–356.

Collavoli, A., Hatcher, C.J., He, J., Okin, D., Deo, R., Basson, C.T. 2003. Tbx5 nuclear localization is mediated by dual cooperative intramolecular signals. J. Mol. Cell. Cardiol. 35, 1191–1195.

Coppen, S.R., Kaba, R.A., Halliday, D., Dupont, E., Skepper, J.N., Elneil, S., Severs, N.J. 2003. Comparison of connexin expression patterns in the developing mouse heart and human foetal heart. Mol. Cell. Biochem. 242, 121–127.

Cripps, R.M., Olson, E.N. 2002. Control of cardiac development by an evolutionarily conserved transcriptional network. Dev. Biol. 246, 14–28.

Davenport, T.G., Jerome-Majewska, L.A., Papaioannou, V.E. 2003. Mammary gland, limb and yolk sac defects in mice lacking *Tbx3*, the gene mutated in human ulnar mammary syndrome. Development 130, 2263–2273.

Davidson, E.H. 2006. *The Regulatory Genome: Gene Regulatory Networks in Development and Evolution*. San Diego: Academic Press.

Firnberg, N., Neubuser, A. 2002. FGF signaling regulates expression of *Tbx2, Erm, Pea3*, and *Pax3* in the early nasal region. Dev. Biol. 247, 237–250.

Fishman, G.I. 2005. Understanding conduction system development: A hop, skip, and jump away? Circ. Res. 96, 809–811.

Garg, V., Yamagishi, C., Hu, T., Kathiriya, I.S., Yamagishi, H., Srivastava, D. 2001. *Tbx1*, a DiGeorge syndrome candidate gene, is regulated by sonic hedgehog during pharyngeal arch development. Dev. Biol. 235, 62–73.

Garg, V., Kathiriya, I.S., Barnes, R., Schluterman, M.K., King, I.N., Butler, C.A., Rothrock, C. R., Eapen, R.S., Hirayama-Yamada, K., Joo, K., Matsuoka, R., Cohen, J.C., et al. 2003. *GATA4* mutations cause human congenital heart defects and reveal an interaction with TBX5. Nature 424, 443–447.

Gibson-Brown, J.J., Agulnik, S.I., Silver, L.M., Niswander, L., Papaioannou, V.E. 1998. Involvement of T-box genes Tbx2-Tbx5 in vertebrate limb specification and development. Development 125, 2499–2509.

Goetz, S.C., Brown, D.D., Conlon, F.L. 2006. TBX5 is required for embryonic cardiac cell cycle progression. Development 133, 2575–2584.

Habets, P.E., Moorman, A.F., Clout, D.E., van Roon, M.A., Lingbeek, M., van Lohuizen, M., Campione, M., Christoffels, V.M. 2002. Cooperative action of Tbx2 and Nkx2.5 inhibits ANF expression in the atrioventricular canal: Implications for cardiac chamber formation. Genes Dev. 16, 1234–1246.

Harrelson, Z., Kelly, R.G., Goldin, S.N., Gibson-Brown, J.J., Bollag, R.J., Silver, L.M., Papaioannou, V.E. 2004. *Tbx2* is essential for patterning the atrioventricular canal and for morphogenesis of the outflow tract during heart development. Development 131, 5041–5052.

Hatcher, C.J., Kim, M.S., Mah, C.S., Goldstein, M.M., Wong, B., Mikawa, T., Basson, C.T. 2001. TBX5 transcription factor regulates cell proliferation during cardiogenesis. Dev. Biol. 230, 177–188.

Hatcher, C.J., Diman, N.Y., Kim, M.S., Pennisi, D., Song, Y., Goldstein, M.M., Mikawa, T., Basson, C.T. 2004. A role for Tbx5 in proepicardial cell migration during cardiogenesis. Physiol. Genom. 18, 129–140.

Heicklen-Klein, A., Evans, T. 2004. T-box binding sites are required for activity of a cardiac GATA-4 enhancer. Dev. Biol. 267, 490–504.

Herrmann, B.G., Labiet, S., Poustka, A., King, T., Lehrach, H. 1990. Cloning of the T gene required in mesoderm formation in the mouse. Nature 343, 617–622.

Hiroi, Y., Kudoh, S., Monzen, K., Ikeda, Y., Yazaki, Y., Nagai, R., Komuro, I. 2001. Tbx5 associates with Nkx2-5 and synergistically promotes cardiomyocyte differentiation. Nat. Genet. 28, 276–280.

Hoogaars, W.M.H., Tessari, A., Moorman, A.F., de Boer, P.A.J., Hagoort, J., Soufan, A.T., Campione, M., Christoffels, V.M. 2004. The transcriptional repressor Tbx3 delineates the developing central conduction system of the heart. Cardiovasc. Res. 62, 489–499.

Houweling, A.C., Somi, S., Van Den Hoff, M.J., Moorman, A.F., Christoffels, V.M. 2002. Developmental pattern of *ANF* gene expression reveals a strict localization of cardiac chamber formation in chicken. Anat. Rec. 266, 93–102.

Iio, A., Koide, M., Hidaka, K., Morisaki, T. 2001. Expression pattern of novel chick T-box gene, *Tbx20*. Dev. Genes Evol. 211, 559–562.

Jerome, L.A., Papaioannou, V.E. 2001. DiGeorge syndrome phenotype in mice mutant for the T-box gene, *Tbx1*. Nat. Genet. 27, 286–291.

Koshiba-Takeuchi, K., Takeuchi, J.K., Matsumoto, K., Momose, T., Uno, K., Hoepker, V., Ogura, K., Takahashi, N., Nakamura, H., Yasuda, K., Ogura, T. 2000. Tbx5 and the retinotectum projection. Science 287, 134–137.

Krause, A., Zacharias, W., Camarata, T., Linkhart, B., Law, E., Lischke, A., Miljan, E., Simon, H.G. 2004. Tbx5 and Tbx4 transcription factors interact with a new chicken PDZ-LIM protein in limb and heart development. Dev. Biol. 273, 106–120.

Lammer, E.J., Chen, D.T., Hoar, R.M., Agnish, N.D., Benke, P.J., Braun, J.T., Curry, C.J., Fernhoff, P.M., Grix, A.W.Jr., Lott, I.T., Richard, J.M., Sun, S.C. 1985. Retinoic acid embryopathy. N. Engl. J. Med. 313, 837–841.

Leconte, L., Lecoin, L., Martin, P., Saule, S. 2004. Pax6 interacts with cVax and Tbx5 to establish the dorsoventral boundary of the developing eye. J. Biol. Chem. 279, 47272–47277.

Li, Q.Y., Newbury-Ecob, R.A., Terrett, J.A., Wilson, D.I., Curtis, A.R., Yi, C.H., Gebuhr, T., Bullen, P.J., Robson, S.C., Strachan, T., Bonnet, D., Lyonnet, S., et al. 1997. Holt-Oram syndrome is caused by mutations in *TBX5*, a member of the *Brachyury* (T) gene family. Nat. Genet. 15, 21–29.

Liberatore, C.M., Searcy-Schrick, R.D., Yutzey, K.E. 2000. Ventricular expression of *tbx5* inhibits normal heart chamber development. Dev. Biol. 223, 169–180.

Lickert, H., Takeuchi, J.K., Von Both, I., Walls, J.R., McAuliffe, F., Adamson, S.L., Henkelman, R.M., Wrana, J.L., Rossant, J., Bruneau, B.G. 2004. Baf60c is essential for function of BAF chromatin remodeling complexes in heart development. Nature 432, 107–112.

Lindsay, E.A., Vitelli, F., Su, H., Morishima, M., Huynh, T., Pramparo, T., Jurecic, V., Ogunrinu, G., Sutherland, H.F., Scambler, P., Bradley, A., Baldini, A. 2001. *Tbx1* haploinsufficiency in the DiGeorge syndrome region causes aortic arch defects in mice. Nature 410, 97–101.

Linhares, V.L., Almeida, N.A., Menezes, D.C., Elliott, D.A., Lai, D., Beyer, E.C., Campos de Carvalho, A.C., Costa, M.W. 2004. Transcriptional regulation of the murine *Connexin40* promoter by cardiac factors Nkx2-5, GATA4 and Tbx5. Cardiovasc. Res. 64, 402–411.

Ma, L., Lu, M.F., Schwartz, R.J., Martin, J.F. 2005. Bmp2 is essential for cardiac cushion epithelial-mesenchymal transition and myocardial patterning. Development 132, 5601–5611.

Manner, J., Perez-Pomares, J.M., Macias, D., Munoz-Chapuli, R. 2001. The origin, formation and developmental significance of the epicardium: A review. Cells Tissues Organs 169, 89–103.

Merscher, S., Funke, B., Epstein, J.A., Heyer, J., Puech, A., Lu, M.M., Xavier, R.J., Demay, M.B., Russell, R.G., Factor, S., Tokooya, K., Jore, B.S., et al. 2001. *TBX1* is responsible for cardiovascular defects in velo-cardio-facial/DiGeorge syndrome. Cell 104, 619–629.

Minguillon, C., Buono, J.D., Logan, M.P. 2005. *Tbx5* and *Tbx4* are not sufficient to determine limb-specific morphologies but have common roles in initiating limb outgrowth. Dev. Cell 8, 75–84.

Moon, A.M., Guris, D.L., Seo, J.H., Li, L., Hammond, J., Talbot, A., Imamoto, A. 2006. *Crk1* deficiency disrupts Fgf8 signaling in a mouse model of 22q11 deletion syndromes. Dev. Cell 10, 71–80.

Moskowitz, I.P.G., Pizard, A., Patel, V.V., Bruneau, B.G., Kim, J.B., Kupershmidt, S., Roden, D., Berul, C.I., Seidman, C.E., Seidman, J.G. 2004. The T-box transcription factor Tbx5 is required for the patterning and maturation of the murine cardiac conduction system. Development 131, 4107–4116.

Murakami, M., Nakagawa, M., Olson, E.N., Nakagawa, O. 2005. A WW domain protein TAZ is a critical coactivator for TBX5, a transcription factor implicated in Holt-Oram syndrome. Proc. Natl. Acad. Sci. USA 102, 18034–18039.

Naiche, L.A., Harrelson, Z., Kelly, R.G., Papaioannou, V.E. 2005. T-box genes in vertebrate development. Annu. Rev. Genet. 39, 219–239.

Nakamura, H., Katahira, T., Sato, T., Watanabe, Y., Funahashi, J. 2004. Gain- and loss-of-function in chick embryos by electroporation. Mech. Dev. 121, 1137–1143.

Ng, J.K., Kawakami, Y., Buscher, D., Raya, A., Itoh, T., Koth, C.M., Rodriguez Esteban, C., Rodriguez-Leon, J., Garrity, D.M., Fishman, M.C., Izpisua Belmonte, J.C. 2002. The limb identity gene Tbx5 promotes limb initiation by interacting with Wnt2b and Fgf10. Development 129, 5161–5170.

Niederreither, K., Vermot, J., Messaddeq, N., Schuhbaur, B., Chambon, P., Dolle, P. 2001. Embryonic retinoic acid synthesis is essential for heart morphogenesis in the mouse. Development 128, 1019–1031.

Nowotschin, S., Liao, J., Gage, P.J., Epstein, J.A., Campione, M., Morrow, B.E. 2006. Tbx1 affects asymmetric cardiac morphogenesis by regulating Pitx2 in the secondary heart field. Development 133, 1565–1573.

Packham, E.A., Brook, J.D. 2003. T-box genes in human disorders. Hum. Mol. Genet. 12, R37–R44.

Papaioannou, V.E. 2001. T-box genes in development: From hydra to humans. Int. Rev. Cytol. 207, 1–70.

Paxton, C., Zhao, H., Chin, Y., Langner, K., Reecy, J. 2002. Murine Tbx2 contains domains that activate and repress gene transcription. Gene 283, 117–124.

Plageman, T.F.Jr., Yutzey, K.E. 2004. Differential expression and function of Tbx5 and Tbx20 in cardiac development. J. Biol. Chem. 279, 19026–19034.

Plageman, T.F.Jr., Yutzey, K.E. 2005. T-box genes and heart development: Putting the "T" in heart. Dev. Dyn. 232, 11–20.

Plageman, T.F.Jr., Yutzey, K.E. 2006. Microarray analysis of Tbx5-induced genes expressed in the developing heart. Dev. Dyn. 235, 2868–2880.

Rivera-Feliciano, J., Tabin, C.J. 2006. Bmp2 instructs cardiac progenitors to form the heart-valve-inducing field. Dev. Biol. 295, 580–588.

Roberts, C., Ivins, S.M., James, C.T., Scambler, P.J. 2005. Retinoic acid down-regulates Tbx1 expression in vivo and in vitro. Dev. Dyn. 232, 928–938.

Rodriguez-Esteban, C., Tsukui, T., Yonei, S., Magallon, J., Tamura, K., Izpisua Belmonte, J.C. 1999. The T-box genes Tbx4 and Tbx5 regulate limb outgrowth and identity. Nature 398, 814–818.

Sakiyama, J., Yamagishi, A., Kuroiwa, A. 2003. Tbx4-Fgf10 system controls lung bud formation during chicken embryonic development. Development 130, 1225–1234.

Schlueter, J., Manner, J., Brand, T. 2006. BMP is an important regulator of proepicardial identity in the chick embryo. Dev. Biol. 295, 546–558.

Schott, J.J., Benson, D.W., Basson, C.T., Pease, W., Silberbach, G.M., Moak, J.P., Maron, B.J., Seidman, C.E., Seidman, J.G. 1998. Congenital heart disease caused by mutations in the transcription factor NKX2-5. Science 281, 108–111.

Schultheiss, T.M., Burch, J.B., Lassar, A.B. 1997. A role for bone morphogenetic proteins in the induction of cardiac myogenesis. Genes Dev. 11, 451–462.

Shelton, E.S., Yutzey, K.E. 2007. Tbx20 regulation of endrocardial cushion cell proliferation and extracellular matrix gene expression. Dev. Biol. 302, 376–388.

Singh, M.K., Christoffels, V.M., Dias, J.M., Trowe, M.O., Petry, M., Schuster-Gossler, K., Burger, A., Ericson, J., Kispert, A. 2005. Tbx20 is essential for cardiac chamber differentiation and repression of Tbx2. Development 132, 2697–2707.

Stennard, F.A., Costa, M.W., Elliot, D.A., Rankin, S., Haast, S.J.P., Lai, D., McDonald, L.P.A., Niederreither, K., Dolle, P., Bruneau, B.G., Zorn, A.M., Harvey, R.P. 2003. Cardiac T-box factor Tbx20 directly interacts with Nkx2-5, GATA4 and GATA5 in regulation of gene expression in the developing heart. Dev. Biol. 262, 206–224.

Stennard, F.A., Harvey, R.P. 2005. T-box transcription factors and their roles in regulatory hierarchies in the developing heart. Development 132, 4897–4910.

Stennard, F.A., Costa, M.W., Lai, D., Biben, C., Furtado, M.B., Solloway, M.J., McCulley, D.J., Leimana, C., Preis, J.I., Dunwoodie, S.L., Elliot, D.E., Prall, O.W., et al. 2005. Murine T-box transcription factor Tbx20 acts as a repressor during heart development, and is essential for adult heart integrity, function and adaptation. Development 132, 2451–2462.

Stern, C.D. 2005. The chick: A great model system becomes even greater. Dev. Cell 8, 9–17.

Stoller, J.Z., Epstein, J.A. 2005. Identification of a novel nuclear localization signal in Tbx1 that is deleted in DiGeorge syndrome patients harboring the 1223delC mutation. Hum. Mol. Genet. 14, 885–892.

Sun, G., Lewis, L.E., Huang, X., Nguyen, Q., Price, C., Huang, T. 2004. *TBX5*, a gene mutated in Holt-Oram syndrome, is regulated through a GC box and T-box binding elements (TBEs). J. Cell. Biochem. 92, 189–199.

Suzuki, T., Takeuchi, J., Koshiba-Takeuchi, K., Ogura, T. 2004. Tbx genes specify posterior digit identity through Shh and BMP signaling. Dev. Cell 6, 43–53.

Tada, M., Smith, J.C. 2001. T-targets: Clues to understanding the functions of T-box proteins. Dev. Growth Differ. 43, 1–11.

Takeuchi, J.K., Koshiba-Takeuchi, K., Matsumoto, K., Vogel-Hopker, A., Naitoh-Matsuo, M., Ogura, K., Takahashi, N., Yasuda, K., Ogura, T. 1999. Tbx5 and Tbx4 genes determine the wing/leg identity of limb buds. Nature 398, 810–814.

Takeuchi, J.K., Koshiba-Takeuchi, K., Suzuki, T., Kamimura, M., Ogura, K., Ogura, T. 2003a. Tbx5 and Tbx4 trigger limb initiation through activation of the Wnt/Fgf signaling cascade. Development 130, 2729–2739.

Takeuchi, J.K., Ohgi, M., Koshiba-Takeuchi, K., Shiratori, H., Sakaki, I., Ogura, K., Saijoh, Y., Ogura, T. 2003b. Tbx5 specifies the left/right ventricles and ventricular septum position during cardiogenesis. Development 130, 5953–5964.

Takeuchi, J.K., Mileikovskaia, M., Koshiba-Takeuchi, K., Heidt, A.B., Mori, A.D., Arruda, E. P., Gertsenstein, M., Georges, R., Davidson, L., Mo, R., Hui, C., Henkelman, R.M., et al. 2005. Tbx20 dose-dependently regulates transcription factor networks required for mouse heart and motor neuron development. Development 132, 2463–2474.

Vitelli, F., Taddei, I., Morishima, M., Meyers, E.N., Lindsay, E.A., Baldini, A. 2002. A genetic link between *Tbx1* and fibroblast growth factor signaling. Development 129, 4605–4611.

Vitelli, F., Zhang, Z., Huynh, T., Sobotka, A., Mupo, A., Baldini, A. 2006. *Fgf8* expression in the *Tbx1* domain causes skeletal abnormalities and modifies the aortic arch but not the outflow tract phenotype of *Tbx1* mutants. Dev. Biol. 295, 559–570.

Xu, H., Morishima, M., Wylie, J.N., Schwartz, R.J., Bruneau, B.G., Lindsay, E.A., Baldini, A. 2004. *Tbx1* has a dual role in the morphogenesis of the cardiac outflow tract. Development 131, 3217–3227.

Yamada, M., Revelli, J.P., Eichele, G., Barron, M., Schwartz, R.J. 2000. Expression of chick Tbx-2, Tbx-3, and Tbx-5 genes during early heart development: Evidence for BMP2 induction of Tbx2. Dev. Biol. 228, 95–105.

Yamagishi, T., Nakajima, Y., Nishimatsu, S., Nohno, T., Ando, K., Nakamura, H. 2004. Expression of *tbx20* RNA during chick heart development. Dev. Dyn. 230, 576–580.

Yang, L., Cai, C.L., Lin, L., Qyang, Y., Chung, C., Monteiro, R.M., Mummery, C.L., Fishman, G.I., Cogen, A., Evans, S. 2006. Isl1Cre reveals a common Bmp pathway in hearts and limb development. Development 133, 1575–1585.

Yutzey, K., Gannon, M., Bader, D. 1995. Diversification of cardiomyogenic cell lineages *in vitro*. Dev. Biol. 170, 531–541.

Yutzey, K.E., Rhee, J.T., Bader, D. 1994. Expression of the atrial-specific myosin heavy chain AMHC1 and the establishment of anteroposterior polarity in the developing chicken heart. Development 120, 871–883.

Zhang, Z., Huynh, T., Baldini, A. 2006. Mesodermal expression of *Tbx1* is necessary and sufficient for pharyngeal arch and cardiac outflow tract development. Development 133, 3587–3595.

CHAPTER 4

Transcriptional Control of Cardiac Boundary Formation

Anabel Rojas and **Brian L. Black**

Contents

Abstract

The heart is the first organ to form during embryogenesis and its function is critical for the viability of the mammalian embryo. As development proceeds, the early heart tube is remodeled by the addition of new structures with discrete boundaries. Each of these boundaries contributes to physical and functional compartments in the mature heart and serves to ensure proper unidirectional blood flow and rhythmic contraction. The establishment and maintenance of each of these boundaries is controlled by an evolutionary-conserved network of signaling pathways and transcription factors that function combinatorially to specify distinct divisions within the heart. This chapter focuses on the transcription factors that are involved in the establishment and maintenance of functional boundaries and distinct identities within the heart. The combinatorial interactions between cell-restricted and ubiquitous transcription factors in the establishment of atrial and ventricular restricted

Cardiovascular Research Institute and Department of Biochemistry and Biophysics, University of California San Francisco, San Francisco, California

Advances in Developmental Biology, Volume 18
ISSN 1574-3349, DOI: 10.1016/S1574-3349(07)18004-2

expression patterns are highlighted. In addition, numerous genes and transgenes exhibit expression that is restricted to either the right ventricle and outflow tract or the left ventricle and atria. This unique division is discussed in the context of the recent discovery that two distinct cardiac progenitor populations, the first and second heart fields (FHF and SHF), reside in the mesoderm and contribute to the heart. A transcriptional network for the SHF and its derivatives in the right ventricle and outflow tract is proposed. Finally, distinctions in the regulation of gene expression in primary myocardium versus chamber or working myocardium are discussed.

1. INTRODUCTION

The vertebrate heart is the first functional embryonic organ and has a vital role in the circulation and distribution of nutrients and oxygen. Initially, cardiac progenitors are derived from the mesoderm, which emerges from the primitive streak during gastrulation. Later, precardiac mesoderm migrates to an anterior–lateral position in the embryo and forms two groups of cells on either side of the midline (Tam and Behringer, 1997). These bilateral fields of cardiogenic mesoderm then extend across the midline to form the cardiac crescent, which fuses at the midline to form the early linear heart tube. The linear tube elongates through the addition of cells to the arterial and venous poles to form the inflow and outflow tracts, respectively (De La Cruz et al., 1989; Kelly and Buckingham, 2002). Studies have shown that the precursor cells at the arterial pole of mouse and chick come from a distinct population of cardiac progenitors in the pharyngeal mesoderm (Kelly et al., 2001; Mjaatvedt et al., 2001; Waldo et al., 2001; Buckingham et al., 2005). These cells are known as the second heart field (SHF) and give rise to the outflow tract and right ventricle (Kelly et al., 2001; Mjaatvedt et al., 2001; Waldo et al., 2001; Cai et al., 2003). After the addition of this population of cells, the heart is remodeled to become a four-chambered organ.

The first step in heart remodeling consists of the bending of the tubular heart toward the ventral side and rightward looping, following cues provided by molecular asymmetries that are established by the embryonic left/right axis pathway (Harvey, 1998). Shortly after the onset of looping, the ventricular and atrial chambers are formed in defined zones within the heart tube. The newly forming myocardium of the nascent chambers, known as working myocardium, has morphological aspects of trabecular myocardium, and will provide much of the contractile force of the heart (Christoffels et al., 2000; Sedmera et al., 2000). The remainder of the heart tube, known as primary myocardium, displays a phenotype distinct from the working myocardium and contributes to the atrioventricular canal (AVC), the inflow tract, inner curvature, and out-flow tract (de Jong et al., 1992; Moorman et al., 1998). The primary myocardium also plays a critical role during heart maturation by providing essential signals for the outgrowth and remodeling of the endocardial cushions. The formation of the cardiac cushions requires endocardial–mesenchymal transformation (EMT) of a subset of endothe-lial cells that have been previously specified in the AVC and outflow tract (Markwald et al., 1977; Delot, 2003; Armstrong and Bischoff, 2004). These endothelial cells will form the membranous portions of the septa, the septal wall separating the outflow tract, and the future valves (Wessels et al., 1996; Lincoln et al., 2004).

The septation of the ventricular chambers and the outflow tract also requires cardiac neural crest cells, which migrate from the neural folds into the pharyngeal arches and ultimately into the endocardial cushions of the outflow tract (Farrell et al., 1999; Waldo et al., 1999). Neural crest cells and SHF-derived endocardial cushion cells cooperate to divide the outflow tract into the aorta and the pulmonary artery (Kirby and Waldo, 1995). These two populations of cells also participate together in the process of ventricular septation. The interventricular septum (IVS), which divides both ventricles, is composed of a muscular as well as a mesenchymal component (Harh and Paul, 1975). The muscular component arises as myocardial outgrowth of the ventricular wall, whereas the mesenchymal component originates from fusion of the atrioventricular (AV) and cono-truncal endocardial cushions (Eisenberg and Markwald, 1995). It has been shown that myocardial cells from both right and left ventricle contribute to the formation of the IVS (Verzi et al., 2005; Franco et al., 2006).

The conversion of a laminar flow in the linear tube, which functions as a peristaltic heart, to a synchronous contracting pump with a flow into the ventricles during relaxation and into the outflow tract during contraction is achieved through the formation of discrete physical and functional boundaries within the developing heart. The developing cardiac meso-derm initially appears to be segmented in the anterior lateral mesoderm that forms the cardiac crescent (Fig. 1). After the formation of the linear heart tube, multiple primordial segments become obvious: inflow region (sinus venosus), common atrium, AVC, primitive ventricle, and outflow tract (Harvey, 1999). These segments correspond to functional compart-ments in the mature heart, suggesting that cardiac progenitors obtain positional information along the anterior–posterior axis even prior to the formation of the linear tube (Fig. 1). This information likely comes from discrete sources that differentially affect cells within distinct segments to set up the initial boundaries of gene expression within the nascent heart. For example, retinoic acid helps to specify the cells in the caudal heart-forming region to develop into atrial rather than ventricular myocardium (Xavier-Neto et al., 2001; Rivera-Feliciano et al., 2006). Similarly, other

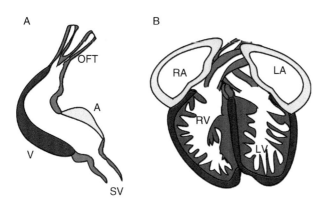

FIGURE 1 Schematic representation of two stages of heart development. Panel A shows a representation of a linear heart tube at E8 in the mouse. Panel B shows a schematic representation of the fully formed, four-chambered heart present at E16 in the mouse. Red and blue colors represent primary and chamber myocardium, respectively. Light blue represents atrial myocardium, while darker blue represents ventricular myocardium. The darkest shade of blue in panel B discriminates the compact zone myocardium from the trabecular working myocardium. Together, the two panels depict the respective contributions of regions within the linear heart tube (Panel A) to the mature heart (Panel B). A, atrial myocardium; LA, left atrium; LV, left ventricle; OFT, outflow tract; RA, right atrium; RV, right ventricle; SV, sinus venosus (inflow tract); V, ventricular myocardium.

morphogens have also been shown to play crucial roles in the establishment and maintenance of boundaries within the developing heart. For example, bone morphogenetic protein (BMP) signaling pathway plays an essential role in the proper development of the structures derived from the primary myocardium, which contribute to the septation of the outflow tract and valve formation (Delot et al., 2003; Liu et al., 2004). Likewise, the FGF ligand Fgf8 is also required for the proper outflow tract formation and remodeling (Ilagan et al., 2006; Park et al., 2006). Hedgehog signaling has also been shown to play an important role in the establishment and maintenance of the SHF-derived structures since ablation of sonic hedgehog or the conditional inactivation of its receptor Smoothened in the SHF leads to malformation of the outflow tract (Washington Smoak et al., 2005; Lin et al., 2006).

Each of these distinct inductive signals activates a unique program of gene expression within the different segments of the heart, including the activation of several transcriptional regulators of the cardiac program, such as Nkx2.5, GATA4, MEF2C, and T-box genes. Many examples of functional and physical interactions among cardiac-restricted factors have been reported in last few years, and the identification of cardiac-specific regulatory elements has facilitated the study of the transcriptional networks involved in the formation of restricted gene expression patterns within the heart. In this chapter, we will discuss how boundaries are recognized and maintained by combinatorial action of cell-restricted as well as ubiquitous transcriptional regulators to activate specific cardiac genetic programs that define these boundaries. We will focus on the AV, right–left side boundaries and on the division between the primary and the working myocardium.

2. AV BOUNDARY

In the mature heart, the physical boundary between the atria and the ventricles is obvious (Fig. 1). Furthermore, these distinct chambers have unique functional properties that allow for the specialized physiological roles and unique hemodynamic properties. Consistent with their specialized functions, the atrial and ventricular chambers express distinct subsets of genes and also direct expression of common genes at different levels, distinct times, or in response to different cues (Small and Krieg, 2004). Understanding how chamber-specific gene expression patterns are initially activated in the absence of a clear physical boundary remains an important challenge. For example, the atrial natriuretic factor (ANF) gene (*Npaa*) is first expressed in progenitor cells that will form the left ventricular chamber. After looping, expression is also observed in the right ventricle, and by E9.5, expression begins to appear in the atria. ANF expression is never seen in the primary myocardium of the inflow tract, AVC, inner curvature, or in the outflow tract, making this gene a good marker for chamber myocardium (Fig. 1). Later in development, ANF expression becomes restricted to the atria with limited expression also in the ventricular trabeculae in the mature heart (Argentin et al., 1994).

The activation of the ANF promoter is one of the best examples of combinatorial transcription factor regulation of spatial and temporal expression. Transgenic analyses have shown that 700bp of upstream sequence from the rat ANF promoter are sufficient to recapitulate endogenous ANF expression *in vivo* (Rockman et al., 1991). Dissection of this region in transactivation experiments revealed the requirement of key cardiac transcription factors binding sites for promoter activity, including serum response factor (SRF), GATA factors, HAND2, MEF2C, Nkx2.5, and T-box factors (Fig. 2). Moreover, functional and physical interactions among these transcription factors have been shown to be required for the transactivation of the ANF promoter in transfection experiments. GATA4 has been shown to physically interact and transcriptionally cooperate with MEF2C, HAND2, and Nkx2.5 (Durocher et al., 1997; Morin et al., 2000; Dai et al., 2002). Likewise, HAND2 synergizes with Nkx2.5 and MEF2C in the transcriptional activation of the ANF promoter (Yamagishi et al., 2001;

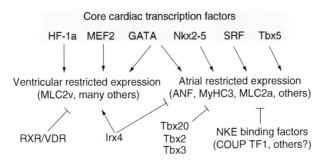

FIGURE 2 Atrioventricular (AV) boundary formation is regulated by core cardiac transcription factors in combination with restricted activators and repressors. Core cardiac-restricted transcription factors (blue shaded box) are essential for the activation of atrial and ventricular restricted gene expression but function combinatorially with each other and with restricted repressors to delimit expression to either atrial or ventricular myocardium.

Thattaliyath et al., 2002; Zang et al., 2004). Some of these interactions are also required for the activation of other genes, including *BNP, cardiac α-actin, myosin heavy chain,* and *myosin light chain* (Chen et al., 1996; Dai et al., 2002; Latinkic et al., 2004).

In vivo analyses of the ANF promoter have shown that mutation of the proximal GATA element or the Nkx element modulated the spatial activity of the ANF promoter such that ANF transcription was no longer restricted to the atria, but persisted in the ventricle and outflow tract during heart development (Small and Krieg, 2003). Thus, GATA and Nkx transcription factors play a role not only in transcriptional activation but also in the spatial control of ANF expression. Interestingly, GATA4 and Nkx2.5 are present in all myocardial cells during development, so it seems highly unlikely that these two transcription factors alone also could account for the restriction of ANF expression to the atria. Rather, the atrial expression of ANF may be achieved through a combination of specific and broadly expressed transcription factors with positive regulatory capabilities in the atria and negative regulatory capabilities in other regions of the heart.

A positive regulatory complex formed by GATA4, Nkx2.5, and Tbx5 could activate the expression of ANF promoter specifically in the atria. Tbx5 is expressed initially throughout the cardiac mesoderm, but its expression pattern is rapidly restricted, first as a posterior–anterior gradient in the linear tube and then during midgestation, to the atria and left ventricle (Bruneau et al., 1999; Liberatore et al., 2000). Tbx5 has been shown to interact physically with Nxk2.5, which results in synergistic activation of the ANF promoter (Hiroi et al., 2001; Plageman and Yutzey, 2004). Similarly, GATA4 and Tbx5 also cooperate to activate the ANF promoter through a T-box element (Plageman and Yutzey, 2004). These interactions may also contribute to the high level of ANF found in the left ventricle (Argentin et al., 1994).

Another model for the spatiotemporal restriction of ANF expression is based on the presence of a repressor that becomes expressed or activated in the ventricles at the time when ANF restriction occurs (Habets et al., 2002). This repressor could recognize the GATA or Nkx sites in the ANF promoter (Fig. 2). The activity of this inhibitor in the ventricle would result in restriction of ANF expression to the atria. Mutation of the inhibitor recognition sequences would therefore result in loss of inhibitor activity and would permit continued ANF transcription in ventricles and outflow tract. This model agrees with the results of mutagenesis studies of the GATA and Nkx sites in the *Xenopus* ANF promoter (Small and Krieg, 2003). In fact, some other transcription factors, including members of the Nkx2 family, Nkx3 family, and COUP-TFI, can recognize Nkx2.5-binding sites and may act as transcriptional repressors (Amendt et al., 1999; Steadman et al., 2000; Guo et al., 2001). This model is also supported by studies that showed that overexpression of Nkx2.5 throughout the myocardium increased the expression of ANF in the ventricle, possibly by competing with a prospective inhibitor for the Nkx site (Takimoto et al., 2000).

The absence of ANF expression in the primary myocardium may also require the inhibitory activity of T-box factors (Fig. 2) (Plageman and Yutzey, 2005; Stennard and Harvey, 2005). For example, Tbx20 can regulate transcription of target genes positively or negatively, depending on the cellular context, and can physically and functionally interact with GATA4 and Nkx2.5 (Stennard et al., 2003, 2005; Plageman and Yutzey, 2004). Tbx2 and Tbx3 are also transcriptional repressors and have been associated with the development of AV canal and conduction system (Habets et al., 2002; Hoogaars et al., 2004). Both transcription factors are expressed in the primary myocardium, more caudally in the linear tube and in the AVC (Christoffels et al., 2004b; Hoogaars et al., 2004; Yamada et al., 2000). On the ANF promoter, Tbx20 binds to a second T-box element to repress transcription in transient transfection experiments (Plageman and Yutzey, 2004). These data agree with the observation that ANF transcripts are not present in regions of strong Tbx20 expression, including the AVC, AVC cushions, and outflow tract (Argentin et al., 1994; Plageman and Yutzey, 2004). Tbx20 also physically and functionally interacts with GATA4 and Nkx2.5 (Stennard et al., 2003; Takeuchi et al., 2005). Similarly, Tbx2 has been shown to form a complex with Nkx2.5 on the T-box/Nk site in the ANF promoter repressing its activation in the AVC (Habets et al., 2002; Christoffels et al., 2004b). Taken together, these observations indicate an interplay between strictly localized repressors (Tbx20, Tbx2, and Tbx3), a restricted activator (Tbx5), and more widely expressed but essential heart-restricted transcription factors such as GATA4/5/6 and Nkx2.5, to repress chamber-specific genes and consequently block chamber differentiation in regions that retain a primary myocardial phenotype.

Another example of how specific transcription factor combinations restrict gene expression is derived from studies of the quail slow *myosin heavy chain* (*MyHC3*) promoter specifically in the atria (Wang et al., 1996, 1998, 2001). *MyHC3* becomes restricted to the future atrial chamber just after formation of the heart tube. Transgenic analyses identified an 840-bp sequence that was able to recapitulate the endogenous pattern of atrial-specific expression (Xavier-Neto et al., 1999). *In vitro* analyses of the same element revealed the requirement of a GATA site for promoter activation, while a vitamin D response element (VDRE), which binds a complex formed by the vitamin D and retinoic acid receptors, was responsible for inhibition of expression (Wang et al., 1998). A similar mechanism for the activation of *MLC2a* has been proposed based on the presence of, MEF2, GATA, SRF, and retinoic acid response elements (Fig. 2) (Dyson et al., 1995; Chen et al., 1998). Again, however, neither retinoic acid receptor family members nor GATA transcription factors exhibit expression restricted exclusively to the atria. Importantly, the atrial-restricted expression of the *MyHC3* gene also appears to require the recruitment of Irx4 to its promoter. Irx4 is an Iroquois homeobox factor expressed exclusively in the ventricles during embryonic and postnatal heart development (Bao et al., 1999). Irx4 is recruited to the VDRE and physically interacts with the retinoic acid receptor, thus building a repressor complex to inhibit expression of *MyHC3* in the ventricles (Wang et al., 2001).

Different atrial-specific genes become chamber restricted at different times, suggesting the involvement of distinct regulatory mechanisms for each gene. However, based on the regulation of the ANF, *MyHC3*, and *MLC2a* promoters, it seems likely that atrial restriction is achieved by repression of the program for ventricular development. In contrast to the atria, candidate transcription factors responsible for ventricular-restricted gene expression remain elusive. Although some *cis*-regulatory elements controlling ventricular expression have been identified, none of the *trans*-activators that regulate expression via these elements are likely to account for the observed restriction of gene expression. Transgenic analyses of the *MLC2v* promoter, for example, have identified a small region of 250bp responsible for the specific expression in the right ventricle, while the regulatory sequences for activation in the left ventricle *in vivo* have yet to be identified (Ross et al., 1996). This region contains binding sites for MEF2 transcription factors and for HF-1A (Ross et al., 1996). As with other examples, however, these two transcription factors are unlikely to account for ventricular-specific expression of MLC2v since both are expressed in other cell types.

In the heart, Irx4 is the first cardiac transcription factor to be identified with early and exclusive expression in ventricular myocytes (Bao et al., 1999; Bruneau et al., 2000). This expression pattern suggests that Irx4 is a key factor involved in the induction of the genetic program for ventricular-specific activation, and it may also be important for repression of atrial genetic program. Indeed, overexpression of *Irx4* in chick leads to aberrant expression of ventricular differentiation markers in the atria (Bao et al., 1999). In addition, Irx4 null mice display reduced expression of ventricular markers and increased activation of atrial genes in the ventricular region, further supporting an essential role for this transcription factor in a genetic program for ventricular specificity (Bruneau et al., 2001a). The identification of targets of Irx4 as well as the regulatory elements controlling its expression will be essential to understand the early signals and transcriptional mechanisms controlling the establishment of the AV boundary during development.

3. LEFT–RIGHT BOUNDARY (FIRST–SECOND HEART FIELDS)

Initial evidence that the outflow tract of the heart was not present at the linear tube stage and was likely derived from an independent progenitor population came from *in vivo* labeling techniques in chick embryos (de la Cruz et al., 1977). Subsequent studies on chick and mouse have determined that the cells that give rise to the outflow tract and right ventricle originate from the splanchnic and pharyngeal mesoderm and are added to the heart after the linear tube stage (Kelly et al., 2001; Mjaatvedt et al., 2001; Waldo et al., 2001). This population of cardiomyocytes is known as the SHF (also referred to as the anterior or secondary heart field). By contrast, the majority of the left ventricle, the atria, and the inflow tract of the developing heart are derived from a region of lateral mesoderm known as the first (or primary) heart field (FHF) (Fig. 3) (Kelly and Buckingham, 2002; Buckingham et al., 2005).

The observation that two distinct populations of cells contribute to the heart provided a likely explanation for the restricted spatial expression and function of some genes (Schwartz and Olson, 1999; Kelly and Buckingham, 2002). For example, *Hand2* is expressed throughout the cardiac crescent but by E9.0 in the mouse, at the time the heart is looping, *Hand2* transcript levels are higher in the outflow tract and in the anterior part of the tube that will form the right ventricle (Firulli et al., 1998). In *Hand2* null mice, development of the right ventricle and outflow tract is initiated, but the expansion of these regions is dramatically impaired as a result of an increase in apoptosis, resulting in the absence of the right ventricle and consequent embryonic lethality (Srivastava et al., 1997). Conversely, Tbx5 expression is restricted to the atria and left ventricle and knockout studies have shown the requirement of this

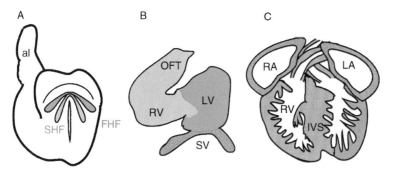

FIGURE 3 Two sources of embryonic mesoderm contribute to the heart. In this schematic, the first heart field (FHF) and its derivatives are depicted in green and the second heart field (SHF) and its derivatives are depicted in yellow. Neural crest cells, which also contribute to the outflow tract, ventricular septum, and cardiac cushions are not depicted in this schematic, which is focused only on mesodermal progenitors. Panel A depicts a schematic of a mouse embryo at E7.5, where both FHF and SHF are evident in the cardiac crescent region of anterior lateral mesoderm. The SHF resides medial, dorsal, and caudal to the FHF. Panel B depicts a mouse heart at E9 showing the contributions of the FHF and SHF to the looping heart. Note that the inflow region at the sinus venosus (SV) appears to be a mixture of cells from both fields and that SHF cells cross into the future left ventricle (LV). Panel C shows a schematic representation of the fully formed, four-chambered mouse heart at E16. Note that the myocardium of the outflow tracts and the interventricular septum (IVS) is derived from the SHF. al; allantois; LA, left atrium; OFT, outflow tract; RA, right atrium; RV, right ventricle.

T-box gene for the proper formation of these structures, while Tbx5 function is dispensable for outflow tract and right ventricle development (Bruneau et al., 1999, 2001b). Thus, it appears that distinct transcriptional programs function in the FHF and SHF and that restricted transcription factor expression patterns are likely to contribute to spatially distinct gene expression pattern and the right–left ventricular boundary in the heart.

A key regulator in the transcriptional hierarchy of SHF development is the LIM-homeodomain transcription factor Isl1 (Pfaff et al., 1996; Ahlgren et al., 1997; Cai et al., 2003). During early development in mice, *Isl1* is expressed in the splanchnic mesoderm in a medial dorsal position relative to the cardiac crescent. As cardiac development proceeds, *Isl1*-expressing cells within the mesoderm are localized anterior and dorsal to the linear heart tube (Cai et al., 2003). At E10, *Isl1* mRNA is present in the ventral endoderm and splanchnic mesoderm near the outflow tract, but not in the myocardium of the heart itself, and mice that are homozygous null for *Isl1* die at approximately E10.5 (Cai et al., 2003). The hearts of *Isl1* mutant embryos fail to undergo proper looping and lack a morphologically distinct outflow tract or right ventricle (Cai et al., 2003), indicating that *Isl1*-expressing cells are important for the proper formation of the SHF and its derivatives. Several growth factors and transcription factors have been shown to be direct or indirect targets of Isl1 (Fig. 4) (Cai et al., 2003; Dodou et al., 2004; Takeuchi et al., 2005). In Isl1 mutant embryos, the levels of BMP4, BMP7, and FGF10 mRNA are drastically reduced (Cai et al., 2003). These factors have been shown to be involved in the development of the outflow tract by recruitment of myocardial precursors from the SHF (Kelly and Buckingham, 2002; Kirby, 2002). Thus, proper activation of BMP- and FGF-signaling pathways via Isl1 activity are likely to be responsible for the establishment and maintenance of the SHF and its derivatives.

A combination of Isl1 and GATA factors may represent a common mechanism for restricted spatial expression in the SHF, as both transcription factors regulate expression of *Mef2c* and *Nkx2.5* via enhancers (Fig. 4) (Dodou et al., 2004; Takeuchi et al., 2005). A proximal *Nkx2.5* enhancer directs expression to the SHF and its derivatives, and requires GATA-binding sites and an Isl1 site for enhancer activity (Searcy et al., 1998; Lien et al., 1999, 2002; Takeuchi et al., 2005). Similarly, identified regulatory sequences in the *Mef2c* gene direct expression at E7.5 in the precardiac mesoderm and continue functioning in the pharyngeal mesoderm, outflow tract, and right ventricle throughout embryonic and fetal development. Analysis of *Mef2c* enhancer function in transgenic mouse

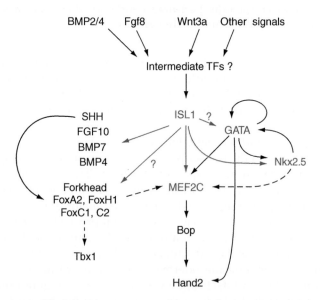

FIGURE 4   A simplified model transcriptional network functioning in SHF development. In this model, the LIM-homeodomain transcription factor Isl1 functions as a key early regulator of this cascade by activating MEF2C, Nkx2.5, and GATA transcription factors, which function at the core of this network (depicted in red).

embryos demonstrated that *Mef2c* is a direct transcriptional target of GATA factors and Isl1 (Dodou et al., 2004). Interestingly, the cardiac phenotype in *Mef2c* null embryos is very similar to the cardiac phenotype of *Isl1* null embryos (Lin et al., 1997; Cai et al., 2003). The hearts of *Mef2c* null embryos fail to undergo looping, have gross abnormalities in the outflow tract, and the right ventricle is not properly developed (Lin et al., 1997), suggesting an important role for this transcription factor in the development of the SHF and its derivatives.

Some forkhead proteins, including FoxH1, FoxA2, FoxC1, and FoxC2, have also been shown to be required for the proper development of the SHF (Hu et al., 2004; von Both et al., 2004; Seo and Kume, 2006). Interestingly, *Mef2c* may also be a direct target of FoxH1, which interacts with Nkx2.5 to mediate the activation of a separate TGF-β response element in the *Mef2c* gene (von Both et al., 2004). Taken together, these results suggest the existence of two parallel pathways for the establishment and maintenance of SHF development via *Mef2c* activation, one involving GATA factors and Isl1, and another involving forkhead factors and Nkx2.5 (Fig. 4) (Dodou et al., 2004; von Both et al., 2004). Subsequently, *Mef2c* is required for the activation of the transcription factor BOP, which in turn is required for activation of *Hand2* (Gottlieb et al., 2002; Phan et al., 2005). Loss-of-function mutations in either Bop or Hand2 results in loss of right ventricle and outflow tract development, suggesting a role for *MEF2C*-dependent pathways in the derivatives of the SHF (Fig. 4) (Srivastava et al., 1997; Gottlieb et al., 2002).

The contribution of the SHF to the heart raises several important issues surrounding the definition of this boundary. A fate map of Isl1 expressing cells showed that the cells from the SHF give rise to the outflow tract and right ventricle and also much of the atria and the majority of the left ventricle (Cai et al., 2003). By contrast, a fate map generated using the *Mef2c* anterior heart field enhancer labeled the right ventricle, outflow tract, and intraventricular septum, but not the atria or the majority of the left ventricle, suggesting that perhaps the *Mef2c* enhancer marks only a subpopulation of the SHF (Verzi et al., 2005). Comparison of *Fgf10* and *Fgf10-lacZ* transgene expression combined with dye-labeling techniques in cultured mouse embryos suggest that myocardial precursor cells migrate from the pharyngeal mesoderm to the arterial pole of the heart during the outflow tract formation (Kelly et al., 2001). Other studies using retrospective clonal analyses have shown that the SHF gives rise to the right ventricle as well as the outflow tract, or even extends to other regions of the heart, including the atria (Meilhac et al., 2004a; Zaffran et al., 2004). The observation that cells from the SHF form the intraventricular septum is also supported by other labeling studies that have shown that myocytes from right and left ventricles contribute to the ventricular septum (Fig. 3) (Meilhac et al., 2004b; Verzi et al., 2005;

Franco et al., 2006). Fgf8 has been implicated as an upstream regulator of Isl1, suggesting an essential role for this pathway in SHF development (Park et al., 2006). It will be essential to define the intermediate steps in the pathway linking Fgf8 to Isl1.

If indeed Isl1 functions as a master transcriptional regulator in the SHF, it will be essential to determine its partners and activators. It will also be important to determine the master regulators of cardiac development functioning in the FHF, but these have remained elusive. Nkx2.5 is clearly a critical early factor in the transcriptional hierarchy of the FHF, based on the studies in invertebrates. In *Drosophila*, the Nkx2.5 ortholog *tinman* is the earliest inducer of the cardiac program and is required for the specification of cardiomyocytes in the dorsal vessel (Azpiazu and Frasch, 1993; Bodmer, 1993; Frasch, 1995, 1999). By contrast, in mice lacking Nkx2.5, the initial events of heart formation occur normally, but embryos die due to malformations in the heart tube and failure in the left ventricle development (Lyons et al., 1995). This phenotype in *Nkx2.5* null embryos could be interpreted as a selective failure in the development of structures derived from the FHF. In this regard, Nkx2.5 may serve to maintain the identity of the FHF and its derivatives, the left ventricle and atria, by activating left-restricted genes, such as *Hand1* and *Pitx2* (Tanaka et al., 1999; Shiratori et al., 2001) and by interacting with other left-specific transcription factors, such as Tbx5 (Tanaka et al., 1999; Bruneau et al., 2001b).

4. PRIMARY MYOCARDIUM–CHAMBER (OR WORKING) MYOCARDIUM BOUNDARY

The early embryonic heart tube has a physiological phenotype characterized by slow conduction, slow growth, and low contractility (Moorman and Christoffels, 2003; Sedmera et al., 2003). After the onset of looping, at specific regions within the linear tube, the myocardium differentiates into chamber or working myocardium that ultimately will give rise to the atria and ventricles. This nascent chamber myocardium changes during embryonic development, and two different morphologies become apparent at midgestation in mice: a compact myocardium lining the outer myocardial wall and a trabecular myocardium filling the ventricular space (Fig. 1) (Sedmera et al., 2000).

Differences in proliferation rate, conduction velocities, contractility, and gene expression between the primary myocardium and the working or chamber myocardium reflect differences in function that have their origins in development (Sedmera et al., 2000; Christoffels et al., 2004a). Chamber myocardium provides the contractile force of the heart and distributes the electrical impulse for contraction throughout the ventricles.

By contrast, the primary myocardium is not the major working muscle of the heart, but the morphogenetic events that lead to cardiac septation and valve formation are largely confined to the primary myocardium (Lamers and Moorman, 2002). The primary myocardium consists of the inner curvature, outflow tract, inflow tract, and AVC myocardium (Fig. 1). The myocardial and endocardial cell layers in the linear heart tube are separated by extracellular matrix (ECM) know as cardiac jelly. In the region known as the cardiac cushions, endothelial cells migrate into the cardiac jelly and undergo endothelial–mesenchymal transformation (EMT), and will ultimately give rise to the valves (Armstrong and Bischoff, 2004). Although some of the transcription factors required to induce and pattern the primary myocardium in the cardiac crescent and linear heart tube have been identified as well as some others that are important for differentiation in specific areas of the primary myocardium, little is known about how the primary myocardium is established and maintained.

BMP signaling has been shown to be essential for the morphogenesis of the AV cushion and the AVC, suggesting a role in the development of the primary myocardium (Sugi et al., 2004). Analyses of conditional knockout mice for the BMP receptors Alk3 and Alk2 in endothelial cells have shown that these receptors mediated the EMT in the AVC (Gaussin et al., 2005; Wang et al., 2005; Song et al., 2007). Another component of BMP signaling, the ligand BMP2, appears to induce the expression of Tbx2, a known repressor of chamber genes in areas of the primitive myocardium (Yamada et al., 2000). There are very few defined *cis*-regulatory elements for the primary myocardium, which has limited the study of the transcriptional networks functioning in this domain. So far, only a few examples of control regions that restrict the expression to the AV conduction system have been identified. A proximal enhancer from the *cardiac troponin I* gene directs expression throughout the linear heart tube at early stages and becomes restricted to the AVC and to the AV node of the conduction system later in development (Di Lisi et al., 1998). However, the transcriptional pathways that control this enhancer *in vivo* have not been defined. A chicken *Gata6* enhancer also directs expression to a subset of cells from the primary myocardium that gives rise to the AVC and part of the conduction system. This enhancer requires a GATA site and a GC-rich element for activation (Davis et al., 2000; Adamo et al., 2004). Interestingly, another study has shown the requirement of another GATA factor, GATA4, for the proper formation of the valves as embryos lacking GATA4 function in the endothelial cells fail to undergo EMT to form the progenitors of the AV valves (Rivera-Feliciano et al., 2006). Taken together, these studies suggest a prominent role of GATA transcription factors in the development of the primary myocardium. As with other regions of the heart, it is likely that GATA factors will cooperate with other restricted transcription factors to maintain the primary myocardial phenotype.

5. COMPACT–TRABECULAR MYOCARDIUM BOUNDARY

The working myocardium undergoes important changes during embryonic development as it diversifies into distinct compact and trabecular myocardial layers (Sedmera et al., 2000). The trabecular myocardium is apparent early in development, just after heart looping, and it begins to be remodeled as ventricular septation occurs. The trabecular myocardium has a high rate of proliferation at this stage. As ventricular volumes increase, the trabeculations become compressed within the ventricular wall, resulting in a significant increase in thickness of the compact myocardium (Sedmera et al., 2000). The thickness of the compact myocardium accounts for the contraction of the more mature heart, and it will proliferate faster than the trabecular myocardium after chamber maturation has occurred (Wessels and Sedmera, 2003). The importance of a proper organization and assembly of the myocardium is apparent in the phenotype of mice with gene disruption that affect the architecture of the myocardium. For example, the EGF signaling pathway is essential for trabecular formation as the inactivation of the ligand neuregulin or the genes that encode two neuregulin receptors erbB4 or erbB2 leads to embryonic lethality due to lack of myocardial trabeculation (Gassmann et al., 1995; Lee et al., 1995; Meyer and Birchmeier, 1995).

Other genes have been implicated in the maturation of the ventricular chamber, including early inducers of cardiogenesis, such as Nkx2.5 and GATA4 (Pashmforoush et al., 2004; Zeisberg et al., 2005). Conditional inactivation of *Nkx2.5* in the ventricular myocardium results in hypertrabeculation and noncompaction of myocardium (Pashmforoush et al., 2004). Similarly, conditional inactivation of *Gata4* in the chamber myocardium leads to ventricular myocardial hypoplasia (Zeisberg et al., 2005). This ventricular hypoplasia is a result of proliferation defects, although the mechanism by which GATA4 is involved in cardiomyocyte proliferation is still unclear (Zeisberg et al., 2005). The phenotypes of mice with mutations in cell cycle genes, including N-myc, D type cyclins, and cyclin-dependent kinases (Cdks) also show defects in trabecular compaction and are perhaps easier to understand as these genes are known to be positive regulators of the cell cycle (Moens et al., 1993; Evan and Littlewood, 1998; Kozar et al., 2004; Kozar and Sicinski, 2005; Berthet et al., 2006). Whether the key cardiac transcription factors such as Nkx2.5 and GATA4 are part of a common transcriptional pathway with cell cycle genes seems plausible but remains to be determined.

6. CONCLUSIONS AND FUTURE DIRECTIONS

A vast network of complex transcriptional pathways and combinatorial interactions are involved in the formation of the developing vertebrate heart. The extremely dynamic expression pattern of key cardiac transcription factors throughout embryonic and fetal development and their different activator or repressor functions contribute to the establishment and maintenance of molecular boundaries within the heart. The identification of transcription factors implicated in specific developmental events has begun to be elucidated but still many questions remain to be answered.

Several key cardiac transcription factors have been shown to be controlled by modular regulatory elements that each direct expression to distinct populations of cardiomyocytes at different stages during development. The selective advantage of this modularity is still unknown, but perhaps it serves to tune the spatiotemporal and quantitative control of transcription factor expression, ensuring the proper regulation of downstream targets. The identification of regulatory elements controlling the expression of key cardiac genes is therefore essential to define the transcriptional networks controlling heart formation and the establishment of distinct boundaries within the heart. For example, Nkx2.5 and GATA4 are crucial factors that govern the establishment and maintenance of different boundaries within the heart, and both genes are regulated by multiple modular elements (Searcy et al., 1998; Lien et al., 1999; Reecy et al., 1999; Tanaka et al., 1999; Brown et al., 2004; Heicklen-Klein and Evans, 2004; Lee et al., 2004; Takeuchi et al., 2005;). It remains to be determined how each of these elements interact to regulate the fine control of cardiac gene expression necessary to subspecify discrete regions within the myocardium.

Over the past several years, some of the signaling pathways involved in cardiac induction in the primary heart field have been identified, but the identity of the essential inducers of cardiac differentiation in the mesoderm is still not known. Ablation of any of early transcription factors does not appear to affect cardiac induction, as all cardiac knockout mice analyzed to date possess differentiated cardiomyocytes. This is probably due to redundancy among member of the same family of transcription factors. The analyses of the phenotype of compound mutations in different cardiac genes will be required to assess a role to these transcription factor families early in development. For example, the compound mutation of *Gata4* and *Gata6* results in the complete absence of cardiomyocyte differentiation in the mouse embryo, indicating the redundancy of GATA transcription factors in the early steps of cardiac induction (Duncan, S., personal communication).

The FHF and SHF share common regulators and likely common signaling pathways; however, a more extensive study of the signaling pathways governing the induction of the SHF is necessary to understand the cross talk between these two populations of cardiomyocytes. Some studies place Isl1 as the earliest marker of the SHF (Cai et al., 2003). Thus, it will be essential to define the hierarchy of molecular events upstream of Isl1 and determine how these earliest pathways may intersect to specify multiple cardiac lineages, including the FHF and SHF, to dictate the form of the mature heart with its distinct physical and functional boundaries.

ACKNOWLEDGMENTS

A.R. was supported by a fellowship from the American Heart Association, Western Affiliate. Work in the BLB laboratory was supported by grants HL64658 and AR52130 from the NIH.

REFERENCES

Adamo, R.F., Guay, C.L., Edwards, A.V., Wessels, A., Burch, J.B. 2004. GATA-6 gene enhancer contains nested regulatory modules for primary myocardium and the embedded nascent atrioventricular conduction system. Anat. Rec. A Discov. Mol. Cell. Evol. Biol. 280, 1062–1071.

Ahlgren, U., Pfaff, S.L., Jessell, T.M., Edlund, T., Edlund, H. 1997. Independent requirement for ISL1 information of pancreatic mesenchyme and islet cells. Nature 385, 257–260.

Amendt, B.A., Sutherland, L.B., Russo, A.F. 1999. Transcriptional antagonism between Hmx1 and Nkx2.5 for a shared DNA-binding site. J. Biol. Chem. 274, 11635–11642.

Argentin, S., Ardati, A., Tremblay, S., Lihrmann, I., Robitaille, L., Drouin, J., Nemer, M. 1994. Developmental stage-specific regulation of atrial natriuretic factor gene transcription in cardiac cells. Mol. Cell. Biol. 14, 777–790.

Armstrong, E.J., Bischoff, J. 2004. Heart valve development: Endothelial cell signaling and differentiation. Circ. Res. 95, 459–470.

Azpiazu, N., Frasch, M. 1993. Tinman and bagpipe: Two homeo box genes that determine cell fates in the dorsal mesoderm of Drosophila. Genes Dev. 7, 1325–1340.

Bao, Z.Z., Bruneau, B.G., Seidman, J.G., Seidman, C.E., Cepko, C.L. 1999. Regulation of chamber-specific gene expression in the developing heart by Irx4. Science 283, 1161–1164.

Berthet, C., Klarmann, K.D., Hilton, M.B., Suh, H.C., Keller, J.R., Kiyokawa, H., Kaldis, P. 2006. Combined loss of Cdk2 and Cdk4 results in embryonic lethality and Rb hypophosphorylation. Dev. Cell 10, 563–573.

Bodmer, R. 1993. The gene tinman is required for specification of the heart and visceral muscles in Drosophila. Development 118, 719–729.

Brown, C.O., 3rd, Chi, X., Garcia-Gras, E., Shirai, M., Feng, X.H., Schwartz, R.J. 2004. The cardiac determination factor, Nkx2–5, is activated by mutual cofactors GATA-4 and Smad1/4 via a novel upstream enhancer. J. Biol. Chem. 279, 10659–10669.

Bruneau, B.G., Logan, M., Davis, N., Levi, T., Tabin, C.J., Seidman, J.G., Seidman, C.E. 1999. Chamber-specific cardiac expression of Tbx5 and heart defects in Holt-Oram syndrome. Dev. Biol. 211, 100–108.

Bruneau, B.G., Bao, Z.Z., Tanaka, M., Schott, J.J., Izumo, S., Cepko, C.L., Seidman, J.G., Seidman, C.E. 2000. Cardiac expression of the ventricle-specific homeobox gene Irx4 is modulated by Nkx2–5 and dHand. Dev. Biol. 217, 266–277.

Bruneau, B.G., Bao, Z.Z., Fatkin, D., Xavier-Neto, J., Georgakopoulos, D., Maguire, C.T., Berul, C.I., Kass, D.A., Kuroski-de Bold, M.L., de Bold, A.J., et al. 2001a. Cardiomyopathy in Irx4-deficient mice is preceded by abnormal ventricular gene expression. Mol. Cell. Biol. 21, 1730–1736.

Bruneau, B.G., Nemer, G., Schmitt, J.P., Charron, F., Robitaille, L., Caron, S., Conner, D.A., Gessler, M., Nemer, M., Seidman, C.E., Seidman, J.G. 2001b. A murine model of Holt-Oram syndrome defines roles of the T-box transcription factor Tbx5 in cardiogenesis and disease. Cell 106, 709–721.

Buckingham, M., Meilhac, S., Zaffran, S. 2005. Building the mammalian heart from two sources of myocardial cells. Nat. Rev. Genet. 6, 826–835.

Cai, C.L., Liang, X., Shi, Y., Chu, P.H., Pfaff, S.L., Chen, J., Evans, S. 2003. Isl1 identifies a cardiac progenitor population that proliferates prior to differentiation and contributes a majority of cells to the heart. Dev. Cell 5, 877–889.

Chen, C.Y., Croissant, J., Majesky, M., Topouzis, S., McQuinn, T., Frankovsky, M.J., Schwartz, R.J. 1996. Activation of the cardiac alpha-actin promoter depends upon serum response factor, Tinman homologue, Nkx-2.5, and intact serum response elements. Dev. Genet. 19, 119–130.

Chen, J., Kubalak, S.W., Chien, K.R. 1998. Ventricular muscle-restricted targeting of the RXRalpha gene reveals a non-cell-autonomous requirement in cardiac chamber morphogenesis. Development 125, 1943–1949.

Christoffels, V.M., Habets, P.E., Franco, D., Campione, M., de Jong, F., Lamers, W.H., Bao, Z. Z., Palmer, S., Biben, C., Harvey, R.P., Moorman, A.F. 2000. Chamber formation and morphogenesis in the developing mammalian heart. Dev. Biol. 223, 266–278.

Christoffels, V.M., Burch, J.B., Moorman, A.F. 2004a. Architectural plan for the heart: Early patterning and delineation of the chambers and the nodes. Trends Cardiovasc. Med. 14, 301–307.

Christoffels, V.M., Hoogaars, W.M., Tessari, A., Clout, D.E., Moorman, A.F., Campione, M. 2004b. T-box transcription factor Tbx2 represses differentiation and formation of the cardiac chambers. Dev. Dyn. 229, 763–770.

Dai, Y.S., Cserjesi, P., Markham, B.E., Molkentin, J.D. 2002. The transcription factors GATA4 and dHAND physically interact to synergistically activate cardiac gene expression through a p300-dependent mechanism. J. Biol. Chem. 277, 24390–24398.

Davis, D.L., Wessels, A., Burch, J.B. 2000. An Nkx-dependent enhancer regulates cGATA-6 gene expression during early stages of heart development. Dev. Biol. 217, 310–322.

de Jong, F., Opthof, T., Wilde, A.A., Janse, M.J., Charles, R., Lamers, W.H., Moorman, A.F. 1992. Persisting zones of slow impulse conduction in developing chicken hearts. Circ. Res. 71, 240–250.

de la Cruz, M.V., Sanchez Gomez, C., Arteaga, M.M., Arguello, C. 1977. Experimental study of the development of the truncus and the conus in the chick embryo. J. Anat. 123, 661–686.

de la Cruz, M.V., Sanchez-Gomez, C., Palomino, M.A. 1989. The primitive cardiac regions in the straight tube heart (Stage 9) and their anatomical expression in the mature heart: An experimental study in the chick embryo. J. Anat. 165, 121–131.

Delot, E.C. 2003. Control of endocardial cushion and cardiac valve maturation by BMP signaling pathways. Mol. Genet. Metab. 80, 27–35.

Delot, E.C., Bahamonde, M.E., Zhao, M., Lyons, K.M. 2003. BMP signaling is required for septation of the outflow tract of the mammalian heart. Development 130, 209–220.

Di Lisi, R., Millino, C., Calabria, E., Altruda, F., Schiaffino, S., Ausoni, S. 1998. Combinatorial cis-acting elements control tissue-specific activation of the cardiac troponin I gene *in vitro* and *in vivo*. J. Biol. Chem. 273, 25371–25380.

Dodou, E., Verzi, M.P., Anderson, J.P., Xu, S.M., Black, B.L. 2004. Mef2c is a direct transcriptional target of Isl-1 and GATA factors in the anterior heart field during mouse embryonic development. Development 131, 3931–3942.

Durocher, D., Charron, F., Warren, R., Schwartz, R.J., Nemer, M. 1997. The cardiac transcription factors Nkx2–5 and GATA-4 are mutual cofactors. EMBO J. 16, 5687–5696.

Dyson, E., Sucov, H.M., Kubalak, S.W., Schmid-Schonbein, G.W., DeLano, F.A., Evans, R.M., Ross, J., Jr., Chien, K.R. 1995. Atrial-like phenotype is associated with embryonic ventricular failure in retinoid X receptor alpha $-/-$ mice. Proc. Natl. Acad. Sci. USA 92, 7386–7390.

Eisenberg, L.M., Markwald, R.R. 1995. Molecular regulation of atrioventricular valvuloseptal morphogenesis. Circ. Res. 77, 1–6.

Evan, G., Littlewood, T. 1998. A matter of life and cell death. Science 281, 1317–1322.

Farrell, M., Waldo, K., Li, Y.X., Kirby, M.L. 1999. A novel role for cardiac neural crest in heart development. Trends Cardiovasc. Med. 9, 214–220.

Firulli, A.B., McFadden, D.G., Lin, Q., Srivastava, D., Olson, E.N. 1998. Heart and extra-embryonic mesodermal defects in mouse embryos lacking the bHLH transcription factor Hand1. Nat. Genet. 18, 266–270.

Franco, D., Meilhac, S.M., Christoffels, V.M., Kispert, A., Buckingham, M., Kelly, R.G. 2006. Left and right ventricular contributions to the formation of the interventricular septum in the mouse heart. Dev. Biol. 294, 366–375.

Frasch, M. 1995. Induction of visceral and cardiac mesoderm by ectodermal Dpp in the early *Drosophila* embryo. Nature 374, 464–467.

Frasch, M. 1999. Intersecting signalling and transcriptional pathways in *Drosophila* heart specification. Semin. Cell Dev. Biol. 10, 61–71.

Gassmann, M., Casagranda, F., Orioli, D., Simon, H., Lai, C., Klein, R., Lemke, G. 1995. Aberrant neural and cardiac development in mice lacking the ErbB4 neuregulin receptor. Nature 378, 390–394.

Gaussin, V., Morley, G.E., Cox, L., Zwijsen, A., Vance, K.M., Emile, L., Tian, Y., Liu, J., Hong, C., Myers, D., Conway, S.J., Depre, C., et al. 2005. Alk3/Bmpr1a receptor is required for development of the atrioventricular canal into valves and annulus fibrosus. Circ. Res. 97, 219–226.

Gottlieb, P.D., Pierce, S.A., Sims, R.J., Yamagishi, H., Weihe, E.K., Harriss, J.V., Maika, S.D., Kuziel, W.A., King, H.L., Olson, E.N., Nakagawa, O., Srivastava, D. 2002. Bop encodes a muscle-restricted protein containing MYND and SET domains and is essential for cardiac differentiation and morphogenesis. Nat. Genet. 31, 25–32.

Guo, L., Lynch, J., Nakamura, K., Fliegel, L., Kasahara, H., Izumo, S., Komuro, I., Agellon, L. B., Michalak, M. 2001. COUP-TFl antagonizes Nkx2.5-mediated activation of the calreti-culin gene during cardiac development. J. Biol. Chem. 276, 2797–2801.

Habets, P.E., Moorman, A.F., Clout, D.E., van Roon, M.A., Lingbeek, M., van Lohuizen, M., Campione, M., Christoffels, V.M. 2002. Cooperative action of Tbx2 and Nkx2.5 inhibits ANF expression in the atrioventricular canal: Implications for cardiac chamber forma-tion. Genes Dev. 16, 1234–1246.

Harh, J.Y., Paul, M.H. 1975. Experimental cardiac morphogenesis. I. Development of the ventricular septum in the chick. J. Embryol. Exp. Morphol. 33, 13–28.

Harvey, R.P. 1998. Links in the left/right axial pathway. Cell 94, 273–276.

Harvey, R.P. 1999. Seeking a regulatory roadmap for heart morphogenesis. Semin. Cell Dev. Biol. 10, 99–107.

Heicklen-Klein, A., Evans, T. 2004. T-box binding sites are required for activity of a cardiac GATA-4 enhancer. Dev. Biol. 267, 490–504.

Hiroi, Y., Kudoh, S., Monzen, K., Ikeda, Y., Yazaki, Y., Nagai, R., Komuro, I. 2001. Tbx5 associates with Nkx2-5 and synergistically promotes cardiomyocyte differentiation. Nat. Genet. 28, 276–280.

Hoogaars, W.M., Tessari, A., Moorman, A.F., de Boer, P.A., Hagoort, J., Soufan, A.T., Campione, M., Christoffels, V.M. 2004. The transcriptional repressor Tbx3 delineates the developing central conduction system of the heart. Cardiovasc. Res. 62, 489–499.

Hu, T., Yamagishi, H., Maeda, J., McAnally, J., Yamagishi, C., Srivastava, D. 2004. Tbx1 regulates fibroblast growth factors in the anterior heart field through a reinforcing auto-regulatory loop involving forkhead transcription factors. Development 131, 5491–5502.

Ilagan, R., Abu-Issa, R., Brown, D., Yang, Y.P., Jiao, K., Schwartz, R.J., Klingensmith, J., Meyers, E.N. 2006. Fgf8 is required for anterior heart field development. Development 133, 2435–2445.

Kelly, R.G., Buckingham, M.E. 2002. The anterior heart-forming field: Voyage to the arterial pole of the heart. Trends Genet. 18, 210–216.

Kelly, R.G., Brown, N.A., Buckingham, M.E. 2001. The arterial pole of the mouse heart forms from Fgf10-expressing cells in pharyngeal mesoderm. Dev. Cell 1, 435–440.

Kirby, M.L. 2002. Molecular embryogenesis of the heart. Pediatr. Dev. Pathol. 5, 516–543.

Kirby, M.L., Waldo, K.L. 1995. Neural crest and cardiovascular patterning. Circ. Res. 77, 211–215.

Kozar, K., Sicinski, P. 2005. Cell cycle progression without cyclin D-CDK4 and cyclin D-CDK6 complexes. Cell Cycle 4, 388–391.

Kozar, K., Ciemerych, M.A., Rebel, V.I., Shigematsu, H., Zagozdzon, A., Sicinska, E., Geng, Y., Yu, Q., Bhattacharya, S., Bronson, R.T., Akashi, K., Sicinski, P. 2004. Mouse development and cell proliferation in the absence of D-cyclins. Cell 118, 477–491.

Lamers, W.H., Moorman, A.F. 2002. Cardiac septation: A late contribution of the embryonic primary myocardium to heart morphogenesis. Circ. Res. 91, 93–103.

Latinkic, B.V., Cooper, B., Smith, S., Kotecha, S., Towers, N., Sparrow, D., Mohun, T.J. 2004. Transcriptional regulation of the cardiac-specific MLC2 gene during Xenopus embryonic development. Development 131, 669–679.

Lee, K.F., Simon, H., Chen, H., Bates, B., Hung, M.C., Hauser, C. 1995. Requirement for neuregulin receptor erbB2 in neural and cardiac development. Nature 378, 394–398.

Lee, K.H., Evans, S., Ruan, T.Y., Lassar, A.B. 2004. SMAD-mediated modulation of YY1 activity regulates the BMP response and cardiac-specific expression of a GATA4/5/6-dependent chick Nkx2.5 enhancer. Development 131, 4709–4723.

Liberatore, C.M., Searcy-Schrick, R.D., Yutzey, K.E. 2000. Ventricular expression of tbx5 inhibits normal heart chamber development. Dev. Biol. 223, 169–180.

Lien, C.L., Wu, C., Mercer, B., Webb, R., Richardson, J.A., Olson, E.N. 1999. Control of early cardiac-specific transcription of Nkx2–5 by a GATA-dependent enhancer. Development 126, 75–84.

Lien, C.L., McAnally, J., Richardson, J.A., Olson, E.N. 2002. Cardiac-specific activity of an Nkx2–5 enhancer requires an evolutionarily conserved Smad binding site. Dev. Biol. 244, 257–266.

Lin, Q., Schwarz, J., Bucana, C., Olson, E.N. 1997. Control of mouse cardiac morphogenesis and myogenesis by transcription factor MEF2C. Science 276, 1404–1407.

Lin, L., Bu, L., Cai, C.L., Zhang, X., Evans, S. 2006. Isl1 is upstream of sonic hedgehog in a pathway required for cardiac morphogenesis. Dev. Biol. 295, 756–763.

Lincoln, J., Alfieri, C.M., Yutzey, K.E. 2004. Development of heart valve leaflets and supporting apparatus in chicken and mouse embryos. Dev. Dyn. 230, 239–250.

Liu, W., Selever, J., Wang, D., Lu, M.F., Moses, K.A., Schwartz, R.J., Martin, J.F. 2004. Bmp4 signaling is required for outflow-tract septation and branchial-arch artery remodeling. Proc. Natl. Acad. Sci. USA 101, 4489–4494.

Lyons, I., Parsons, L.M., Hartley, L., Li, R., Andrews, J.E., Robb, L., Harvey, R.P. 1995. Myogenic and morphogenetic defects in the heart tubes of murine embryos lacking the homeo box gene Nkx2–5. Genes Dev. 9, 1654–1666.

Markwald, R.R., Fitzharris, T.P., Manasek, F.J. 1977. Structural development of endocardial cushions. Am J. Anat. 148, 85–119.

Meilhac, S.M., Esner, M., Kelly, R.G., Nicolas, J.F., Buckingham, M.E. 2004a. The clonal origin of myocardial cells in different regions of the embryonic mouse heart. Dev. Cell 6, 685–698.

Meilhac, S.M., Esner, M., Kerszberg, M., Moss, J.E., Buckingham, M.E. 2004b. Oriented clonal cell growth in the developing mouse myocardium underlies cardiac morphogenesis. J. Cell Biol. 164, 97–109.

Meyer, D., Birchmeier, C. 1995. Multiple essential functions of neuregulin in development. Nature 378, 386–390.

Mjaatvedt, C.H., Nakaoka, T., Moreno-Rodriguez, R., Norris, R.A., Kern, M.J., Eisenberg, C. A., Turner, D., Markwald, R.R. 2001. The outflow tract of the heart is recruited from a novel heart-forming field. Dev. Biol. 238, 97–109.

Moens, C.B., Stanton, B.R., Parada, L.F., Rossant, J. 1993. Defects in heart and lung development in compound heterozygotes for two different targeted mutations at the N-myc locus. Development 119, 485–499.

Moorman, A.F., Christoffels, V.M. 2003. Cardiac chamber formation: Development, genes, and evolution. Physiol. Rev. 83, 1223–1267.

Moorman, A.F., de Jong, F., Denyn, M.M., Lamers, W.H. 1998. Development of the cardiac conduction system. Circ. Res. 82, 629–644.

Morin, S., Charron, F., Robitaille, L., Nemer, M. 2000. GATA-dependent recruitment of MEF2 proteins to target promoters. EMBO J. 19, 2046–2055.

Park, E.J., Ogden, L.A., Talbot, A., Evans, S., Cai, C.L., Black, B.L., Frank, D.U., Moon, A.M. 2006. Required, tissue-specific roles for Fgf8 in outflow tract formation and remodeling. Development 133, 2419–2433.

Pashmforoush, M., Lu, J.T., Chen, H., Amand, T.S., Kondo, R., Pradervand, S., Evans, S.M., Clark, B., Feramisco, J.R., Giles, W., Ho, S.Y., Benson, D.W., et al. 2004. Nkx2–5 pathways and congenital heart disease; loss of ventricular myocyte lineage specification leads to progressive cardiomyopathy and complete heart block. Cell 117, 373–386.

Pfaff, S.L., Mendelsohn, M., Stewart, C.L., Edlund, T., Jessell, T.M. 1996. Requirement for LIM homeobox gene Isl1 in motor neuron generation reveals a motor neuron-dependent step in interneuron differentiation. Cell 84, 309–320.

Phan, D., Rasmussen, T.L., Nakagawa, O., McAnally, J., Gottlieb, P.D., Tucker, P.W., Richardson, J.A., Bassel-Duby, R., Olson, E.N. 2005. BOP, a regulator of right ventricular heart development, is a direct transcriptional target of MEF2C in the developing heart. Development 132, 2669–2678.

Plageman, T.F., Jr., Yutzey, K.E. 2004. Differential expression and function of Tbx5 and Tbx20 in cardiac development. J. Biol. Chem. 279, 19026–19034.

Plageman, T.F., Jr., Yutzey, K.E. 2005. T-box genes and heart development: Putting the "T" in heart. Dev. Dyn. 232, 11–20.

Reecy, J.M., Li, X., Yamada, M., DeMayo, F.J., Newman, C.S., Harvey, R.P., Schwartz, R.J. 1999. Identification of upstream regulatory regions in the heart-expressed homeobox gene Nkx2–5. Development 126, 839–849.

Rivera-Feliciano, J., Lee, K.H., Kong, S.W., Rajagopal, S., Ma, Q., Springer, Z., Izumo, S., Tabin, C.J., Pu, W.T. 2006. Development of heart valves requires Gata4 expression in endothelial-derived cells. Development 133, 3607–3618.

Rockman, H.A., Ross, R.S., Harris, A.N., Knowlton, K.U., Steinhelper, M.E., Field, L.J., Ross, J., Jr., Chien, K.R. 1991. Segregation of atrial-specific and inducible expression of an atrial natriuretic factor transgene in an *in vivo* murine model of cardiac hypertrophy. Proc. Natl. Acad. Sci. USA 88, 8277–8281.

Ross, R.S., Navankasattusas, S., Harvey, R.P., Chien, K.R. 1996. An HF-1a/HF-1b/MEF-2 combinatorial element confers cardiac ventricular specificity and established an anterior-posterior gradient of expression. Development 122, 1799–1809.

Schwartz, R.J., Olson, E.N. 1999. Building the heart piece by piece: Modularity of cis-elements regulating Nkx2–5 transcription. Development 126, 4187–4192.

Searcy, R.D., Vincent, E.B., Liberatore, C.M., Yutzey, K.E. 1998. A GATA-dependent nkx–2.5 regulatory element activates early cardiac gene expression in transgenic mice. Development 125, 4461–4470.

Sedmera, D., Pexieder, T., Vuillemin, M., Thompson, R.P., Anderson, R.H. 2000. Developmental patterning of the myocardium. Anat. Rec. 258, 319–337.

Sedmera, D., Reckova, M., DeAlmeida, A., Coppen, S.R., Kubalak, S.W., Gourdie, R.G., Thompson, R.P. 2003. Spatiotemporal pattern of commitment to slowed proliferation in the embryonic mouse heart indicates progressive differentiation of the cardiac conduction system. Anat. Rec. A Discov. Mol. Cell. Evol. Biol. 274, 773–777.

Seo, S., Kume, T. 2006. Forkhead transcription factors, Foxc1 and Foxc2, are required for the morphogenesis of the cardiac outflow tract. Dev. Biol. 296, 421–436.

Shiratori, H., Sakuma, R., Watanabe, M., Hashiguchi, H., Mochida, K., Sakai, Y., Nishino, J., Saijoh, Y., Whitman, M., Hamada, H. 2001. Two-step regulation of left-right asymmetric expression of Pitx2: Initiation by nodal signaling and maintenance by Nkx2. Mol. Cell 7, 137–149.

Small, E.M., Krieg, P.A. 2003. Transgenic analysis of the atrialnatriuretic factor (ANF) promoter: Nkx2–5 and GATA-4 binding sites are required for atrial specific expression of ANF. Dev. Biol. 261, 116–131.

Small, E.M., Krieg, P.A. 2004. Molecular regulation of cardiac chamber-specific gene expression. Trends Cardiovasc. Med. 14, 13–18.

Song, L., Fassler, R., Mishina, Y., Jiao, K., Baldwin, H.S. 2007. Essential functions of Alk3 during AV cushion morphogenesis in mouse embryonic hearts. Dev. Biol. 301, 276–286.

Srivastava, D., Thomas, T., Lin, Q., Kirby, M.L., Brown, D., Olson, E.N. 1997. Regulation of cardiac mesodermal and neural crest development by the bHLH transcription factor, dHAND. Nat. Genet. 16, 154–160.

Steadman, D.J., Giuffrida, D., Gelmann, E.P. 2000. DNA-binding sequence of the human prostate-specific homeodomain protein NKX3.1. Nucleic Acids Res. 28, 2389–2395.

Stennard, F.A., Harvey, R.P. 2005. T-box transcription factors and their roles in regulatory hierarchies in the developing heart. Development 132, 4897–4910.

Stennard, F.A., Costa, M.W., Elliott, D.A., Rankin, S., Haast, S.J., Lai, D., McDonald, L.P., Niederreither, K., Dolle, P., Bruneau, B.G., Zorn, A.M., Harvey, R.P. 2003. Cardiac T-box factor Tbx20 directly interacts with Nkx2–5, GATA4, and GATA5 in regulation of gene expression in the developing heart. Dev. Biol. 262, 206–224.

Stennard, F.A., Costa, M.W., Lai, D., Biben, C., Furtado, M.B., Solloway, M.J., McCulley, D.J., Leimena, C., Preis, J.I., Dunwoodie, S.L., Elliott, D.E., Prall, O.W., et al. 2005. Murine T-box transcription factor Tbx20 acts as a repressor during heart development, and is essential for adult heart integrity, function and adaptation. Development 132, 2451–2462.

Sugi, Y., Yamamura, H., Okagawa, H., Markwald, R.R. 2004. Bone morphogenetic protein-2 can mediate myocardial regulation of atrioventricular cushion mesenchymal cell formation in mice. Dev. Biol. 269, 505–518.

Takeuchi, J.K., Mileikovskaia, M., Koshiba-Takeuchi, K., Heidt, A.B., Mori, A.D., Arruda, E.P., Gertsenstein, M., Georges, R., Davidson, L., Mo, R., Hui, C.C., Henkelman, R.M., et al. 2005. Tbx20 dose-dependently regulates transcription factor networks required for mouse heart and motoneuron development. Development 132, 2463–2474.

Takimoto, E., Mizuno, T., Terasaki, F., Shimoyama, M., Honda, H., Shiojima, I., Hiroi, Y., Oka, T., Hayashi, D., Hirai, H., Kudoh, S., Toko, H., et al. 2000. Up-regulation of natriuretic peptides in the ventricle of Csx/Nkx2–5 transgenic mice. Biochem. Biophys. Res. Commun. 270, 1074–1079.

Tam, P.P., Behringer, R.R. 1997. Mouse gastrulation: The formation of a mammalian body plan. Mech. Dev. 68, 3–25.

Tanaka, M., Wechsler, S.B., Lee, I.W., Yamasaki, N., Lawitts, J.A., Izumo, S. 1999. Complex modular cis-acting elements regulate expression of the cardiac specifying homeobox gene Csx/Nkx2.5. Development 126, 1439–1450.

Thattaliyath, B.D., Firulli, B.A., Firulli, A.B. 2002. The basic-helix-loop-helix transcription factor HAND2 directly regulates transcription of the atrial naturetic peptide gene. J. Mol. Cell. Cardiol. 34, 1335–1344.

Verzi, M.P., McCulley, D.J., De Val, S., Dodou, E., Black, B.L. 2005. The right ventricle, outflow tract, and ventricular septum comprise a restricted expression domain within the secondary/anterior heart field. Dev. Biol. 287, 134–145.

von Both, I., Silvestri, C., Erdemir, T., Lickert, H., Walls, J.R., Henkelman, R.M., Rossant, J., Harvey, R.P., Attisano, L., Wrana, J.L. 2004. Foxh1 is essential for development of the anterior heart field. Dev. Cell 7, 331–345.

Waldo, K.L., Lo, C.W., Kirby, M.L. 1999. Connexin 43 expression reflects neural crest patterns during cardiovascular development. Dev. Biol. 208, 307–323.

Waldo, K.L., Kumiski, D.H., Wallis, K.T., Stadt, H.A., Hutson, M.R., Platt, D.H., Kirby, M.L. 2001. Conotruncal myocardium arises from a secondary heart field. Development 128, 3179–3188.

Wang, G.F., Nikovits, W., Schleinitz, M., Stockdale, F.E. 1996. Atrial chamber-specific expression of the slow myosin heavy chain 3 gene in the embryonic heart. J. Biol. Chem. 271, 19836–19845.

Wang, G.F., Nikovits, W., Jr., Schleinitz, M., Stockdale, F.E. 1998. A positive GATA element and a negative vitamin D receptor-like element control atrial chamber-specific expression of a slow myosin heavy-chain gene during cardiac morphogenesis. Mol. Cell. Biol. 18, 6023–6034.

Wang, G.F., Nikovits, W., Jr., Bao, Z.Z., Stockdale, F.E. 2001. Irx4 forms an inhibitory complex with the vitamin D and retinoic X receptors to regulate cardiac chamber-specific slow MyHC3 expression. J. Biol. Chem. 276, 28835–28841.

Wang, J., Sridurongrit, S., Dudas, M., Thomas, P., Nagy, A., Schneider, M.D., Epstein, J.A., Kaartinen, V. 2005. Atrioventricular cushion transformation is mediated by ALK2 in the developing mouse heart. Dev. Biol. 286, 299–310.

Washington Smoak, I., Byrd, N.A., Abu-Issa, R., Goddeeris, M.M., Anderson, R., Morris, J., Yamamura, K., Klingensmith, J., Meyers, E.N. 2005. Sonic hedgehog is required for cardiac outflow tract and neural crest cell development. Dev. Biol. 283, 357–372.

Wessels, A., Sedmera, D. 2003. Developmental anatomy of the heart: A tale of mice and man. Physiol. Genomics 15, 165–176.

Wessels, A., Markman, M.W., Vermeulen, J.L., Anderson, R.H., Moorman, A.F., Lamers, W. H. 1996. The development of the atrioventricular junction in the human heart. Circ. Res. 78, 110–117.

Xavier-Neto, J., Neville, C.M., Shapiro, M.D., Houghton, L., Wang, G.F., Nikovits, W., Jr., Stockdale, F.E., Rosenthal, N. 1999. A retinoic acid-inducible transgenic marker of sino-atrial development in the mouse heart. Development 126, 2677–2687.

Xavier-Neto, J., Rosenthal, N., Silva, F.A., Matos, T.G., Hochgreb, T., Linhares, V.L. 2001. Retinoid signaling and cardiac anteroposterior segmentation. Genesis 31, 97–104.

Yamada, M., Revelli, J.P., Eichele, G., Barron, M., Schwartz, R.J. 2000. Expression of chick Tbx-2, Tbx-3, and Tbx-5 genes during early heart development: Evidence for BMP2 induction of Tbx2. Dev. Biol. 228, 95–105.

Yamagishi, H., Yamagishi, C., Nakagawa, O., Harvey, R.P., Olson, E.N., Srivastava, D. 2001. The combinatorial activities of Nkx2.5 and dHAND are essential for cardiac ventricle formation. Dev. Biol. 239, 190–203.

Zaffran, S., Kelly, R.G., Meilhac, S.M., Buckingham, M.E., Brown, N.A. 2004. Right ventricular myocardium derives from the anterior heart field. Circ. Res. 95, 261–268.

Zang, M.X., Li, Y., Xue, L.X., Jia, H.T., Jing, H. 2004. Cooperative activation of atrial naturetic peptide promoter by dHAND and MEF2C. J. Cell. Biochem. 93, 1255–1266.

Zeisberg, E.M., Ma, Q., Juraszek, A.L., Moses, K., Schwartz, R.J., Izumo, S., Pu, W.T. 2005. Morphogenesis of the right ventricle requires myocardial expression of Gata4. J. Clin. Invest. 115, 1522–1531.

CHAPTER 5

Signaling Pathways in Embryonic Heart Induction

Rosa M. Guzzo,* Ann C. Foley,*
Yessenia M. Ibarra* and Mark Mercola*,[1]

Contents

Abstract

It is well established that the heart has only a limited ability to repair damage caused by heart disease. However, the observation that human embryonic stem (ES) cells differentiate into beating cardiomyocytes raises the possibility that these cells may be used to regenerate damaged cardiac tissue. Unfortunately, limited progress has been made in developing protocols for the efficient differentiation of human ES cells

* Burnham Institute for Medical Research, La Jolla, California
[1] Present Address: Weill Cornell Medical College, New York, New York

Advances in Developmental Biology, Volume 18
ISSN 1574-3349, DOI: 10.1016/S1574-3349(07)18005-4

into ventricular cardiomyocytes. Studies in various animal models have provided significant insights into the signaling molecules that are necessary for heart induction, making it clear that both the spatial and temporal activation and suppression of defined signaling pathways are critical for the development of the myocardial cells in both embryos and ES cell cultures. We propose that there are at least four distinct steps that lead to the development of cardiac lineages in the embryo, namely: (1) establishment of organizing centers, (2) mesendoderm induction, (3) establishment of cardiac precursors, and (4) terminal differentiation of beating cardiomyocytes. Here, we provide a description of each stage, the signaling pathways that dominate each stage and present evidence that those signaling pathways also govern cardiac differentiation in ES cell models. Further identification of soluble growth factors and signaling cascades that promote heart formation is of great importance for an improved understanding of cardiac development and function and may reveal new paradigms for the treatment of cardiovascular disease.

1. INTRODUCTION

1.1. Significance and cardiovascular relevance

Heart disease is the leading cause of mortality and decline in the quality of life in the developed world. The poor prognoses for heart disease (Thom et al., 2006) reflect the inability of current therapies to restore function to damaged heart tissue and underscore the critical need to develop alternative therapeutic strategies. The demonstrated ability of human embryonic stem (ES) cells to form cardiomyocytes (Kehat et al., 2001; Xu et al., 2002a; He et al., 2003) has spawned widespread hope that these cells may be used as a cell source to replace damaged myocardium. Despite their ability to form cardiomyocytes, efficient and controlled cardiomyogenesis in ES cell cultures has not been achieved due to an incomplete understanding of the pathways that regulate cardiac development. Embryologic studies have provided a laundry list of factors involved in the early steps of cardiomyogenesis and yet simply adding these factors or even cocktails of various factors does not significantly enhance cardiac differentiation of ES cells. Rather, it has become clear that both the spatial and temporal regulation of these signals are critical for the proper development of the myocardium. We propose that there are at least four distinct embryonic steps that lead to the development of cardiac lineages in the embryo, namely: (1) establishment of organizing centers, (2) mesendoderm induction, (3) establishment of cardiac precursors, and (4) terminal differentiation of beating cardiomyocytes. Here, we provide a description of each stage, the signaling pathways that dominate each stage, and present evidence that those signaling pathways also govern cardiac differentiation in ES cell models.

1.2. ES cell differentiation into cardiomyocytes

ES cells, isolated from the inner cell mass of preimplantation blastocysts (Evans and Kaufman, 1981; Martin, 1981; Thomson et al., 1998), retain the capacity for virtually unlimited self-renewal and multilineage differentiation. Differentiation of ES cells is initiated by the removal of factors known to support stemness (Smith et al., 1988; Niwa et al., 1998; Thomson et al., 1998; Chambers et al., 2003), concomitant with the formation of suspension aggregates termed as embryoid bodies (EBs) (Doetschman et al., 1985) (Fig. 1). It is believed that the differentiation of multiple lineages in EBs occurs because these cell aggregates permit cell–cell interactions that recapitulate normal embryonic development, including the formation of all three germ layers (ectoderm, endoderm, mesoderm) and the development of rhythmically contracting cardiomyocytes (reviewed in Boheler et al., 2002; Gadue et al., 2005; Wei et al., 2005; Wobus and Boheler, 2005).

In many respects, cardiomyogenesis in ES cells mimics the developmental dynamics of embryonic cardiogenesis (Fig. 2). Typically, *brachyury* expressing mesoderm forms 3–4 days after EB formation, followed by the generation of cardiac progenitors (5–6 days), which give rise to differentiated, spontaneously contracting cardiomyocytes (7 days) situated between an epithelial layer and a mesenchymal cell layer (Hescheler et al., 1997; reviewed by Wei et al., 2005). In a similar manner, pluripotent P19 cells derived from embryonal carcinoma cells display morphological and

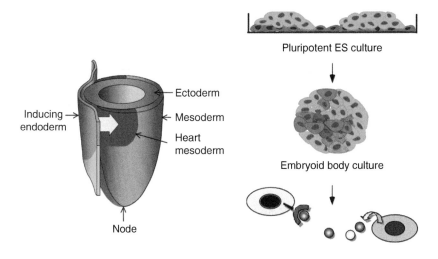

FIGURE 1   Heart induction in mouse embryos and EBs. Stochastic interactions within the EB recapitulate normal interactions in the embryo. EB model of differentiation recapitulates the early events of embryonic development, including formation of the primary germ layers (endoderm, mesoderm, and ectoderm) that give rise to all lineages. This figure was adapted from (Foley and Mercola, 2004).

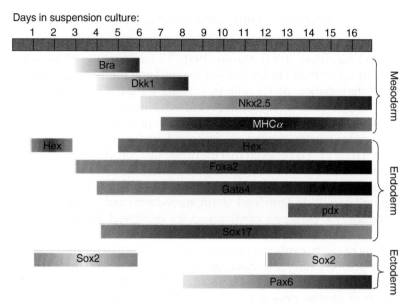

FIGURE 2   Expression of lineage-specific marker genes in mouse embyroid bodies.

molecular evidence of cardiomyocyte development when cultured as suspension aggregates (McBurney et al., 1982). Several groups have demonstrated that differentiated cardiomyocytes derived from mouse and human ES cells display functional (ion channel development and gap junction formation) and structural (myofibrillogenesis and maturation) properties resembling those of embryonic ventricular, atrial, and nodal cardiomyocytes (Kehat et al., 2001; Fijnvandraat et al., 2003; He et al., 2003; Mummery et al., 2003; Reppel et al., 2005; Dolnikov et al., 2006).

ES cells are also amenable to large-scale biochemical studies that would otherwise be prohibitive using embryologic models. Thus, in addition to their potential therapeutic applications, ES cell assays represent a validated experimental system for dissecting the complex networks of signaling pathways that regulate cardiogenesis.

2. EMBRYOLOGY OF HEART INDUCTION

2.1. Where are heart cells derived in the embryo?

All vertebrate embryos start off as a single cell that divides several times to form an undifferentiated layer of cells, referred to as the epiblast or primitive ectoderm in amniotes. Embryos then begin the process of gastrulation in which the primary germ layers are formed: (1) the ectoderm, which gives rise to the skin and nervous system, (2) the mesoderm, which gives rise to the muscles, and (3) endoderm, which gives rise to the gut and associated organs. Fate mapping and explant studies in chick and mouse have shown that cells with heart-forming capacity are scattered throughout the pre-gastrula epiblast (Olivo, 1928; Butler, 1935; Hatada and Stern, 1994; Tam and Behringer, 1997). During gastrulation, cells of the prospective heart-forming region migrate through a structure called the primitive streak in mouse and chick, or the dorsal lip in amphibians. Heart cells are among the first derivatives of embryonic mesoderm to differentiate during gastrulation [chick (Rawles, 1936; Rudnick, 1938; Spratt, 1942; Rosenquist and DeHaan, 1966; Garcia-Martinez and Schoenwolf, 1993; Psychoyos and Stern, 1996; reviewed in Rudnick, 1944), mouse (Kinder et al., 1999), and frog (Keller and Tibbetts, 1989)]. Experiments described below have demonstrated that heart cells have acquired a significant ability to self-differentiate before the overall process of mesoderm formation is complete, suggesting that cardiac specification occurs just before and during early gastrulation. A thorough understanding of the mechanisms underlying these early processes will be essential for the development of strategies that improve on the efficacy of cardiomyocyte differentiation in ES cells.

So what is happening in the embryo at the onset of gastrulation? All embryos are initially radially symmetrical, and this symmetry must be broken to orient the body axes in a process that determines the site where the cell movements of gastrulation will begin. The site of cell involution or ingression is called, variously, the streak in amniotes or the dorsal lip in amphibians. As cells pass through the streak or dorsal lip, they receive signals that contribute to the elaboration of the overall axial body plan. Indeed among the most famous experiments in experimental embryology were those of Spemann and Mangold (1924) who transplanted early dorsal lip cells to recipient embryos and found that they could re-specify host cells to form a secondary body axis mirroring that of the recipient, giving rise to the term Spemann's Organizer for the dorsal lip cells at the onset of gastrulation, which are now known to secrete a variety of soluble molecules (see below) that specify axial pattern. Analogous structures exist in the chick (Hensen's node) (Waddington, 1932) and mouse (node) (Beddington, 1994). Once cells involute over the dorsal lip or ingress through the streak, they will coalesce to form two of the three definitive cell lineages in the embryo (endoderm and mesoderm). The third lineage, ectoderm, arises from cells that do not involute or ingress during this process. Below we summarize a large number of embryo studies demonstrating that signals from both the endoderm and mesoderm are required for heart induction during this time.

2.2. Requirement for an organizer signal

In amphibians, signals for the dorsal lip midline mesoderm are required for cardiac specification (Sater and Jacobson, 1990; Nascone and Mercola, 1995). In zebrafish, signals that regulate the size and shape of the heart field are found in the prechordal plate (Goldstein and Fishman, 1998; Serbedzija et al., 1998). Interestingly in mouse, it was found through an elegant series of transplantation experiments that migration through the streak is not required for epiblast cells to adopt a cardiac fate (Tam et al., 1997) suggesting that signals from the dorsal midline may not be strictly required for cardiac differentiation. Instead it appears that in mouse, cells adopt commitment to a cardiac fate once they have reached the heart field and Tam suggests that since the endoderm shows similar anterior displacement (Lawson et al., 1986; Lawson and Pedersen, 1987) that it might be an early association of cardiac mesoderm with anterior endoderm that is required for heart induction. [See Tam and Behringer (1997) for a discussion]. This does not rule out the possibility that there is an indirect requirement for dorsal midline signals to pattern the endoderm, which in turn secretes signals that pattern the cardiac mesoderm. Our findings in *Xenopus*, described in some detail below, are consistent with the notion that organizer-derived signals induce heart indirectly by patterning the early endoderm (Foley and Mercola, 2005; Foley et al., 2006).

2.3. Requirement for an endodermal signal

An endodermal signal has been shown to be either essential for or supportive of cardiac differentiation in vertebrate species as well as in ES cells. In amphibians, endoderm has been shown to be necessary for cardiac differentiation (Jacobson, 1960; Jacobson and Duncan, 1968; Fullilove, 1970; Sater and Jacobson, 1989; Nascone and Mercola, 1995; Schneider and Mercola, 1999) with a requirement ending by mid-gastrulation (Nascone and Mercola, 1995). After gastrulation, the endoderm appears to play a supportive role that increases the rate of self-differentiation and decreases the time it takes for explants of mesoderm to develop beating (Jacobson, 1960; Jacobson and Duncan, 1968; Fullilove, 1970). In mouse, it has been shown that proper differentiation of the cardiac mesoderm requires the presence of a layer of cells known as visceral endoderm (VE) (Arai et al., 1997a). VE and its avian equivalent (discussed below) is termed extraembryonic to signify that it does not contribute to the embryo proper, and is thus not related by lineage to the definitive endoderm of the embryo.

The chick, as with mouse, has a distinct extraembryonic endoderm that forms prior to the onset of gastrulation known as the hypoblast. While the hypoblast and definitive endoderm of the chick share the ability to influence the formation of cardiac mesoderm, there are also distinct differences between them.

The hypoblast that covers the epiblast prior to gastrulation is able to induce expression of cardiac markers from undifferentiated epiblast

(Yatskievych et al., 1997; Matsui et al., 2005) and can respecify mesodermal cells from posterior primitive streak at mid-gastrula to a cardiac fate (Yatskievych et al., 1997). Hypoblast is thought, in many respects, to be the equivalent of the anterior visceral endoderm (AVE) of mouse [see Foley et al. (2000) for a review], which has also been shown to be a heart inducer (Arai et al., 1997b). The hypoblast expresses many of the molecular signals, such as Hex, Dkk-1, and Cerberus, that have been shown to play an important role in the endoderm's ability to induce cardiac tissue in other systems (Foley et al., 2000; Chapman et al., 2003). The notion that an endodermal-like signal is required in birds just before or during the early stages of gastrulation and not after is consistent with fate mapping studies and embryological studies in other organisms described above.

By mid-gastrulation, the hypoblast has been almost completely displaced by the definitive endoderm, which now underlies the heart field [see Foley et al. (2000) for a review]. The definitive endoderm retains many of the heart-forming properties of the hypoblast; for example, it is able to respecify mesoderm from the posterior primitive streak to a cardiac fate (Schultheiss et al., 1995; Yatskievych et al., 1997) and its removal during the early stages of heart induction results in rapid downregulation of *Nkx2.5* and *Mef2C* (Alsan and Schultheiss, 2002), even if removed after initiation of *Nkx2.5* expression. In addition, it has been reported to be essential to stimulate formation of the myofibrillar apparatus in cardiomyocytes (Gannon and Bader, 1995) and proliferation (Sugi and Lough, 1994; Sugi et al., 1995) and might therefore contribute morphogenetic signals needed to form heart tubes. However, it does not induce cardiac marker expression from undifferentiated epiblast (Yatskievych et al., 1997). It is unclear whether definitive endoderm is required for continued maintenance of cardiac markers at mid-gastrula stages since removal of endoderm from the heart field results in a rapid decrease in cardiac markers (Alsan and Schultheiss, 2002), but explant studies of the precardiac mesoderm's ability to self-differentiate suggest that the requirement for an endodermal signal is complete by mid-gastrula stages (Gannon and Bader, 1995; Yutzey et al., 1995). Although mesoderm can maintain expression of cardiac markers in the absence of endoderm, these studies do not rule out the possibility that cardiac endoderm might also provide signals that maintain heart specification in the face of antagonizing signals from the axial mesoderm or ectoderm (Climent et al., 1995; Goldstein and Fishman, 1998; Raffin et al., 2000; Tzahor and Lassar, 2001a).

By analogy to embryonic cardiogenesis, it seems likely that stochastic interactions between the endoderm and mesoderm induce cardiogenesis within clusters of differentiating ES cells. To date, few studies have specifically addressed the role of endoderm in EB cardiac differentiation, primarily because of the difficulties in manipulating cell–cell interactions within differentiating EBs as well as the lack of lineage-specific reporters to identify anterior visceral or anterior definitive endoderm in living cells. To this end, the inductive role of endoderm during *in vitro* cardiac development has been largely investigated via heterogeneous EB cocultures assays. Rudy-Reil and Lough (2004) demonstrated that paracrine signals derived from monolayer explants of avian precardiac endoderm robustly increase the expression of various cardiac-specific genes, as well as the incidence of beating cardiomyocytes in mouse EBs (Rudy-Reil and Lough, 2004). Christine Mummery and colleagues further demonstrated that mouse VE-like cells (END2) mimic this effect. Spontaneously contracting cardiomyocytes with functional and electrophysiological properties resembling fetal ventricular cells develop from human ES cells when cocultured with END2 cells or grown in END2-conditioned media (Mummery et al., 1991, 2003; Passier et al., 2005). Since spontaneous cardiomyocyte development is considerably less efficient in human than in mouse EBs, this strategy may be useful for identification and characterization of inducing signals and their mode of action, including downstream signaling cascades.

2.4. Signals that inhibit heart development

It should be noted that in addition to providing signals that induce cardiac mesoderm, the embryo also secretes factors that inhibit heart formation. For example, in zebrafish, the notochord appears to inhibit the size of the heart field (Goldstein and Fishman, 1998), and inclusion of neural tissues in explants of amphibian cardiac tissues completely blocks expression of cardiac markers (Raffin et al., 2000) and beating (Jacobson, 1960; Jacobson and Duncan, 1968) in *Xenopus* and chick (Climent et al., 1995; Tzahor and Lassar, 2001a) and this blockade to heart induction may be mediated by Wnt signaling (Tzahor and Lassar, 2001a). Importantly, the embryologic experiments reveal that mesoderm adjacent to the heart field will differentiate as heart if the source of the signal is removed, indicating that the purpose of the inhibitory factors is to constrain the size of the heart field.

3. STEP 1: ESTABLISHING THE ORGANIZING CENTERS: CANONICAL WNTS

The first embryological step required for heart formation is the establishment of a dorsal organizing center in the pre-gastrula stage embryo. This important signaling center determines the site where gastrulation will be initiated and over time will secrete factors that both induce the formation of endoderm and mesoderm and pattern these tissues as they form.

Molecularly, these early organizing centers are marked by the expression of genes, including *goosecoid* (*gsc*), *twins* (*twns*), and *brachury* (*bra*).

The formation of the pre-gastrula organizing center is best studied in *Xenopus* embryos where early on, activation of the canonical Wnt/ β-catenin pathway results in the localized nuclear accumulation of β-catenin that is required to establish the early signaling centers that both induce the various germ layers and pattern the embryonic axis. In *Xenopus*, embryonic depletion of maternal β-catenin blocks axis formation (Heasman et al., 1994) and overexpression of β-catenin results in the formation of secondary axes and the expression of markers for the dorsal mesendoderm (Wylie et al., 1996; Larabell et al., 1997; Crease et al., 1998). These findings have been extended to other animal models. For example, overexpression of β-catenin in zebrafish (Kelly et al., 1995) results in the formation of secondary axes and mice possessing a homozygous deletion of β-catenin lack a primitive streak and form no mesoderm (Huelsken et al., 2000). β-catenin localization also plays a key role in inducing and patterning expression of the gene Nodal (see below) (Xanthos et al., 2002; Hashimoto-Partyka et al., 2003), which induces and patterns both endoderm and mesoderm.

One of the most interesting findings of this research has been that while signaling through the Wnt/β-catenin pathway is required for formation of the early dorsal organizing centers, none of the Wnts shown to activate this pathway are present in the right part of the early embryo at the relevant stage. It has been suggested that nuclear accumulation of β-catenin may be attributed to the activation of some other pathway in the early embryo (reviewed in Harland and Gerhart, 1997). More recently, Tao et al. have suggested that Wnt11, which in most settings signals independently of β-catenin to activate a noncanonical Wnt signaling pathway characterized by JNK phosphorylation, might cooperate with the glycosyl transferase X.ETX1 and the EGF-CFC protein FRL1 to activate the canonical Wnt/β-catenin pathway and induce the pre-gastrula organizer in the early embryo (Tao et al., 2005).

At the onset of gastrulation, the dorsal organizing centers secrete factors that act on ingressing cells to pattern the body axis. Known as the node, Hensen's node, or the Spemann's organizer in mouse, avians, and amphibians, respectively, each of these early gastrula stage organizers will re-specify recipient tissue to form an ectopic body axis when transplanted to a host embryo [frog (Spemann and Mangold, 1924), chick (Waddington, 1932), and mouse (Beddington, 1994)].

At this time, however, it appears that Wnt signaling is no longer required and, in fact, inhibits the specific formation of cardiac mesoderm as well as other functions of the early gastrula stage organizer. For example, it has been shown in both frog and chick that cardiac induction is blocked either by misexpression of Wnt3a or Wnt8c (Marvin et al., 2001;

Schneider and Mercola, 2001) or by morpholino-mediated knockdown of the Wnt antagonist Frodo (Brott and Sokol, 2005). In addition, it has been demonstrated in chick that endogenous expression of Wnt1 and Wnt3a in the neural tube also functions to inhibit cardiac differentiation during normal development (Tzahor and Lassar, 2001b). It has also been proposed that the transcription factor Goosecoid (gsc), which has long been considered to be the best marker for the dorsal organizer, acts to inhibit Wnt signaling within the organizer (Yao and Kessler, 2001). Finally, several antagonists of the canonical Wnt/β-catenin pathway have been shown to be potent inducers of cardiac markers. These will be discussed in a separate section below.

In ES and embryonical carcinoma (EC) cell culture systems, stimulation of mammalian cardiogenesis by the canonical Wnt signaling pathway has underscored a pivotal role for transient, stage-specific activation of the canonical Wnt pathway during differentiation. Activation of the Wnt/β-catenin signaling pathway is initiated at the onset of cardiac differentiation in P19 EC cells, and stimulation of canonical Wnt/β-catenin signaling cascade by treatment with Wnt3a-conditioned media at the early stages of differentiation was sufficient to promote P19 cardiomyocyte differentiation. (Nakamura et al., 2003; Naito et al., 2005). Chronic activation of the Wnt3/β-catenin pathway, on the other hand, blocked P19 cardiogenesis (Naito et al., 2005). Furthermore, inhibition of the Wnt/β-catenin pathway at the stage when canonical Wnt activity is required for cardiomyocyte differentiation suppressed cardiac marker gene expression such as *Nkx2.5, GATA4, MEF2C, Tbx5, and αMHC*; however, cardiomyocyte differentiation was unaffected when P19 cells were treated with inhibitors of canonical Wnt signaling after the formation of *brachury* expressing mesoderm (Naito et al., 2005). Taken together, these observations are consistent with the biphasic role of Wnt/β-catenin signaling in embryonic development (Marvin et al., 2001; Schneider and Mercola, 2001; Tzahor and Lassar, 2001a) and support a critical role for the transient activation of canonical Wnt pathway during the early stage of mesoderm induction with subsequent requirement for Wnt inhibition for cardiac specification (Nakamura et al., 2003; Naito et al., 2005). *brachyury*, whose expression is initiated at the onset of gastrulation and is restricted to the nascent and migrating mesoderm, node, notochordal plate, and notochord (Wilkinson et al., 1990), was shown to be induced by Wnt/β-catenin signaling in mouse ES cells, and TCF/LEF-1-binding elements have been identified within the *brachyury* promoter, upstream of transcriptional start site (Arnold et al., 2000). However, it is not clear whether canonical Wnts signal directly to mesoderm or act indirectly by signaling to the endoderm to generate factors with cardiogenic activity. Further studies are necessary to address the molecular mechanism by which canonical Wnts stimulate cardiomyogenesis in ES cells.

While a direct interaction is not yet established, it is has been suggested that signaling by canonical Wnt acts downstream of the phosphatylinositol 3-kinase (PI3-K)/Akt-dependent pathway and that these pathways operate synergistically at the early stages of *in vitro* differentiation to induce cardiogenesis (Naito et al., 2005). PI3-K, situated downstream of receptor tyrosine kinases and G-protein–coupled receptors, phosphorylates phosphatidylinositol, which functions as a second messenger to activate downstream proteins such as the serine/threonine kinase Akt (reviewed by Vanhaesebroeck et al., 2005). Temporal activation of PI3-K/Akt signaling pathway in differentiating P19 cells coincides with the early stage-specific activation of the canonical Wnt pathway during cardiogenesis and *in vitro* studies have demonstrated that activation of PI3-K pathway suppresses GSK3β activity to maintain a level of canonical Wnt signaling that is sufficient to support cardiac differentiation (Naito et al., 2005). Inactivation of PI3-K signaling at the stage when canonical Wnt signaling is required for cardiomyocyte differentiation in P19 cells attenuated cardiomyocyte differentiation without affecting general mesoderm differentiation, in support of a specific function for PI3-K signaling during cardiac commitment (Klinz et al., 1999; Naito et al., 2003; Nakamura et al., 2003).

4. STEP 2: MESODERM INDUCTION

Nodal, a secreted factor belonging to the TGFβ superfamily, has been implicated as a potential heart inducer as a consequence of its role in the induction of axial mesoderm in *Xenopus* (Jones et al., 1995; Agius et al., 2000), zebrafish (Toyama et al., 1995; Feldman et al., 1998), mouse (Zhou et al., 1993; Conlon et al., 1994), and chick (Bertocchini and Stern, 2002; Bertocchini et al., 2004) and in the induction and patterning of the endoderm (Henry et al., 1996; Alexander and Stainier, 1999; Agius et al., 2000; Chang et al., 2000; David and Rosa, 2001) (Fig. 3). In addition, several members of the *Xenopus* Nodal-related (XNr) family, including Xnr1 (Reissmann et al., 2001), the closely related TGFβ family member Activin (Logan and Mohun, 1993; Mangiacapra et al., 1995; Yatskievych et al., 1997; Ladd et al., 1998), and a constitutively active form of the Nodal receptor, Alk4 (Takahashi et al., 2000) have been shown to induce cardiac tissue in *Xenopus* and chick. Aberrant Nodal signaling in the mouse embryo prevents formation of primitive streak and inhibits AVE formation (Zhou et al., 1993; Conlon et al., 1994). Consistent with its role in mesoderm induction in amphibians (Smith et al., 1990; Thomsen et al., 1990; van den Eijnden-van Raaij et al., 1990; Dale et al., 1992), Nodal signaling influences cardiac differentiation by operating in a binary cell fate decision between mesendoderm and ectoderm development in

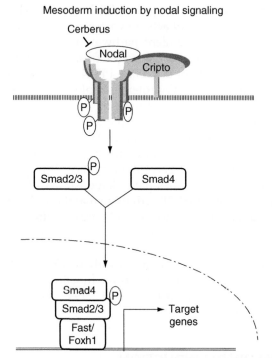

Mesoderm induction by nodal signaling

FIGURE 3 Nodal signaling pathway plays a key role in mesoderm induction. Signaling by Nodal is transduced to the cytoplasm through a heteromeric receptor complex consisting of the activin type II receptor, the Alk4 type I receptor, and the EGF-CFC protein Cripto, a GPI-anchored protein that directly binds Nodal via its EGF domain. Receptor activation leads to phosphorylation of Smad2/3 and association with co-Smad4. The pSmad2/3–Smad4 complex translocates to the nucleus and activates target genes such as *Brachyury*. Nodal signaling is inhibition by the secreted factor, Cerberus. FGF (not shown) is thought to act as a competence factor for Nodal during this stage of cardiogenesis.

ES cells (Johansson and Wiles, 1995; Parisi et al., 2003; Pfendler et al., 2005; Tada et al., 2005).

   A critical role for Nodal-dependent signaling in vertebrate mesendo-derm development has also been confirmed by homozygous deletion of the gene encoding the Nodal coreceptor Cripto in mice, which results in embryonic lethality at day 7.5 due to lack of primitive streak formation, absence of a node, and defects in embryonic mesoderm formation (Ding et al., 1998; Xu et al., 1999; Liguori et al., 2003; reviewed in Minchiotti et al., 2002; Minchiotti, 2005; Strizzi et al., 2005). These studies also revealed that Cripto$^{-/-}$ mice display severe defects in cardiogenesis while exhibiting precocious differentiation of anterior neuroectoderm (Ding et al., 1998).

Interestingly, chimeric embryos generated from the combination of wild-type ES cells and Cripto$^{-/-}$ ES cells develop normally (Xu et al., 1999; Parisi et al., 2003), which suggests that rescue of mutant cells is mediated either by a soluble, functionally active form of Cripto cleaved from the cell membrane of wild-type cells or by paracrine factors acting downstream of Nodal signaling.

The temporal requirements for Nodal-dependent Cripto signaling in cardiogenesis as well as the downstream signaling cascades that mediate cardiac development have been elucidated using mouse ES cells (Parisi et al., 2003). Consistent with the developmental defects observed *in vivo*, Cripto$^{-/-}$ mES cells lack the ability to differentiate to the cardiac lineage; however, the paucity of cardiac differentiation does not reflect a general differentiation defect as Cripto$^{-/-}$ mES cells undergo extensive differentiation to the neuroectodermal lineage. Furthermore, cardiogenesis could be restored by the administration of a soluble, recombinant version of Cripto during a defined stage of EB formation (day 0–2) (Parisi et al., 2003; Minchiotti, 2005). Collectively, these observations support a model whereby Nodal-dependent Cripto signaling acts in a dose-dependent, temporally regulated manner to activate the Smad2/3 pathway during the first few days of differentiation in order to initiate cardiogenesis while negatively regulating neuroectoderm differentiation (Parisi et al., 2003; Minchiotti, 2005; Parish et al., 2005).

Although activation of the Nodal signaling pathway by Nodal or the constitutively active, ligand independent Nodal receptor (ALK4) is capable of inducing cardiac mesoderm, a thorough understanding of Nodal signaling remains to be established. While Nodal is clearly necessary for cardiac differentiation, it has been difficult to completely separate its cardiac-inducing activity from its more general role as mesendodermal inducer. In addition, studies in zebrafish suggest that Nodal signaling may act primarily in the endoderm (David and Rosa, 2001) and our studies have indicated that activation of the signaling pathway acts cell nonautonomously to heart induction, suggesting that its role in mesoderm, and therefore heart induction, is indirect (Foley *et al.*, 2006). In addition, Activin (Green and Smith, 1990; Symes et al., 1994), as well as Nodal and its homologues, can act in a concentration and/or duration-sensitive manner to specify fate and behavioral differences of induced mesendoderm in zebrafish (Gritsman et al., 1999; Reiter et al., 2001), *Xenopus* (Agius et al., 2000; Chang et al., 2000; Faure et al., 2000; Lee et al., 2001; Bourillot et al., 2002), and mouse (Meno et al., 1999; Lowe et al., 2001; Norris et al., 2002; for reviews, see Schier and Shen, 2000; Gurdon and Bourillot, 2001; Whitman, 2001) and ES cells (Kubo et al., 2004; Ng et al., 2005). It is unclear how the embryo establishes and maintains proper timing and/or levels of Nodal that are required for cardiac or other developmental outputs. We have shown that these issues at least in part can be addressed by evoking the role of the secreted multiple antagonist Cerberus in the endoderm that underlies the heart (Foley et al., 2006). Cerberus is a multiple antagonist of Wnt, Bmp, and Nodal signaling (Bouwmeester et al., 1996; Piccolo et al., 1999), and it is the Cerberus-expressing deep endoderm that is required for heart induction (Schneider and Mercola, 1999). In addition, Cerberus is induced by ectopic Nodal in a temporal and spatial pattern similar to its endogenous expression (Osada and Wright, 1999; Piccolo et al., 1999; Yamamoto et al., 2003). Since Cerberus overexpression induces expression of cardiac field markers and attenuation of Cerberus by morpholino knockdown significantly reduces both endogenous heart field markers and the ability of Nodal to induce hearts, Cerberus is likely to be involved in blocking Nodal signaling in the heart field at a time when it would otherwise inhibit heart induction.

Cerberus might also integrate signaling between the Nodal and the Wnt/β-catenin signaling pathways in mesoderm (Nishita et al., 2000; Xanthos et al., 2002; Yamamoto et al., 2003) and endoderm formation (Zorn et al., 1999). In particular, it has been shown that activation of the transcription factor Siamois by Wnt works in conjunction with activation of Mix-1, Xlim-1 and Xotx2 by Xnr1 to induce the expression of Cerberus (Yamamoto et al., 2003), suggesting that Cerberus may act to coordinate the activities of these two important signaling pathways.

Although there is little direct evidence linking the fibroblast growth factor (FGF) signaling cascade to the early stages of heart induction, there is ample evidence suggesting that FGF signaling is required for early induction of mesoderm. More recently, it has been suggested that FGF may serve as a competence factor during mesoderm formation by Nodal and Activin and may also play a role in mesodermal maintenance. [For a more thorough review of the FGF signaling pathway, see Bottcher and Niehrs (2005).]

5. STEP 3: ESTABLISHING THE PRECARDIAC MESODERM

As explained above, signaling through the Wnt/β-catenin pathway is required for establishment of the early dorsal organizing centers; however, after gastrulation, canonical Wnts block production of the dorsal mesoderm, including cardiac mesoderm (Fig. 4A). Indeed it has now been shown in chick (Marvin et al., 2001) and *Xenopus* (Schneider and Mercola, 2001) that Wnt antagonists such as Dickkopf (Dkk-1), Crescent, and XDbf4 (Brott and Sokol, 2005) play an important role in establishing the early heart field. Few studies have specifically addressed the role of Wnt antagonists during mammalian cardiogenesis; however, transient activation of the Wnt/β-catenin signaling cascade in P19 cells (see above) is

Establishng the precardiac mesoderm

A Inhibition of Wnt/β-cat   B Noncanoncial Wnt pathway

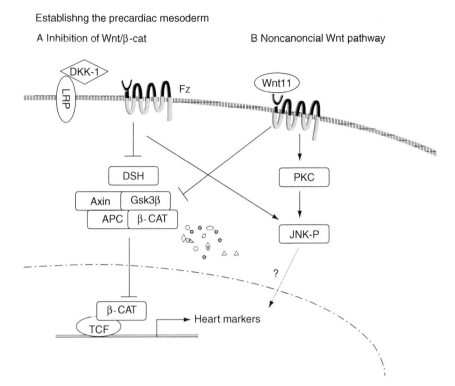

FIGURE 4   Development of precardiac mesoderm is characterized by inhibition of the canonical Wnt signaling and activation of the noncanonical Wnt pathway. Functional inhibition of canonical Wnt is mediated by Dkk-1 binding to LRP6 coreceptor at the cell membrane. Activation of the so-called "destruction box" (APC, axin Gsk3β, disheveled) degrades β-catenin, thus preventing it from entering the nucleus and activating target genes. Induction of noncanonical Wnt signaling initiated by binding of Wnt11 to the frizzled (Fz) receptor, activates the "destruction box" and inhibits downstream targets of canonical Wnt signaling. In addition, both Wnt11 and Dkk-1 have been shown to phosphorylate JNK and activate the planar cell polarity pathway, and possibly heart markers downstream of phospho-JNK.

necessary to support cardiogenesis (Nakamura et al., 2003; Naito et al., 2005). It is not clear from these P19 studies if the activity of endogenous Wnt antagonists, such as Dkk-1, turns off Wnt/β-catenin signaling, although the induction of Dkk-1 expression in differentiating mouse EBs does coincide with the downregulation of canonical Wnt activity that occurs after the formation of *brachyury*-expressing mesoderm (Fig. 2). Furthermore, inhibition of canonical Wnt signaling in mesoderm lineage selected (Flk1+) mouse ES cells by either Dkk-1 or the soluble canonical Wnt antagonist Frizzled-8/Fc enhances differentiation to αMHC positive cardiomyocytes (Yamashita et al., 2005). As these studies suggest that biphasic Wnt/β-catenin signaling is an important component of the cardiac developmental program, culture protocols that integrate the transient activation of canonical Wnt, followed by Wnt inhibition at the relevant stage of ES cell differentiation, are predicted to augment the number of ES cell-derived cardiomyocytes.

Inhibition of canonical Wnt signaling by Dkk-1 also induces secondary signals that promote heart formation. We have shown that Dkk-1, through inhibition of β-catenin/TCF complex, induces expression of the homeodomain transcription factor Hex in the endoderm, which then controls production of a diffusible factor that signals back to mesoderm to induce expression of early heart field markers (Foley and Mercola, 2005). Evidence that such a mechanism might also operate in the mouse comes from the finding that conditional depletion of β-catenin from the endoderm causes the formation of multiple beating hearts in the mouse midline (Lickert et al., 2002).

Taken together, these findings suggest that inhibition of canonical Wnt signaling in the endoderm is both sufficient and necessary for the formation of heart tissue. What positive heart-inducing signals are turned on or what heart-repressing signals are turned off in response to the Wnt antagonists has yet to be discovered. While this is a nice model to explain the initiation of cardiogenesis, induction of *Hex* in the endoderm is insufficient for heart formation since Hex alone induces only early but not late heart markers and never induced beating cardiomyocytes. This is in contrast to overexpression of Dkk-1 that induces formation of well-formed beating heart tubes in *Xenopus* assays. We are testing how divergent pathways upstream of β-catenin/TCF complex are involved in later stages of differentiation, including morphogenesis of the heart tube and control of myofibrillogenesis (see Step 4 below).

One likely pathway that may cooperate with antagonists of the canonical Wnt/β-catenin pathway in heart induction is the Wnt5a class of secreted factors, which include Wnt4, Wnt5a, and Wnt11 (Fig. 4B). These so-called noncanonical Wnts regulate intracellular calcium (Kuhl et al., 2000; Kohn and Moon, 2005) planar cell polarity and convergent extension movements (reviewed by Mlodzik, 2002) and do not induce secondary axes in *Xenopus* assays. Gain- and loss-of-function studies in chick and *Xenopus* have demonstrated that signaling by Wnt11 acts to establish the early heart field (Eisenberg and Eisenberg, 1999; Pandur et al., 2002). In mammalian systems, a role for noncanonical Wnt signaling pathway in cardiac development has been inferred based on the coincident expression of Wnt11 and cardiac-specific transcription factors such as *Nkx2.5* in the embryonic mouse myocardium and in differentiating mouse EBs (Kispert et al., 1996; Terami et al., 2004). Consistent with findings in *Xenopus* and chick, exogenous administration of Wnt11 stimulates the induction of early cardiac markers such as *Nkx2.5* and *Gata4* in mouse ES cells as well as in P19 cells (Pandur et al., 2002; Terami et al., 2004), and may also enhance cardiomyocyte differentiation of human circulating progenitor cells (CPC) cocultured with rat neonatal cardiomyocytes in a PKC-dependent manner (Koyanagi et al., 2005). Interestingly, the induction of endogenous Wnt11 expression correlates with formation of cardiac progenitors and downregulation of canonical Wnt activity in differentiating mouse EBs (Terami et al., 2004). Thus, dissection of the precise temporal requirements for canonical and noncanonical Wnt signaling during the course of ES cell cardiac differentiation warrants further investigation as a plausible strategy to improve the efficiency of cardiogenesis.

Inhibition of canonical Wnt signaling by noncanoncial Wnt11 has been put forth as a possible mechanism to explain the ability of the noncanonical Wnt11 to stimulate cardiogenesis (Pandur et al., 2002; Maye et al., 2004) (Fig. 4B). It has been proposed that repression of canonical Wnt signaling by Wnt11 occurs by either: (1) intracellular calcium-mediated repression of β-catenin-dependent transcriptional activity (Kuhl et al., 2001; Li and Iyengar, 2002; Maye et al., 2004) or (2) inhibition at the level of the cell surface, possibly via receptor (Frizzled) competition (Maye et al., 2004). Since there is cross talk between the canonical and noncanonical Wnt signaling pathways, it may be technically difficult to separate these pathways as activating one inhibits the other and vice versa. Evidence that inhibition of the canonical Wnt signaling pathway serves a primary role in the establishment of the precardiac mesoderm is that blocking Wnt at any level including down to the transcriptional response induces gene markers of the early heart field (Foley and Mercola, 2005). A second possible mechanism to explain the role of noncanonical Wnts in cardiogenesis is the ability of Wnt11 to activate the downstream phospho-JNK pathway (Pandur et al., 2002). Interestingly, while Dkk-1 induces the formation of well-organized beating heart tubes in *Xenopus*, downstream antagonists of the Wnt/β-catenin pathway give either poorly organized cardiac tissue that rarely beat or early markers only, suggesting that the ability of Dkk-1 to activate JNK may be essential for the formation of beating heart tissue. Taken together, these studies raise the possibility that inhibition of canonical Wnt/β-catenin signaling is probably critical for early heart induction whereas activation of the noncanonical phospho-JNK pathway is necessary for the formation of beating heart tissue and possibly tube formation for the heart.

In addition to noncanonical Wnts and Wnt antagonists, the evolutionarily conserved Notch-dependent receptor signaling pathway that is involved in regulating various cell fate decisions critical for embryonic and adult developmental processes (reviewed by Artavanis-Tsakonas et al., 1999) also plays a role at this stage of cardiac development. Notch receptors and ligands (Delta, Serrate/Jagged), for example, have been detected in cardiac tissue of vertebrate embryos (Myat et al., 1996; Westin and Lardelli, 1997; Loomes et al., 1999; Rones et al., 2000; Loomes et al., 2002). In *Xenopus*, Activin/TGFβ signaling has been shown to activate the Notch pathway in *Xenopus* animal caps (Abe et al., 2004) and signaling by Notch controls cardiac cell fate after Nkx2.5-expressing heart field has been specified (Rones et al., 2000). Mouse knockout studies have been employed to further examine the requirement for Notch signaling during development; however, the absence of Notch signaling results in gross developmental abnormalities leading to early embryonic death (Oka et al., 1995; Gridley, 1997, 2003), thus confounding *in vivo* analysis of Notch signaling during early heart development. In the absence of Notch signaling (Notch1$^{-/-}$ or RBP-Jκ$^{-/-}$), mouse ES cells display enhanced differentiation to the cardiac lineage (Schroeder et al., 2003; Nemir et al., 2006), without affecting differentiation to other mesodermal lineages such as endothelial and hematopoietic lineages (Schroeder et al., 2003, 2006). These results corroborate an antagonistic role for Notch in mesoderm development, acting before and at the specification of cardiac mesoderm. Subsequently, Notch components exhibit highly dynamic patterns of expression indicating complex roles such as delineating the border of chamber with atrioventricular canal and inner curvature myocardium (Rutenberg et al., 2006).

Mitogen-activated protein kinases (p38MAPK) are downstream effectors of various signaling pathways that have been shown to play multiple roles in early cell fate decisions, including determination of the cardiac lineage (Davidson and Morange, 2000; Davis, 2000; Nebreda and Porras, 2000; Aouadi et al., 2006). Functional inactivation of p38αMAPK signaling by gene knockout in mice leads to mid-gestational lethality characterized by multiple developmental defects (Tamura et al., 2000); however, the cardiovascular defects are rescued by placental expression of p38 indicating that this defect is due to insufficient oxygen and nutrient transfer rather than a specific role for cardiac specification (Adams et al., 2000). On the other hand, inactivation of p38MAPK activity by either deletion of the p38α isoform gene in mouse ES cells or chemical inhibition of endogenous p38MAPK activity results in a significant block or delay in mesoderm differentiation and severe impairment of *in vitro* cardiomyogenesis in favor of neuronal differentiation (Davidson and Morange, 2000; Aouadi et al., 2006). Although p38MAPK is activated at a stage that corresponds to the development of *brachyury*-expressing mesoderm and cardiac progenitors in mouse EBs, it is not clear from these studies whether signaling by p38MAPK acts at the level of mesoderm formation or specifically affects the development of cardiac progenitors since expression of gene markers for other mesodermal lineages have not been examined. Pharmacological inhibition of other MAPK pathways, such as ERK1/ERK2 or JNK, did not appear to significantly impair *in vitro* cardiogenesis (Davidson and Morange, 2000; Bost et al., 2002). Using a biochemical approach to examine the downstream phosphorylation substrates that mediate p38MAPK signaling during P19 cardiac differentiation, Eriksson and Leppa (2002) demonstrated that induction of AP-1 transcription factor by p38 kinase is necessary for the cardiac differentiation response; however, little is known of the downstream transcriptional targets that integrate these signals and influence cardiac-specific gene expression.

Several groups have demonstrated that reactive oxygen species (ROS)-dependent signaling promotes cardiogenesis and cardiac myofibrillogenesis in mouse ES cells (Sauer et al., 2000; Li et al., 2006) and this effect may be elicited by activation of the downstream p38MAPK pathway (Li et al., 2006; Schmelter et al., 2006). Whereas treatment with hydrogen peroxide promotes cardiac differentiation in mouse EBs, ROS scavengers or functional inhibition of NAPDH oxidase NOX4 prevents phosphorylation of p38MAPK and suppresses cardiac differentiation (Li et al., 2006). Furthermore, it is thought that p38MAPK activation by ROS signaling mediates the activity of Mef2C, a transcription factor that is critical for commitment to the cardiac lineage. Whether ROS is produced endogenously by cardiac-committed cells or other lineages within the developing EBs has not been investigated (Li et al., 2006); however, it has been speculated that ROS levels in EBs may be regulated by phosphatidylinositiol-3 (PI-3) kinase (Sauer et al., 2000). Collectively, these studies indicate the intracellular redox state influences the activity of signaling cascades that participate in commitment to the cardiac lineage.

6. STEP 4: DIFFERENTIATION OF THE CARDIAC LINEAGE INTO BEATING CARDIOMYOCYTES

Bone morphogenetic proteins (BMPs) have been implicated as potential heart-inducing molecules (Fig. 5) due, in part, to the observation that, in chick, the heart field develops in a region having high BMP signaling and low Wnt signaling (Marvin et al., 2001) and in zebrafish, Swirl/BMP2b mutants have reduced or absent expression of *Nkx2.5* (Kishimoto et al., 1997). However, parsing out the role of BMPs in heart development has proven to be particularly problematic because it has been difficult to separate BMPs' role in heart from their overall role in patterning the early embryo and the early lethality observed in mice possessing homozygous deletions of either the BMP receptor BMPRII, ALK3/BMPR1A, or ALK2/ActRIA prevents the analysis of BMP signaling at later stages (Mishina et al., 1995; Gu et al., 1999; Beppu et al., 2000).

Several groups have shown that the heart field of birds can be expanded by grafting pellets of cells expressing BMP or BMP-soaked

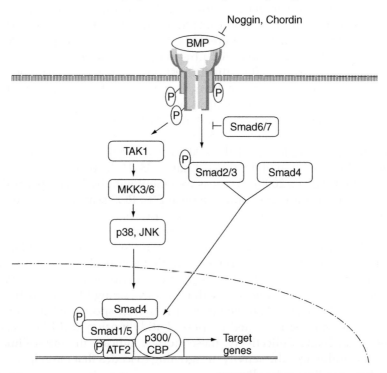

Cardiomyocyte differentiaton by BMP signaling

FIGURE 5   BMP signaling influences cardiomyocyte differentiation. BMPs bind serine/ threonine kinase receptors consisting of the type I receptor (ALK3) and type II receptor (BMPR-II). Activation of receptor serine/threonine kinase activity phosphorylates Smad1/5, which associates with Smad4 in the cytoplasm. Smad1/5-Smad4 complex is translocated to the nucleus where it associates with DNA-binding proteins and transcriptional cofactors (i.e., ATF2) and influences gene expression. Smad6/7 are inhibitory Smads that interfere with BMP signaling by associating with BMP type I receptor, preventing SMAD1/5 activation. TAK1, a mitogen-activated protein kinase (MAPKKK), is activated by BMP and phosphorylates ATF-2 transcription factor. Activated ATF-2 binds directly to heterooligomers of Smad. Natural BMP inhibitors Noggin or Chordin can interfere with BMP signaling by directly interacting with BMP2/4.

beads lateral to the heart field and contracted by similar treatment with the BMP antagonist Noggin. (Schultheiss et al., 1997; Andree et al., 1998; Schlange et al., 2000; Yamada et al., 2000; Nakajima et al., 2002). In addition, isolated explants of the mesendoderm that lies just medial to the normal heart field can turn on heart markers in response to BMP (Schultheiss et al., 1997; Schlange et al., 2000). However, other types of mesoderm that are competent to respond to heart-inducing factors such as the posterior lateral plate mesoderm only form heart tissue in response to BMP when combined with other factors such as FGF (Ladd et al., 1998; Barron et al., 2000), and addition of BMP at early streak stages or prestreak stages of the chick embryo blocks cardiogenesis (Ladd et al., 1998; Matsui et al., 2005). In *Xenopus*, blocking BMP signaling by overexpression of dominant negative forms of downstream effectors, such as dominant negative ALK3 (dnALK3), truncated BMP receptor (tBRII), and the inhibitory SMAD6, does not block the initial induction of cardiac markers but rather prevents progression to a state of terminal differentiation (Shi et al., 2000; Walters et al., 2001). Mice possessing a homozygous deletion of BMP2 (Zhang and Bradley, 1996), BMP4 (Winnier et al., 1995), SMAD5 (Chang et al., 1999), or double mutant for BMP 5/7 (Solloway and Robertson, 1999) specify heart correctly but have later heart defects associated with cardiac morphogenesis. In addition, the specific deletion of the type I serine-threonine kinase receptor ALK3 within the mouse epiblast reveals that BMP signaling is not required for the initial induction of cardiogenic markers but is necessary to maintain cardiac-specific gene expression (Mishina et al., 2002). Together these findings suggest that BMP is required for the maintenance of cardiac tissue but not for its early induction.

Stem cell-based assays also suggest that transient application of BMP at later stages enhances cardiac differentiation. Activation of BMP signaling promotes cardiac differentiation of mouse ES cells as well as P19 cells by upregulating the expression of the mesodermal marker *brachyury* and cardiac transcription factors (Johansson and Wiles, 1995; Monzen et al., 1999, 2001, 2002; Behfar et al., 2002). Furthermore, the chronic inhibition of BMP signaling by Noggin inhibits or significantly attenuates cardiomyocyte differentiation in P19 and mES cells (Monzen et al., 1999, 2001, 2002; Kawai et al., 2004). Administration of BMP to differentiating human EBs, however, did not improve the efficiency of cardiac differentiation (Kehat et al., 2002; Xu et al., 2002b; Mummery et al., 2003) and thus, may point to complications regarding dose, timing, as well as a requirement for other signals or confounding influences of undefined media components, as were discussed as early as 1995 by Johansson et al. In one study, Yuasa et al. (2005) addressed the importance of temporal regulation of BMP signaling during key stages of cardiomyocyte differentiation in mouse ES cells. For example, transient inhibition of BMP signaling by Noggin or Chordin in undifferentiated mouse ES cells, at least 3 days prior to EB formation and in early stage EBs (day 0), robustly induced expression of the cardiac-specific transcription factors *Nkx2.5, GATA4, Tbx5*, and *MEF2C* as well as markers of terminal cardiomyocyte differentiation such as *MLC, αMHC, ANF*, and *sarcomeric actinin*, indicating that BMP signaling must be inhibited during mesoderm formation during initial cardiac induction but then provided later during differentiation for either specification or maintenance of cardiac markers. This regimen is in agreement with the spatiotemporal dyamics of BMP signaling that occurs during mammalian cardiogenesis, as Noggin is expressed transiently in the mouse cardiac crescent (E7.5) and late crescent stage (E8) (anterolateral plate) in an overlapping pattern with Nkx2.5 (Yuasa et al., 2005).

To gain insight into the molecular mechanism by which BMP signaling promotes cardiogenesis, several groups have examined the intracellular signaling pathways situated downstream of the BMP pathway (Fig. 5). Using P19CL6 cells, a clonal derivative of P19, Komuro and colleagues demonstrated that downstream BMP effectors phospho-Smad1/4 acts cooperatively with the TAK1 MAPKKK-ATF2 signal transduction pathway to positively regulate cardiomyocyte differentiation. For instance, administration of recombinant BMP or activation of either Smad1/4 or the TAK1 MAPKKK-ATF2 pathways rescued the Noggin-induced block in P19 cardiomyocyte differentiation (Monzen et al., 1999, 2001, 2002). Results of these studies support a model whereby ATF-2, Smads, and TAK1 cooperate to regulate the expression of cardiac-specific factors. Furthermore, several Smad response elements were identified with the regulatory region of Nkx2.5 upstream of its transcriptional start site and activation by BMP is necessary for early expression of Nkx2.5 (Brown et al., 2004). For a more comprehensive review of complex transcriptional networks that regulate cardiac development, the reader is encouraged to refer to Cripps and Olson (2002), Firulli and Thattaliyath (2002), and Olson (2002).

In addition to evidence described above suggesting a role for FGF functioning as a competence factor in mesoderm induction, data is now accumulating from chick and mouse studies suggesting that FGF may play a specific role in the differentiation of cardiac lineages. Mouse ES cells that lack FGFR1 show defective cardiomyocyte differentiation due to deregulation of mesodermal genes as demonstrated by the persistent upregulation of brachyury and a delay or reduction in the expression of cardiac-specific markers and formation of beating regions within EBs (Dell'Era et al., 2003). This study also addressed the role of downstream effectors of the FGF pathway in the context of cardiac differentiation. For example, chemical activation of protein kinase C (PKC) partially rescues cardiomyocyte differentiation in $FGFR1^{-/-}$ EBs and this rescue was abolished by cotreatment with inhibitors of ERK/MEK1/2. These results suggest that PKC acts upstream of ERK and downstream of FGF in cardiomyocyte differentiation (Dell'Era et al., 2003). As described above, FGF is thought to be a competence factor for mesoderm induction and therefore it is tempting to suggest that the lack of cardiac mesoderm could be a consequence of general failure of mesoderm. However, this cannot be the case since other mesodermal lineages are not affected in FGFR1 mutant EBs. Taken together these observations suggest that FGF signaling plays a more specific role in the development of cardiac lineages. Indeed, embryology data suggests that FGF may cooperate with other factors present in the anterior endoderm to maintain the expression of cardiac markers (Barron et al., 2000).

The presence of FGF8 and the chick FGF receptor, *Cek-1* in the heart field endoderm (Parlow et al., 1991; Sugi et al., 1995) suggests an endodermal role for this factor. This idea is further supported by the findings that the reduction in cardiogenesis caused by endoderm ablation after initial *Nkx2.5* expression could be rescued by administration of FGF8, but not by administration of BMP2 (Alsan and Schultheiss, 2002). Interestingly, FGF8 was only capable of rescuing the loss of cardiac gene expression when placed in a region of the embryo expressing BMPs. Similarly, FGF has been shown to support cardiac differentiation of cardiac mesoderm (Sugi and Lough, 1995) and to induce expression of early cardiac markers in noncardiac mesoderm, but only when acting in combination with either BMP or Activin (Lough et al., 1996; Barron et al., 2000).

In fact, strategies based on coadministration of FGF2 and BMP2 have been utilized to improve the efficiency of cardiomyocyte development in ES cells (Kawai et al., 2004). In differentiating mouse ES cells cultures, addition of both BMP2 and FGF to the culture media enhanced the expression of both early and late cardiac markers, including *Nkx2.5*, *MEF2C*, *GATA4*, and α*MHC* (Kawai et al., 2004). Combination of the two growth factors (treatment at day 0–3) was more effective at promoting the cardiomyocyte differentiation compared to treatment with either factor alone corroborating a synergy between these pathways.

7. CONCLUSIONS AND FUTURE DIRECTIONS

Studies in various vertebrate systems have revealed a remarkable conservation in the temporal program of signaling pathways used to specify cardiomyogenesis, even if the tissue sources of the factors vary depending on developmental anatomy. As might be expected, many of these pathways operate with an analogous time course during ES cell cardiomyogenesis (Fig. 6).

We have characterized cardiomyogenesis as occurring in four steps. In the first, nuclear accumulation of β-catenin, probably mediated by Wnt signaling, is required for the establishment of dorsal organizing center. Second, signals from the dorsal organizing center such as Nodal, in cooperation with FGF, act to induce and pattern mesoderm. Shortly after mesoderm development has been initiated, canonical Wnt signaling must be blocked by antagonists such as Dkk-1 and noncanonical Wnts. At least Dkk-1 acts on endoderm to secrete as yet uncharacterized signals that specify cardiac mesoderm. In the fourth step, a series of signaling molecules, including BMP, possibly noncanonical Wnts, and FGF act on the newly

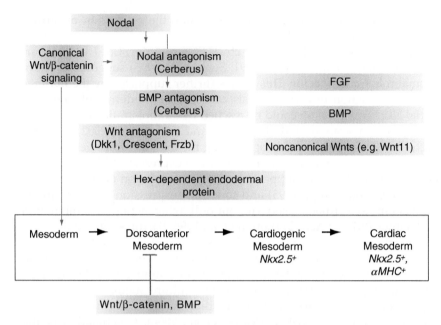

FIGURE 6   Development of the cardiac lineage depends on temporal regulation of key signaling pathways.

specified cardiac mesoderm to maintain expression of cardiac markers and establish patterns of myocardial proliferation and regionalization.

How the various pathways described above interact and integrate signals to influence cardiac development is of great importance not only for an improved understanding of cardiac development but also to realize the promise that ES cells or endogenous cardiac stem cells hold for regenerative therapy through the development of novel, efficient regimens for the controlled differentiation of mature human ventricular cardiomyocytes. Of equal importance, the mechanisms that regulate the processes of myofibrillogenesis, ion channel development, and gap junction formation during cardiac differentiation of ES cells warrant further investigation to ensure transplantation or differentiation of myocytes with normal ventricular function. A combined approach of embryologic and cell-based studies should provide an excellent approach to the study of these problems.

ACKNOWLEDGMENTS

We are grateful to Dr. Shankar Subramaniam, Dr. Mano Maurya, and Frederick Lo (University of California, San Diego) for insights into bioinformatics analyses; and Dr. Ramon Diaz-Trelles (Burnham Institute for Medical Research) for helpful discussions in the preparation of this chapter. This work was supported by NIH grants (R01 HL059502, R01 HL067079, and R01 HL083463) to M.M. and American Heart Association Fellowships to A.C.F. and R.M.G.

REFERENCES

Abe, T., Furue, M., Myoishi, Y., Okamoto, T., Kondow, A., Asashima, M. 2004. Activin-like signaling activates Notch signaling during mesodermal induction. Int. J. Dev. Biol. 48, 327–332.

Adams, R.H., Porras, A., Alonso, G., Jones, M., Vintersten, K., Panelli, S., Valladares, A., Perez, L., Klein, R., Nebreda, A.R. 2000. Essential role of p38alpha MAP kinase in placental but not embryonic cardiovascular development. Mol. Cell 6, 109–116.

Agius, E., Oelgeschlager, M., Wessely, O., Kemp, C., De Robertis, E.M. 2000. Endodermal Nodal-related signals and mesoderm induction in *Xenopus*. Development 127, 1173–1183.

Alexander, J., Stainier, D.Y. 1999. A molecular pathway leading to endoderm formation in zebrafish. Curr. Biol. 9, 1147–1157.

Alsan, B.H., Schultheiss, T.M. 2002. Regulation of avian cardiogenesis by Fgf8 signaling. Development 129, 1935–1943.

Andree, B., Duprez, D., Vorbusch, B., Arnold, H.-H., Brand, T. 1998. BMP-2 induces ectopic expression of cardiac lineage markers and interferes with somite formation in chicken embryos. Mech. Dev. 70, 119–131.

Aouadi, M., Bost, F., Caron, L., Laurent, K., Le Marchand Brustel, Y., Binetruy, B. 2006. p38MAPK activity commits embryonic stem cells to either neurogenesis or cardiomyogenesis. Stem Cells 24(5), 1399–1406.

Arai, A., Yamamoto, K., Toyama, J. 1997a. Murine cardiac progenitor cells require visceral embryonic endoderm and primitive streak for terminal differentiation. Dev. Dyn. 210, 344–353.

Arai, A., Yamamoto, K., Toyama, J. 1997b. Murine cardiac progenitor cells require visceral embryonic endoderm and primitive streak for terminal differentiation. Dev. Dyn. 210, 344–353.

Arnold, S.J., Stappert, J., Bauer, A., Kispert, A., Herrmann, B., Kemler, R. 2000. Brachyury is a target gene of the Wnt/β-catenin signaling pathway. Mech. Dev. 91, 249–258.

Artavanis-Tsakonas, S., Rand, M.D., Lake, R.J. 1999. Notch signaling: Cell fate control and signal integration in development. Science 284, 770–776.

Barron, M., Gao, M., Lough, J. 2000. Requirement for BMP and FGF signaling during cardiogenic induction in non-precardiac mesoderm is specific, transient, and cooperative. Dev. Dyn. 218, 383–393.

Beddington, R.S. 1994. Induction of a second neural axis by the mouse node. Development 120, 613–620.

Behfar, A., Zingman, L.V., Hodgson, D.M., Rauzier, J.M., Kane, G.C., Terzic, A., Puceat, M. 2002. Stem cell differentiation requires a paracrine pathway in the heart. FASEB J. 16, 1558–1566.

Beppu, H., Kawabata, M., Hamamoto, T., Chytil, A., Minowa, O., Noda, T., Miyazono, K. 2000. BMP type II receptor is required for gastrulation and early development of mouse embryos. Dev. Biol. 221, 249–258.

Bertocchini, F., Stern, C.D. 2002. The hypoblast of the chick embryo positions the primitive streak by antagonizing nodal signaling. Dev. Cell 3, 735–744.

Bertocchini, F., Skromne, I., Wolpert, L., Stern, C.D. 2004. Determination of embryonic polarity in a regulative system: Evidence for endogenous inhibitors acting sequentially during primitive streak formation in the chick embryo. Development 131, 3381–3390.

Boheler, K.R., Czyz, J., Tweedie, D., Yang, H.T., Anisimov, S.V., Wobus, A.M. 2002. Differentiation of pluripotent embryonic stem cells into cardiomyocytes. Circ. Res. 91, 189–201.

Bost, F., Caron, L., Marchetti, I., Dani, C., Le Marchand-Brustel, Y., Binetruy, B. 2002. Retinoic acid activation of the ERK pathway is required for embryonic stem cell commitment into the adipocyte lineage. Biochem. J. 361, 621–627.

Bottcher, R.T., Niehrs, C. 2005. Fibroblast growth factor signaling during early vertebrate development. Endocr. Rev. 26, 63–77.

Bourillot, P.Y., Garrett, N., Gurdon, J.B. 2002. A changing morphogen gradient is interpreted by continuous transduction flow. Development 129, 2167–2180.

Bouwmeester, T., Kim, S., Sasai, Y., Lu, B., De Robertis, E.M. 1996. Cerberus is a head-inducing secreted factor expressed in the anterior endoderm of Spemann's organizer. Nature 382, 595–601.

Brott, B.K., Sokol, S.Y. 2005. A vertebrate homolog of the cell cycle regulator Dbf4 is an inhibitor of Wnt signaling required for heart Development. Dev. Cell 8, 703–715.

Brown, C.O.3rd, Chi, X., Garcia-Gras, E., Shirai, M., Feng, X.H., Schwartz, R.J. 2004. The cardiac determination factor, Nkx2-5, is activated by mutual cofactors GATA-4 and Smad1/4 via a novel upstream enhancer. J. Biol. Chem. 279, 10659–10669.

Butler, E. 1935. The developmental capacity of regions of the unincubated chick blastoderm as tested in chorio-allantoic grafts. J. Exp. Zool. 70, 357–389.

Chambers, I., Colby, D., Robertson, M., Nichols, J., Lee, S., Tweedie, S., Smith, A. 2003. Functional expression cloning of Nanog, a pluripotency sustaining factor in embryonic stem cells. Cell 113, 643–655.

Chang, H., Huylebroeck, D., Verschueren, K., Guo, Q., Matzuk, M.M., Zwijsen, A. 1999. Smad5 knockout mice die at mid-gestation due to multiple embryonic and extraembryonic defects. Development 126, 1631–1642.

Chang, H., Zwijsen, A., Vogel, H., Huylebroeck, D., Matzuk, M.M. 2000. Smad5 is essential for left-right asymmetry in mice. Dev. Biol. 219, 71–78.

Chapman, S.C., Schubert, F.R., Schoenwolf, G.C., Lumsden, A. 2003. Anterior identity is established in chick epiblast by hypoblast and anterior definitive endoderm. Development 130, 5091–5101.

Climent, S., Sarasa, M., Villar, J.M., Murillo-Ferrol, N.L. 1995. Neurogenic cells inhibit the differentiation of cardiogenic cells. Dev. Biol. 171, 130–148.

Conlon, F.L., Lyons, F.M., Takaesu, N., Barth, K.S., Kispert, A., Herrmann, B., Robertson, E.J. 1994. A primary requirement for nodal in the formation and maintenance of the primitive streak in the mouse. Development 120, 1919–1928.

Crease, D.J., Dyson, S., Gurdon, J.B. 1998. Cooperation between the activin and Wnt pathways in the spatial control of organizer gene expression. Proc. Natl. Acad. Sci. USA 95, 4398–4403.

Cripps, R.M., Olson, E.N. 2002. Control of cardiac development by an evolutionarily conserved transcriptional network. Dev. Biol. 246, 14–28.

Dale, L., Howes, G., Price, B.M., Smith, J.C. 1992. Bone morphogenetic protein 4: A ventralizing factor in early *Xenopus* development. Development 115, 573–585.

David, N.B., Rosa, F.M. 2001. Cell autonomous commitment to an endodermal fate and behaviour by activation of Nodal signalling. Development 128, 3937–3947.

Davidson, S.M., Morange, M. 2000. Hsp25 and the p38 MAPK pathway are involved in differentiation of cardiomyocytes. Dev. Biol. 218, 146–160.

Davis, R.J. 2000. Signal transduction by the JNK group of MAP kinases. Cell 103, 239–252.

Dell'Era, P., Ronca, R., Coco, L., Nicoli, S., Metra, M., Presta, M. 2003. Fibroblast growth factor receptor-1 is essential for *in vitro* cardiomyocyte development. Circ. Res. 93, 414–420.

Ding, J., Yang, L., Yan, Y.T., Chen, A., Desai, N., Wynshaw-Boris, A., Shen, M.M. 1998. Cripto is required for correct orientation of the anterior-posterior axis in the mouse embryo. Nature 395, 702–707.

Doetschman, T.C., Eistetter, H., Katz, M., Schmidt, W., Kemler, R. 1985. The *in vitro* development of blastocyst-derived embryonic stem cell lines: Formation of visceral yolk sac, blood islands and myocardium. J. Embryol. Exp. Morphol. 87, 27–45.

Dolnikov, K., Shilkrut, M., Zeevi-Levin, N., Gerecht-Nir, S., Amit, M., Danon, A., Itskovitz-Eldor, J., Binah, O. 2006. Functional properties of human embryonic stem cell-derived cardiomyocytes: Intracellular Ca^{2+} handling and the role of sarcoplasmic reticulum in the contraction. Stem Cells 24, 236–245.

Eisenberg, C.A., Eisenberg, L.M. 1999. WNT11 promotes cardiac tissue formation of early mesoderm. Dev. Dyn. 216, 45–58.

Eriksson, M., Leppa, S. 2002. Mitogen-activated protein kinases and activator protein 1 are required for proliferation and cardiomyocyte differentiation of P19 embryonal carcinomal cells. J. Biol. Chem. 277, 15992–16001.

Evans, M., Kaufman, M.H. 1981. Establishment in culture of pluripotential cells from mouse embryos. Nature 292, 154–155.

Faure, S., Lee, M.A., Keller, T., ten Dijke, P., Whitman, M. 2000. Endogenous patterns of TGF-β superfamily signaling during early *Xenopus* development. Development 127, 2917–2931.

Feldman, B., Gates, M.A., Egan, E.S., Dougan, S.T., Rennebeck, G., Sirotkin, H.I., Schier, A.F., Talbot, W.S. 1998. Zebrafish organizer development and germ-layer formation require nodal-related signals. Nature 395, 181–185.

Fijnvandraat, A.C., van Ginneken, A.C., de Boer, P.A., Ruijter, J.M., Christoffels, V.M., Moorman, A.F., Lekanne Deprez, R.H. 2003. Cardiomyocytes derived from embryonic stem cells resemble cardiomyocytes of the embryonic heart tube. Cardiovasc. Res. 58, 399–409.

Firulli, A.B., Thattaliyath, B.D. 2002. Transcription factors in cardiogenesis: The combinations that unlock the mysteries of the heart. Int. Rev. Cytol. 214, 1–62.

Foley, A., Mercola, M. 2004. Heart Induction: Embryology to cardiomyocyte regeneration. Trends cardiovasc. Med. 14(3), 121–125.

Foley, A., Mercola, M. 2005. Heart induction by Wnt antagonists depends on the homeodomain transcription factor Hex. Genes Dev. 19, 387–396.

Foley, A., Korol, O., Timmer, A.M., Mercola, M. 2006. Nodal and Cerberus Coopeerate in Heart Induction. Dev. Biol. 303(1), 57–65.

Foley, A.C., Skromne, I., Stern, C.D. 2000. Reconciling different models of forebrain induction and patterning: A dual role for the hypoblast. Development 127, 3839–3854.

Fullilove, S.L. 1970. Heart induction: Distribution of active factors in newt endoderm. J. Exp. Zool. 175, 323–326.

Gadue, P., Huber, T.L., Nostro, M.C., Kattman, S., Keller, G.M. 2005. Germ layer induction from embryonic stem cells. Exp. Hematol. 33, 955–964.

Gannon, M., Bader, D. 1995. Initiation of cardiac differentiation occurs in the absence of anterior endoderm. Development 121, 2439–2450.

Garcia-Martinez, V., Schoenwolf, G.C. 1993. Primitive-streak origin of the cardiovascular system in avian embryos. Dev. Biol. 159, 706–719.

Goldstein, A.M., Fishman, M.C. 1998. Notochord regulates cardiac lineage in zebrafish embryos. Dev. Biol. 201, 247–252.

Green, J.B.A., Smith, J.C. 1990. Graded changes in dose of a *Xenopus* activin A homologue elicit stepwise transitions in embryonic cell fate. Nature 347, 391–394.

Gridley, T. 1997. Notch signaling in vertebrate development and disease. Mol. Cell Neurosci. 9, 103–108.

Gridley, T. 2003. Notch signaling and inherited disease syndromes. Hum. Mol. Genet. 12(Spec No 1), R9–R13.

Gritsman, K., Zhang, J., Cheng, S., Heckscher, E., Talbot, W.S., Schier, A.F. 1999. The EGF-CFC protein one-eyed pinhead is essential for nodal signaling. Cell 97, 121–132.

Gu, Z., Reynolds, E.M., Song, J., Lei, H., Feijen, A., Yu, L., He, W., MacLaughlin, D.T., van den Eijnden-van Raaij, J., Donahoe, P.K., Li, E. 1999. The type I serine/threonine kinase receptor ActRIA (ALK2) is required for gastrulation of the mouse embryo. Development 126, 2551–2561.

Gurdon, J.B., Bourillot, P.Y. 2001. Morphogen gradient interpretation. Nature 413, 797–803.

Harland, R., Gerhart, J. 1997. Formation and function of Spemann's organizer. Annu. Rev. Cell Dev. Biol. 13, 611–667.

Hashimoto-Partyka, M.K., Yuge, M., Cho, K.W. 2003. Nodal signaling in *Xenopus* gastrulae is cell-autonomous and patterned by beta-catenin. Dev. Biol. 253, 125–138.

Hatada, Y., Stern, C.D. 1994. A fate map of the epiblast of the early chick embryo. Development 120, 2879–2889.

He, J.Q., Ma, Y., Lee, Y., Thomson, J.A., Kamp, T.J. 2003. Human embryonic stem cells develop into multiple types of cardiac myocytes: Action potential characterization. Circ. Res. 93, 32–39.

Heasman, J., Crawford, A., Goldstone, K., Garner-Hamrick, P., Gumbiner, B., McCrea, P., Kintner, C., Noro, C.Y., Wylie, C. 1994. Overexpression of cadherins and underexpression of beta-catenin inhibit dorsal mesoderm induction in early *Xenopus* embryos. Cell 79, 791–803.

Henry, G.L., Brivanlou, I.H., Kessler, D.S., Hemmati-Brivanlou, A., Melton, D.A. 1996. TGF-b signals and a prepattern in *Xenopus laevis* endodermal development. Development 122, 1007–1015.

Hescheler, J., Fleischmann, B.K., Lentini, S., Maltsev, V.A., Rohwedel, J., Wobus, A.M., Addicks, K. 1997. Embryonic stem cells: A model to study structural and functional properties in cardiomyogenesis. Cardiovasc. Res. 36, 149–162.

Huelsken, J., Vogel, R., Brinkmann, V., Erdmann, B., Birchmeier, C., Birchmeier, W. 2000. Requirement for beta-catenin in anterior–posterior axis formation in mice. J. Cell Biol. 148, 567–578.

Jacobson, A.G. 1960. Influences of ectoderm and endoderm on heart differentiation in the newt. Dev. Biol. 2, 138–154.

Jacobson, A.G., Duncan, J.T. 1968. Heart induction in salamanders. J. Exp. Zool. 167, 79–103.

Johansson, B.M., Wiles, M.V. 1995. Evidence for involvement of activin A and bone morphogenetic protein 4 in mammalian mesoderm and hematopoietic development. Mol. Cell Biol. 15, 141–151.

Jones, C.M., Kuehn, M.R., Hogan, B.L., Smith, J.C., Wright, C.V. 1995. Nodal-related signals induce axial mesoderm and dorsalize mesoderm during gastrulation. Development 121, 3651–3662.

Kawai, T., Takahashi, T., Esaki, M., Ushikoshi, H., Nagano, S., Fujiwara, H., Kosai, K. 2004. Efficient cardiomyogenic differentiation of embryonic stem cell by fibroblast growth factor 2 and bone morphogenetic protein 2. Circ. J. 68, 691–702.

Kehat, I., Kenyagin-Karsenti, D., Snir, M., Segev, H., Amit, M., Gepstein, A., Livne, E., Binah, O., Itskovitz-Eldor, J., Gepstein, L. 2001. Human embryonic stem cells can differentiate into myocytes with structural and functional properties of cardiomyocytes. J. Clin. Invest. 108, 407–414.

Kehat, I., Gepstein, A., Spira, A., Itskovitz-Eldor, J., Gepstein, L. 2002. High-resolution electrophysiological assessment of human embryonic stem cell-derived cardiomyocytes: A novel *in vitro* model for the study of conduction. Circ. Res. 91, 659–661.

Keller, R., Tibbetts, P. 1989. Mediolateral cell intercalation in the dorsal, axial mesoderm of *Xenopus laevis*. Dev. Biol. 131, 539–549.

Kelly, G.M., Erezyilmaz, D.F., Moon, R.T. 1995. Induction of a secondary embryonic axis in zebrafish occurs following the overexpression of beta-catenin. Mech. Dev. 53, 261–273.

Kinder, S.J., Tsang, T.E., Quinlan, G.A., Hadjantonakis, A.K., Nagy, A., Tam, P.P. 1999. The orderly allocation of mesodermal cells to the extraembryonic structures and the anteroposterior axis during gastrulation of the mouse embryo. Development 126, 4691–4701.

Kishimoto, Y., Lee, K.-H., Zon, L., Hammerschmidt, M., Schulte-Merker, S. 1997. The molecular nature of zebrafish swirl: BMP2 function is essential during early dorsoventral patterning. Development 124, 4457–4466.

Kispert, A., Svainio, S., Shen, L., Rowitch, D.H., McMahon, A.P. 1996. Proteoglycans are required for maintenance of Wnt-11 expression in the ureter tips. Development 122, 3627.

Klinz, F., Bloch, W., Addicks, K., Hescheler, J. 1999. Inhibition of phosphatidylinositol-3-kinase blocks development of functional embryonic cardiomyocytes. Exp. Cell Res. 247, 79–83.

Kohn, A.D., Moon, R.T. 2005. Wnt and calcium signaling: Beta-Catenin-independent pathways. Cell Calcium 38, 439–446.

Koyanagi, M., Urbich, C., Chavakis, E., Hoffmann, J., Rupp, S., Badorff, C., Zeiher, A.M., Starzinski-Powitz, A., Haendeler, J., Dimmeler, S. 2005. Differentiation of circulating endothelial progenitor cells to a cardiomyogenic phenotype depends on E-cadherin. FEBS Lett. 579, 6060–6066.

Kubo, A., Shinozaki, K., Shannon, J.M., Kouskoff, V., Kennedy, M., Woo, S., Fehling, H.J., Keller, G. 2004. Development of definitive endoderm from embryonic stem cells in culture. Development 131, 1651–1662.

Kuhl, M., Sheldahl, L.C., Park, M., Miller, J.R., Moon, R.T. 2000. The Wnt/Ca^{2+} pathway: A new vertebrate Wnt signaling pathway takes shape. Trends Genet. 16, 279–283.

Kuhl, M., Geis, K., Sheldahl, L.C., Pukrop, T., Moon, R.T., Wedlich, D. 2001. Antagonistic regulation of convergent extension movements in *Xenopus* by Wnt/beta-catenin and Wnt/Ca2 + signaling. Mech. Dev. 106, 61–76.

Ladd, A., Yatskievych, T.A., Antin, P.B. 1998. Regulation of avian cardiac myogenesis by activin/TGFβ and bone morphogenetic proteins. Dev. Biol. 204, 407–419.

Larabell, C.A., Torres, M., Rowning, B.A., Yost, C., Miller, J.R., Wu, M., Kimelman, D., Moon, R.T. 1997. Establishment of the dorso-ventral axis in *Xenopus* embryos is presaged by early asymmetries in beta-catenin that are modulated by the Wnt signaling pathway. J. Cell Biol. 136, 1123–1136.

Lawson, K.A., Pedersen, R.A. 1987. Cell fate, morphogenetic movement and population kinetics of embryonic endoderm at the time of germ layer formation in the mouse. Development 101, 627–652.

Lawson, K.A., Meneses, J.J., Pedersen, R.A. 1986. Cell fate and cell lineage in the endoderm of the presomite mouse embryo, studied with an intracellular tracer. Dev. Biol. 115, 325–339.

Lee, M.A., Heasman, J., Whitman, M. 2001. Timing of endogenous activin-like signals and regional specification of the *Xenopus* embryo. Development 128, 2939–2952.

Li, G., Iyengar, R. 2002. Calpain as an effector of the Gq signaling pathway for inhibition of Wnt/beta-catenin-regulated cell proliferation. Proc. Natl. Acad. Sci. USA 99, 13254–13259.

Li, J., Stouffs, M., Serrander, L., Banfi, B., Bettiol, E., Charnay, Y., Steger, K., Krause, K.H., Jaconi, M.E. 2006. The NADPH oxidase NOX4 drives cardiac differentiation: Role in regulating cardiac transcription factors and MAP kinase activation. Mol. Biol. Cell. 17 (9), 3978–3988.

Lickert, H., Kutsch, S., Kanzler, B., Tamai, Y., Taketo, M.M., Kemler, R. 2002. Formation of multiple hearts in mice following deletion of beta-catenin in the embryonic endoderm. Dev. Cell 3, 171–181.

Liguori, G.L., Echevarria, D., Improta, R., Signore, M., Adamson, E., Martinez, S., Persico, M.G. 2003. Anterior neural plate regionalization in cripto null mutant mouse embryos in the absence of node and primitive streak. Dev. Biol. 264, 537–549.

Logan, M., Mohun, T. 1993. Induction of cardiac muscle differentiation in isolated animal pole explants of *Xenopus laevis* embryos. Development 118, 865–875.

Loomes, K.M., Underkoffler, L.A., Morabito, J., Gottlieb, S., Piccoli, D.A., Spinner, N.B., Baldwin, H.S., Oakey, R.J. 1999. The expression of Jagged1 in the developing mammalian heart correlates with cardiovascular disease in Alagille syndrome. Hum. Mol. Genet. 8, 2443–2449.

Loomes, K.M., Taichman, D.B., Glover, C.L., Williams, P.T., Markowitz, J.E., Piccoli, D.A., Baldwin, H.S., Oakey, R.J. 2002. Characterization of Notch receptor expression in the developing mammalian heart and liver. Am. J. Med. Genet. 112, 181–189.

Lough, J., Barron, M., Brogley, M., Sugi, Y., Bolender, D.L., Xhu, X. 1996. Combined BMP-2 and FGF-4, but neither factor alone, induces cardiogenesis in non-precardiac embryonic mesoderm. Dev. Biol. 178, 198–202.

Lowe, L.A., Yamada, S., Kuehn, M.R. 2001. Genetic dissection of nodal function in patterning the mouse embryo. Development 128, 1831–1843.

Mangiacapra, F.J., Fransen, M.E., Lemanski, L.F. 1995. Activin A and transforming growth factor-beta stimulate heart formation in axolotls but not rescue cardiac lethal mutants. Cell Tissue Res. 282, 227–236.

Martin, G.R. 1981. Isolation of a pluripotent cell line from early mouse embryos cultured in medium conditioned by teratocarcinoma stem cells. Proc. Natl. Acad. Sci. USA 78, 7634–7638.

Marvin, M.J., Di Rocco, G., Gardiner, A., Bush, S.M., Lassar, A.B. 2001. Inhibition of Wnt activity induces heart formation from posterior mesoderm. Genes Dev. 15, 316–327.

Matsui, H., Ikeda, K., Nakatani, K., Sakabe, M., Yamagishi, T., Nakanishi, T., Nakajima, Y. 2005. Induction of initial cardiomyocyte alpha-actin-smooth muscle alpha-actin-in cultured avian pregastrula epiblast: A role for nodal and BMP antagonist. Dev. Dyn. 233, 1419–1429.

Maye, P., Zheng, J., Li, L., Wu, D. 2004. Multiple mechanisms for Wnt11-mediated repression of the canonical Wnt signaling pathway. J. Biol. Chem. 279, 24659–24665.

McBurney, M.W., Jones-Villeneuve, E.M., Edwards, M.K., Anderson, P.J. 1982. Control of muscle and neuronal differentiation in a cultured embryonal carcinoma cell line. Nature 299, 165–167.

Meno, C., Gritsman, K., Ohishi, S., Ohfuji, Y., Heckscher, E., Mochida, K., Shimono, A., Kondoh, H., Talbot, W.S., Robertson, E.J., Schier, A.F., Hamada, H. 1999. Mouse Lefty2 and zebrafish antivin are feedback inhibitors of nodal signaling during vertebrate gastrulation. Mol. Cell 4, 287–298.

Minchiotti, G. 2005. Nodal-dependant Cripto signaling in ES cells: From stem cells to tumor biology. Oncogene 24, 5668–5675.

Minchiotti, G., Parisi, S., Liguori, G.L., D'Andrea, D., Persico, M.G. 2002. Role of the EGF-CFC gene cripto in cell differentiation and embryo development. Gene 287, 33–37.

Mishina, Y., Suzuki, A., Ueno, N., Behringer, R.R. 1995. Bmpr encodes a type I bone morphogenetic protein receptor that is essential for gastrulation during mouse embryogenesis. Genes Dev. 9, 3027–3037.

Mishina, Y., Hanks, M.C., Miura, S., Tallquist, M.D., Behringer, R.R. 2002. Generation of Bmpr/Alk3 conditional knockout mice. Genesis 32, 69–72.

Mlodzik, M. 2002. Planar cell polarization: Do the same mechanisms regulate *Drosophila* tissue polarity and vertebrate gastrulation?Trends Genet. 18, 564–571.

Monzen, K., Shiojima, I., Hiroi, Y., Kudoh, S., Oka, T., Takimoto, E., Hayashi, D., Hosoda, T., Habara-Ohkubo, A., Nakaoka, T., Fujita, T., Yazaki, Y., et al. 1999. Bone morphogenetic proteins induce cardiomyocyte differentiation through the mitogen-activated protein kinase kinase kinase TAK1 and cardiac transcription factors Csx/Nkx2.5 and Gata-4. Mol. Cell. Biol. 19(10), 7096–7105.

Monzen, K., Hiroi, Y., Kudoh, S., Akazawa, H., Oka, T., Takimoto, E., Hayashi, D., Hosoda, T., Kawabata, M., Miyazono, K., Ishii, S., Yazaki, Y., et al. 2001. Smads, TAK1 and their common target ATF-2 play a critical role in cardiomyocyte differentiation. J. Cell Biol. 153, 687–698.

Monzen, K., Nagai, R., Komuro, I. 2002. A role for bone morphogenetic protein signaling in cardiomyocyte differentiation. Trends Cardiovasc. Med. 12, 263–269.

Mummery, C., Ward-van Oostwaard, D., Doevendans, P., Spijker, R., van den Brink, S., Hassink, R., van der Heyden, M., Opthof, T., Pera, M., de la Riviere, A.B., Passier, R.,

Tertoolen, L. 2003. Differentiation of human embryonic stem cells to cardiomyocytes: Role of coculture with visceral endoderm-like cells. Circulation 107, 2733–2740.

Mummery, C.L., van Achterberg, T.A., van den Eijnden-van Raaij, A.J., van Haaster, L., Willemse, A., de Laat, S.W., Piersma, A.H. 1991. Visceral-endoderm-like cell lines induce differentiaiton of murine P19 embryonal carcinoma cells. Differentiation 46, 51–60.

Myat, A., Henrique, D., Ish-Horowicz, D., Lewis, J. 1996. A chick homologue of Serrate and its relationship with Notch and Delta homologues during central neurogenesis. Dev. Biol. 174, 233–247.

Naito, A.T., Tominaga, A., Oyamada, M., Oyamada, Y., Shiraishi, I., Monzen, K., Komuro, I., Takamatsu, T. 2003. Early stage-specific inhibitions of cardiomyocyte differentiation and expression of Csx/Nkx-2.5 and GATA-4 by phosphatidylinositol 3-kinase inhibitor LY294002. Exp. Cell Res. 291, 56–69.

Naito, A.T., Akazawa, H., Takano, H., Minamino, T., Nagai, T., Aburatani, H., Komuro, I. 2005. Phosphatidylinositol 3-kinase-Akt pathway plays a critical role in early cardiomyogenesis by regulating canonical Wnt signaling. Circ. Res. 97, 144–151.

Nakajima, Y., Yamagishi, T., Ando, K., Nakamura, H. 2002. Significance of bone morphogenetic protein-4 function in the initial myofibrillogenesis of chick cardiogenesis. Dev. Biol. 245, 291–303.

Nakamura, T., Sano, M., Songyang, Z., Schneider, M.D. 2003. A Wnt- and beta-catenin-dependent pathway for mammalian cardiac myogenesis. Proc. Natl. Acad. Sci. USA 100, 5834–5839.

Nascone, N., Mercola, M. 1995. An inductive role for the endoderm in *Xenopus* cardiogenesis. Development 121, 515–523.

Nebreda, A.R., Porras, A. 2000. p38 MAP kinases: Beyond the stress response. Trends Biochem. Sci. 25, 257–260.

Nemir, M., Croquelois, A., Pedrazzini, T., Radtke, F. 2006. Induction of cardiogenesis in embryonic stem cells via downregulation of Notch1 signaling. Circ. Res. 98, 1471–1478.

Ng, E.S., Azzola, L., Sourris, K., Robb, L., Stanley, E.G., Elefanty, A.G. 2005. The primitive streak gene Mixl1 is required for efficient haematopoiesis and BMP4-induced ventral mesoderm patterning in differentiating ES cells. Development 132, 873–884.

Nishita, M., Hashimoto, M.K., Ogata, S., Laurent, M.N., Ueno, N., Shibuya, H., Cho, K.W. 2000. Interaction between Wnt and TGF-beta signalling pathways during formation of Spemann's organizer. Nature 403, 781–785.

Niwa, H., Burdon, T., Chambers, I., Smith, A. 1998. Self-renewal of pluripotent embryonic stem cells is mediated via activation of STAT3. Genes Dev. 12, 2048–2060.

Norris, D.P., Brennan, J., Bikoff, E.K., Robertson, E.J. 2002. The Foxh1-dependent autoregulatory enhancer controls the level of Nodal signals in the mouse embryo. Development 129, 3455–3468.

Oka, C., Nakano, T., Wakeham, A., de la Pompa, J.L., Mori, C., Sakai, T., Okazaki, S., Kawaichi, M., Shiota, K., Mak, T.W., Honjo, T. 1995. Disruption of the mouse RBP-J kappa gene results in early embryonic death. Development 121, 3291–3301.

Olivo, O.M. 1928. Précoce Détermination de l'Ébauche du Coeur dans l'Embryon de Poulet et Sa Différenciation Histologique et Physiologique *in vitro*. Comptes Rendus Association Anatomique de Prague 23, 357–374.

Olson, E.N. 2002. A genetic blueprint for growth and development of the heart. Harvey Lect 98, 41–64.

Osada, S.I., Wright, C.V. 1999. *Xenopus* nodal-related signaling is essential for mesendodermal patterning during early embryogenesis. Development 126, 3229–3240.

Pandur, P., Lasche, M., Eisenberg, L.M., Kuhl, M. 2002. Wnt-11 activation of a non-canonical Wnt signalling pathway is required for cardiogenesis. Nature 418, 636–641.

Parish, C.L., Parisi, S., Persico, M.G., Arenas, E., Minchiotti, G. 2005. Cripto as a target for improving embryonic stem cell-based therapy in Parkinson's disease. Stem Cells 23, 471–476.

Parisi, S., D'Andrea, D., Lago, C.T., Adamson, E.D., Persico, M.G., Minchiotti, G. 2003. Nodal-dependent Cripto signaling promotes cardiomyogenesis and redirects the neural fate of embryonic stem cells. J. Cell Biol. 163, 303–314.

Parlow, M.H., Bolender, D.L., Kokan-Moore, N.P., Lough, J. 1991. Localization of bFGF-like proteins as punctate inclusions in the preseptation myocardium of the chicken embryo. Dev. Biol. 146, 139–147.

Passier, R., Oostwaard, D.W., Snapper, J., Kloots, J., Hassink, R.J., Kuijk, E., Roelen, B., de la Riviere, A.B., Mummery, C. 2005. Increased cardiomyocyte differentiation from human embryonic stem cells in serum-free cultures. Stem Cells 23, 772–780.

Pfendler, K.C., Catuar, C.S., Meneses, J.J., Pederson, R.A. 2005. Overexpression of Nodal promotes differentiation of mouse embryonic stem cells into mesoderm and endoderm at the expense of neuroectoderm formation. Stem Cells Dev. 14, 162–172.

Piccolo, S., Agius, E., Leyns, L., Bhattacharyya, S., Grunz, H., Bouwmeester, T., De Robertis, E.M. 1999. The head inducer Cerberus is a multifunctional antagonist of Nodal, BMP and Wnt signals. Nature 397, 707–710.

Psychoyos, D., Stern, C.D. 1996. Fates and migratory routes of primitive streak cells in the chick embryo. Development 122, 1523–1534.

Raffin, M., Leong, L.M., Rones, M.S., Sparrow, D., Mohun, T., Mercola, M. 2000. Subdivision of the cardiac Nkx2.5 expression domain into myogenic and non-myogenic compartments. Dev. Biol. 218, 326–340.

Rawles, M.E. 1936. A study in the localization of organ-forming areas in the chick blastoderm of the head process stage. J. Exp. Zool. 138, 505–555.

Reissmann, E., Jornvall, H., Blokzijl, A., Andersson, O., Chang, C., Minchiotti, G., Persico, M. G., Ibanez, C.F., Brivanlou, A.H. 2001. The orphan receptor ALK7 and the Activin receptor ALK4 mediate signaling by Nodal proteins during vertebrate development. Genes Dev. 15, 2010–2022.

Reiter, J.F., Verkade, H., Stainier, D.Y. 2001. Bmp2b and Oep promote early myocardial differentiation through their regulation of gata5. Dev. Biol. 234, 330–338.

Reppel, M., Pillekamp, F., Brockmeier, K., Matzkies, M., Bekcioglu, A., Lipke, T., Nguemo, F., Bonnemeier, H., Hescheler, J. 2005. The electrocardiogram of human embryonic stem cell-derived cardiomyocytes. J. Electrocardiol. 38, 166–170.

Rones, M.S., McLaughlin, K.A., Raffin, M., Mercola, M. 2000. Serrate and Notch specify cell fates in the heart field by suppressing cardiomyogenesis. Development 127, 3865–3876.

Rosenquist, G.C., DeHaan, R.L. 1966. Migration of precardiac cells in the chick embryo: A radioautographic study. Carnegie Institute of Washington Publication 625, Contributions to Embryology 263, 113–121.

Rudnick, D. 1938. Differentiation of culture of pieces of early chick blastoderm. II short primitive streak stages. J. Exp. Zool. 79, 399–425.

Rudnick, D. 1944. Early history and mechanics of the chick blastoderm. Q. Rev. Bio. 19, 187–212.

Rudy-Reil, D., Lough, J. 2004. Avian precardiac endoderm/mesoderm induces cardiac myocyte differentiation in murine embryonic stem Cell. Circ. Res. 94, 107–116.

Rutenberg, J.B., Fischer, A., Jia, H., Gessler, M., Mercola, M. 2006. Developmental patterning of the cardiac atrioventricular canal by Notch and Hairy-related transcription factors. Development 133(21), 4381–4390.

Sater, A.K., Jacobson, A.G. 1989. The specification of heart mesoderm occurs during gastrulation in *Xenopus laevis*. Development 105, 821–830.

Sater, A.K., Jacobson, A.G. 1990. The role of the dorsal lip in the induction of heart mesoderm in *Xenopus laevis*. Development 108, 461–470.

Sauer, H., Rahimi, G., Hescheler, J., Wartenberg, M. 2000. Role of reactive oxygen species and phosphatidylinositol 3-kinase in cardiomyocyte differentiation of embryonic stem cells. FEBS Lett. 476, 218–223.

Schier, A.F., Shen, M.M. 2000. Nodal signalling in vertebrate development. Nature 403, 385–389.

Schlange, T., Andree, B., Arnold, H.H., Brand, T. 2000. BMP2 is required for early heart development during a distinct time period. Mech. Dev. 91, 259–270.

Schmelter, M., Ateghang, B., Helmig, S., Wartenberg, M., Sauer, H. 2006. Embryonic stem cells utilize reactive oxygen species as transducers of mechanical strain-induced cardiovascular differentiation. FASEB J. 20, 1182–1184.

Schneider, V.A., Mercola, M. 1999. Spatially distinct head and heart inducers within the *Xenopus* organizer region. Curr. Biol. 9, 800–809.

Schneider, V.A., Mercola, M. 2001. Wnt antagonism initiates cardiogenesis in *Xenopus laevis*. Genes Dev. 15, 304–315.

Schroeder, T., Fraser, S.T., Ogawa, M., Nishikawa, S., Oka, C., Bornkamm, G.W., Honjo, T., Just, U. 2003. Recombination signal sequence-binding protein Jkappa alters mesodermal cell fate decisions by suppressing cardiomyogenesis. Proc. Natl. Acad. Sci. USA 100, 4018–4023.

Schroeder, T., Meier-Stiegen, F., Schwanbeck, R., Eilken, H., Nishikawa, S., Hasler, R., Schreiber, S., Bornkamm, G.W., Nishikawa, S.I., Just, U. 2006. Activated Notch1 alters differentiation of embryonic stem cells into mesodermal cell lineages at multiple stages of development. Mech. Dev. 123, 570–579.

Schultheiss, T.M., Xydas, S., Lassar, A.B. 1995. Induction of avian cardiac myogenesis by anterior endoderm. Development 121, 4203–4214.

Schultheiss, T.M., Burch, J.B., Lassar, A.B. 1997. A role for bone morphogenetic proteins in the induction of cardiac myogenesis. Genes Dev. 11, 451–462.

Serbedzija, G.N., Chen, J.N., Fishman, M.C. 1998. Regulation in the heart field of zebrafish. Development 125, 1095–1101.

Shi, Y., Katsev, S., Cai, C., Evans, S. 2000. BMP signaling is required for heart formation in vertebrates. Dev. Biol. 224, 226–237.

Smith, A.G., Heath, J.K., Donaldson, D.D., Wong, G.G., Moreau, J., Stahl, M., Rogers, D. 1988. Inhibition of pluripotential embryonic stem cell differentiation by purified polypeptides. Nature 336, 688–690.

Smith, J.C., Price, B.M., Van Nimmen, K., Huylebroeck, D. 1990. Identification of a potent *Xenopus* mesoderm-inducing factor as a homologue of activin A. Nature 345, 729–731.

Solloway, M.J., Robertson, E.J. 1999. Early embryonic lethality in Bmp5;Bmp7 double mutant mice suggests functional redundancy within the 60A subgroup. Development 126, 1753–1768.

Spemann, H., Mangold, H. 1924. Über Induktion von Embryonalanlagen durch Implantation artfremder Organisatoren. Arch. mikr. Entwicklungsmech. Org. 100, 599–638.

Spratt, N.T. 1942. Location of organ-specific regions and their relationship to the development of the primitve streak in the early chick blastoderm. J. Exp. Zool. 114, 69–101.

Strizzi, L., Bianco, C., Normanno, N., Salomon, D. 2005. Cripto-1: A multifunctional modulator during embryogenesis and oncogenesis. Oncogene 24, 5731–5741.

Sugi, Y., Lough, J. 1994. Anterior endoderm is a specific effector of terminal cardiac myocyte differentiation of cells from the embryonic heart forming region. Dev. Dyn. 200, 155–162.

Sugi, Y., Lough, J. 1995. Activin-A and FGF-2 mimic the inductive effects of anterior endoderm on terminal cardiac myogenesis *in vitro*. Dev. Biol. 168, 567–574.

Sugi, Y., Sasse, J., Barron, M., Lough, J. 1995. Developmental expression of fibroblast growth factor receptor-1 (cek-1; flg) during heart development. Dev. Dyn. 202, 115–125.

Symes, K., Yordán, C., Mercola, M. 1994. Morphological differences in *Xenopus* embryonic mesodermal cells are specified as an early response to distinct threshold concentrations of activin. Development 120, 2339–2346.

Tada, S., Era, T., Furusawa, C., Sakurai, H., Nishikawa, S., Kinoshita, M., Nakao, K., Chiba, T. 2005. Characterization of mesendoderm: A diverging point of the definitive endoderm and mesoderm in embryonic stem cell differentiation culture. Development 132, 4363–4374.

Takahashi, S., Yokota, C., Takano, K., Tanegashima, K., Onuma, Y., Goto, J., Asashima, M. 2000. Two novel nodal-related genes initiate early inductive events in *Xenopus* Nieuwkoop center. Development 127, 5319–5329.

Tam, P.P., Behringer, R.R. 1997. Mouse gastrulation: The formation of a mammalian body plan. Mech. Dev. 68, 3–25.

Tam, P.P., Parameswaran, M., Kinder, S.J., Weinberger, R.P. 1997. The allocation of epiblast cells to the embryonic heart and other mesodermal lineages: The role of ingression and tissue movement during gastrulation. Development 124, 1631–1642.

Tamura, K., Sudo, T., Senftleben, U., Dadak, A.M., Johnson, R., Karin, M. 2000. Requirement for p38alpha in erythropoietin expression: A role for stress kinases in erythropoiesis. Cell 102, 221–231.

Tao, Q., Yokota, C., Puck, H., Kofron, M., Birsoy, B., Yan, D., Asashima, M., Wylie, C.C., Lin, X., Heasman, J. 2005. Maternal wnt11 activates the canonical wnt signaling pathway required for axis formation in *Xenopus* embryos. Cell 120, 857–871.

Terami, H., Hidaka, K., Katsumata, T., Iio, A., Morisaki, T. 2004. Wnt11 facilitates embryonic stem cell differentiation to Nkx2.5-positive cardiomyocytes. Biochem. Biophys. Res. Commun. 325, 968–975.

Thom, T., Haase, N., Rosamond, W., Howard, V.J., Rumsfeld, J., Manolio, T., Zheng, Z.J., Flegal, K., O'Donnell, C., Kittner, S., Lloyd-Jones, D., Goff, D.C., Jr., et al. 2006. Heart disease and stroke statistics-2006 update: A report from the American Heart Association Statistics Committee and Stroke Statistics Subcommittee. Circulation 113, 85–151.

Thomsen, G., Woolf, T., Whitman, M., Sokol, S., Vaughan, J., Vale, W., Melton, D.A. 1990. Activins are expressed early in *Xenopus* embryogenesis and can induce axial mesoderm and anterior structures. Cell 63, 485–493.

Thomson, J.A., Itskovitz-Eldor, J., Shapiro, S.S., Waknitz, M.A., Swiergiel, J.J., Marshall, V.S., Jones, J.M. 1998. Embryonic stem cell lines derived from human blastocysts. Science 282, 1145–1147.

Toyama, R., O'Connell, M.L., Wright, C.V., Kuehn, M.R., Dawid, I.B. 1995. Nodal induces ectopic goosecoid and lim1 expression and axis duplication in zebrafish. Development 121, 383–391.

Tzahor, E., Lassar, A.B. 2001a. Wnt signals from the neural tube block ectopic cardiogenesis. Genes Dev. 15, 255–260.

Tzahor, E., Lassar, A.B. 2001b. Wnt signals from the neural tube block ectopic cardiogenesis. Genes Dev. 15, 255–260.

van den Eijnden-van Raaij, A.J.M., van Zoelent, E.J.J., van Nimmen, K., Koster, C.H., Snoek, G.T., Durston, A.J., Huylebroeck, D. 1990. Activin-like factor from a *Xenopus laevis* cell line responsible for mesoderm induction. Nature 345, 819–822.

Vanhaesebroeck, B., Ali, K., Bilancio, A., Geering, B., Foukas, L.C. 2005. Signalling by PI3K isoforms: Insights from gene-targeted mice. Trends Biochem. Sci. 30, 194–204.

Waddington, C.H. 1932. Experiments on the development of chick and duck embryos, cultivated *in vitro*. Philos. Trans. R. Soc. Lond. B 221, 179–230.

Walters, M.J., Wayman, G.A., Christian, J.L. 2001. Bone morphogenetic protein function is required for terminal differentiation of the heart but not for early expression of cardiac marker genes. Mech. Dev. 100, 263–273.

Wei, H., Juhasz, O., Li, J., Tarasova, Y.S., Boheler, K.R. 2005. Embryonic stem cells and cardiomyocyte differentiation: Phenotypic and molecular analyses. J. Cell. Mol. Med. 9, 804–817.

Westin, J., Lardelli, M. 1997. Three novel Notch genes in zebrafish: Implications for vertebrate Notch gene evolution and function. Dev. Genes Evol. 207, 51–63.

Whitman, M. 2001. Nodal signaling in early vertebrate embryos: Themes and variations. Dev. Cell 1, 605–617.

Wilkinson, D.G., Bhatt, S., Herrmann, B.G. 1990. Expression pattern of the mouse T gene and its role in mesoderm formation. Nature 343, 657–659.

Winnier, G., Blessing, M., Labosky, P.A., Hogan, B.L. 1995. Bone morphogenetic protein-4 is required for mesoderm formation and patterning in the mouse. Genes Dev. 9, 2105–2116.

Wobus, A.M., Boheler, K.R. 2005. Embryonic stem cells: Prospects for developmental biology and cell therapy. Physiol. Rev. 85, 635–678.

Wylie, C., Kofron, M., Payne, C., Anderson, R., Hosobuchi, M., Joseph, E., Heasman, J. 1996. Maternal beta-catenin establishes a 'dorsal signal' in early *Xenopus* embryos. Development 122, 2987–2996.

Xanthos, J.B., Kofron, M., Tao, Q., Schaible, K., Wylie, C., Heasman, J. 2002. The roles of three signaling pathways in the formation and function of the Spemann Organizer. Development 129, 4027–4043.

Xu, C., Liguori, G., Persico, M.G., Adamson, E.D. 1999. Abrogation of the Cripto gene in mouse leads to failure of postgastrulation morphogenesis and lack of differentiation of cardiomyocytes. Development 126, 483–494.

Xu, C., Police, S., Rao, N., Carpenter, M.K. 2002a. Characterization and enrichment of cardiomyocytes derived from human embryonic stem cells. Circ. Res. 91, 501–508.

Xu, R.H., Chen, X., Li, D.S., Li, R., Addicks, G.C., Glennon, C., Zwaka, T.P., Thomson, J.A. 2002b. BMP4 initiates human embryonic stem cell differentiation to trophoblast. Nat. Biotechnol. 20, 1261–1264.

Yamada, M., Revelli, J.P., Eichele, G., Barron, M., Schwartz, R.J. 2000. Expression of chick Tbx-2, Tbx-3, and Tbx-5 genes during early heart development: Evidence for BMP2 induction of Tbx2. Dev. Biol. 228, 95–105.

Yamamoto, S., Hikasa, H., Ono, H., Taira, M. 2003. Molecular link in the sequential induction of the Spemann organizer: Direct activation of the cerberus gene by Xlim-1, Xotx2, Mix.1, and Siamois, immediately downstream from Nodal and Wnt signaling. Dev. Biol. 257, 190–204.

Yamashita, J.K., Takano, M., Hiraoka-Kanie, M., Shimazu, C., Peishi, Y., Yanagi, K., Nakano, A., Inoue, E., Kita, F., Nishikawa, S. 2005. Prospective identification of cardiac progenitors by a novel single cell-based cardiomyocyte induction. FASEB J. 19, 1534–1536.

Yao, J., Kessler, D.S. 2001. Goosecoid promotes head organizer activity by direct repression of Xwnt8 in Spemann's organizer. Development 128, 2975–2987.

Yatskievych, T., Ladd, A., Antin, P. 1997. Induction of cardiac myogenesis in avian pregastrula epiblast: The role of the hypoblast and activin. Development 124, 2561–2570.

Yuasa, S., Itabashi, Y., Koshimizu, U., Tanaka, T., Sugimura, K., Kinoshita, M., Hattori, F., Fukami, S., Shimazaki, T., Okano, H., Ogawa, S., Fukuda, K. 2005. Transient inhibition of BMP signaling by Noggin induces cardiomyocyte differentiation of mouse embryonic stem cells. Nat. Biotechnol. 23(5), 607–611.

Yutzey, K., Gannon, M., Bader, D. 1995. Diversification of cardiomyogenic cell lineages *in vitro*. Dev. Biol. 170, 531–541.

Zhang, H., Bradley, A. 1996. Mice deficeint for BMP2 are nonviable and have defects in amnion/chorion and cardiac development. Development 122, 2977–2986.

Zhou, X., Sasaki, H., Lowe, L., Hogan, B.L., Kuehn, M.R. 1993. Nodal is a novel TGF-beta like gene expressed in mouse node during gastrulation. Nature 361, 543–547.

Zorn, A.M., Butler, K., Gurdon, J.B. 1999. Anterior endomesoderm specification in *Xenopus* by Wnt/beta-catenin and TGF-beta signalling pathways. Dev. Biol. 209, 282–297.

CHAPTER 6

Islet1 Progenitors in Developing and Postnatal Heart

Yunfu Sun, Xingqun Liang, and **Sylvia M. Evans**

Contents

Abstract

The LIM-homeodomain transcription factor Islet1 (Isl1) is a marker of pluripotential undifferentiated cardiovascular progenitors that will give rise to myocytes, endothelial cells, and smooth muscle cells within the heart and the vasculature. During development, Isl1 is actively expressed in cardiac progenitors prior to their differentiation. Isl1 function is also required for normal heart development, as Isl1 mutant hearts are completely lacking the outflow tract, right ventricle, and have severely reduced atrial tissue. Isl1 is expressed both in

Skaggs School of Pharmacy and Pharmaceutical Sciences, University of California, San Diego, La Jolla, California

Advances in Developmental Biology, Volume 18
ISSN 1574-3349, DOI: 10.1016/S1574-3349(07)18006-6

cardiogenic mesoderm and in endoderm, the latter a tissue that is required for heart induction. It will also be important to understand tissue-specific roles for Isl1 in this context. The early expression of Isl1 within multipotential cardiac progenitors and its essential role in heart development suggest that understanding factors that are both upstream and downstream of Isl1 will give critical insights into genetic pathways required for specification of distinct cardiovascular lineages and for normal heart development.

1. INTRODUCTION

Congenital heart disease is the most common birth defect, affecting one in one hundred live births (Olson and Schneider, 2003). To understand root causes of congenital heart disease, it is critical to gain insight into pathways underlying development of the heart. Developmental pathways may also come into play during cardiac repair or regeneration, through activation of intrinsic or extrinsic progenitor cells with cardiovascular potential. In recent years, new understanding of cardiac progenitor populations has emerged, which has been comprehensively treated in several review articles (Abu-Issa et al., 2004; Buckingham et al., 2005). In this chapter, we will focus on the role of Islet1 (Isl1), a LIM-homeodomain transcription factor, in cardiac development and as a marker for cardiovascular progenitor populations in both developing and postnatal heart.

Early microscopic studies had described development of the heart from cardiogenic plate mesoderm, with continuous transformation of splanchnic mesoderm into myocardium at arterial and venous poles of the heart (Viragh and Challice, 1973; deVries, 1981). Radiolabel lineage tracing in chick embryos by de la Cruz et al. (1977) demonstrated that in the early linear heart tube, outflow tract progenitors were not yet present, suggesting a later migration of these progenitors into the forming heart. More DiI labeling studies in chick and mouse embryos indicated that some outflow tract progenitors were present in pharyngeal mesoderm, dorsal and anterior to the forming heart (Kelly et al., 2001; Mjaatvedt et al., 2001; Waldo et al., 2001). Outflow tract progenitors in this region were deemed to the secondary or anterior heart field, each of these defined by specific regions of the pharyngeal mesoderm which were labeled in respective studies (Abu-Issa et al., 2004). Expression of an Fgf10-lacZ transgene in mouse embryos suggested that cells in the right ventricle might also originate in the anterior heart field (Kelly and Buckingham, 2002). How extensive the contribution of the secondary/ anterior heart field to the outflow tract, and/or right ventricle, and the overall location of these progenitors remained unknown.

2. THE LIM-HOMEODOMAIN TRANSCRIPTION FACTOR ISL1 MARKS CARDIAC PROGENITORS AND IS REQUIRED FOR DEVELOPMENT OF THE EARLY HEART TUBE

Studies on chick embryos had described expression of the LIM-homeodomain transcription factor Isl1 in cardiogenic mesoderm (Yuan and Schoenwolf, 2000). Initial analysis of Isl1 null mice had demonstrated an essential role in motor neuron and pancreatic development (Pfaff et al., 1996; Ahlgren et al., 1997), with null mutants dying at E9.5. Although the cause of death was speculated as being cardiovascular in origin (Pfaff et al., 1996), the hearts of Isl1 null mice had not been examined. Morphological and marker analysis revealed that hearts of Isl1 mutants were missing the outflow tract, right ventricle, and had severely reduced amounts of atrial tissue (Cai et al., 2003). Expression analysis of Isl1 in mouse embryos revealed expression in pharyngeal endoderm and pharyngeal/splanchnic mesoderm, beginning at the cardiac crescent stage and continuing through stages of early cardiac looping morphogenesis. Double label RNA *in situ* hybridization demonstrated that Isl1 and a marker of differentiated cardiomyocytes, MLC2a, were never coexpressed.

Expression of Isl1 in pharyngeal mesoderm and the lack of outflow tract in Isl1 null mouse embryos suggested that Isl1 might mark and be required for the secondary/anterior heart field. To investigate this, Isl1 lineage studies were performed by crossing Isl1-cre mice (Soriano, 1999) to R26R-lacZ indicator mice (Soriano, 1999). Results of this lineage analysis demonstrated that Isl1 expressing progenitors contributed a majority of cells to the outflow tract and the right ventricle, suggesting that Isl1 was a marker for the secondary/anterior heart field. More surprisingly, perhaps, Isl1 progenitors were also found to contribute some cells to the left ventricle and a majority of cells to both atria. These results suggested that Isl1 progenitors not only gave rise to the secondary/anterior heart field, but also marked a broader population of progenitors, which contributed most cells to the early heart, with the exception of the majority of cells in the early left ventricle. Importantly, Isl1 was downregulated as these progenitors differentiated, rendering Isl1 a marker for undifferentiated progenitors. These lineage studies were consistent with the observed phenotype of Isl1 null mutants. Further analysis demonstrated that Isl1 was required for proliferation, survival, and migration of Isl1 expressing progenitors into developing heart (Cai et al., 2003).

Recent retrospective clonal analysis in mouse embryos has revealed two cardiac lineages, the first and second cardiac lineage. The second lineage corresponds to the contribution of the second heart field, as shown with Isl1-lineage traced cells (Buckingham et al., 2005).

Lineage labeling studies have demonstrated that cranial paraxial mesodermal cells can give rise to both myocardial and endocardial cells within
the outflow tract (Noden, 1991; Cai et al., 2003; Tirosh-Finkel et al., 2006).
In both mouse and chick, Isl1 is expressed in cranial paraxial mesoderm
at the time during which migration from the cranial paraxial mesoderm into
the arterial pole of the heart is occurring. These observations suggest that the
second heart field as marked by Isl1 may have its anterior boundary within
cranial paraxial mesoderm (see Fig. 1).

More lineage analysis with an inducible Isl1-cre mouse line, performed by a series of inductions at distinct embryonic stages from E7 to

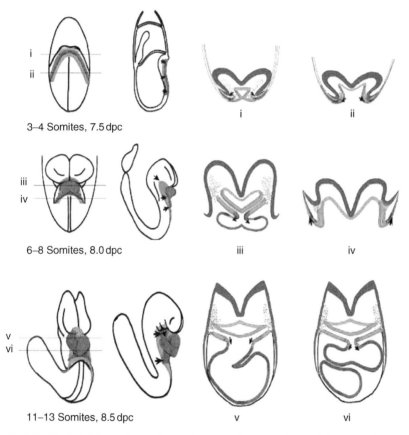

3–4 Somites, 7.5 dpc

6–8 Somites, 8.0 dpc

11–13 Somites, 8.5 dpc

FIGURE 1 Modified from Cai et al. (2003). Isl1 marks progenitors of the second heart
field. Isl1 expressing cells are medial and dorsal to differentiating cells in the cardiac
crescent, and at later stages dorsal to the forming heart, entering the heart both
anteriorly and posteriorly. The anterior extent of Isl1 expression overlaps with cranial
paraxial mesoderm, which contributes to myocardial and endocardial cells of the
outflow tract (Noden, 1991; Tirosh-Finkel et al., 2006). Isl1 expression in this domain
may define the anterior extent of the second heart field.

E11, has demonstrated that most Isl1 progenitors have migrated into the heart by E9 (Sun et al., 2007).

3. ISL1 AND TRANSCRIPTIONAL NETWORKS REQUIRED FOR HEART DEVELOPMENT

Several factors downstream and upstream of Isl1 in heart development have been identified. These are diagramed in Fig. 2, and described in more detail later.

3.1. Factors downstream of Isl1

The first identified direct downstream target of Isl1 in the anterior heart field is mef2c, which encodes a MADS box transcription factor required for early outflow tract and right ventricular morphogenesis (Lin et al., 1997).

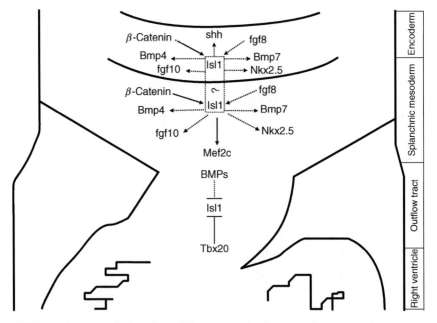

FIGURE 2   Diagram of Isl1 and regulatory networks during cardiogenesis. Isl1 expressed in endoderm and splanchnic mesoderm of the second heart lineage is both downstream and upstream of multiple genetic pathways which regulate overall cardiac and outflow tract development (refer to text for references). Whether Isl1 is required in endoderm or mesoderm for regulation of these pathways is not yet known, as indicated in this diagram by the question mark. Isl1 is downregulated as second lineage cells differentiate, and both BMP signaling and Tbx20 contribute to its downregulation. Dotted arrows or lines indicate direct or indirect action of Isl1, whereas solid arrows or lines indicate direct action of Isl1 on target promoter sequences.

An anterior heart field enhancer for mef2c has been identified which is activated in response to GATA factors and Isl1, and requires homeodomain consensus elements for expression (Dodou et al., 2004). Mice which are null for the winged helix transcription factor Foxh1 exhibit a cardiac phenotype which resembles that of Mef2c mutants in some aspects (von Both et al., 2004). Mef2c is downregulated in the ventricle of Foxh1 mutants, and a mef2c intronic enhancer which drives expression in the anterior heart field is a direct target of Foxh1 and the homeodomain transcription factor Nkx2.5, which is also required for heart development (Lyons et al., 1995). Isl1 regulates expression of Nkx2.5 in the second heart field, as this domain of expression is decreased in Isl1 null mice (Chenleng Cai and S.E., unpublished observations), suggesting that Isl1 regulates Mef2c expression directly and indirectly. Isl1 is not downregulated in Foxh1 mutants, suggesting that it may be upstream of Foxh1 (von Both et al., 2004).

Isl1 is critical for both early and later events in outflow tract morphogenesis. Isl1 mutants have no apparent outflow tract tissue. Additionally, studies in Isl1 null mice have demonstrated that several growth factors required for later cardiac outflow tract morphogenesis, BMP4, BMP7, Fgf10, and Shh (Gruber and Epstein, 2004; Liu et al., 2004; Washington Smoak et al., 2005; Lin et al., 2006; Marguerie et al., 2006) are selectively downregulated in regions overlapping with Isl1, including pharyngeal endoderm, pharyngeal mesoderm, or both (Cai et al., 2003). These observations suggest that these genes are direct or indirect targets of Isl1. They also suggest a pivotal role for Isl1 in coordinating expression of genes required for outflow tract morphogenesis, a suggestion supported by analysis of a hypomorphic mutant of Isl1 which exhibits defects in outflow tract remodeling (Y.S., X.L., S.E., in preparation).

3.2. Factors upstream of Isl1

Results of experiments have demonstrated a key role for β-catenin signaling in maintaining Isl1 expression in cardiac progenitors (Lin et al., 2007). Analysis of a LEF/TCF-lacZ reporter (Mohamed et al., 2004) demonstrated active β-catenin signaling in early pharyngeal mesoderm and endoderm, overlapping with Isl1 expression. Ablation of β-catenin with Isl1-cre results in lethality at E12.5, with mutants exhibiting persistent truncus arteriosus. Isl1 mRNA and protein expression are severely affected in Isl1-cre;β-catenin mutants. An evolutionarily conserved LEF/TCF-binding site within the Isl1 promoter binds LEF/TCF *in vivo* and is required for the activation by LEF/TCF/β-catenin *in vitro*, demonstrating that the Isl1 promoter is a direct target of β-catenin. Expression of other genes required for outflow tract morphogenesis are also decreased in Isl1-cre;β-catenin mutants, including shh, wnt11, and Tbx2 (Harrelson et al., 2004; Washington Smoak et al., 2005; Lin et al., 2006; Zhou et al., in press). Decreased Shh is likely to be at least in part secondary to decreased Isl1 expression, and it will be of future interest to examine whether Tbx2 and wnt11 are also downstream of Isl1.

Isl1 expression is also downstream of Fgf8 signaling within cardiac progenitors. Ablation of Fgf8 in early cardiogenic mesoderm, second heart field progenitors, or their derivatives, utilizing Mesp1-cre, Isl1-cre, Nkx2.5-cre, TnT-cre, or mef2c anterior heart field enhancer-cre mice, have demonstrated that Fgf8 signaling is required within early splanchnic mesoderm for early growth and morphogenesis of the outflow tract and right ventricle (Ilagan et al., 2006; Park et al., 2006). Expression of both Isl1 and wnt11 is downregulated in Mesp1-cre or Isl1-cre;Fgf8 mutants (Park et al., 2006).

Isl1 mRNA expression is downregulated as Isl1 progenitors migrate into the forming heart and differentiate. The T-box transcription factor Tbx20 and BMP signaling through the Type1 BMP receptor ALK3 may cooperatively downregulate expression of Isl1 as cardiac progenitors differentiate. Both Tbx20 null mice and mice with BMP signaling ablated in the Isl1 domain, Isl1-cre;ALK3 mice, exhibit increased expression of Isl1 mRNA in the outflow tract (Cai et al., 2005; Yang et al., 2006). In Tbx20 mutants, Isl1 mRNA is slightly upregulated throughout the heart including the left ventricle (Cai et al., 2005). The Isl1 promoter appears to be a direct target of Tbx20, as a conserved Tbx element within the Isl1 promoter binds Tbx20 in developing heart, and is required for repression of Isl1 by Tbx20 in cell transfection studies.

4. ISL1 MARKS A DIVERSITY OF CARDIOVASCULAR LINEAGES

Lineage studies of Isl1 expressing cells have revealed that Isl1 progenitors contribute to a number of cardiovascular lineages (Fig. 3). Early fate mapping studies with Isl1-cre mice demonstrated that Isl1 expressing cells migrate into the forming heart, contributing to most cells of outflow tract, right ventricle, atria, and some cells of the left ventricle (Cai et al., 2003). Within the heart, Isl1 progenitors gave rise to both myocytes and a subset of endocardial cells. In addition to contributing to both myocardium and endocardium, Isl1-lineage traced cells were also observed within endothelium of the aorta (Cai et al., 2003).

More fate mapping studies, examining later stages, have demonstrated that Isl1 descendents also contribute to smooth muscle in the proximal portion of the aorta and pulmonary artery, consistent with results which have demonstrated that this population of smooth muscle arises in part from the anterior/secondary heart field (Verzi et al., 2005; Waldo et al., 2005). Lineage labeling and chick/quail chimeric studies have demonstrated that the proximal smooth muscle of the great vessels derives from a late-migrating population of cells from the secondary heart field (Waldo et al., 2005). Isl1-expressing cells also give rise to a subset of coronary vascular

Isl1 + progenitors

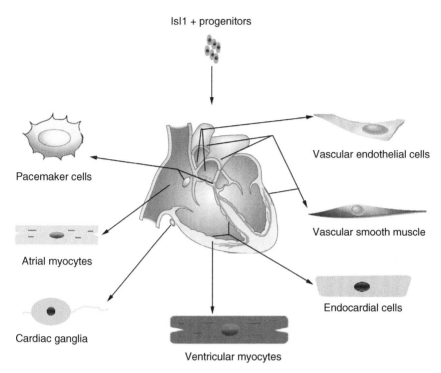

Pacemaker cells

Vascular endothelial cells

Vascular smooth muscle

Atrial myocytes

Endocardial cells

Cardiac ganglia

Ventricular myocytes

FIGURE 3   Isl1 marks progenitors of diverse cardiovascular lineages. Isl1 is expressed in progenitors of multiple cardiovascular lineages, as shown.

smooth muscle cells (Sun et al., 2007). These smooth muscle cells appear to be distinct from those which arise from the epicardium (Reese et al., 2002; Chenleng Cai, Jody Martin, Y.S., S.E., in preparation), demonstrating previously unsuspected embryonic lineage heterogeneity for coronary vascular smooth muscle.

Isl1 is expressed in several neural crest derivatives, including neurons of the autonomic nervous system and sensory neurons in sensory ganglia (Thor et al., 1991). Isl1 is also expressed in cardiac ganglia which derive from cardiac neural crest (Kirby and Stewart, 1983; Sun et al., 2007). However, Isl1 does not appear to be expressed in the cardiac neural crest population that contributes to outflow tract septation, and to subsets of smooth muscle of the outflow tract and aortic arch arteries, as evidenced by coimmunostaining of Isl1 protein and wnt1-cre;R26R lineage traced tissue (Sun et al., 2007). Additionally, ablation of a floxed allele of Isl1 with wnt1-cre does not give a cardiac outflow tract phenotype (Y.S. and S.E., unpublished observations), suggesting that if Isl1 is expressed in progenitors of the cardiac neural crest it is not required within them for outflow tract remodeling. Further evidence that Isl1 is not expressed in cardiac neural crest, which contributes smooth muscle to the aortic arch arteries and outflow tract, arises from the observation that ablation of several genes with wnt1-cre gives a cardiac outflow tract phenotype, whereas ablation of the same genes with Isl1-cre does not (Lizhu Lin and S.E., unpublished observations).

5. ISL1 PROTEIN PERSISTS IN OUTFLOW TRACT AND NODAL CONDUCTION TISSUE OF THE MID-GESTATION HEART

Although Isl1 mRNA is downregulated as Isl1 progenitors enter the developing heart and differentiate, studies with an Isl1-lacZ mouse line and coimmunostaining for Isl1 protein have demonstrated persistent expression of Isl1 protein in the myocardium of the outflow tract and in the pacemaker cells of the sinus and atrioventricular node until E11.5 (Sun et al., 2007). Outflow tract myocardium is relatively undifferentiated, exhibiting persistent expression of α-smooth muscle actin (Kruithof et al., 2003). It will be of interest to examine whether persistence of Isl1 protein in the outflow tract myocardium is causally associated with the less differentiated state of outflow tract myocardium, which may be required to effect the extensive remodeling process and which must occur during this time to arrive at concordant connections between each of the great vessels and their respective ventricular chambers. Similarly, conduction tissue has been observed to express genes associated with early cardiac morphogenesis, and may be relatively undifferentiated (Christoffels et al., 2004). Persistence of Isl1 protein in the two pacemaking centers of the heart suggests a critical role for Isl1 in the pacemaking function of the early and mid-gestation heart. These observations also suggest that pacemaking tissue derives from the second heart field as marked by Isl1. Consistent with dependence of Isl1 expression on β-catenin signaling, β-catenin signaling remains active in early outflow tract myocardium and in the region of the sinus and atrioventricular nodes (Lin et al., 2007).

6. ISL1 AS A MARKER OF A PLURIPOTENT CARDIOVASCULAR PROGENITOR

Expression of Isl1 in the second heart field, its downregulation as progenitors differentiate, and differentiation of those progenitors into endothelial, myocardial, endocardial, and smooth muscle lineages suggest that Isl1 may mark a pluripotent cardiovascular progenitor state (Sun et al., 2007). Inducible Isl1-cre;R26R-lacZ lineage experiments and Isl1 immunostaining have demonstrated the presence of Isl1 expressing cells within postnatal myocardium, in rodent and human tissue (Laugwitz et al., 2005). Isolation and gene profiling of these cells demonstrated expression of cardiac lineage markers, but no expression of differentiated cardiomyocyte markers. Following expansion on cardiac mesenchymal feeder layers, Isl1 expressing cells were cocultured with differentiated myocytes and demonstrated a high efficiency of conversion into cardiomyocytes.

These studies suggest that Isl1 will be a useful marker for cardiovascular progenitors in embryonic stem cells, or in other potential cardiovascular progenitor populations, including humans. This will allow for their isolation and further characterization, including their potential therapeutic utility.

7. CONCLUSIONS

The discovery of Isl1 as a marker for an undifferentiated cardiovascular progenitor has allowed visualization of progenitors in development, and in postnatal heart. Isl1 may also be utilized as a marker of cardiovascular progenitors in other contexts, including embryonic stem cells and potentially extracardiac adult progenitors, allowing for their enrichment and utilization both *in vitro* and *in vivo* studies.

Isl1 has utility as a marker, but is also required for multiple aspects of cardiac lineage development, including proliferation, survival, and migration of cardiac progenitors during early heart formation. Isl1 is also required for expression of a number of genes required for later stages of outflow tract morphogenesis. Together, these observations place Isl1 upstream in diverse aspects of heart and cardiovascular development.

8. FUTURE CHALLENGES

Owing to the pivotal role of Isl1 in cardiovascular progenitors, insights into Isl1 regulation will provide important information on initial specification of cardiovascular progenitors. Similarly, it will be important to understand mechanisms by which Isl1 controls multiple aspects of cardiovascular progenitor biology. Although some progress has been made in defining factors directly upstream and downstream of Isl1, much remains to be understood, including a more comprehensive understanding of direct targets of Isl1 action. The potential role of persistent expression of Isl1 in outflow tract and pacemaker tissues of mid-gestation heart, and later in postnatal progenitors requires further investigation.

Isl1 is expressed both in pharyngeal endoderm and in cardiogenic mesoderm. Pharyngeal endoderm is an inducer of cardiogenic mesoderm; therefore, it is likely that Isl1 is required in each of these lineages for distinct aspects of heart formation. Studying the role of Isl1 in each of these tissues may shed further light on distinct roles of endoderm and mesoderm in cardiogenesis, and the cross talk that occurs between them.

Isl1 progenitors adopt multiple distinct cardiovascular fates, including that of myocytes, pacemaker cells, smooth muscles, and endothelial lineages. The requirement for Isl1 in adoption of each of these cell fates remains to be explored.

REFERENCES

Abu-Issa, R., Waldo, K., Kirby, M.L. 2004. Heart fields: One, two or more? Dev. Biol. 272, 281–285.

Ahlgren, U., Pfaff, S.L., Jessell, T.M., Edlund, T., Edlund, H. 1997. Independent requirement for ISL1 in formation of pancreatic mesenchyme and islet cells. Nature 385, 257–260.

Buckingham, M., Meilhac, S., Zaffran, S. 2005. Building the mammalian heart from two sources of myocardial cells. Nat. Rev. Genet. 6, 826–835.

Cai, C.L., Liang, X., Shi, Y., Chu, P.H., Pfaff, S.L., Chen, J., Evans, S. 2003. Isl1 identifies a cardiac progenitor population that proliferates prior to differentiation and contributes a majority of cells to the heart. Dev. Cell 5, 877–889.

Cai, C.L., Zhou, W., Yang, L., Bu, L., Qyang, Y., Zhang, X., Li, X., Rosenfeld, M.G., Chen, J., Evans, S. 2005. T-box genes coordinate regional rates of proliferation and regional specification during cardiogenesis. Development 132, 2475–2487.

Christoffels, V.M., Burch, J.B., Moorman, A.F. 2004. Architectural plan for the heart: Early patterning and delineation of the chambers and the nodes. Trends Cardiovasc. Med. 14, 301–307.

de la Cruz, M.V., Sanchez Gomez, C., Arteaga, M.M., Arguello, C. 1977. Experimental study of the development of the truncus and the conus in the chick embryo. J. Anat. 123, 661–686.

deVries, P.A. 1981. Evolution of precardiac and splanchnic mesoderm in relationship to the infundibulum and truncus. In: *Mechanisms of Cardiac Morphogenesis and Teratogenesis* (T. Pexeider, Ed.), pp. 31–47Raven Press, .

Dodou, E., Verzi, M.P., Anderson, J.P., Xu, S.M., Black, B.L. 2004. Mef2c is a direct transcriptional target of ISL1 and GATA factors in the anterior heart field during mouse embryonic development. Development 131, 3931–3942.

Gruber, P.J., Epstein, J.A. 2004. Development gone awry: Congenital heart disease. Circ. Res. 94, 273–283.

Harrelson, Z., Kelly, R.G., Goldin, S.N., Gibson-Brown, J.J., Bollag, R.J., Silver, L.M., Papaioannou, V.E. 2004. Tbx2 is essential for patterning the atrioventricular canal and for morphogenesis of the outflow tract during heart development. Development 131, 5041–5052.

Ilagan, R., Abu-Issa, R., Brown, D., Yang, Y.P., Jiao, K., Schwartz, R.J., Klingensmith, J., Meyers, E.N. 2006. Fgf8 is required for anterior heart field development. Development 133, 2435–2445.

Kelly, R.G., Brown, N.A., Buckingham, M.E. 2001. The arterial pole of the mouse heart forms from Fgf10-expressing cells in pharyngeal mesoderm. Dev. Cell 1, 435–440.

Kelly, R.G., Buckingham, M.E. 2002. The anterior heart-forming field: Voyage to the arterial pole of the heart. Trends Genet. 18, 210–216.

Kirby, M.L., Stewart, D.E. 1983. Neural crest origin of cardiac ganglion cells in the chick embryo: Identification and extirpation. Dev. Biol. 97, 433–443.

Kruithof, B.P., Van Den Hoff, M.J., Tesink-Taekema, S., Moorman, A.F. 2003. Recruitment of intra- and extracardiac cells into the myocardial lineage during mouse development. Anat. Rec. A Discov. Mol. Cell. Evol. Biol. 271, 303–314.

Laugwitz, K.L., Moretti, A., Lam, J., Gruber, P., Chen, Y., Woodard, S., Lin, L.Z., Cai, C.L., Lu, M.M., Reth, M., et al. 2005. Postnatal isl1 + cardioblasts enter fully differentiated cardiomyocyte lineages. Nature 433, 647–653.

Lin, L., Bu, L., Cai, C.L., Zhang, X., Evans, S. 2006. Isl1 is upstream of sonic hedgehog in a pathway required for cardiac morphogenesis. Dev. Biol. 295, 756–763.

Lin, L., Cui, L., Zhou, W., Dufort, D., Zhang, X., Cai, C.L, Bu, L., Yang, L., Martin, J., Kemler, R., Rosenfeld, M.G., Chen, J., et al. 2007. Beta-catenin directly regulates Islet1 expression in cardiovascular progenitors and is required for multiple aspects of cardiogenesis. Proc. Natl. Acad. Sci. USA 104, 9313–9318.

Lin, Q., Schwarz, J., Bucana, C., Olson, E.N. 1997. Control of mouse cardiac morphogenesis and myogenesis by transcription factor MEF2C. Science 276, 1404–1407.

Liu, W., Selever, J., Wang, D., Lu, M.F., Moses, K.A., Schwartz, R.J., Martin, J.F. 2004. Bmp4 signaling is required for outflow-tract septation and branchial-arch artery remodeling. Proc. Natl. Acad. Sci. USA 101, 4489–4494.

Lyons, I., Parsons, L.M., Hartley, L., Li, R., Andrews, J.E., Robb, L., Harvey, R.P. 1995. Myogenic and morphogenetic defects in the heart tubes of murine embryos lacking the homeo box gene Nkx2-5. Genes Dev. 9, 1654–1666.

Marguerie, A., Bajolle, F., Zaffran, S., Brown, N.A., Dickson, C., Buckingham, M.E., Kelly, R. G. 2006. Congenital heart defects in Fgfr2-IIIb and Fgf10 mutant mice. Cardiovasc. Res. 71, 50–60.

Mjaatvedt, C.H., Nakaoka, T., Moreno-Rodriguez, R., Norris, R.A., Kern, M.J., Eisenberg, C.A., Turner, D., Markwald, R.R. 2001. The outflow tract of the heart is recruited from a novel heart-forming field. Dev. Biol. 238, 97–109.

Mohamed, O.A., Clarke, H.J., Dufort, D. 2004. Beta-catenin signaling marks the prospective site of primitive streak formation in the mouse embryo. Dev. Dyn. 231, 416–424.

Noden, D.M. 1991. Origins and patterning of avian outflow tract endocardium. Development 111, 867–876.

Olson, E.N., Schneider, M.D. 2003. Sizing up the heart: Development redux in disease. Genes Dev. 17, 1937–1956.

Park, E.J., Ogden, L.A., Talbot, A., Evans, S., Cai, C.L., Black, B.L., Frank, D.U., Moon, A.M. 2006. Required, tissue-specific roles for Fgf8 in outflow tract formation and remodeling. Development 133, 2419–2433.

Pfaff, S.L., Mendelsohn, M., Stewart, C.L., Edlund, T., Jessell, T.M. 1996. Requirement for LIM homeobox gene Isl1 in motor neuron generation reveals a motor neuron-dependent step in interneuron differentiation. Cell 84, 309–320.

Reese, D.E., Mikawa, T., Bader, D.M. 2002. Development of the coronary vessel system. Circ. Res. 91, 761–768.

Soriano, P. 1999. Generalized lacZ expression with the ROSA26 Cre reporter strain. Nat. Genet. 21, 70–71.

Sun, Y., Liang, X., Najafi, N., Cass, M., Lin, L, Cai, C.L., Chen, J., Evans, S.M. 2007. Islet1 is expressed in distinct cardiovascular lineages, including pacemaker and coronary vascular cells. Dev. Biol. 304, 286–296.

Thor, S., Ericson, J., Brannstrom, T., Edlund, T. 1991. The homeodomain LIM protein Isl-1 is expressed in subsets of neurons and endocrine cells in the adult rat. Neuron 7, 881–889.

Tirosh-Finkel, L., Elhanany, H., Rinon, A., Tzahor, E. 2006. Mesoderm progenitor cells of common origin contribute to the head musculature and the cardiac outflow tract. Development 133, 1943–1953.

Verzi, M.P., McCulley, D.J., De Val, S., Dodou, E., Black, B.L. 2005. The right ventricle, outflow tract, and ventricular septum comprise a restricted expression domain within the secondary/anterior heart field. Dev. Biol. 287, 134–145.

Viragh, S., Challice, C.E. 1973. Origin and differentiation of cardiac muscle cells in the mouse. J. Ultrastruct. Res. 42, 1–24.

von Both, I., Silvestri, C., Erdemir, T., Lickert, H., Walls, J.R., Henkelman, R.M., Rossant, J., Harvey, R.P., Attisano, L., Wrana, J.L. 2004. Foxh1 is essential for development of the anterior heart field. Dev. Cell 7, 331–345.

Waldo, K.L., Hutson, M.R., Ward, C.C., Zdanowicz, M., Stadt, H.A., Kumiski, D., Abu-Issa, R., Kirby, M.L. 2005. Secondary heart field contributes myocardium and smooth muscle to the arterial pole of the developing heart. Dev. Biol. 281, 78–90.

Waldo, K.L., Kumiski, D.H., Wallis, K.T., Stadt, H.A., Hutson, M.R., Platt, D.H., Kirby, M.L. 2001. Conotruncal myocardium arises from a secondary heart field. Development 128, 3179–3188.

Washington Smoak, I., Byrd, N.A., Abu-Issa, R., Goddeeris, M.M., Anderson, R., Morris, J., Yamamura, K., Klingensmith, J., Meyers, E.N. 2005. Sonic hedgehog is required for cardiac outflow tract and neural crest cell development. Dev. Biol. 283, 357–372.

Yang, L., Cai, C.L., Lin, L., Qyang, Y., Chung, C., Monteiro, R.M., Mummery, C.L., Fishman, G.I., Cogen, A., Evans, S. 2006. Isl1Cre reveals a common Bmp pathway in heart and limb development. Development 133, 1575–1585.

Yuan, S., Schoenwolf, G.C. 2000. Islet-1 marks the early heart rudiments and is asymmetrically expressed during early rotation of the foregut in the chick embryo. Anat. Rec. 260, 204–207.

Zhou, W., Lin, L., Majumdar, A., Li, X., Zhang, X., Liu, W., Etheridge, L., Shi, Y., Martin, J., Van de Ven, W., Kaartinen, V., Wynshaw-Boris, A., et al. (in press). Modulation of morphogenesis by non-canonical Wnt signaling requires ATF/CREB-mediated transcriptional activation of TGF-β. Nature Genetics.

CHAPTER 7

Role of microRNAs in Cardiovascular Biology

Eva Samal*,† and Deepak Srivastava*,†

Abstract   Remarkable advances in cardiac developmental biology over the last decade have revealed elegant networks of signaling and transcriptional cascades that regulate cardiomyocyte commitment and organogenesis. Differentiation is often reinforced via positive and negative feedback loops that serve to irreversibly control sequential cell fate decisions. As pluripotent cells become progressively

* Gladstone Institute of Cardiovascular Disease, University of California, San Francisco, California
† Departments of Pediatrics and Biochemistry and Biophysics, University of California, San Francisco, California

Advances in Developmental Biology, Volume 18
ISSN 1574-3349, DOI: 10.1016/S1574-3349(07)18007-8

specialized, it may be equally important to promote specific gene programs and also suppress unwanted pathways to allow cell type-specific protein expression. This is likely accomplished in part by differential transcription, but recent reports suggest that this process is also regulated at the level of translation. microRNAs (miR-NAs) have emerged as endogenous and widely used mediators of translational repression and likely influence most important cellular processes. Here, we will review the emerging role of miRNAs in translational regulation during cardiac development and disease and consider the potentially broad impact of this rapidly evolving field on cardiovascular research.

1. INTRODUCTION

microRNAs (miRNAs) are ~20- to 22-nucleotide RNAs that are encoded by the genome and regulate either translation or stability of mRNAs via a sequence-dependent interaction (Ambros, 2001; Pasquinelli and Ruvkun, 2002; Cullen, 2006). miRNA-mediated regulation of protein dosage is a phylogenetically conserved mechanism used to control the flow of genetic information that ultimately results in a given cell's proteome. Over 400 miRNAs are known in humans, and each likely "fine-tunes" translation of numerous mRNA targets, with estimates ranging from the tens to hundreds. As such, it is likely that most important mRNAs are titrated to some degree by one or more miRNAs. During the last few years, several muscle-specific miRNAs have been described that appear to regulate pivotal decisions during cardiogenesis and skeletal myogenesis (Zhao et al., 2005; Chen et al., 2006; Kim et al., 2006; Rao et al., 2006). A brief introduction to miRNA biology is provided below followed by a review of the current findings regarding miRNA-mediated events during cardiogenesis and cardiac disease.

2. miRNA BIOGENESIS AND REPRESSION OF TARGET mRNAs

miRNAs are initially transcribed as long RNA transcripts known as pri-miRNAs that have capped structures and poly(A) tails (Fig. 1) (Lee et al., 2004). Similar to mRNAs, their synthesis is often controlled by lineage-specific transcription factors, thus allowing for tissue-specific expression and participation in the lineage commitment decisions (Kwon et al., 2005; Zhao et al., 2005; Rao et al., 2006). Pri-miRNAs are processed in the nucleus into ~70-nucleotide pre-miRNAs with extensive stem-loop structures by the RNase III enzyme, Drosha, and the double-stranded RNA-binding protein, Pasha (Gregory et al., 2004). The resulting pre-miRNAs are exported to the cytoplasm by exportin-5, a RAN GTP-dependent

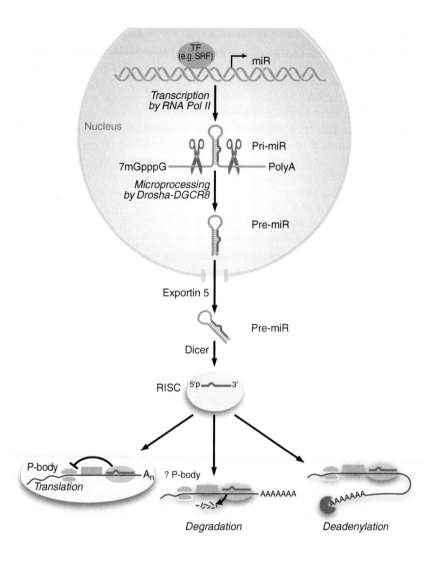

FIGURE 1   Current model of miRNA biogenesis and function. The initial RNA transcript is typically transcribed by RNA polymerase II as primary miRNAs (pri-miRNAs), which range from a few hundred to thousands of nucleotides (nt) in length. The pri-miRNA of each miRNA has a characteristic stem-loop structure that can be recognized and cleaved by the ribonuclease III (RNase III) endonuclease Drosha within the nucleus. Efficient pri-miRNA cleavage by Drosha requires a protein partner, Pasha/DGCR8, which has a double-stranded RNA-binding domain (dsRBD). The cleavage product, a ~70-nt stem-loop RNA called pre-miRNA, is exported from the nucleus to the cytoplasm by exportin-5/RanGTP. In the cytoplasm, another evolutionarily conserved RNase III enzyme, Dicer, together with its dsRBD protein partners, TAR RNA-binding protein (TRBP) and PACT, further process pre-miRNA into mature miRNA (~21 nt). The mature miRNA is then unwound, and a single strand is incorporated into a protein complex known as the RNA-induced silencing complex

transporter, where another RNase III enzyme, Dicer, cleaves them into ~20- to 22-nucleotide RNA duplexes (reviewed in Matranga and Zamore, 2004; Cullen, 2006). A single strand of this duplex, the mature miRNA, is preferentially incorporated into the RNA-induced silencing complex (RISC) where miRNA–mRNA interactions can result in translational repression (Schwarz et al., 2003).

miRNAs negatively regulate mRNA translation through several mechanisms. Interaction of a miRNA with an mRNA target, typically in the 3'-untranslated region (UTR), was initially reported to repress translation of target mRNAs at the level of translation initiation (Pillai et al., 2005). Curiously, limited Watson-Crick base-pairing between the 5' end of the miRNA and the mRNA was sufficient for this repression, with sequence matching among nucleotides 2–7 of the miRNA being most important. More recently, it has become clear that, in addition to translational repression, miRNAs can regulate mRNA stability (reviewed in Pillai et al., 2007). A high degree of sequence matching results in cleavage of mRNAs similar to siRNA-mediated events, but it appears that even limited complementarity can affect stability of mRNA targets through unknown mechanisms. For example, overexpression of miR-1, a heart and skeletal muscle-restricted miRNA in HeLa cells shifted the mRNA expression profile toward that of a myogenic lineage (Lim et al., 2005). Furthermore, the 3' UTRs of miR-1 repressed transcripts were enriched for the 5'-seed motif CAUUCC, corresponding to the 2- to 7-nucleotide sequences of miR-1. Hence, miRNAs potentially promote and maintain a cell type-specific identity directly by acting as negative regulators of alternate differentiation fates. In agreement with a repressive role for miRNAs in promoting a tissue-specific signature gene expression, targets of miRNAs are often expressed at lower levels exclusively in tissues where the corresponding miRNA is present (Farh et al., 2005). Specificity of miRNA effects is evident from the observation that genes upregulated or unchanged after overexpression of miR-1 did not possess seed motifs for miRNA binding in their annotated 3' UTRs (Lim et al., 2005). Identifying targets of myogenic-specific miRNAs should shed light on how mesodermal progenitor cells take on and maintain a cardiac, smooth or skeletal muscle lineage. The ability of a miRNA to stimulate distinctive cell-type mRNA levels could be exploited to coax embryonic stem cells toward a certain lineage. A detailed understanding of parallel cascades of transcription factors and their downstream effector promoters in combination with knowledge of myogenic-specific miRNAs and their mRNA targets could help in the identification and treatment of muscle diseases.

(RISC), which represses mRNA translation or destabilizes mRNA transcripts through cleavage or deadenylation. TF, transcription factor; SRF, (Reproduced with permission from Zhao and Srivastava, *TIBS*, 2007).

3. EXPRESSION AND REGULATION OF MUSCLE-SPECIFIC miRNAs

Several miRNAs are expressed in a muscle-specific fashion, including miR-1, miR-133, miR-206, miR-208, and miR-181 (Lagos-Quintana et al., 2002; Zhao et al., 2005; Kim et al., 2006; Naguibneva et al., 2006; Rao et al., 2006). miR-1, one of the most conserved miRNAs, is expressed specifically in cardiac and skeletal muscle cells (Zhao et al., 2005). It is transcribed as a bicistronic message with miR-133. A genomic duplication appears to have resulted in two identical mature miR-1 and miR-133 genes on separate chromosomes (miR-1-1, miR-1-2, miR-133a-1, miR-133a-2). An enhancer controlling miR-1-1 expression is active in both atrial and ventricular precursors at E9.5 in mouse embryos, whereas expression from a miR-1-2 enhancer is restricted to the ventricles (Fig. 2) (Zhao et al., 2005). However, their expression expands, and both are expressed ubiquitously in the developing heart by E10.5. Based on transgenic lacZ expression from the miR-1 enhancer, expression in skeletal muscle progenitors of the somites begins by E9.5 (Zhao et al., 2005). Transcription of the miR-1/miR-133 bicistronic precursor is under the direct regulation of the major myogenic DNA-binding proteins. In the heart, serum response factor (SRF) binds and activates the promoter regions of miR-1-1 and miR-1-2 *in vitro* and *in vivo* through a highly conserved serum response element (SRE) (Zhao et al., 2005). Consistent with this, mice with targeted SRF deletion in the heart lacked endogenous miR-1-1 and miR-1-2 expression. Similar findings supported miR-1/-133 as direct *in vivo* targets of MyoD and Mef2 in skeletal muscle. The transcriptional modules appear to be ancient, as they are conserved even in fruit flies, where an SRF-like *cis* element is essential for cardiac expression, and the basic helix-loop-helix (bHLH) transcription factor twist and mef2 regulate somatic muscle expression of miR-1 (Kwon et al., 2005).

4. MUSCLE-SPECIFIC miRNAs AS DEVELOPMENTAL REGULATORS

4.1. miR-1

The regulation of miRNAs by master regulators of muscle differentiation programs, such as MyoD and SRF, suggests that miRNAs promote and/or maintain differentiation of the lineage in which they are expressed. Consistent with such a role, restricted overexpression of miR-1 in the embryonic heart with a β-MHC promoter leads to an early exit from the cell cycle and, consequently, poor trabeculation of the myocardial wall (Zhao et al., 2005). Similarly, overexpression of miR-1 in C2C12 myoblasts promotes differentiation of myoblasts into myotubes (Chen et al., 2006).

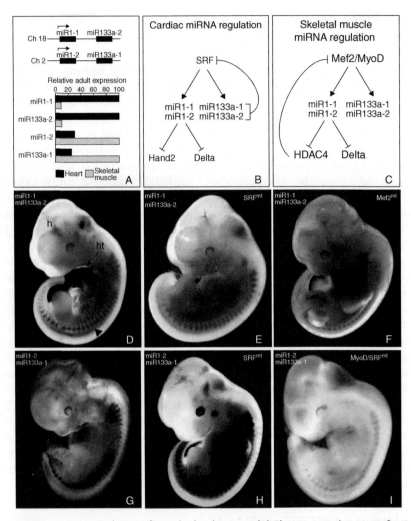

FIGURE 2   miRNA regulation of muscle development. (A) Chromosome location of miR-1 and miR-133a family members. Combinations of miR-1 and miR-133a are transcribed as single polycistronic mRNAs and therefore share common regulation. Relative levels of adult expression of each in the heart and skeletal muscle are shown. (B) and (C) represent simplified schematics of our current understanding of miR-1 and miR-133a pathways in cardiac or skeletal muscle. (D) and (G) represent lacZ expression in the heart (ht) and somites (arrowhead) directed by the upstream enhancer of the miR-1-1/miR-133a-2 cluster or the miR-1-2/miR-133a-1 cluster, respectively. (E) and (H) indicate loss of heart expression in the presence of a mutation of an SRF binding site. (F) and (I) demonstrate decreased somatic expression in the setting of a mutation in a Mef2 or MyoD binding site within the enhancers. (Embyronic images adapted from Zhao et al., 2005).

The biologic functions of miRNAs will be dictated by the spectrum of mRNAs that they directly regulate. Because of the limited amount of sequence matching necessary for miRNA–mRNA interactions, identification of miRNA targets has been difficult and relatively few targets have been validated at the level of protein expression. Using an approach that incorporated mRNA target site sequence matching and physical accessibility, we identified Hand2, a bHLH transcription factor involved in ventricular cardiomyocyte expansion, as a target of miR-1 (Zhao et al., 2005). The functions of Hand2 during cardiac morphogenesis are dose dependent and consistent with its potential role in mediating miR-1 effects on the balance of ventricular myocyte proliferation versus differentiation.

In skeletal muscle, miR-1 targets HDAC4, a histone deacetylase that represses Mef2-dependent activation of muscle-specific genes (Chen et al., 2006). Translational inhibition of HDAC4 by miR-1 may promote activation of Mef2c-dependent promoters, such as myosin heavy chain (MyHC) and other markers of terminally differentiated muscle. This observation is consistent with a function for miR-1 in maintaining the differentiated myogenic state and its abundance in terminally differentiated muscle.

While miRNA overexpression studies have been informative, loss-of-function studies promise to reveal the endogenous requirement of specific miRNAs. Disruption of the single fly orthologue of miR-1 had catastrophic consequences, resulting in uniform lethality at embryonic or larval stages with a frequent defect in maintaining cardiac gene expression (Kwon et al., 2005). In a subset of flies lacking miR-1, a severe defect of cardiac progenitor cell differentiation provided loss-of-function evidence that miR-1 participates in muscle differentiation events (Kwon et al., 2005). Embryonic lethality was also observed by knocking down fly miR-1 activity by O-methyl-modified antisense oligonucleotides (Leaman et al., 2005), although a milder miR-1 phenotype has also been described, suggesting a highly variable phenotype (Sokol and Ambros, 2005).

miR-1 in flies regulates the Notch signaling pathway by directly targeting mRNA of the Notch ligand, Delta (Kwon et al., 2005), suggesting miR-1 functions to induce differentiated cardiac cells from an equivalency group of progenitor cells. Whether miR-1 functions similarly in mammalian cells awaits targeted deletion of the miR-1-1 and miR-1-2 alleles.

4.2. miR-133

Although miR-133 is transcribed with miR-1, it appears, at least in some contexts, to function in a very different manner compared to miR-1. Overexpression of miR-133 in C2C12 skeletal myoblasts antagonizes differentiation into myotubes and promotes continued proliferation of the muscle progenitors (Chen et al., 2006). One mechanism by which miR-133 might mediate the repression of differentiation is by inhibiting the translation of SRF, which itself regulates miR-1 and miR-133 expression. This negative feedback loop may serve to control the proper dosage and timing of this key regulatory network as differentiation proceeds (Chen et al., 2006). Interestingly, comparison of changes in global gene expression after transfection of C2C12 myoblasts with miR-1 or miR-133 showed many overlapping gene changes, suggesting that miR-1 and miR-133 may function antagonistically in some settings but also may cooperate in regulation of subsets of target genes (Kim et al., 2006).

4.3. miR-206

miR-206 is a skeletal muscle-specific miRNA that is not expressed in cardiac muscle, yet shares a high degree of sequence identity with miR-1 (Kim et al., 2006). The mature forms of miR-206 and miR-1 have identical 5' seed matches and differ by only three nucleotides outside of the seed match. However, miR-206 is expressed earlier than miR-1 in differentiating skeletal muscle. Given the similarity in sequence to miR-1, it is not surprising that miR-206 can also promote myogenesis of C2C12 myoblasts. As with miR-1, overexpression of miR-206 in C2C12 myoblasts leads to increased numbers of myogenin and MyHC positive cells (Kim et al., 2006). Several targets of miR-206 have been described, including Connexin43, a gap junction protein that is absent in adult muscle (Anderson et al., 2006). Connexin43 protein reappears after muscle injury to establish gap junctions involved in incorporation of activated muscle progenitor cells known as satellite cells. MyoD also directly regulates miR-206, which may downregulate markers of other lineages including Utropin (Utrn) and follistatin-like 1 (Fstl1), during formation of adult muscle (Rosenberg et al., 2006).

4.4. miR-181

miR-181 is expressed in a pattern opposite to that of miR-1, miR-133, and miR-206. It is present in muscle precursors but is undetectable in differentiated muscle. After muscle injury, miR-181 is induced in regenerating muscle as satellite cells become activated, but subsequently regresses as differentiation proceeds (Naguibneva et al., 2006). This timing suggests a potential role for miR-181 in activating satellite cells and providing a permissive environment for expansion of muscle precursor cells. The trans-factors regulating miR-181 expression and induction following injury may reveal pathways involved in activation of myogenic progenitors.

5. MYOGENIC miRNAs AS CELL CYCLE REGULATORS

A mechanism by which skeletal muscle-specific miRNAs can increase the efficacy of MyoD-mediated differentiation is by coordinating the exit from cell cycle with the onset of MyoD-dependent differentiation.

To this end, both miR-1 and miR-206 target DNA polymerase α, a key component required for DNA replication (Kim et al., 2006). Overexpression of either miR-1 or miR-206 leads to rapid block of DNA replication and exit from cell cycle (Kim et al., 2006). miR-206 decreases DNA Pol α by directly decreasing mRNA levels rather than repressing translation. This is consistent with a need for permanent exit from the cell cycle during the conversion of myoblasts into myotubes. However, both these myogenic-promoting miRNAs potentially target more than one mRNA involved in cell cycle to facilitate coordinated inhibition of multiple stages of the cell cycle.

The role for miR-1 and miR-206 in facilitating cardioblast and myoblast differentiation and their upregulation in terminally differentiated cells strongly suggest that their ablation may lead to excess proliferation reminiscent of cancers. In fact, even the very limited expression of all three myogenic miRNAs, miR-1, miR-133, and miR-206 in prostrate tissue was downregulated in prostrate tumor (Kim et al., 2006). This is consistent with similar observations of other miRNAs in many different human cancers (Lu et al., 2005).

6. miRNAs IN THE ADULT HEART

The ability to match metabolic demand with energy substrate availability is indispensable for all living cells. The heart is required to perpetually respond to dynamic changes in myocardial workload, often in very short time frames. By acting as negative regulators at the last step before translation, miRNAs are well-suited to play a major role in responding rapidly to cardiac stress. Levels of several miRNAs that are not specific for the heart are altered after changes in cardiac workload or injury, providing a miRNA signature to cardiac hypertrophy or failure (van Rooij et al., 2006). Several of these miRNAs could independently induce hypertrophy, suggesting that their upregulation may be involved in the cardiac response to stress. As a corollary, modulation of specific miRNAs could be useful in preventing some of the pathologic consequences of cardiac stress.

7. FUTURE DIRECTIONS

7.1. Determining miRNA function via identification of targets

Similar to transcription factors, miRNAs mediate their function by binding in *trans* to their targets. Delineating the function of a miRNA, therefore, requires the identification of the mRNAs to which it binds. This has been hindered by the lack of consistent complementarity between the two interacting RNA moieties. It has now been demonstrated that seed matches thought to be critical for miRNA–mRNA stable interaction need not be restricted to just the 5′ end of the miRNA. A poor seed match or G:U wobbles within the 5′ seed match region of the miRNA may be compensated by extensive base pairing of other regions of the miRNA (Brennecke et al., 2005). Developing techniques to directly assay binding between the miRNA and its target transcript, akin to establishing binding of a transcription factor to a promoter or enhancer, would be desirable. This seems especially necessary as miRNAs might function *in vivo* in a manner conceptually similar to transcription factors, where the presence of a miRNA binding site is not sufficient for its functionality but might require the equivalent of cofactors or even other miRNAs, thus introducing additional specificity. A *cis* parameter that might restrict the access of a miRNA is the secondary structure of the target transcript. Another *cis* factor that can have major consequences on the availability of an mRNA for repression is posttranscriptional RNA editing of the miRNA target site or regions affecting its accessibility. Thus, miRNA target validation in the future will need integration of computational approaches with "context-dependent" validation using experimental approaches.

7.2. Implications of miRNA biology to cardiovascular disease

The identification of muscle-specific miRNAs introduces a new level of regulation in developing an adult muscle function. Similar to mutations in transcription factors or their binding sites, it is likely that mutations in miRNAs or their binding sites within mRNAs could lead to human congenital or acquired heart disease. The role of miRNAs in cardiac hypertrophy suggests that there may be an opportunity to offset the progression toward heart failure using agonists or inhibitors of specific miRNAs. As miRNA targets are identified, downstream pathways may be modulated via miRNAs or other novel approaches to treat cardiac and muscle diseases. With the maturation of this nascent field, we will likely find numerous miRNAs that control cardiac lineage commitment, morphogenesis, and maintenance. Integration of miRNA-dependent pathways with established signaling and transcriptional networks will provide a more complete view of cardiac regulation.

NOTES ADDED IN PROOF

During the publication phase of this review, several important papers regarding cardiac miRNAs were published that further highlight the significance of miRNAs in both cardiogenesis and postnatal cardiac maintenance. Of particular relevance was the reporting of the first two targeted deletions of miRNAs in mice, one involving miR-1-2 and the other targeting miR-208. miR-1-2 mutant mice displayed cardiac morphogenetic defects in a subset and those surviving to adulthood had cardiac arrhythmias and perturbation of the exquisite cell cycle control typical of cardiomyocytes (Zhao et al., Cell, 129:303–317, 2007). Mice lacking miR-208, which is expressed only in the heart and embedded in a myosin heavy chain gene, survived to adulthood but failed to develop hypertrophy in response to pressure-overload stress (van Rooij et al., Science, 316:575–579, 2007).

ACKNOWLEDGMENTS

Authors thank S. Ordway and G. Howard for editorial assistance. D.S. was supported by grants from NHLBI/NIH, March of Dimes Birth Defects Foundation, and is an established investigator of the American Heart Association. This work was also supported by NIH/NCRR grant (C06 RR018928) to the Gladstone Institutes.

REFERENCES

Ambros, V. 2001. microRNAs: Tiny regulators with great potential. Cell 107, 823–826.

Anderson, C., Catoe, H., Werner, R. 2006. MIR-206 regulates connexin43 expression during skeletal muscle development. Nucleic Acids Res. 34, 5863–5871.

Brennecke, J., Stark, A., Russell, R.B., Cohen, S.M. 2005. Principles of microRNA-target recognition. PLoS Biol. 3, e85.

Chen, J.F., Mandel, E.M., Thomson, J.M., Wu, Q., Callis, T.E., Hammond, S.M., Conlon, F.L., Wang, D.Z. 2006. The role of microRNA-1 and microRNA-133 in skeletal muscle proliferation and differentiation. Nat. Genet. 38, 228–233.

Cullen, B.R. 2006. Viruses and microRNAs. Nat. Genet. 38(Suppl.), S25–S30.

Farh, K.K., Grimson, A., Jan, C., Lewis, B.P., Johnston, W.K., Lim, L.P., Burge, C.B., Bartel, D.P. 2005. The widespread impact of mammalian MicroRNAs on mRNA repression and evolution. Science 310, 1817–1821.

Gregory, R.I., Yan, K.P., Amuthan, G., Chendrimada, T., Doratotaj, B., Cooch, N., Shiekhattar, R. 2004. The Microprocessor complex mediates the genesis of microRNAs. Nature 432, 235–240.

Kim, H.K., Lee, Y.S., Sivaprasad, U., Malhotra, A., Dutta, A. 2006. Muscle-specific microRNA miR-206 promotes muscle differentiation. J. Cell Biol. 174, 677–687.

Kwon, C., Han, Z., Olson, E.N., Srivastava, D. 2005. MicroRNA1 influences cardiac differentiation in Drosophila and regulates Notch signaling. Proc. Natl. Acad. Sci. USA 102, 18986–18991.

Lagos-Quintana, M., Rauhut, R., Yalcin, A., Meyer, J., Lendeckel, W., Tuschl, T. 2002. Identification of tissue-specific microRNAs from mouse. Curr. Biol. 12, 735–739.

Leaman, D., Chen, P.Y., Fak, J., Yalcin, A., Pearce, M., Unnerstall, U., Marks, D.S., Sander, C., Tuschl, T., Gaul, U. 2005. Antisense-mediated depletion reveals essential and specific functions of microRNAs in Drosophila development. Cell 121, 1097–1108.

Lee, Y., Kim, M., Han, J., Yeom, K.H., Lee, S., Baek, S.H., Kim, V.N. 2004. MicroRNA genes are transcribed by RNA polymerase II. EMBO J. 23, 4051–4060.

Lim, L.P., Lau, N.C., Garrett-Engele, P., Grimson, A., Schelter, J.M., Castle, J., Bartel, D.P., Linsley, P.S., Johnson, J.M. 2005. Microarray analysis shows that some microRNAs downregulate large numbers of target mRNAs. Nature 433, 769–773.

Lu, J., Getz, G., Miska, E.A., Alvarez-Saavedra, E., Lamb, J., Peck, D., Sweet-Cordero, A., Ebert, B.L., Mak, R.H., Ferrando, A.A., Downing, J.R., Jacks, T., et al. 2005. MicroRNA expression profiles classify human cancers. Nature 435, 834–838.

Matranga, C., Zamore, P.D. 2004. Plant RNA interference *in vitro*. Cold Spring Harb. Symp. Quant. Biol. 69, 403–408.

Naguibneva, I., Ameyar-Zazoua, M., Polesskaya, A., Ait-Si-Ali, S., Groisman, R., Souidi, M., Cuvellier, S., Harel-Bellan, A. 2006. The microRNA miR-181 targets the homeobox protein Hox-A11 during mammalian myoblast differentiation. Nat. Cell. Biol. 8, 278–284.

Pasquinelli, A.E., Ruvkun, G. 2002. Control of developmental timing by micrornas and their targets. Annu. Rev. Cell. Dev. Biol. 18, 495–513.

Pillai, R.S., Bhattacharyya, S.N., Artus, C.G., Zoller, T., Cougot, N., Basyuk, E., Bertrand, E., Filipowicz, W. 2005. Inhibition of translational initiation by Let-7 MicroRNA in human cells. Science 309, 1573–1576.

Pillai, R.S., Bhattacharyya, S.N., Filipowicz, W. 2007. Repression of protein synthesis by miRNAs: How many mechanisms? Trends. Cell Biol. 17, 118–126.

Rao, P.K., Kumar, R.M., Farkhondeh, M., Baskerville, S., Lodish, H.F. 2006. Myogenic factors that regulate expression of muscle-specific microRNAs. Proc. Natl. Acad. Sci. USA 103, 8721–8726.

Rosenberg, M.I., Georges, S.A., Asawachaicharn, A., Analau, E., Tapscott, S.J. 2006. MyoD inhibits Fstl1 and Utrn expression by inducing transcription of miR-206. J. Cell Biol. 175, 77–85.

Schwarz, D.S., Hutvagner, G., Du, T., Xu, Z., Aronin, N., Zamore, P.D. 2003. Asymmetry in the assembly of the RNAi enzyme complex. Cell 115, 199–208.

Sokol, N.S., Ambros, V. 2005. Mesodermally expressed *Drosophila* microRNA-1 is regulated by Twist and is required in muscles during larval growth. Genes Dev. 19, 2343–2354.

van Rooij, E., Sutherland, L.B., Liu, N., Williams, A.H., McAnally, J., Gerard, R.D., Richardson, J.A., Olson, E.N. 2006. A signature pattern of stress-responsive microRNAs that can evoke cardiac hypertrophy and heart failure. Proc. Natl. Acad. Sci. USA 103, 18255–18260.

van Rooij, E., Sutherland, L.B., Qi, X., Richardson, J.A., Hill, J., Olson, E.N., 2007. Control of stress-dependent cardiac growth and gene expression by a microRNA. Science 316, 575–579.

Zhao, Y., Samal, E., Srivastava, D. 2005. Serum response factor regulates a muscle-specific microRNA that targets Hand2 during cardiogenesis. Nature 436, 214–220.

Zhao, Y., Ransom, J.F., Li, A., Vedantham, V., von Drehle, M., Muth, A.N., Tsuchihashi, T., McManus, M.T., Schwartz, R.J., Srivastava, D. 2007. Dysregulation of cardiogenesis, cardiac conduction, and cell cycle in mice lacking miRNA-1-2. Cell 129, 303–317.

Zhao, Y., Srivastava, D. 2007. A developmental view of microRNA function. Trends Biochem. Science 32, 189–197.

CHAPTER 8

Divergent Roles of Hedgehog and Fibroblast Growth Factor Signaling in Left–Right Development

Judith M. Neugebauer and **H. Joseph Yost**

Contents

Abstract

Left–right (LR) patterning of the vertebrate embryo regulates heart, gut, and brain development. Three transient signaling centers in the embryo contribute to LR patterning: ciliated cells in the node/organizer/shield, the floorplate and notochord in the embryonic midline, and the lateral plate mesoderm. A cassette of asymmetrically expressed genes in the left lateral plate mesoderm is evolutionarily conserved and upstream of organ asymmetry. This cassette includes members of the TGF-β signaling pathway family (Nodal

Department of Neurobiology and Anatomy, University of Utah, Salt Lake City, Utah

Advances in Developmental Biology, Volume 18
ISSN 1574-3349, DOI: 10.1016/S1574-3349(07)18008-X

and Lefty) and the transcription factor Pitx2. Here, we review two other signaling pathways hedgehog and fibroblast growth factor, which have been implicated in LR patterning but appear to have divergent roles in distinct classes of vertebrates.

1. INTRODUCTION: BRIEF HISTORICAL PERSPECTIVE ON LEFT–RIGHT ASYMMETRY

External structures such as limbs and eyes are positioned at specific coordinates along the anterior–posterior (AP) (head to tail) axis and dorsal–ventral (DV) (back to front) axis. Typically, these exterior structures are bilaterally symmetric, so that they are equally sized and positioned from an imaginary plane that separates the left and right mirror-image halves of the animal. In contrast, there is very little mirror-image symmetry inside the animal. Internal organs are often asymmetric along the left–right (LR) axis. For example, the chambers of the heart, the major vessels of the cardiovascular system, the coiling of the gut, and placement of liver, gall bladder, pancreas, stomach, and spleen are all LR asymmetric. Two questions arise from the examination of internal organ morphology. First, what is the function of organ asymmetry; does it have advantages over bilateral symmetry? Second, how are LR asymmetries oriented with respect to the other body axes?

From a developmental perspective, it is striking that asymmetric organs arise from bilaterally symmetric primordial. The complex four-chambered LR asymmetric heart in mammals develops from a pair of primitive, bilaterally symmetric cardiac tubes or groups of cells that fuse across the embryonic midline during early development (Olson and Srivastava, 1996). The LR asymmetric gut arises from a primitive, bilaterally symmetric tube that coils in a stereotypic fashion and buds off asymmetric organs (Branford et al., 2000). It is likely that LR asymmetry provides mechanical advantages over symmetry. An unlooped, bilaterally symmetric cardiac tube forms a single, unidirectional pump, resembling the dorsal–vessel (analogous to the heart) that circulates hemolymph in insects (Bodmer and Venkatesh, 1998). By looping the cardiac tube on itself, the tube becomes septated into multiple chambers. The four-chamber heart is essentially two parallel pumps, so that a single heartbeat efficiently and simultaneously sends deoxygenated blood (through the right atrium and ventricle) to the pulmonary system and oxygenated blood (through the left atrium and ventricle) to systemic circulation. Looping and coiling allow the intestinal tract to have extensive surface area for nutrient and waste exchange, by packing a length of 6–8 m of intestine in humans into in a small space.

The organs that are asymmetric, for example the heart, gut, and brain, arise from embryologically distinct primordial layers, the mesoderm, endoderm, and ectoderm, respectively. The morphogenic processes by which they form asymmetries appear to be divergent, and certainly the functions of the asymmetries are unique to each organ. This led to the assumption that asymmetries arise by distinct mechanisms during the development of each organ system. However, embryological manipulations in the twentieth century found that early perturbations during embryogenesis, before the formation of organ primordia, resulted in alterations of orientation of multiple organs (Spemann and Falkenberg, 1919; Yost, 1991). Similarly, congenital defects in organ placement of visceral organs in humans are often coincident with complex congenital heart defects. While there are manipulations that block asymmetric morphogenesis, so that the embryo retains a bilaterally symmetric heart tube (Yost, 1990), most manipulations in early embryos that alter LR development result in organs that are asymmetric but randomly oriented with respect to each other. This led to the proposal that there is a common mechanism that establishes the "LR axis." The major tenets of the LR axis are that: (1) it is orthogonal to the embryologic AP and DV axes; (2) LR axial information is established early in development, concurrent with establishment of the other body axes; (3) LR axis information is transmitted to all developing organ primordial so that their orientations throughout the embryonic body plan are coordinated. Thus, embryo-wide LR axial information is superimposed on the subsequent mechanisms that drive morphogenesis in individual organ primordia, providing a bias to the orientation of LR organ asymmetry. In this model, embryos that have lost normal LR axial information would still develop asymmetric organs, but the LR orientation of each organ would be randomly assigned.

1.1. Molecular asymmetry

The foundations described in the field's "premolecular era" were based on embryological manipulations and a few uncloned mouse mutants [see Ciba Foundation Symposium (1991) and Yost (1995) for reviews]. A major breakthrough occurred in 1995 with the report of asymmetric gene expression patterns in chick during gastrulation and early somitogenesis, including asymmetric expression of sonic hedgehog (Shh) near the node (an early organizing center) and the TGF-β superfamily member Nodal (cNR1 in chick) in the left lateral plate mesoderm (LPM) (Levin et al., 1995). A series of manipulations indicated that a molecular pathway involving Shh and Nodal establishes LR asymmetric gene expression in left LPM, and that this pathway regulates organ asymmetry (Levin et al., 1995). This provided molecular markers for analysis of LR development, gave molecular evidence to the previously described embryologic idea that LR patterning occurs significantly earlier in development than organ formation, and implicated cells in or near the node as a source for signals that establish asymmetry in LPM.

1.2. Is LR asymmetry conserved among classes of vertebrates?

1.2.1. Asymmetric gene expression in LPM: phylotypic stages?

Asymmetric expression of Nodal in left LPM was rapidly found in other vertebrates, including Xnr1 in *Xenopus*, Nodal in mice, and two members, cyclops and southpaw, in zebrafish (Zhou et al., 1993; Lowe et al., 1996; Lustig et al., 1996; Sampath et al., 1997; Rebagliati et al., 1998; Long et al., 2003). Two other genes, Lefty, a divergent member of the TGF-β family, and the transcription factor Pitx2 are downstream of Nodal. Together, Nodal, Lefty, and Pitx2 make a highly conserved cassette that is asymmetrically expressed in left LPM in all vertebrates examined to date. *cis*-regulatory elements in the Lefty and Pitx2 genes allow asymmetric expression and appear to be conserved and downstream of Nodal signaling (Shiratori et al., 2006). In addition, Pitx2 is asymmetrically expressed in ascidians and amphioxus (Boorman and Shimeld, 2002a,b), indicating that at least one component of LR patterning is highly conserved throughout chordates.

1.2.2. The roles of cilia in LR asymmetry

The next breakthrough in the field occurred with the serendipitous discovery that motile cilia in the embryonic node contribute to an early step in LR patterning (Nonaka et al., 1998). Professor Hirokowa's group had been systematically knocking out murine members of the kinesin gene family. The goal was to understand the roles of these microtubule-dependent motors in vesicle transport, predominantly in neural development and function. While neural development was difficult to assess due to embryonic lethality, Nonaka et al. (1998) discovered that the kinesis KIF3B mutants lost normal nodal cilia and normal LR asymmetric expression of Lefty (Nonaka et al., 1998). Because the node cilia are predominantly 9 + 0, the previous assumption was that these cilia were nonmotile. This study demonstrated that the cilia generate a strong asymmetric flow of extracellular fluid across the node, capable of moving fluorescent beads from the right to the left side of the embryo. Loss of KIF3B prevents cells from building cilia on their cell surfaces. The resulting "Nodal flow" model proposed that this asymmetric fluid flow was responsible for establishing a LR gradient of an unknown signaling molecule. Motile cilia and asymmetric fluid flow have been documented in zebrafish, *Medaka*, rabbit, and *Xenopus*, and shown at least in zebrafish to provide an functionally obligatory role in LR development (Essner et al., 2005; Kramer-Zucker et al., 2005; Okada et al., 2005; Schweickert et al., 2007). Thus, it is likely that cilia-dependent generation of extracellular asymmetric fluid flow is a conserved step in LR development, although there are some alternative explanations (Wagner and Yost, 2000).

The recurrent theme—laboratories that study other developmental processes knocking out their favorite genes only to discover a LR phenotype—has brought many exciting genes into the field. This trend will likely increase, since hundreds of genes are predicted to be utilized in building a functional cilium (Inglis et al., 2006), and mutations that alter cilia formation or function are likely to also result in LR defects.

1.2.3. Do other steps precede the roles of cilia in LR development?

It is becoming clear that cilia have a critical role in LR development and in several other aspects of development (Bisgrove and Yost, 2006), but it is still debated whether cilia control the initial step that converts bilateral symmetry into asymmetry. While the requirement for normal cilia function for LR patterning appears to be conserved in vertebrates, and the models by which nodal flow establish LR axis information are elegant and parsimonious, it is not formally known whether cilia control the first step in LR patterning. Are they the prime mover, or do earlier molecular decisions influence LR axis formation? Several lines of evidence indicate that molecular asymmetries across the prospective LR axis exist prior to cilia formation or function (Adams et al., 2006), including asymmetric RNA localization (Levin et al., 2002) and PKCγ-dependent asymmetric phosphorylation of the heparin sulfate proteoglycan syndecan-2 (Kramer and Yost, 2002; Kramer et al., 2002). These asymmetries are best described in amphibians. An important question in the field is whether these asymmetries are conserved in other classes of vertebrates, and whether these asymmetries control parallel pathways or converge on cilia function in LR patterning.

2. ARE THE FUNCTIONS OF MAJOR CELL–CELL SIGNALING PATHWAYS, HEDGEHOG AND FGF, CONSERVED IN LR DEVELOPMENT?

Two signaling pathways, hedgehog (HH) and fibroblast growth factor (FGF), have been implicated in LR development in multiple classes of vertebrates. Strikingly, it is not clear whether they have conserved, multiple or divergent functions in LR development. This chapter will summarize and attempt to resolve some of the apparent contradictions in the roles of HH and FGF signaling in LR development.

2.1. Hedgehog

First discovered in *Drosophila*, HH signaling proteins are a conserved family of secreted factors responsible for diverse roles in embryogenesis. In vertebrates, there are three known HH members Shh, Indian hedgehog (Ihh), and Desert hedgehog (Dhh), although additional homologues are present in both zebrafish and *Xenopus* (Sampath et al., 1997; Schilling et al., 1999).

Shh, the most widely studied HH, plays a role in the central nervous system, limb development, and axial patterning (Echelard et al., 1993; Riddle et al., 1993; Chiang et al., 1996). Ihh and Dhh, less well-studied family members, participate in cartilage differentiation and developing germ line cells, respectively (Bitgood et al., 1996; Lanske et al., 1996; Vortkamp et al., 1996).

HH proteins are cleaved, yielding an active N-terminus 19-kDa fragment, called HH-N (Lee et al., 1994; Bumcrot et al., 1995; Porter et al., 1995), which undergoes further modifications including a choles-terol moiety and palmitoylation, creating a membrane-associated protein (Porter et al., 1995; Pepinsky et al., 1998). HH proteins bind to and are sequestered by a large membrane-spanning protein, Patched (Ptc) (Ingham et al., 1991; Chen and Struhl, 1996; Marigo et al., 1996a). In the absence of HH protein, Ptc inhibits Smoothened (Smo), a membrane-associated member of the G-protein–coupled receptor superfamily (Alcedo et al., 1996; van den Heuvel and Ingham, 1996). Through an unknown mechanism, HH interaction with Ptc releases Smo and allows the nuclear import of the transcription factor cubitus interruptus (Ci) which activates transcription of HH response genes, including Wnt, TGF-β family members, and Ptc (Ingham and McMahon, 2001). In verte-brates, the gli family of transcription factors are downstream targets of Smo (Marigo et al., 1996b; Lee et al., 1997).

2.1.1. Chick hedgehog

Shh was the first LR asymmetrically expressed gene to be discovered. In chick, Shh expression initiates in Hensen's node and expands to become asymmetrically expressed with stronger expression on the left side of Hensen's node at Hamburger and Hamilton stage 4+ (Levin et al., 1995). Shh is thought to act as a left-sided determinant capable of initiat-ing a cassette of left-side genes (Fig. 1). Implantation of Shh-expressing cells into right LPM induces Nodal expression and randomizes subsequent heart looping (Levin et al., 1995). Implantation of a Shh protein bead on the right side of the node induced reversal of gut looping and embryo rotation (Levin et al., 1997). Similarly, implantation of Nodal in the right LPM randomize heart looping, suggesting that Shh affects heart looping via the Nodal pathway (Levin et al., 1997). Expression of Ptc, which is indicative of sustained HH signaling, becomes strongly asymmetric around the node until stage 5+. Inhibition of Shh with a function-blocking antibody results in decreased Nodal and Ptc expression on the left side and bilateral expression of chick Snail-related (cSnR), a right-sided marker (Fig. 1; Pagan-Westphal and Tabin, 1998). These stud-ies indicate that Shh acts as a left-sided determinant to activate left-sided gene expression and antagonize expression of right-sided genes.

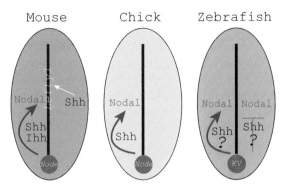

FIGURE 1   Activity of HH signaling diagrams of dorsal views of embryos with anterior (cranial) up, posterior down, left to the left, showing relative positions of the major signaling centers for LR patterning: node/ciliated cells, midline, and LPM. The proposed positive (arrows) and negative (bars) functions of signaling pathways are superimposed on the diagrams. Mouse: HH as a left determinant; both Shh and Ihh are necessary, but not sufficient, to activate Nodal in LPM. In addition Shh turns on Lefty1 in the midline, which maintains midline barrier function. Chick: Shh acts as a left-determinant, is expressed asymmetrically on the left side of the node, and can induce Nodal. Zebrafish: 'The class of determinant to which HH belongs in zebrafish is unknown. However, over-expression of Shh protein in zebrafish causes randomization of heart looping.

The midline is thought to provide a barrier separating left- and right-asymmetric signals (Danos and Yost, 1996; Levin et al., 1996). Lefty1 is expressed in the midline. When Lefty1 is mutated in mice or knocked down in zebrafish, Nodal expression becomes bilateral (Meno et al., 1998; Bisgrove et al., 1999). In the chick, expression of Shh and Lefty1 overlaps in the midline. Implantation of Shh protein bead on the right side of Hensen's node induces Lefty1 expression, and function-blocking anti-Shh antibodies block endogenous Lefty1 expression (Tsukui et al., 1999). To determine whether Shh alone was capable of inducing Nodal expression Pagan-Westphal and Tabin used explants of the LPM with and without associated tissues. When the right LPM was explanted without midline and paraxial tissue, there was no induction of Nodal. However, if the right LPM was explanted and cocultured with these tissues, ectopic expression of Nodal was induced. Shh beads were not capable of inducing Nodal expression in isolated LPM, but could induce Nodal expression when LPM was cocultured with paraxial and midline tissue. At later stages (stage 6), Nodal could be induced in LPM in the absence of paraxial and midline tissue (Pagan-Westphal and Tabin, 1998). These experiments suggest that HH in the chick could be acting to turn on Nodal signaling by activating signals from the midline.

In contrast to the asymmetric expression of Shh near the node in chick (Levin et al., 1995), and despite efforts by several laboratories, members of the HH family have not been found to be asymmetrically expressed in other vertebrates. Knockout of Shh in mice has a very subtle and late laterality phenotype and mutants in HH-related genes in zebrafish have normal cardiac laterality (Chiang et al., 1996; Schilling et al., 1999). This suggests that HH is a LR pathway component in aves but not in other classes of vertebrates, that components of HH signaling yet to be discovered in the genomes of other vertebrates replace the function of HH seen in chick, or that the LR pathway is initiated by divergent mechanisms in distinct classes of vertebrates.

2.1.2. Mouse hedgehog

Manipulations of the mouse HH-signaling pathway do not result in early laterality phenotypes seen dramatically in chick. Unlike in chick, HH-associated genes, including Shh, Ihh, Ptc, SMO, are all symmetrically expressed around the node (Zhang et al., 2001). Several studies of mouse mutant have examined possible roles for Shh in laterality. Chiang et al. (1996) determined that in $Shh^{-/-}$ mutants have abnormalities in heart, lung, and gut development and have no distinguishable midline, but they did not detect any asymmetrical defects of early heart and gut looping. A subsequent analysis of $Shh^{-/-}$ mutants reveals that the tail points to the left in these mutants, whereas in normal littermates the tails point to the right, indicating a randomized body axis turning. In addition, $Shh^{-/-}$ mutants had severe heart abnormalities with delayed or incomplete heart looping, and displayed left pulmonary isomerism (Tsukui et al., 1999). In $Shh^{-/-}$ mutants, Nodal, Lefty2, and Pitx2 are all bilaterally expressed in both left and right LPM, presumably due to midline defects in $Shh^{-/-}$ mutants (Meyers and Martin, 1999; Tsukui et al., 1999). The absence of midline may be due to absence of Lefty1 expression in the left prospective floor plate and absence of Lefty2 expression in the midline. Lefty1 was also absent in the left LPM in $Shh^{-/-}$ mutants (Tsukui et al., 1999). Shh and Lefty1 are expressed in similar domains in the mouse and Shh expression is unaffected in $Lefty1^{-/-}$ embryos (Tsukui et al., 1999). Together, these results suggest that Shh controls Lefty1 expression in mouse embryos (Fig. 1).

The laterality phenotype seen in Shh mouse mutants is apparently due to the loss of a normal midline barrier. To determine whether the HH family controlled initiation of expression of laterality genes, an attempt to abrogate all HH function was performed by creating a $Smo^{-/-}$ knockout mouse (Table 1). $Smo^{-/-}$ mutants have no embryonic turning, exhibit abnormal heart looping and a cyclopia phenotype characteristic of $Shh^{-/-}$ mutants (Zhang et al., 2001). HH signaling might control initiation of left-sided genes, since Pitx2 and Nodal expression in the LPM is decreased in either $Shh^{-/-}$:$Ihh^{-/-}$ double mutants or $Smo^{-/-}$. In contrast, Pitx2 expression is normal in $Ihh^{-/-}$ mutants. However, in $Ptc1^{-/-}$

TABLE 1 Phenotypes and manipulations of the hedgehog pathway

Organism	Gene	Genetic manipulation	Gross phenotype	LR phenotype	References
Chick	Shh	Cell implantation: Shh expressing cells	ND	Upregulation of Nodal expression, heart Reversals	Levin et al., 1995
Chick	Shh	Shh bead	ND	Heart, gut, embryo reversals, upregulation of Nodal expression	Levin et al., 1997; Pathi et al., 2001
Chick	Shh	Anti-Shh antibody	ND	Absent Nodal, Lefty1, and Ptc1 expression in LPM; bilateral cSnR expression in LPM	Pagan-Westphal and Tabin, 1998; Tsukui et al., 1999; Rodriguez-Esteban et al., 2001
Chick	Ihh	Ihh bead	ND	Upregulation of Nodal expression	Pathi et al., 2001
Chick	Dhh	Dhh bead	ND	Upregulation of Nodal expression	Pathi et al., 2001
Chick	PKA	PKA activator	ND	Heart reversals, ectopic expression of Nodal in right LPM	Rodriguez-Esteban et al., 2001
Mouse	Ihh, Shh	Ihh$^{-/-}$ mutant, Shh$^{-/-}$ mutant	Cyclopia	Absence of Pitx2 in LPM	Zhang et al., 2001
Mouse	Shh	Shh$^{-/-}$ mutant	Cyclopia, heart, lung, gut defects; abnormal midline	None	Chiang et al., 1996

(*continued*)

TABLE 1 (continued)

Organism	Gene	Genetic manipulation	Gross phenotype	LR phenotype	References
Mouse	Shh	Shh$^{-/-}$ mutant	ND	Randomized body axis, incomplete heart looping, left pulmonary isomerism; Absent Lefty1 expression; Bilateral Lefty2, Nodal, Pitx2 expression	Tsukui et al., 1999; Meyers and Martin, 1999; Zhang et al., 2001
Mouse	Smo	Smo$^{-/-}$ mutant	Cyclopia	No embryonic turning, abnormal heart looping, absent Nodal, lefty1, lefty2, and Pitx2 expression in LPM	Zhang et al., 2001
Mouse	Ihh	Ihh$^{-/-}$ mutant	None	none	Zhang et al., 2001
Mouse	Ihh, Shh	Ihh$^{-/-}$ mutant, Shh$^{-/-}$ mutant	Cyclopia	Absence of Pitx2 in LPM	Zhang et al., 2001
Zebrafish	Shh-a	Overexpression: Shh mRNA	ND	Heart and gut reversals	Schilling et al., 1999
Zebrafish	PKA	Overexpression: dominant negative; PKA mRNA	ND	Heart reversals	Schilling et al., 1999
Zebrafish	Smo	Mutant line: Smo	Heart and craniofacial defects, absence of pectoral fins	none	Chen et al., 2001

ND, not determined.

knockout mice (which results in constitutively active HH signaling), both Pitx2 and Nodal expression is normal, suggesting that HH signaling is not sufficient to induce ectopic gene expression. In $Smo^{-/-}$, both Lefty1 and Lefty2 were not expressed, but the LPM was grossly normal as seen by other LPM markers (Zhang et al., 2001). Cryptic, an EGF-CFC protein that serves as a Nodal coreceptor to turn on Lefty2, is reduced in $Smo^{-/-}$ embryos and upregulated in $Ptc1^{-/-}$ mutants, suggesting that Cryptic and its downstream targets are controlled by HH signaling (Zhang et al., 2001). This suggests that HH signaling may be necessary but not sufficient to induce left-sided markers in the mouse embryo.

2.1.3. Zebrafish hedgehog
Analysis of HH signaling in the establishment of laterality in zebrafish has not revealed a clear mechanism. Although HH appears to play a role in laterality, it is unclear whether it is a left or right signaling determinant. Due to a partial genome duplication several additional paralogues of HH exist, for example zebrafish Shh-a and Shh-b (also known as twiggy-winkle HH) (Ekker et al., 1995). A large-scale screen to identify heart-looping mutants in zebrafish characterized the Shh mutant [sonic you (syu)] as having normal heart jogging, indicative of normal LR patterning. However, BMP4, which is normally enriched on the left side of the developing heart tube, is symmetrically expressed in syu mutants. To further explore the role of Shh in zebrafish heart development, Schilling et al. overexpressed Shh mRNA by injection in 1-cell of a 2-cell embryo. Overexpression of Shh on the right side induced heart reversals, whereas overexpression on the left side did not. Protein kinase A (PKA) is capable of inhibiting downstream competitors of HH signaling, therefore injection of a PKA-dominant negative (dnPKA) would create constitutively active HH signaling. Injection of this dnPKA increased heart looping defects, with stronger effects than overexpression of Shh mRNA (Schilling et al., 1999). Knockdown of Smo in the zebrafish, which should abrogate HH signaling, resulted in abnormal heart development but never exhibiting reversed heart looping, although many hearts had high amounts of edema which may have affected heart looping (Table 1; Chen et al., 2001). In order to determine whether and where HH signaling affects laterality, and whether members of the HH family serve as right-side and/or left-side determinants, knockdown or mutant analysis of multiple family members and analysis of molecular-laterality markers will be necessary.

2.1.4. Possible explanations for divergent HH signaling
At first glace it appears that HH signaling plays divergent roles in chick and mouse LR development. However, the discrepancies might simply be due to differing roles of HH family members in the control of laterality. One potential explanation of why Shh is only needed in chick versus both

Ihh and Shh in mouse is that the blocking antibody was capable of inhibiting both Shh and Ihh (Zhang et al., 2001). This is possible given that the antibody cross-reacts with IHH (Ericson et al., 1996; Vortkamp et al., 1996). Although the expression patterns of Ihh in the chick embryo have not been published, Ihh has been shown to induce Nodal expression in the right LPM of chick embryos (Pathi et al., 2001). In the chick, Shh protein was shown to be necessary and sufficient to induce Nodal signaling. In mouse embryos, HH signaling is necessary, but not sufficient to induce Nodal. This conclusion was based on a $Ptc^{-/-}$ mouse mutant, and since there are two known vertebrate Ptc proteins, Smo activity may not have been fully activated in the $Ptc^{-/-}$ mutant.

Considering that HH proteins are conserved, do their activities differ in activating cellular targets? Pathi et al. showed that each HH protein could bind to Ptc with comparable affinities; however, the cellular responses to each HH protein were very different. At a given protein concentration, Shh was more effective than Ihh, and Ihh was more effective than Dhh (Pathi et al., 2001). This was determined in several different context, including a human cell line induction assay, Nodal induction in the chick LPM, and digit duplication of the chick wing bud. In the case of the wing bud, Dhh induced duplication more efficiently than Ihh and at levels near Shh (Pathi et al., 2001). These results suggest that the limiting factor for HH signaling is both expression pattern and levels of expressed protein required to induce a cellular response.

Zebrafish HH family creates an interesting argument for the divergence of HH protein activity. Contrary to the mouse, knockdown of both Shh-a and Shh-b does not abolish floor plate development (Etheridge et al., 2001). Also distinct from results in mouse, loss of all HH signaling by knocking down Smo protein results in loss of floor plate structure but ectopic expression of certain floor plate markers (Placzek et al., 2000; Chen et al., 2001). Further elucidations of laterality control by HH will advance our understanding of both conserved and divergent mechanisms of HH signaling in vertebrates.

2.2. Fibroblast growth factors

The question of LR pathway divergence also arises when considering the roles of FGFs, a highly conserved family of secreted signaling molecules. Diffusible FGF ligands bind cell surface receptors and induce a variety of downstream signaling events including activation of the MAPK cascade, anti-apoptosis cascade, and Ca^{2+} release, regulating differentiation, cell migration, and cell survival (Dailey et al., 2005; Thisse and Thisse, 2005). There are 4 known transmembrane FGF receptors (FGFR) and 22 different FGF ligands, with both ligands and receptors capable of forming multiple ligand–receptor complexes (Ornitz et al., 1996; Ornitz and Itoh, 2001; Itoh and Ornitz, 2004). Ligand–receptor specificity is controlled in part by association with cell surface heparan sulfate proteoglycans (HSPGs) (Yayon et al., 1991; Schlessinger et al., 2000). FGF family members are found in mouse, chicken, zebrafish, *Drosophila*, *C. elegans*, and planaria. FGF signaling has roles in mesoderm development, neurulation, brain, and limb development (Klambt et al., 1992; DeVore et al., 1995; Ornitz and Itoh, 2001; Cebria et al., 2002; Itoh and Ornitz, 2004; Thisse and Thisse, 2005). Although FGFs have been widely studied, the extent to which each ligand–receptor pair induces a unique intracellular response in any given cell type is unclear. Thus, it might be unsurprising that FGFs have been implicated in multiple roles for LR patterning.

2.2.1. FGF control of node formation/function

One of the first steps to establishing bilateral symmetry and subsequent LR asymmetry is formation of the node. FGF signaling has a critical role in both formation and function of the node in mouse, chick, and zebrafish. Several groups have examined FGF signaling within Hensen's node, but no studies have focused on the role of FGF signaling in Hensen's node in relation to LR development. However, studies have looked at the control of cells within Hensen's node by FGF signaling in other arenas. Of particular interest is the finding that FGF controls the migration and proliferation of cells in the node. By expressing a dominant-negative FGFR (lacking the cytoplasmic domain) in three regions within and surrounding Hensen's node, Mathis et al. found that cells lacking FGF signaling prematurely exit the region of the node, creating neural tube defects and preventing elongation of the embryonic axis (Mathis et al., 2001). This effect was cell autonomous, indicated by the lack of effects on nontransformed cells in the same region. Based on the ability of FGF bead implantation to mimic Hensen's node transplantation in the ability to induce cardiac tissue, it is suggested that FGF is produced and secreted by Hensen's node (Lopez-Sanchez et al., 2002). TGF-β family members could only induce cardiac tissue in the presence of FGF beads, where together the two signaling molecules had an increased induction of cardiac markers (Lopez-Sanchez et al., 2002). These two studies suggest that FGF signaling is active in Hensen's node both cell autonomously and noncell autonomously.

In the mouse, FGF signaling is necessary for nodal flow. As correlative evidence, three of four FGFRs were detected on mouse node monocilia. However, FGFR activation, as detected by high levels of phosphorylation of tyrosines on FGFRs, was detected symmetrically around the node (Tanaka et al., 2005). Treatment of embryos with an FGFR inhibitor, SU5402, decreased phosphotyrosine levels. The mechanoreceptor model of nodal flow postulates that asymmetric flow results in asymmetrical Ca^{2+} spikes on the left side of the node. In the mouse, inhibition of FGF signaling by SU5402 treatment abolished asymmetric Ca^{2+} activity (Tanaka et al., 2005). A controversial mechanism for nodal flow postulates movement and asymmetric accumulation of "nodal vesicular parcels" (NVPs), lipophilic parcels that are moved from the floor of the node to the left by node cilia. Shh and retinoic acid (RA) are thought to be associated with the NVPs. Pharmalogical inhibition of FGF signaling results in a decrease in the speed and frequency of NVPs moving across the mouse node (Tanaka et al., 2005).

In zebrafish, Kupffer's vesicle (KV) is ciliated and is necessary for the induction of LR asymmetry (Essner et al., 2002, 2005). One study focusing on FGF8 in craniofacial development uncovered a role for FGF8 in the development of the KV (Albertson and Yelick, 2005). KV is absent or diminished in a hypomorphic strain of FGF8 zebrafish (ace; acerebellar) (Reifers et al., 1998; Draper et al., 2001; Albertson and Yelick, 2005), which correlated with perturbation of laterality marker expression. In contrast, ace mutants that had apparently normal KV formation displayed abnormal, asymmetric craniofacial development occurring in the absence of FGF8 signaling (Albertson and Yelick, 2005). This suggests that FGF8 plays an early role in KV formation and asymmetric patterning and a later role in craniofacial symmetry.

2.2.2. FGF8 acts divergently for correct LPM gene expression

After symmetry breaking at the node, LR patterning is propagated by asymmetric gene expression of Nodal in the left LPM. In the chick, FGF8 appears to act as a right determinant, suppressing the expression of left determinants in the right LPM (Fig. 2). In chick, FGF8 is expressed asymmetrically around Hensen's node starting at the Hamburger and Hamilton stage six and persisting to the four-somite stage (Boettger et al., 1999; Stolte et al., 2002). This expression is thought to be controlled by ActivinA, since ActivinA bead implantation adjacent to the left side of Hensen's node increases FGF8 expression. When FGF8 beads were placed in the left LPM, Nodal expression was decreased, even over relatively large distances, but expression of Shh in the node was unaffected (Boettger et al., 1999). This affect can be mimicked by FGF4 and FGF1 but not FGF7. In addition, FGF8 beads implanted into left LPM induce another right determinant, chicken–snail related (cSnR), downregulate Pitx2c, and randomize the orientation of subsequent heart looping (Boettger et al., 1999).

Asymmetric expression of FGF8 appears to be controlled by ActivinA, as discussed earlier, and by another FGF family member FGF18. FGF18 is predominantly expressed on the right-side of Hensen's node from stages 4 to 6, when FGF8 expression becomes more prominent on the right side of the node. Implantation of FGF18 beads in the left LPM induces FGF8 expression and cSnR expression (Ohuchi et al., 2000). Implantation of FGF18 beads in right LPM inhibits expression of Pitx2 in left LPM, suggesting long-range effects. Although the possibility of a feedback loop between FGF8 and FGF18 was not tested, the earlier asymmetric expression of FGF18 suggests that FGF18 is responsible for the asymmetric expression of FGF8 on the right side of the node.

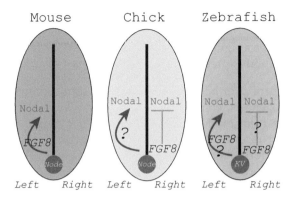

FIGURE 2   Activity of FGF signaling diagrams as in Fig. 1. Mouse: FGF8 acts as a left determinant. Hypomorphic FGF8 mouse mutants lack Nodal signaling in the LPM. Nodal is upregulated in LPM by ectopic FGF8 protein. Chick: In contrast to mouse, FGF8 acts as a right determinant. Ectopic FGF8 signaling on the left causes downregulation of Nodal in left LPM and upregulation of right-sided markers. Zebrafish: Knockdown of FGF8 activity in zebrafish causes "randomization" of Nodal; however, this could be due to disruption of Kupffer's vesicle (analogous to node ciliated cells). Further studies are necessary to characterize the role of FGF signaling in zebrafish LR development.

Contrary to the activity in chick, FGF8 is hypothesized to be a left determinant in mouse, turning on left-sided markers in the left LPM (Fig. 2). FGF8 null mice embryos die early in gastrulation due to cell migration defects (Meyers et al., 1998; Sun et al., 1999; Moon and Capecchi, 2000); however, FGF8$^{neo/-}$, an FGF8 hypomorphic mouse line, displays LR patterning defects. Approximately half of these embryos displayed abnormal heart looping and right pulmonary isomerism (both lungs had right identity). In FGF8$^{neo/-}$ embryos, Nodal and other left-sided markers Lefty2 and Pitx2 were absent from the left LPM. FGF8 protein beads implanted into right LPM during early somite stages induce expression of Nodal, presumably as a left LPM marker (Meyers and Martin, 1999). FGF8 is symmetrically expressed around the node, but appears to be acting in an asymmetrical manner (Tanaka et al., 2005).

In rabbit and zebrafish, FGF8 is symmetrically expressed around the node, or in the case of the zebrafish, KV (Fischer et al., 2002; Albertson and Yelick, 2005). In zebrafish *ace* mutants, *southpaw* (*spaw*; zebrafish nodal homologue) have "randomized" expression in the LPM, with embryos displaying either bilateral or right-sided expression. Downstream of *spaw*, *Pitx2c* expression in the brain is either left, bilateral, or absent, with no right-side only expression (Albertson and Yelick, 2005). *Ace* mutants have altered heart jogging and heart lopping, in addition to having midline or reversed placement of the gall bladder and pancreas. As discussed earlier, this control over laterality appears to stem from FGF8's control over KV formation/function, therefore this study does not ascertain whether FGF8 is acting as a left or right determinant in zebrafish (Fig. 2). In contrast, FGF8 appears to be a right determinant in rabbit—FGF8 beads implanted into left LPM inhibit expression of Nodal (Fischer et al., 2002).

2.2.3. FGF8 control over bilateral symmetry of somite formation

During embryogenesis, somite formation is consistent with a bilateral clock and wave front model, partially controlled by FGF8 (Baker et al., 2006). However, recent studies have shown that in mouse, chick, and zebrafish somite formation becomes asymmetrical when RA signaling is disrupted. Somite development in RALDH2$^{-/-}$ mouse mutants (mice deficient in RA) is asymmetric and FGF8 acquires an asymmetric distribution pattern in the right presomitic mesoderm (Vermot and Pourquie, 2005). This suggests that FGF8 might be responsible for an underlying asymmetric progression of the somite clock that is only revealed in the absence of RA signaling (Vermot et al., 2005). Inhibition of FGF8 signaling in zebrafish does not lead to asymmetric somite formation, but disruption of RA signaling resulted in both asymmetric somite formation and asymmetric FGF8 expression (Kawakami et al., 2005). In both the zebrafish and mice, inhibiting RA signaling created more somites on the left side, in contrast, inhibiting RA signaling in chick induced more somites on the right (Kawakami et al., 2005; Vermot and Pourquie, 2005; Vermot et al., 2005). This lead to the hypothesis that perhaps somite progression was also being controlled by the LR pathway, and indeed asymmetric somite formation was randomized when components known to control LR development were disrupted in conjunction with inhibition of RA signaling in both the mouse and chick (Vermot and Pourquie, 2005). Vermot and Pourquie hypothesize that the sided somite bias seen in chick and mouse might be due to the differential control of FGF8 over the LR pathway; however, FGF8 expression remained symmetrical in the chick when RA signaling was inhibited. Hamade et al. (2006) have suggested that FGF8 is a direct target of RA signaling for regulation of somitogenesis with upregulation of FGF8 expression by local induction of RA. That study conflicts with another mouse study where FGF8 expression was shown to be expanded in RALDH$^{-/-}$ embryos, and RA is proposed to confer somite symmetry by setting the anterior limit of FGF8 expression (Sirbu and Duester, 2006). Although no direct evidence demonstrates that FGF8 regulates bilateral somite progression, it appears that FGF8 might play opposite roles in the chick and mouse somitogenesis.

2.2.4. Explanation of divergent usage of FGF signaling

During embryo development, the roles of FGF signaling in node formation or function appear to be conserved. However, FGF signaling has divergent role in the initiation of asymmetric gene expression in the LPM. Several mechanisms have been proposed, including the hypothesis that in mouse FGF8 might asymmetrically activate Nodal as a morphogen, which is then moved across the node by cilia flow (Meyers and Martin, 1999). FGFR phosphotyrosine levels are symmetrical around the node, suggesting that asymmetry is not directly conferred by the FGF pathway (Tanaka et al., 2005). Fischer et al. proposed that divergence of embryo shape among classes of vertebrates correlates with the distinction of FGF8 as a right-side or left-side determinant. Mouse develops as a cup-shaped embryo, whereas chick, zebrafish, and rabbits develop as flat blastodiscs (Fischer et al., 2002). This difference is potentially important when exploring the role of gap junctional communication (GJC) around the node, which appear to connect differently around the blastodisc node versus the cup-shaped mouse embryos (Mercola and Levin, 2001; Fischer et al., 2002).

Another possible explanation of the divergent roles of FGF8 is differential specificities and activities of FGF ligands and FGF receptors. The FGFRs have several different isoforms of the third Ig domain, which is responsible for ligand specificities (Johnson et al., 1991; Duan et al., 1992; Chellaiah et al., 1994; Ornitz et al., 1996; Yeh et al., 2003). Zhang et al. (2006) cataloged all the FGF ligands binding to all splice forms of FGFR to determine binding specificity. They concluded that the FGF8 subfamily of FGF ligands activates FGFR1c, FGFR2c, FGFR3c, and both FGFR4 isoforms. However, the FGF8 subfamily of ligands did not activate the –b isoforms of these receptors, as detected by mitogenic activity (Zhang et al., 2006). Different FGF receptors have distinct levels of tyrosine kinase activity (Raffioni et al., 1999; Dailey et al., 2005). Therefore, if different FGFR isoforms are expressed in the LPM of the chick, mouse, and zebrafish, different ligand specificity along with variations of receptor activity could explain the activity of FGF8 as a Nodal-inducer or Nodal-inhibitor.

It is clear that we are only beginning to understand the complexities of signaling by FGFs. There are several other known activities of FGF signaling which are important to consider in the context of LR development. Several studies have explored the role of dosage dependence on FGF8 signaling; surprisingly, a hypomorphic phenotype is opposite a null phenotype (Storm et al., 2003; Jaskoll et al., 2004). Elucidating the mechanisms by which FGF8 signaling controls LR development might provide a deeper understanding of FGF signaling.

3. CONCLUSIONS

It is becoming clear that both the HH and FGF signaling pathways have roles in the LR pathway, but the roles for each pathway are either poorly defined or appear to be distinct in different classes of vertebrates.

Compounding the uncertainties for each pathway is a growing number of observations indicating that HH and FGF pathways have functional intersections in a signaling network that controls LR patterning. For example, asymmetric Ca^{2+} around the node is dependent on FGF signaling and can be partially rescued by the addition of Shh protein around the node. Ihh treatment also raises Ca^{2+} levels, but in a symmetrical manner around the node (Tanaka et al., 2005). This suggests that FGF signaling around the node is upstream of HH signaling, which in turn might be necessary for LR asymmetric gene expression. Clearly, it will be important to simultaneously consider multiple signaling pathways as components of a complex network that controls LR patterning in vertebrates. Similarly, it will be important to discern the roles of HH and FGF signaling in each of the three embryonic signaling centers that contribute to LR patterning: ciliated cells in the node/organizer/shield, the floorplate and notochord in the embryonic midline, and the LPM.

REFERENCES

Adams, D.S., Robinson, K.R., Fukumoto, T., Yuan, S., Albertson, R.C., Yelick, P., Kuo, L., McSweeney, M., Levin, M. 2006. Early, H+-V-ATPase-dependent proton flux is necessary for consistent left-right patterning of non-mammalian vertebrates. Development 133(9), 1657–1671.

Albertson, R.C., Yelick, P.C. 2005. Roles for fgf8 signaling in left-right patterning of the visceral organs and craniofacial skeleton. Dev. Biol. 283(2), 310–321.

Alcedo, J., Ayzenzon, M., Von Ohlen, T., Noll, M., Hooper, J.E. 1996. The *Drosophila* smoothened gene encodes a seven-pass membrane protein, a putative receptor for the hedgehog signal. Cell 86(2), 221–232.

Baker, R.E., Schnell, S., Maini, P.K. 2006. A clock and wavefront mechanism for somite formation. Dev. Biol. 293(1), 116–126.

Bisgrove, B.W., Yost, H.J. 2006. The roles of cilia in developmental disorders and disease. Development 133(21), 4131–4143.

Bisgrove, B.W., Essner, J.J., Yost, H.J. 1999. Regulation of midline development by antagonism of lefty and nodal signaling. Development 126(14), 3253–3262.

Bitgood, M.J., Shen, L., McMahon, A.P. 1996. Sertoli cell signaling by Desert hedgehog regulates the male germline. Curr. Biol. 6(3), 298–304.

Bodmer, R., Venkatesh, T.V. 1998. Heart development in *Drosophila* and vertebrates: Conservation of molecular mechanisms. Dev. Genet. 22(3), 181–186.

Boettger, T., Wittler, L., Kessel, M. 1999. FGF8 functions in the specification of the right body side of the chick. Curr. Biol. 9(5), 277–280.

Boorman, C.J., Shimeld, S.M. 2002a. Cloning and expression of a Pitx homeobox gene from the lamprey, a jawless vertebrate. Dev. Genes Evol. 212(7), 349–353.

Boorman, C.J., Shimeld, S.M. 2002b. Pitx homeobox genes in Ciona and amphioxus show left-right asymmetry is a conserved chordate character and define the ascidian adenohypophysis. Evol. Dev. 4(5), 354–365.

Branford, W.W., Essner, J.J., Yost, H.J. 2000. Regulation of gut and heart left-right asymmetry by context-dependent interactions between xenopus lefty and BMP4 signaling. Dev. Biol. 223(2), 291–306.

Bumcrot, D.A., Takada, R., McMahon, A.P. 1995. Proteolytic processing yields two secreted forms of sonic hedgehog. Mol. Cell. Biol. 15(4), 2294–2303.

Cebria, F., Kobayashi, C., Umesono, Y., Nakazawa, M., Mineta, K., Ikeo, K., Gojobori, T., Itoh, M., Taira, M., Sanchez Alvarado, A., Agata, K. 2002. FGFR-related gene nou-darake restricts brain tissues to the head region of planarians. Nature 419(6907), 620–624.

Chellaiah, A.T., McEwen, D.G., Werner, S., Xu, J., Ornitz, D.M. 1994. Fibroblast growth factor receptor (FGFR) 3. Alternative splicing in immunoglobulin-like domain III creates a receptor highly specific for acidic FGF/FGF-1. J. Biol. Chem. 269(15), 11620–11627.

Chen, W., Burgess, S., Hopkins, N. 2001. Analysis of the zebrafish smoothened mutant reveals conserved and divergent functions of hedgehog activity. Development 128(12), 2385–2396.

Chen, Y., Struhl, G. 1996. Dual roles for patched in sequestering and transducing Hedgehog. Cell 87(3), 553–563.

Chiang, C., Litingtung, Y., Lee, E., Young, K.E., Corden, J.L., Westphal, H., Beachy, P.A. 1996. Cyclopia and defective axial patterning in mice lacking sonic hedgehog gene function. Nature 383(6599), 407–413.

Ciba Foundation Symposium 1991. Symposium on the childhood environment and adult disease. Ciba Foundation London, 15–17 May 1990. Ciba Found. Symp. 156, 1–243.

Dailey, L., Ambrosetti, D., Mansukhani, A., Basilico, C. 2005. Mechanisms underlying differential responses to FGF signaling. Cytokine Growth Factor Rev. 16(2), 233–247.

Danos, M.C., Yost, H.J. 1995. Linkage of cardiac left-right asymmetry and dorsal-anterior development in Xenopus. Development 121(5), 1467–1474.

Danos, M.C., Yost, H.J. 1996. Role of notochord in specification of cardiac left-right orientation in zebrafish and Xenopus. Dev. Biol. 177(1), 96–103.

DeVore, D.L., Horvitz, H.R., Stern, M.J. 1995. An FGF receptor signaling pathway is required for the normal cell migrations of the sex myoblasts in C. elegans hermaphrodites. Cell 83(4), 611–620.

Draper, B.W., Morcos, P.A., Kimmel, C.B. 2001. Inhibition of zebrafish fgf8 pre-mRNA splicing with morpholino oligos: A quantifiable method for gene knockdown. Genesis 30(3), 154–156.

Duan, D.S., Werner, S., Williams, L.T. 1992. A naturally occurring secreted form of fibroblast growth factor (FGF) receptor 1 binds basic FGF in preference over acidic FGF. J. Biol. Chem. 267(23), 16076–16080.

Echelard, Y., Epstein, D.J., St-Jacques, B., Shen, L., Mohler, J., McMahon, J.A., McMahon, A.P. 1993. Sonic hedgehog, a member of a family of putative signaling molecules, is implicated in the regulation of CNS polarity. Cell 75(7), 1417–1430.

Ekker, S.C., Ungar, A.R., Greenstein, P., von Kessler, D.P., Porter, J.A., Moon, R.T., Beachy, P. A. 1995. Patterning activities of vertebrate hedgehog proteins in the developing eye and brain. Curr. Biol. 5(8), 944–955.

Ericson, J., Morton, S., Kawakami, A., Roelink, H., Jessell, T.M. 1996. Two critical periods of sonic Hedgehog signaling required for the specification of motor neuron identity. Cell 87(4), 661–673.

Essner, J.J., Vogan, K.J., Wagner, M.K., Tabin, C.J., Yost, H.J., Brueckner, M. 2002. Conserved function for embryonic nodal cilia. Nature 418(6893), 37–38.

Essner, J.J., Amack, J.D., Nyholm, M.K., Harris, E.B., Yost, H.J. 2005. Kupffer's vesicle is a ciliated organ of asymmetry in the zebrafish embryo that initiates left-right development of the brain, heart and gut. Development 132(6), 1247–1260.

Etheridge, L.A., Wu, T., Liang, J.O., Ekker, S.C., Halpern, M.E. 2001. Floor plate develops upon depletion of tiggy-winkle and sonic hedgehog. Genesis 30(3), 164–169.

Fischer, A., Viebahn, C., Blum, M. 2002. FGF8 acts as a right determinant during establishment of the left-right axis in the rabbit. Curr. Biol. 12(21), 1807–1816.

Hamade, A., Deries, M., Begemann, G., Bally-Cuif, L., Genet, C., Sabatier, F., Bonnieu, A., Cousin, X. 2006. Retinoic acid activates myogenesis in vivo through Fgf8 signalling. Dev. Biol. 289(1), 127–140.

Ingham, P.W., McMahon, A.P. 2001. Hedgehog signaling in animal development: Paradigms and principles. Genes Dev. 15(23), 3059–3087.

Ingham, P.W., Taylor, A.M., Nakano, Y. 1991. Role of the *Drosophila* patched gene in positional signalling. Nature 353(6340), 184–187.

Inglis, P.N., Boroevich, K.A., Leroux, M.R. 2006. Piecing together a ciliome. Trends Genet. 22(9), 491–500.

Itoh, N., Ornitz, D.M. 2004. Evolution of the Fgf and Fgfr gene families. Trends Genet. 20(11), 563–569.

Jaskoll, T., Witcher, D., Toreno, L., Bringas, P., Moon, A.M., Melnick, M. 2004. FGF8 dose-dependent regulation of embryonic submandibular salivary gland morphogenesis. Dev. Biol. 268(2), 457–469.

Johnson, D.E., Lu, J., Chen, H., Werner, S., Williams, L.T. 1991. The human fibroblast growth factor receptor genes: A common structural arrangement underlies the mechanisms for generating receptor forms that differ in their third immunoglobulin domain. Mol. Cell. Biol. 11(9), 4627–4634.

Kawakami, Y., Raya, A., Raya, R.M., Rodriguez-Esteban, C., Belmonte, J.C. 2005. Retinoic acid signalling links left-right asymmetric patterning and bilaterally symmetric somito-genesis in the zebrafish embryo. Nature 435(7039), 165–171.

Klambt, C., Glazer, L., Shilo, B.Z. 1992. Breathless, a *Drosophila* FGF receptor homolog, is essential for migration of tracheal and specific midline glial cells. Genes Dev. 6(9), 1668–1678.

Kramer, K.L., Barnette, J.E., Yost, H.J. 2002. PKCgamma regulates syndecan-2 inside-out signaling during xenopus left-right development. Cell 111(7), 981–990.

Kramer, K.L., Yost, H.J. 2002. Ectodermal syndecan-2 mediates left-right axis formation in migrating mesoderm as a cell-nonautonomous Vg1 cofactor. Dev. Cell 2(1), 115–124.

Kramer-Zucker, A.G., Olale, F., Haycraft, C.J., Yoder, B.K., Schier, A.F., Drummond, I.A. 2005. Cilia-driven fluid flow in the zebrafish pronephros, brain and Kupffer's vesicle is required for normal organogenesis. Development 132(8), 1907–1921.

Lanske, B., Karaplis, A.C., Lee, K., Luz, A., Vortkamp, A., Pirro, A., Karperien, M., Defize, L. H., Ho, C., Mulligan, R.C., Abou-Samra, A.B., Juppner, H., et al. 1996. PTH/PTHrP receptor in early development and Indian hedgehog-regulated bone growth. Science 273(5275), 663–666.

Lee, J., Platt, K.A., Censullo, P., Ruiz i Altaba, A. 1997. Gli1 is a target of sonic hedgehog that induces ventral neural tube development. Development 124(13), 2537–2552.

Lee, J.J., Ekker, S.C., von Kessler, D.P., Porter, J.A., Sun, B.I., Beachy, P.A. 1994. Autoproteo-lysis in hedgehog protein biogenesis. Science 266(5190), 1528–1537.

Levin, M., Johnson, R.L., Stern, C.D., Kuehn, M., Tabin, C. 1995. A molecular pathway determining left-right asymmetry in chick embryogenesis. Cell 82(5), 803–814.

Levin, M., Roberts, D.J., Holmes, L.B., Tabin, C. 1996. Laterality defects in conjoined twins. Nature 384(6607), 321.

Levin, M., Pagan, S., Roberts, D.J., Cooke, J., Kuehn, M.R., Tabin, C.J. 1997. Left/right patterning signals and the independent regulation of different aspects of situs in the chick embryo. Dev. Biol. 189(1), 57–67.

Levin, M., Thorlin, T., Robinson, K.R., Nogi, T., Mercola, M. 2002. Asymmetries in H+/K+-ATPase and cell membrane potentials comprise a very early step in left-right patterning. Cell 111(1), 77–89.

Long, S., Ahmad, N., Rebagliati, M. 2003. The zebrafish nodal-related gene southpaw is required for visceral and diencephalic left-right asymmetry. Development 130(11), 2303–2316.

Lopez-Sanchez, C., Climent, V., Schoenwolf, G.C., Alvarez, I.S., Garcia-Martinez, V. 2002. Induction of cardiogenesis by Hensen's node and fibroblast growth factors. Cell Tissue Res. 309(2), 237–249.

Lowe, L.A., Supp, D.M., Sampath, K., Yokoyama, T., Wright, C.V., Potter, S.S., Overbeek, P., Kuehn, M.R. 1996. Conserved left-right asymmetry of nodal expression and alterations in murine situs inversus. Nature 381(6578), 158–161.

Lustig, K.D., Kroll, K., Sun, E., Ramos, R., Elmendorf, H., Kirschner, M.W. 1996. A Xenopus nodal-related gene that acts in synergy with noggin to induce complete secondary axis and notochord formation. Development 122(10), 3275–3282.

Marigo, V., Davey, R.A., Zuo, Y., Cunningham, J.M., Tabin, C.J. 1996a. Biochemical evidence that patched is the Hedgehog receptor. Nature 384(6605), 176–179.

Marigo, V., Johnson, R.L., Vortkamp, A., Tabin, C.J. 1996b. Sonic hedgehog differentially regulates expression of GLI and GLI3 during limb development. Dev. Biol. 180(1), 273–283.

Mathis, L., Kulesa, P.M., Fraser, S.E. 2001. FGF receptor signalling is required to maintain neural progenitors during Hensen's node progression. Nat. Cell Biol. 3(6), 559–566.

Meno, C., Shimono, A., Saijoh, Y., Yashiro, K., Mochida, K., Ohishi, S., Noji, S., Kondoh, H., Hamada, H. 1998. Lefty-1 is required for left-right determination as a regulator of lefty-2 and nodal. Cell 94(3), 287–297.

Mercola, M., Levin, M. 2001. Left-right asymmetry determination in vertebrates. Annu. Rev. Cell Dev. Biol. 17, 779–805.

Meyers, E.N., Martin, G.R. 1999. Differences in left-right axis pathways in mouse and chick: Functions of FGF8 and SHH. Science 285(5426), 403–406.

Meyers, E.N., Lewandoski, M., Martin, G.R. 1998. An Fgf8 mutant allelic series generated by Cre- and Flp-mediated recombination. Nat. Genet. 18(2), 136–141.

Moon, A.M., Capecchi, M.R. 2000. Fgf8 is required for outgrowth and patterning of the limbs. Nat. Genet. 26(4), 455–459.

Nonaka, S., Tanaka, Y., Okada, Y., Takeda, S., Harada, A., Kanai, Y., Kido, M., Hirokawa, N. 1998. Randomization of left-right asymmetry due to loss of nodal cilia generating leftward flow of extraembryonic fluid in mice lacking KIF3B motor protein. Cell 95(6), 829–837.

Ohuchi, H., Kimura, S., Watamoto, M., Itoh, N. 2000. Involvement of fibroblast growth factor (FGF)18-FGF8 signaling in specification of left-right asymmetry and brain and limb development of the chick embryo. Mech. Dev. 95(1–2), 55–66.

Okada, Y., Takeda, S., Tanaka, Y., Belmonte, J.C., Hirokawa, N. 2005. Mechanism of nodal flow: A conserved symmetry breaking event in left-right axis determination. Cell 121(4), 633–644.

Olson, E.N., Srivastava, D. 1996. Molecular pathways controlling heart development. Science 272(5262), 671–676.

Ornitz, D.M., Itoh, N. 2001. Fibroblast growth factors. Genome Biol. 2(3), REVIEWS3005, 1–12.

Ornitz, D.M., Xu, J., Colvin, J.S., McEwen, D.G., MacArthur, C.A., Coulier, F., Gao, G., Goldfarb, M. 1996. Receptor specificity of the fibroblast growth factor family. J. Biol. Chem. 271(25), 15292–15297.

Pagan-Westphal, S.M., Tabin, C.J. 1998. The transfer of left-right positional information during chick embryogenesis. Cell 93(1), 25–35.

Pathi, S., Pagan-Westphal, S., Baker, D.P., Garber, E.A., Rayhorn, P., Bumcrot, D., Tabin, C.J., Blake Pepinsky, R., Williams, K.P. 2001. Comparative biological responses to human sonic, Indian, and Desert hedgehog. Mech. Dev. 106(1–2), 107–117.

Pepinsky, R.B., Zeng, C., Wen, D., Rayhorn, P., Baker, D.P., Williams, K.P., Bixler, S.A., Ambrose, C.M., Garber, E.A., Miatkowski, K., Taylor, F.R., Wang, E.A., et al. 1998. Identification of a palmitic acid-modified form of human sonic hedgehog. J. Biol. Chem. 273(22), 14037–14045.

Placzek, M., Dodd, J., Jessell, T.M. 2000. Discussion point. The case for floor plate induction by the notochord. Curr. Opin. Neurobiol. 10(1), 15–22.

Porter, J.A., von Kessler, D.P., Ekker, S.C., Young, K.E., Lee, J.J., Moses, K., Beachy, P.A. 1995. The product of hedgehog autoproteolytic cleavage active in local and long-range signalling. Nature 374(6520), 363–366.

Raffioni, S., Thomas, D., Foehr, E.D., Thompson, L.M., Bradshaw, R.A. 1999. Comparison of the intracellular signaling responses by three chimeric fibroblast growth factor receptors in PC12 cells. Proc. Natl. Acad. Sci. USA 96(13), 7178–7183.

Rebagliati, M.R., Toyama, R., Haffter, P., Dawid, I.B. 1998. Cyclops encodes a nodal-related factor involved in midline signaling. Proc. Natl. Acad. Sci. USA 95(17), 9932–9937.

Reifers, F., Bohli, H., Walsh, E.C., Crossley, P.H., Stainier, D.Y., Brand, M. 1998. Fgf8 is mutated in zebrafish acerebellar (ace) mutants and is required for maintenance of midbrain-hindbrain boundary development and somitogenesis. Development 125(13), 2381–2395.

Riddle, R.D., Johnson, R.L., Laufer, E., Tabin, C. 1993. Sonic hedgehog mediates the polarizing activity of the ZPA. Cell 75(7), 1401–1416.

Rodriguez-Esteban, C., Capdevila, J., Kawamami, Y., Izpisúa Belmonte, J.C. 2001. Wnt signaling and PKA control Nodal expression and left-right determination in the chick embryo. Development 128, 3189–3195.

Sampath, K., Cheng, A.M., Frisch, A., Wright, C.V. 1997. Functional differences among Xenopus nodal-related genes in left-right axis determination. Development 124(17), 3293–3302.

Schilling, T.F., Concordet, J.P., Ingham, P.W. 1999. Regulation of left-right asymmetries in the zebrafish by Shh and BMP4. Dev. Biol. 210(2), 277–287.

Schlessinger, J., Plotnikov, A.N., Ibrahimi, O.A., Eliseenkova, A.V., Yeh, B.K., Yayon, A., Linhardt, R.J., Mohammadi, M. 2000. Crystal structure of a ternary FGF-FGFR-heparin complex reveals a dual role for heparin in FGFR binding and dimerization. Mol. Cell 6(3), 743–750.

Schweickert, A., Weber, T., Beyer, T., Vick, P., Bogusch, S., Feistel, K., Blum, M. 2007. Cilia-driven leftward flow determines laterality in Xenopus. Curr. Biol. 17(1), 60–66.

Shiratori, H., Yashiro, K., Shen, M.M., Hamada, H. 2006. Conserved regulation and role of Pitx2 in situs-specific morphogenesis of visceral organs. Development 133(15), 3015–3025.

Sirbu, I.O., Duester, G. 2006. Retinoic-acid signalling in node ectoderm and posterior neural plate directs left-right patterning of somitic mesoderm. Nat. Cell Biol. 8(3), 271–277.

Spemann, H., Falkenberg, H. 1919. Über asymmetriche Entwicklung und situs inversus viscerum bei Zwillingen und Doppelbildungen. Arch. Entwickmech. Org. 45, 371–422.

Stolte, D., Huang, R., Christ, B. 2002. Spatial and temporal pattern of Fgf-8 expression during chicken development. Anat. Embryol. (Berl.) 205(1), 1–6.

Storm, E.E., Rubenstein, J.L., Martin, G.R. 2003. Dosage of Fgf8 determines whether cell survival is positively or negatively regulated in the developing forebrain. Proc. Natl. Acad. Sci. USA 100(4), 1757–1762.

Sun, X., Meyers, E.N., Lewandoski, M., Martin, G.R. 1999. Targeted disruption of Fgf8 causes failure of cell migration in the gastrulating mouse embryo. Genes Dev. 13(14), 1834–1846.

Tanaka, Y., Okada, Y., Hirokawa, N. 2005. FGF-induced vesicular release of sonic hedgehog and retinoic acid in leftward nodal flow is critical for left-right determination. Nature 435(7039), 172–177.

Thisse, B., Thisse, C. 2005. Functions and regulations of fibroblast growth factor signaling during embryonic development. Dev. Biol. 287(2), 390–402.

Tsukui, T., Capdevila, J., Tamura, K., Ruiz-Lozano, P., Rodriguez-Esteban, C., Yonei-Tamura, S., Magallon, J., Chandraratna, R.A., Chien, K., Blumberg, B., Evans, R.M., Belmonte, J.C. 1999. Multiple left-right asymmetry defects in Shh($-/-$) mutant mice unveil a convergence of the shh and retinoic acid pathways in the control of Lefty-1. Proc. Natl. Acad. Sci. USA 96(20), 11376–11381.

van den Heuvel, M., Ingham, P.W. 1996. Smoothened encodes a receptor-like serpentine protein required for hedgehog signalling. Nature 382(6591), 547–551.

Vermot, J., Pourquie, O. 2005. Retinoic acid coordinates somitogenesis and left-right patterning in vertebrate embryos. Nature 435(7039), 215–220.

Vermot, J., Gallego Llamas, J., Fraulob, V., Niederreither, K., Chambon, P., Dolle, P. 2005. Retinoic acid controls the bilateral symmetry of somite formation in the mouse embryo. Science 308(5721), 563–566.

Vortkamp, A., Lee, K., Lanske, B., Segre, G.V., Kronenberg, H.M., Tabin, C.J. 1996. Regulation of rate of cartilage differentiation by Indian hedgehog and PTH-related protein. Science 273(5275), 613–622.

Wagner, M.K., Yost, H.J. 2000. Left-right development: The roles of nodal cilia. Curr. Biol. 10(4), R149–R151.

Yayon, A., Klagsbrun, M., Esko, J.D., Leder, P., Ornitz, D.M. 1991. Cell surface, heparin-like molecules are required for binding of basic fibroblast growth factor to its high affinity receptor. Cell 64(4), 841–848.

Yeh, B.K., Igarashi, M., Eliseenkova, A.V., Plotnikov, A.N., Sher, I., Ron, D., Aaronson, S.A., Mohammadi, M. 2003. Structural basis by which alternative splicing confers specificity in fibroblast growth factor receptors. Proc. Natl. Acad. Sci. USA 100(5), 2266–2271.

Yost, H.J. 1990. Inhibition of proteoglycan synthesis eliminates left-right asymmetry in Xenopus laevis cardiac looping. Development 110, 865–874.

Yost, H.J. 1991. Development of the left-right axis in amphibians. Ciba Found. Symp. 162, 165–176; discussion 176–181.

Yost, H.J. 1995. Vertebrate left-right development. Cell 82(5), 689–692.

Zhang, X., Ibrahimi, O.A., Olsen, S.K., Umemori, H., Mohammadi, M., Ornitz, D.M. 2006. Receptor specificity of the fibroblast growth factor family. The complete mammalian FGF family. J. Biol. Chem. 281(23), 15694–15700.

Zhang, X.M., Ramalho-Santos, M., McMahon, A.P. 2001. Smoothened mutants reveal redundant roles for Shh and Ihh signaling including regulation of L/R asymmetry by the mouse node. Cell 105(6), 781–792.

Zhou, X., Sasaki, H., Lowe, L., Hogan, B.L., Kuehn, M.R. 1993. Nodal is a novel TGF-beta-like gene expressed in the mouse node during gastrulation. Nature 361(6412), 543–547.

CHAPTER **9**

Development of the Conduction System: Picking up the Pace

Dina Myers Stroud and **Gregory E. Morley**

Contents		

Abstract   The cardiac conduction system initiates and maintains electrical activity that must pass through every cell of the heart to trigger contraction. How are these "specialized" cells differentiated from working cardiomyocytes? The question is simple, the story complex, and the answers remain elusive. Discrepancies between morphological studies and functional data leave genetics and molecular biology to fill the gaps in our understanding. In this chapter, we travel the path from the early gross anatomy discoveries to high-resolution optical mapping studies, focusing on the latest work from both murine and avian systems. We speculate on the future direction of investigation into conduction system development and highlight the clinical significance of this work.

Advances in Developmental Biology, Volume 18
ISSN 1574-3349, DOI: 10.1016/S1574-3349(07)18009-1

1. INTRODUCTION

One definition of a system is the condition of harmonious, orderly interaction. It is a fitting description for the cardiac conduction system (CCS), whose independent parts must function together to create and maintain perfect rhythm. The CCS consists of several molecularly and electrophysiologically distinct components including the sino-atrial (SA) and atrio-ventricular (AV) nodes, the His bundle, left and right bundle branches, and finally the extensive Purkinje fiber network. We study CCS development in the hope that by elucidating the signaling and genetic pathways that direct CCS formation, we can better understand the sources of conduction defects in congenital heart disease and arrhythmogenesis. Initial work on the conduction system was focused on detailed anatomical descriptions. With the advent of high-resolution optical mapping, we and others were able to investigate global CCS function (Komuro et al., 1985; Kamino et al., 1988; Rentschler et al., 2001, 2002). These studies revealed that conduction system activity preceded morphologically distinct structures (Van Mierop, 1967; Hirota et al., 1979; Fujii et al., 1980; Sawanobori et al., 1981). The focus is now to identify the molecular and genetic distinctions between conduction system cells and working cardiomyocytes. For the future, the challenge will be the integration of these fields, anatomy, electrophysiology, and genetics, to generate a global picture of CCS development. In this chapter, we put into context the historical origin of the conduction system with more recent work addressing the electrophysiological changes during embryogenesis.

2. HISTORICAL BACKGROUND AND BASIC ELECTROPHYSIOLOGY PRINCIPLES

The question of how the heart beats was addressed anatomically in the late nineteenth and early twentieth centuries. In a relatively short time, most of what we know about the structure of the CCS was described in detail. Interestingly, the components of the conduction system were actually discovered in the opposite direction to the normal sequence of activation during sinus rhythm. Figure 1 shows historical photos of the investigators who have been given credit for identifying the major components of the CCS. The accomplishments of these investigators have been reviewed in more detail (Suma, 2001; Roguin, 2006; Silverman et al., 2006).

Beginning in 1839, Jan Evangelista Purkinje described a network of fibers that lined the ventricular endocardial surface of the sheep heart. The morphology of the cells was such that Purkinje initially failed to

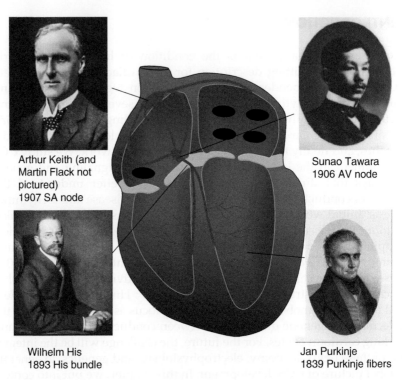

Arthur Keith (and
Martin Flack not
pictured)
1907 SA node

Sunao Tawara
1906 AV node

Wilhelm His
1893 His bundle

Jan Purkinje
1839 Purkinje fibers

FIGURE 1   Investigators credited with describing the major components of the CCS.
Jan Evengelista Purkinje originally described the Purkinje fibers of the heart in 1839.
In 1893, Wilhelm His reported a muscular connection between the atria and ventricles.
Sunao Tawara described the AV node in 1906. Sir Author Keith and Martin Flack (not
pictured) are credited with the first description of the SA node.

recognize these fibers as important to the activation of the heart; rather he
believed they were a cartilaginous network. Only a few years later did he
conclude that they were indeed muscle cells. In 1890, His and colleagues
proposed it was muscle, not innervation that regulated heart rhythm
based on embryological studies but the structures had not been defined
(Bast and Gardner, 1949). Examining serial sections of the AV groove, His
described a continuous sheet of connective tissue which separated the
atria from the ventricles, now known as the annulus fibrosus or cardiac
skeleton. He identified a single muscular connection which originated
near the posterior aspect of the right atria and interatrial septum and
connected to the top of the interventricular septum. The muscular bundle
continued along the ventricular septum toward the front of the heart.
Discovery of the bundle was followed by the work of Tawara (1906) and
the description of the AV node. Tracing both ends of the bundle of His,

Tawara identified the AV node, and where the bundle bifurcated into the bundle branches. He also recognized the bundle branches linked with the network of cells described earlier by Purkinje. As a result, Tawara is credited with providing the first complete description of the functional connections that begin in the atria and lead to the ventricles giving rise to the modern concept of the CCS. Finally, one year later in 1907, Sir Author Keith and Martin Flack defined the location of the SA node (Keith and Flack, 1907). They identified a group of cells near the junction of the superior vena cava and right atria structurally similar to the AV nodal cells described by Tawara.

Cellular electrophysiology is a relatively new discipline compared with the anatomical research of the heart. Early microelectrode studies were used to characterize the electrophysiological phenotypes of different regions of the CCS (Draper and Weidmann, 1951; Weidmann, 1951, 1952, 1955; de Carvalho et al., 1959; de Carvalho and de Almeida, 1960; Spach et al., 1971; Alanis and Benitez, 1975; Bleeker et al., 1980; Masson-Pevet et al., 1984). These studies established important regional variations in action potential shape and the speed electrical depolarization travels. Figure 2 shows characteristic action potentials from different regions of the CCS. Major differences in action potential shape include regional distinctions in upstroke and conduction velocity, action potential duration, and whether the membrane potential remains stable at rest or slowly depolarizes. Early patch clamp studies identified the mechanisms for these differences as being due to varying levels and types of ion channels

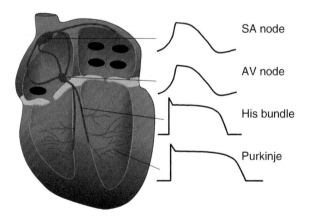

FIGURE 2   Action potential morphology from different regions of the CCS. Action potential morphology varies depending on location within the CCS. Action potentials recorded from the SA and AV nodes propagate slowly and have slow rise time upstrokes. Action potentials that originate in the His bundle and bundle branches propagate more rapidly, are longer, and have more rapid rates of depolarization.

(Nakayama et al., 1984; Noma et al., 1984; DiFrancesco, 1986; Shibasaki, 1987; Anumonwo et al., 1992).

Within the pacemaking regions of the heart including the SA and AV nodes, the maximum diastolic potentials are much less depolarized. This is due to a lower resting permeability to potassium which is associated with the lack of the inward rectifier current I_{K1} (Noma et al., 1984). In addition, these cells have slower upstrokes which are mediated primarily by voltage-gated calcium channels instead of the voltage-gated sodium channels found in atrial and ventricular myocytes. Most importantly, the resting membrane potential is not stable in SA and AV nodal cells. Between action potentials in these cells, the membrane potential slowly depolarizes which is a critical feature for rhythmic pacemaking. This is referred to as slow diastolic depolarization and is due to a dynamic imbalance between inward and outward currents. The inwardly directed hyperpolarization-activated current, I_f provides depolarizing current, while outwardly directed currents such as the delayed rectifier current, I_K act in opposition to return the cell membrane to rest. Conduction velocities in these regions are also slower due to lower levels of gap junction proteins and slower rates of depolarization (Anumonwo et al., 1992; Saffitz et al., 1997).

Fast response action potentials are principally found in the atria, specialized conducting fibers, and ventricular myocytes. These cells express high levels of the inward rectifier current I_{K1} and are more permeable to potassium (Noma et al., 1984). The high resting permeability to potassium clamps the membrane potential close to the reversal potential for potassium thus limiting or eliminating slow diastolic depolarization. Rapid depolarizing current in these cells is mediated by the inward movement of sodium ions across the cell membrane. Therefore, while we think of the CCS as system due to its anatomical continuity and coordinated function, it is electrophysiologically and cellularly heterogeneous. Work identifying the molecular and genetic pathways that underlie these differences and direct the specification of the CCS is just beginning (Myers and Fishman, 2004).

3. DEVELOPMENT OF THE CCS

Although initial anatomical and electrophysiological work typically made use of larger animal models such as sheep, rabbits, and dogs, the pioneering studies into CCS development have employed chick and mice (Van Mierop, 1967; Arguello et al., 1988; Kamino et al., 1989; de Jong et al., 1992; Cheng et al., 1999; Rentschler et al., 2001, 2002). The avian and murine systems are particularly important for developmental experiments as their gestational periods are less than three weeks. They have structurally similar four-chamber hearts and recent work has determined the stage at which electrical activity is initiated. In the chick heart, electrical impulses begin between embryonic days 7 and 7.5, before a morphologically definable sinus node is present or mechanical activity initiated (Fujii et al., 1981). At this stage, the heart is a simple tubular structure and lacks valves, vascular and connective tissue; however, regional distinctions are present. The inflow and primitive atria are positioned below the ventricles and outflow tract. Chamber-specific gene expression is associated with a cooperative interaction of transcription factors including Tbx5 and Nkx2.5 (Jay et al., 2004; Moskowitz et al., 2004). These transcription factors drive the chamber myocardium to rapidly proliferate, become trabeculated, and increase gap junction channel density resulting in faster conduction velocities. However, within the AV canal, these functions are suppressed (Christoffels et al., 2004a). The cells of the AV canal proliferate more slowly; they preserve automaticity and have relatively poor intercellular coupling and slow conduction velocities (de Jong et al., 1992). These properties are important for delaying impulse propagation between the atria and ventricles. It has been suggested that Tbx2 and 3 play a critical role by repressing chamber gene expression in the AV canal (Christoffels et al., 2004b; Hoogaars et al., 2004; Rutenberg et al., 2006). The process to begin to form a four-chambered organ begins with an asymmetric elongation of the right side of the heart to produce an S-shaped tube. As the heart grows, the walls of the atria and ventricles become trabeculated and thicken. The atrial chambers shift posteriorly to their mature position above the ventricles. The chambers of the heart tube become separated by endocardial cushions that give rise to valves and atrial and ventricular septation is completed in the mouse at embryonic day 14.

Primitive nodal tissue was first described by Viragh and Challice at embryonic day 9.5 as a group of distinct, glycogen-filled cells that form along the dorsal wall of the AV canal. Accordingly, electrophysiological studies in our lab and others have demonstrated that conduction in the AV junction at this stage is already significantly slower than either the developing atria or ventricles (Rentschler et al., 2002). However, morphological evidence of the SA node, His bundle, and bundle branches was not reported until embryonic day 10.5 or later (Viragh and Challice, 1977a,b, 1980, 1982; Oosthoek et al., 1993). Thus, these regions functionally differentiate before they are morphologically definable.

4. ELECTRICAL ACTIVATION OF THE DEVELOPING HEART

All adult mammalian hearts share a similar ventricular activation sequence during sinus rhythm, determined by the His–Purkinje system (Durrer, 1970; Kanai and Salama, 1995; Tamaddon et al., 2000; Ramanathan et al., 2006; Azarov et al., 2007). Electrical activation begins in the conducting cells lining the endocardial surface. The right bundle branch is a more compact structure responsible for exciting the right ventricular free wall. The left bundle has a more branched morphology and activates the interventricular septum and left ventricular free wall. Assessment of electrical function within the CCS of adult hearts is possible using both surface electrocardiographic and optical techniques; however, these approaches are not scalable to the embryonic heart (Tamaddon et al., 2000; Zhu et al., 2005).

Chuck et al. (1997) were the first to investigate ventricular activation patterns in the chick heart during embryonic development. Their studies relied on extracellular contact electrodes which were positioned to record electrical activity from four locations. Figure 3 shows a summary of their findings. At early stages of heart development, prior to looping, electrical activation propagated at a constant rate from the inflow toward the outflow region without evidence of an AV delay. Panel B shows the pattern of activation following looping. At this stage, conduction in the AV junction was substantially slower than either the atria or ventricular myocardium. However, the pattern of electrical activation within the ventricles remained essentially the same, beginning near the AV junction and propagating toward the outflow. At relatively late stages of heart development, following closure of the interventricular septum, activation was seen to switch to the mature apex to base pattern. From these studies, Chuck et al. (1997) concluded in the chick heart that the ventricular conduction system begins to function following closure of the interventricular septum.

More recently, high-resolution optical mapping studies have provided detailed activation patterns in the developing chick heart (Reckova et al., 2003; Chuck et al., 2004). Figure 4 shows electrical activation patterns

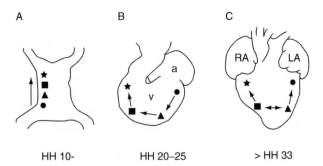

FIGURE 3 Developmental changes in ventricular activation in the chick heart. At early stages of heart development, electrical activation was seen to move at a constant rate from inflow to outflow. Following looping, the electrical activation in the ventricle continues to begin near the AV junction. Following closure of the interventricular septum, the ventricular pattern of activation changes and begins near the apex. Reprinted with permission from Chuck et al. (1997).

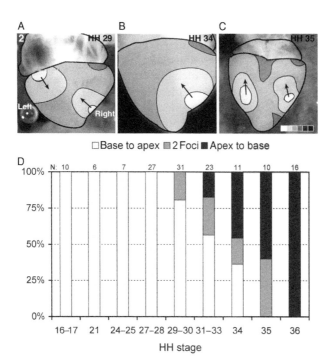

FIGURE 4 Optical mapping of ventricular activation in the developing chick heart. Optical mapping studies were confirmed at the early stage of heart development and electrical activation of the ventricles begins near the AV junction. At HH stage 29, evidence of a functional CCS with electrical activation begins near the apex. By stage 36, all of the hearts showed a mature activation pattern. Reprinted with permission from Reckova et al. (2003).

recorded from late stages of chick heart development. These findings were consistent with the earlier microelectrode studies (Chuck et al., 1997). However, they suggested that the CCS becomes functionally mature over a broader period of time instead of a simple switch generated by "linking up" the components of the conduction system. Beginning at Hamburger–Hamilton (HH) stage 29 and continuing through stage 36, ventricular activation changes from the immature pattern to the mature apex-to-base pattern. These data demonstrated, at least in some hearts, a mature apex-to-base pattern of ventricular activation was present at stages prior to the completion of ventricular septation (Reckova et al., 2003; Chuck et al., 2004).

Questions remained regarding the morphological changes of the CCS and whether alterations in electrical activation occurred at the same developmental stages in the mammalian and avian hearts. We have described a line of mice in which a random insertion of *Engrailed-2* regulatory elements and the *lacZ* reporter gene led to serendipitous expression in the developing and adult CCS (Rentschler et al., 2001; Myers and Fishman, 2003; Myers and Fishman, 2004). The insertion resulted in a chromosomal rearrangement such that the specific regulatory elements are not easily discernible (Stroud et al., 2007). However, with these mice for the first time, all of the major components of the CCS were revealed in unprecedented detail. Figure 5 illustrates the typical pattern of expression that is found in a chemically cleared X-gal stained neonatal CCS-*lacZ* heart. Using this line of mice, we were able to relate structural changes in the anatomy of the CCS with ventricular activation patterns during heart development.

Ventricular myocardial activation patterns from looping to the end of septation are shown in Fig. 6. At embryonic day 10.5, interventricular septation has not yet begun (Fig 6A). At this stage, a dark region of *lacZ* positive cells was present at the junction of the future right and left ventricles. Optical mapping from either the anterior or posterior surfaces revealed that impulse propagation in the ventricles occurred from apex to base. Activation begins as a single breakthrough and spreads through the

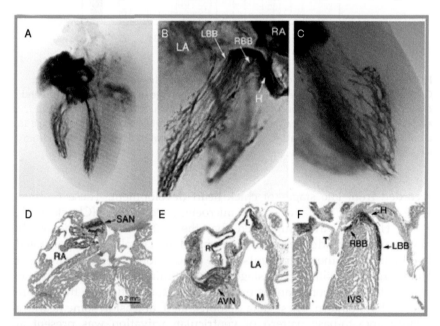

FIGURE 5 Visualization of the CCS in the CCS-*lacZ* mouse. (A–C) Expression of the beta-galactosidase (β-gal) reporter in the neonatal heart was found throughout the entire CCS from the sinoatrial node (SAN) located within the RA to the His bundle, the LBB (left bundle branches) and RBB (right bundle branches), and distal Purkinje fiber network. (D–F) Although not apparent in the intact heart, sectioning showed *lacZ* expression labeled distinct structures within atria. Reprinted with permission from Rentschler et al. (2001).

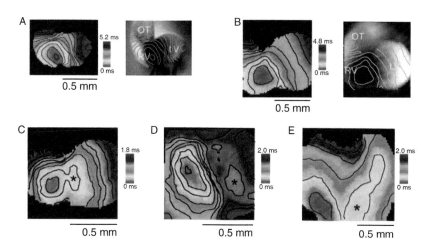

FIGURE 6 Developmental changes in ventricular activation patterns. Patterns of ventricular activation (left panels) and *lacZ* staining (right panels) from 10.5 to 15.5 dpc hearts positioned in approximately the same orientation as they were mapped. Reprinted with permission from Rentschler et al. (2001).

rest of the ventricles in ~5 ms. This epicardial breakthrough is due to a single endocardial activation site within the field of view, suggesting even at this very early developmental stage, specialized fibers deliver the impulse to the apex of the heart. Note the correlation between the site of earliest activation and the *lacZ* positive regions of the heart. The relationship becomes more apparent by superimposing 0.5 ms isochronal lines on the *lacZ* stained heart.

Activation patterns and *lacZ* expression from embryonic day 11.5 hearts are compared in panels B and C. At this stage, both patterns are similar to those described for embryonic day 10.5, yet in some hearts two epicardial breakthroughs were observed, arguing the bundle branches have functionally bifurcated. By embryonic day 12.5, distinct bundle branch function is manifested as two consistent epicardial breakthroughs. In agreement with the optical mapping data, chemical clearing of the embryonic day 12.5 heart reveals independent left and right bundle branches. Impulse propagation is complete within 2 ms, considerably faster than at embryonic day 11.5. Near the end of the gestational period, at embryonic day 15.5, the size of the heart has increased considerably. The activation pattern is essentially the same as that recorded at embryonic day 12.5, beginning near the apex of the heart on both the RV and LV and spreading toward the base. Thus, the mature base-to-apex pattern of activation is evident before septation is complete in the mouse as was also described for the rabbit (Fig. 7). These findings support the notion that conduction system development occurs at a different rate in the mammalian heart compared with the avian heart.

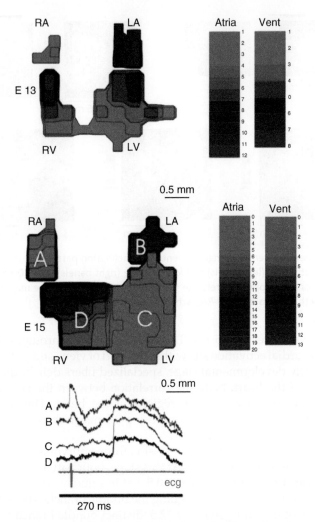

FIGURE 7 Activation maps from the dorsal surface of rabbit embryo hearts. A mature apex-to-base pattern of activation precedes closure of the interventricular septum in the rabbit providing early evidence of a functional conduction system. Reprinted with permission from Rothenberg et al. 2005.

5. CONCLUSIONS

Heart abnormalities represent the most common type of birth defect, with a reported incidence of ~8 per 1000 live births, and represent a significant source of infant mortality. The majority of these are valvular and ventricular septation defects (Minette and Sahn, 2006). These regions are in part derived from the embryonic AV canal, a critical component of the developing conduction system. Therefore, while new surgical procedures offer significantly better odds for survival; successful correction of the morphological anomalies is presenting new challenges as adult patients are at increased risk for arrhythmias and other complications (Khairy et al., 2006; Walsh and Cecchin, 2007).

A long-standing goal of the electrophysiological research field has been to determine the particular substrate changes that occur within the myocardium and identify arrhythmogenic sites. We have known arrhythmias associated with Wolf–Parkinson–White syndrome and atrial tachycardia involve the tricuspid valve annulus; ablation of specific sites can immediately terminate arrhythmic activity. However, more recently we have learned this may also be the case for more complex arrhythmias including atrial fibrillation. Clinical studies have demonstrated in some cases of paroxysmal atrial arrhythmias that regions within the pulmonary veins may play an active role in generating and maintaining arrhythmic activity. The clinical mapping studies of Haissaguerre et al. (1998, 2000) have demonstrated that ectopic atrial activation often originates in cells found in the sleeves of the pulmonary vein region. Additional evidence supports the involvement of other specialized regions of the atria including Bachmann's bundle in maintaining such complex activity (Berenfeld et al., 2002; Ott et al., 2007). Interestingly, there is evidence that specialized cells within the sleeves of the pulmonary veins are indeed present and may share embryonic origins with other regions of the CCS (Jongbloed et al., 2004). These studies highlight the importance of better understanding the genetic and molecular signaling pathways involved in conduction system differentiation.

Conduction system disease poses a major clinical challenge and we know very little regarding the molecular mechanisms that differentiate conduction system cells from the working myocytes. Through a better understanding of the processes that regulate the formation and function of the specialized regions of the heart, we hope to provide new opportunities for the prevention and treatment of conduction system disease. Future studies are needed to better identify the cellular electrophysiological properties from different regions of the CCS. An important question that is to date not adequately addressed is how cells of the CCS are remodeled in response to cardiac disease. Perhaps holding the most promise is the combination of current electrophysiological techniques with anatomical and genetic studies of mouse mutants. These animal models along with detailed gene expression analysis of conduction system tissue may lead the way in our expanding picture of CCS development.

ACKNOWLEDGMENTS

This work was supported by a grant from the National Heart, Lung, and Blood Institute HL76751 (to G.E.M.).

REFERENCES

Alanis, J., Benitez, D. 1975. Two preferential conducting pathways within the bundle of His of the dog heart. Jpn. J. Physiol. 25, 371–385.

Anumonwo, J.M., Wang, H.Z., Trabka-Janik, E., Dunham, B., Veenstra, R.D., Delmar, M., Jalife, J. 1992. Gap junctional channels in adult mammalian sinus nodal cells. Immunolocalization and electrophysiology. Circ. Res. 71, 229–239.

Arguello, C., Alanis, J., Valenzuela, B. 1988. The early development of the atrioventricular node and bundle of His in the embryonic chick heart. An electrophysiological and morphological study. Development 102, 623–637.

Azarov, J.E., Shmakov, D.N., Vityazev, V.A., Roshchevskaya, I.M., Roshchevsky, M.P. 2007. Activation and repolarization patterns in the ventricular epicardium under sinus rhythm in frog and rabbit hearts. Comp. Biochem. Physiol. A Mol. Integr. Physiol. 146, 310–316.

Bast, T.H., Gardner, W.D. 1949. Wilhelm His, Jr. and the bundle of His. J. Hist. Med. 4, 170–187.

Berenfeld, O., Zaitsev, A.V., Mironov, S.F., Pertsov, A.M., Jalife, J. 2002. Frequency-dependent breakdown of wave propagation into fibrillatory conduction across the pectinate muscle network in the isolated sheep right atrium. Circ. Res. 90, 1173–1180.

Bleeker, W.K., Mackaay, A.J., Masson-Pevet, M., Bouman, L.N., Becker, A.E. 1980. Functional and morphological organization of the rabbit sinus node. Circ. Res. 46, 11–22.

Cheng, G., Litchenberg, W.H., Cole, G.J., Mikawa, T., Thompson, R.P., Gourdie, R.G. 1999. Development of the cardiac conduction system involves recruitment within a multipotent cardiomyogenic lineage. Development 126, 5041–5049.

Christoffels, V.M., Burch, J.B.E., Moorman, A.F.M. 2004a. Architectural plan for the heart: Early patterning and delineation of the chambers and the nodes. Trends Cardiovasc. Med. 14, 301–307.

Christoffels, V.M., Hoogaars, W.M.H., Tessari, A., Clout, D.E., Campione, M. 2004b. T-box transcription factor Tbx2 represses differentiation and formation of the cardiac chambers. Dev. Dyn. 229, 763–770.

Chuck, E.T., Freeman, D.M., Watanabe, M., Rosenbaum, D.S. 1997. Changing activation sequence in the embryonic chick heart. Implications for the development of the His-Purkinje system. Circ. Res. 81, 470–476.

Chuck, E.T., Meyers, K., France, D., Creazzo, T.L., Morley, G.E. 2004. Transitions in ventricular activation revealed by two-dimensional optical mapping. Anat. Rec. A Discov. Mol. Cell. Evol. Biol. 280, 990–1000.

de Carvalho, A.P., de Almeida, D.F. 1960. Spread of activity through the atrioventricular node. Circ. Res. 8, 801–809.

de Carvalho, A.P., de Mello, W.C., Hoffman, B.F. 1959. Electrophysiological evidence for specialized fiber types in rabbit atrium. AJP–Legacy 196, 483–488.

de Jong, F., Opthof, T., Wilde, A.A., Janse, M.J., Charles, R., Lamers, W.H., Moorman, A.F. 1992. Persisting zones of slow impulse conduction in developing chicken hearts. Circ. Res. 71, 240–250.

DiFrancesco, D. 1986. Characterization of single pacemaker channels in cardiac sino-atrial node cells. Nature 324, 470–473.

Draper, M.H., Weidmann, S. 1951. Cardiac resting and action potentials recorded with an intracellular electrode. J. Physiol. 115, 74–94.

Durrer, D. 1970. Total excitation of the isolated human heart. Circulation 41, 899–912.

Fujii, S., Hirota, A., Kamino, K. 1980. Optical signals from early embryonic chick heart stained with potential sensitive dyes: Evidence for electrical activity. J. Physiol. 304, 503–518.

Fujii, S., Hirota, A., Kamino, K. 1981. Action potential synchrony in embryonic precontractile chick heart: Optical monitoring with potentiometric dyes. J. Physiol. 319, 529–541.

Haissaguerre, M., Jais, P., Shah, D.C., Takahashi, A., Hocini, M., Quiniou, G., Garrigue, S., Le Mouroux, A., Le Metayer, P., Clementy, J. 1998. Spontaneous initiation of atrial fibrillation by ectopic beats originating in the pulmonary veins. N. Engl. J. Med. 339, 659–666.

Haissaguerre, M., Jais, P., Shah, D.C., Garrigue, S., Takahashi, A., Lavergne, T., Hocini, M., Peng, J.T., Roudaut, R., Clementy, J. 2000. Electrophysiological end point for catheter ablation of atrial fibrillation initiated from multiple pulmonary venous foci. Circulation 101, 1409–1417.

Hirota, A., Fujii, S., Kamino, K. 1979. Optical monitoring of spontaneous electrical activity of 8-somite embryonic chick heart. Jpn. J. Physiol. 29, 635–639.

His, W. 1893. Die thatigkeit des embryonalen herzens und deren bedeutung fur die lehre von der herzbewegung beim erwachsenen. Med. Klin. Leipzig 1, 14–49.

Hoogaars, W.M.H., Tessari, A., Moorman, A.F.M., de Boer, P.A.J., Hagoort, J., Soufan, A.T., Campione, M., Christoffels, V.M. 2004. The transcriptional repressor Tbx3 delineates the developing central conduction system of the heart. Cardiovasc. Res. 62, 489–499.

Jay, P.Y., Harris, B.S., Maguire, C.T., Buerger, A., Wakimoto, H., Tanaka, M., Kupershmidt, S., Roden, D.M., Schultheiss, T.M., O'Brien, T.X., Gourdie, R.G., Berul, C.I., et al. 2004. Nkx2-5 mutation causes anatomic hypoplasia of the cardiac conduction system. J. Clin. Invest. 113, 1130–1137.

Jongbloed, M.R.M., Schalij, M.J., Poelmann, R.E., Blom, N.A., Fekkes, M.L., Wang, Z.H.I.Y., Fishman, G.I., Gittenberger-de Groot, A.C. 2004. Embryonic conduction tissue: A spatial correlation with adult arrhythmogenic areas. J. Cardiovasc. Electrophysiol. 15, 349–355.

Kamino, K., Komuro, H., Sakai, T. 1988. Regional gradient of pacemaker activity in the early embryonic chick heart monitored by multisite optical recording. J. Physiol. 402, 301–314.

Kamino, K., Hirota, A., Komuro, H. 1989. Optical indications of electrical activity and excitation-contraction coupling in the early embryonic heart. Adv. Biophys. 25, 45–93.

Kanai, A., Salama, G. 1995. Optical mapping reveals that repolarization spreads anisotropically and is guided by fiber orientation in guinea pig hearts. Circ. Res. 77, 784–802.

Keith, A., Flack, M.W. 1907. The form and nature of the muscular connections between the primary divisions of the vertebrate heart. J. Anat. Physiol. 41, 172–189.

Khairy, P., Dore, A., Talajic, M., Dubuc, M., Poirier, N., Roy, D., Mercier, L.A. 2006. Arrhythmias in adult congenital heart disease. Expert Rev. Cardiovasc. Ther. 4, 83–95.

Komuro, H., Hirota, A., Yada, T., Sakai, T., Fujii, S., Kamino, K. 1985. Effects of calcium on electrical propagation in early embryonic precontractile heart as revealed by multiple-site optical recording of action potentials. J. Gen. Physiol. 85, 365–382.

Masson-Pevet, M.A., Bleeker, W.K., Besselsen, E., Treytel, B.W., Jongsma, H.J., Bouman, L.N. 1984. Pacemaker cell types in the rabbit sinus node: A correlative ultrastructural and electrophysiological study. J. Mol. Cell. Cardiol. 16, 53–63.

Minette, M.S., Sahn, D.J. 2006. Ventricular septal defects. Circulation 114, 2190–2197.

Moskowitz, I.P.G., Pizard, A., Patel, V.V., Bruneau, B.G., Kim, J.B., Kupershmidt, S., Roden, D., Berul, C.I., Seidman, C.E., Seidman, J.G. 2004. The T-box transcription factor Tbx5 is required for the patterning and maturation of the murine cardiac conduction system. Development 131, 4107–4116.

Myers, D.C., Fishman, G.I. 2003. Molecular and functional maturation of the murine cardiac conduction system. Trends Cardiovasc. Med. 13, 289–295.

Myers, D.C., Fishman, G.I. 2004. Toward an understanding of the genetics of murine cardiac pacemaking and conduction system development. Anat. Rec. A Discov. Mol. Cell. Evol. Biol. 280, 1018–1021.

Nakayama, T., Kurachi, Y., Noma, A., Irisawa, H. 1984. Action potential and membrane currents of single pacemaker cells of the rabbit heart. Pflugers Arch. Eur. J. Physiol. 402, 248–257.

Noma, A., Nakayama, T., Kurachi, Y., Irisawa, H. 1984. Resting K conductances in pacemaker and non-pacemaker heart cells of the rabbit. Jpn. J. Physiol. 34, 245–254.

Oosthoek, P.W., Viragh, S., Mayen, A.E., van Kempen, M.J., Lamers, W.H., Moorman, A.F. 1993. Immunohistochemical delineation of the conduction system. I: The sinoatrial node. Circ. Res. 73, 473–481.

Ott, P., Kirk, M.M., Koo, C., He, D.S., Bhattacharya, B., Buxton, A. 2007. Coronary sinus and fossa ovalis ablation: Effect on interatrial conduction and atrial fibrillation. J. Cardiovasc. Electrophysiol. 18, 310–317.

Ramanathan, C., Jia, P., Ghanem, R., Ryu, K., Rudy, Y. 2006. Activation and repolarization of the normal human heart under complete physiological conditions. Proc. Natl. Acad. Sci. USA 103, 6309–6314.

Reckova, M., Rosengarten, C., deAlmeida, A., Stanley, C.P., Wessels, A., Gourdie, R.G., Thompson, R.P., Sedmera, D. 2003. Hemodynamics is a key epigenetic factor in development of the cardiac conduction system. Circ. Res. 93, 77–85.

Rentschler, S., Vaidya, D.M., Tamaddon, H., Degenhardt, K., Sassoon, D., Morley, G.E., Jalife, J., Fishman, G.I. 2001. Visualization and functional characterization of the developing murine cardiac conduction system. Development 128, 1785–1792.

Rentschler, S., Zander, J., Meyers, K., France, D., Levine, R., Porter, G., Rivkees, S.A., Morley, G.E., Fishman, G.I. 2002. Neuregulin-1 promotes formation of the murine cardiac conduction system. Proc. Natl. Acad. Sci. USA 99, 10464–10469.

Roguin, A. 2006. Wilhelm His Jr. (1863–1934)—The man behind the bundle. Heart Rhythm 3, 480–483.

Rothenberg, F., Nikolski, V.P., Watanabe, M., Efimov, I.R. 2005. Electrophysiology and anatomy of embryonic rabbit hearts before and after septation . Am. J. Physiol. 288, H344–H351.

Rutenberg, J.B., Fischer, A., Jia, H., Gessler, M., Zhong, T.P., Mercola, M. 2006. Developmental patterning of the cardiac atrioventricular canal by Notch and Hairy-related transcription factors. Development 133, 4381–4390.

Saffitz, J.E., Green, K.G., Schuessler, R.B. 1997. Structural determinants of slow conduction in the canine sinus node. J. Cardiovasc. Electrophysiol. 8, 738–744.

Sawanobori, T., Hirota, A., Fujii, S., Kamino, K. 1981. Optical recording of conducted action potential in heart muscle using a voltage-sensitive dye. Jpn. J. Physiol. 31, 369–380.

Shibasaki, T. 1987. Conductance and kinetics of delayed rectifier potassium channels in nodal cells of the rabbit heart. J. Physiol. 387, 227–250.

Silverman, M.E., Grove, D., Upshaw, C.B., Jr. 2006. Why does the heart beat? The discovery of the electrical system of the heart. Circulation 113, 2775–2781.

Spach, M.S., Lieberman, M., Scott, J.G., Barr, R.C., Johnson, E.A., Kootsey, J.M. 1971. Excitation sequences of the atrial septum and the AV node in isolated hearts of the dog and rabbit. Circ. Res. 29, 156–172.

Stroud, D.M., Darrow, B.J., Kim, S.D., Zhang, J., Jongbloed, M.R., Rentschler, S., Moskowitz, I.P., Seidman, J., Fishman, G.I. 2007. Complex genomic rearrangement in CCS-LacZ transgenic mice. Genesis 45, 76–82.

Suma, K. 2001. Sunao Tawara: A father of modern cardiology. Pacing Clin. Electrophysiol. 24, 88–96.

Tamaddon, H.S., Vaidya, D., Simon, A.M., Paul, D.L., Jalife, J., Morley, G.E. 2000. High-resolution optical mapping of the right bundle branch in connexin40 knockout mice reveals slow conduction in the specialized conduction system. Circ. Res. 87, 929–936.

Tawara, S. 1906. Das Reizleitungssystem Des Saugetierherzens. Eine Anatomisch-Histolo-gische Studie Uber Das Atrioventrikularbundel Und Die Purlinjeschen Faden., Jena, Germany: Verlag von Gustav Fischer.

Van Mierop, L.H. 1967. Location of pacemaker in chick embryo heart at the time of initiation of heartbeat. Am. J. Physiol. 212, 407–415.

Viragh, S., Challice, C.E. 1977a. The development of the conduction system in the mouse embryo heart. I. The first embryonic A-V conduction pathway. Dev. Biol. 56, 382–396.

Viragh, S., Challice, C.E. 1977b. The development of the conduction system in the mouse embryo heart. II. Histogenesis of the atrioventricular node and bundle. Dev. Biol. 56, 397–411.

Viragh, S., Challice, C.E. 1980. The development of the conduction system in the mouse embryo heart: III. The development of the sinus muscle and sinoatrial node. Dev. Biol. 80, 28–45.

Viragh, S., Challice, C.E. 1982. The development of the conduction system in the mouse embryo heart: IV. Differentiation of the atrioventricular conduction system. Dev. Biol. 89, 25–40.

Walsh, E.P., Cecchin, F. 2007. Arrhythmias in adult patients with congenital heart disease. Circulation 115, 534–545.

Weidmann, S. 1951. Effect of current flow on the membrane potential of cardiac muscle. J. Physiol. 115, 227–236.

Weidmann, S. 1952. The electrical constants of Purkinje fibers. J. Physiol. 118, 348–360.

Weidmann, S. 1955. Rectifier properties of Purkinje fibers. Am. J. Physiol. 183, 671.

Zhu, W., Saba, S., Link, M.S., Bak, E., Homoud, M.K., Estes, N.A.M.III, Paul, D.L., Wang, P.J. 2005. Atrioventricular nodal reverse facilitation in connexin40-deficient mice. Heart Rhythm 2, 1231–1237.

CHAPTER 10

Transcriptional Control of the Cardiac Conduction System

Shan-Shan Zhang and **Benoit G. Bruneau**

Contents

Department of Pediatrics and Biomedical Sciences Program, Gladstone Institute of Cardiovascular Disease, University of California, San Francisco, California

Advances in Developmental Biology, Volume 18
ISSN 1574-3349, DOI: 10.1016/S1574-3349(07)18010-8

Abstract   Cardiovascular disease, often claiming lives through arrhythmias, is a major cause of mortality in western societies. The cardiac conduction system (CCS) is responsible for the initiation, coordination, and synchronization of the heartbeat. CCS dysfunction, which can result from improper embryonic heart development, directly leads to arrhythmias and sudden cardiac death. Research from the past century has culminated in our current extensive yet incomplete understanding of CCS morphology, function, molecular signature, and regulatory cues governing its formation. In combination with a growing number of molecular markers and sensitive functional assays, mouse models of CCS development, continue to be instrumental in the unraveling of this dynamic process while highlighting the importance of transcriptional control.

1. AN INTRODUCTION TO THE CARDIAC CONDUCTION SYSTEM

1.1. A brief history of cardiac conduction system discovery, function, and anatomy

Since the second century, the driving force behind heart contractions has been a fascinating unknown. It was then that Claudius Galen noted that the denervated heart continued to contract after it had been excised suggesting that the heartbeat was triggered by inherent muscular excitation (Luderitz, 2002). The discovery of the sympathetic and parasympathetic nervous system in the 1830s to 1840s showed that analogous to nervous control of skeletal muscle, the heart, which is intimately associated with nerves and ganglia, could also be under this regulation (Fleming, 1997). These early observations fueled debate over whether the rhythmic contractions were a consequence of muscular or external nervous system control. These two schools of thought are known as the myogenic and the neurogenic theories, respectively.

Heated physiological debate over how the heart beats was not to be settled until the turn of the twentieth century (Silverman, 2006). The initial discovery of the fast-conducting ventricular fibers was made in 1839 by Jan Evangelista Purkinje. He found a peculiar net of gray gelatinous fibers in the subendocardial layer of the sheep heart. Although he had concluded that the fibers resemble embryonic muscular cells that had been stunted in growth, the significance of their pervasive presence in the ventricles was not understood until the work of Sunao Tawara in 1906. Walter Gaskell's work with the tortoise heart in the early 1880s offered significant insight into impulse formation and cardiac conduction through the atrioventricular (AV) junction (Silverman and Upshaw, 2002). Gaskell showed that an isolated strip of tortoise ventricular muscle continued to pulsate at a similar rate as the intact heart. In 1886, he demonstrated that the first excitation point of the heart originates in the atrium, near the sinus venosus, and is subsequently delayed through a muscular bridge at the AV groove. The AV junction appeared critical to heart contraction since various degrees of conduction block between the atria and the ventricles was achievable by ligature of this area. These experiments were the first to show the existence of a specialized muscle tract joining the upper and lower chambers and provided the basis for the discovery of the bundle of His.

The connecting bundle between what was later realized as the atrio-ventricular node (AVN) and the bundle branches was discovered by Wilhelm His Jr. in 1893. Through histological examination of the AV junction of embryonic heart sections, he demonstrated the existence of a ring-like connective tissue sheet. By following serial sections of this region through successive developmental stages, he was the first to describe a penetrating muscular bridge arising from the posterior wall of the right atrium near the septum. Although he did not functionally test the potential of this bundle in playing a role in conduction, his efforts led to the discovery of a crucial cardiac conduction system (CCS) component later known as the bundle of His. Sunao Tawara, through close examination of human, sheep, and other mammalian hearts, traced the bundle of His further toward the base of the atrial septum. Tawara noted the presence of an unsheathed compact node of fibers at the base of the interatrial septum near the endocardium, which he believed to be the anatomical basis of AV delay. This node of Tawara, later known as the AVN, was immediately recognized as being essential for the heart's synchronous and sequential contractions. In the opposite direction, the bundle was seen dividing into branches before terminating in the ventricular wall as a fan-like group of distinct subendocardial muscular fibers. He likened this "AV connecting system" to a tree with its roots in the AV junction before it branches along the interventricular septum (IVS) and reaching out as the fine network discovered by Purkinje. In a 1906 publication, Tawara intro-duced his findings and presented a new understanding of the physiology of the heart's electrical system proposing that each specialized component had a distinct electrical function. In contrast to his contemporaries, Tawara suggested that the sheathed His bundle could act as a part of a fast ventricular conducting pathway that drives the apex-to-base contraction sequence. Tawara's pioneering anatomical description and insightful functional interpretation of the CCS laid the foundation upon which our current understanding of the heartbeat could eventually develop.

On the basis of previous work conducted by Walter Gaskell, H.E. Hering, Karel Wenckebach, and others suggesting that the impulse originates near the junction of the right atrium and the superior vena cava, Arthur Keith and Martin Flack found an unusual structure at this particular region in the heart of a mole. This tissue was seen connecting to the vagal and sympathetic nerves while having its own blood supply through associated arteries. These observations and the striking similarity between this tissue and the node of Tawara led Keith and Flack to propose its function as the dominant automaticity center driving the heartbeat. Functional characterization of the sinuatrial node (SAN) awaited Willem Einthoven's invention of the string galvanometer that records the heart's electrical rhythm. From 1910 to 1915, Thomas Lewis applied this electrocardiography (ECG) technique to verify locations of the SAN, AVN, and the ventricular His–Purkinje system in exposed hearts thus confirming previous findings of the pioneering anatomists.

To summarize, functionally distinct components of the adult CCS include the SAN pacemaker, the impulse delaying AVN, and the electrically insulated fast-conducting tract composed of the His bundle and its bifurcations along the IVS that extend into the ventricles as Purkinje fibers (Fig. 1B). The SAN is a heterogeneous tissue that differs in morphology and degree of embedding in the atrial wall depending on the species of heart examined. Following initiation of a cardiac action potential within the node, the impulse is propagated throughout the atrial muscle before focusing into the AVN. The AVN is located at the junction between the upper and lower

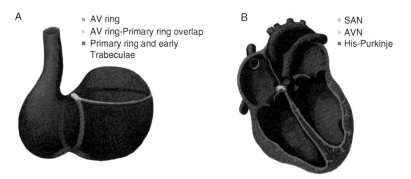

FIGURE 1   The developing and mature CCS. (A) CCS primordia at the cardiac looping stage illustrating early trabeculae, the primary ring, and the AV ring. Newly formed trabecular cells are continuously added to the primary ring separating the future right and left ventricles. The AVN and central AVCS tissue are thought to develop from the overlapping region between these myocardial rings. (B) The mature CCS includes the sinuatrial node (SAN), the atrioventricular node (AVN), and the His–Purkinje network.

chambers and ensures a momentary delay in impulse progression before it reaches the ventricles. After passage through the AVN, the impulse propagates along the His bundle and branches to activate the ventricular chambers via a dense network of Purkinje fibers.

1.2. Anatomical development of the CCS

Since the discoveries of a century ago, the developmental origin of the structures responsible for cardiac impulse generation and propagation remain a heated debate. Some of the critical questions that are left to be answered include: what developmental mechanisms govern the final placement of the adult conduction system components? Furthermore, how are these tissues patterned? What mechanisms are involved during their structural and functional maturation? Finally, what are the consequences when these embryonic pathways go awry?

It is accepted that the anatomical criteria set forth by Tawara still holds true for the histological identification of adult conduction system components. Accordingly, cells of the pathway should be histologically discrete and be tracked continuously from section to section (Anderson, 2004). Moreover, cells within the conducting tracts should be insulated by fibrous sheaths from the adjacent myocardium. The first two criteria are fulfilled by the mature SAN and AVN, while all three can be applied in identifying cells of the fast conducting His–Purkinje pathway. However, such a gold standard does not exist for clearly distinguishing embryonic CCS constituents from the surrounding myocardium. Furthermore, an adult type ECG is observable long before the emergence of the histologically distinct mature tissue. The current challenge for the field is to integrate our emerging understanding of CCS formation with the wealth of established anatomical findings of the formed conduction system.

It is important to appreciate that although not histologically distinguishable, the arrangement of myocyte populations (discernible by their contractile, conductive, and molecular characteristics) underly contraction of the embryonic heart. The linear heart tube that forms at embryonic day (E) 8 of mouse development already contracts peristaltically to ensure unidirectional blood flow. Thereafter, synchronous and fast contracting atrial and ventricular chambers proliferate and differentiate within this slow conducting tube around E8.5. Ventricular chamber myocardium develops at the ventral side of the anterior part of this heart tube. An intermediate stage of ventricular chamber development is the trabecular myocardium. As development proceeds, compact myocardium proliferates exteriorly whereas interiorly located trabeculations are slow proliferating. The ventricular conduction system (VCS) develops from the slowly proliferating trabeculae around the IVS while differentiating toward the heart's periphery. As the ventricular and atrial chambers balloon out, the inflow tract (IFT), AVC, and outflow tract (OFT) remain slow conducting and serve as sphincters to push blood forward. This alternating arrangement ensures that the ventricular portion does not contract before completion of atrial contraction and that the atrium does not relax prematurely to cause regurgitation of blood. Multipotent myocytes within the slowly conducting IFT and AVC are patterned to contribute to the future SAN and AVN, respectively.

Phylogenetic and ontogenetic studies have revealed the presence of four rings of specialized tissue that are potential candidate sites of CCS formation (Wenink, 1976; Anderson et al., 1981). These junctional rings include the ventriculo-arterial (OFT) ring, the inlet–outlet ring (or primary ring), the AV ring, and the sinoatrial ring. These rings, or conduction system primordia, are thought to further elaborate while being brought together to form the definitive CCS of the four-chambered heart. Recent studies of heart-specific transcription factors have revealed their unique requirements and expression patterns within the primordial rings. The AVN is thought to arise from the regression and remodeling of the AV ring. The primary ring, marked by the GLN2 antibody in human hearts, is found between the future right and left ventricles extending into the inner curvature (Wessels et al., 1992). As looping occurs and the two limbs of the heart are brought together, the primary ring overlaps with the rightmost portion of the AV ring. This region of overlap, as indicated in Fig. 1A, is thought to give rise to the central atrioventricular conduction system (AVCS). Although not all rings are necessarily discernable within the same organism or with a particular approach, the concept of rings/segments was the first step toward understanding CCS patterning. This uncomplicated theory has been helpful for interpreting CCS development and can account for areas of ectopic conduction observed in malformed hearts.

A morphologically distinguishable SAN was discovered by Van Meirop, Gessner, and Heintzberger in the mouse at E10 (Van Mierop and Gessner, 1970; Heintzberger, 1974). The SAN forms from the most posterior portion of the developing heart tube, presumably from the progressive fusion of the paired primordia. Upon formation of the sinus venosus, the initially bilateral nodal primordium becomes dominant in the right side. How the dominant pacemaker ends up at its final location and forms a morphologically and molecularly distinct node remains a challenging yet exciting area of investigation. In humans, the SAN is recognizable as early as cardiac looping. The nodal primordium is found at the junction of the superior vena cava and the right atrium. It begins as a horseshoe-shaped structure containing cells that are discrete from the adjacent myocardium in terms of arrangement and staining properties (Ho, 2003). The cells are small in size and are nonuniformly surrounded by transitional cells that have an intermediate appearance that is between nodal and working myocytes. These nodal cells are set within a connective tissue matrix that increases in volume as development proceeds.

The richly innervated SA node is largest in comparison to the atrial working myocardium at early fetal stages. The rest of the SA ring or junction is not always histologically distinct and is folded and remodeled to form the fibrous valves of the sinus venosus.

The AVN primordium develops from the ring of specialized tissue of the AVC. AV-delay is first detectable when the looping heart physically begins to constrict at the future AVC region. As a consequence, the delay gives rise to a biphasic wave of myocardial contraction through the looping heart. The morphological peculiarities of the AVC wall tissue, which allows for this delay in conduction, raised the question of what molecular differences might exist in this region. From extensive histological and microcopy analysis of mouse embryos, the presence of a specialized conduction pathway in the developing heart has been thoroughly described by Virágh and Challice (Viragh and Challice, 1977a,b, 1982; Viragh et al., 1982). They showed that development of the AVN is intricately associated with the formation and remodeling of the AVC, which connects the common atrium to the future left ventricle. At E10, there exists two morphologically distinct cell layers within the AVC region. The inner layer contains a histologically distinguishable tract of cells along the dorsal wall. These myocytes were described as being larger in size, containing a smaller number of myofibrils, and a large amount of glycogen. The AVN is thought to mature from these myocytes within the distinctive tract at the intersection of the AV and primary rings (Fig. 1A).

During later stages, the area of AV and primary ring colocalization expands to the right resulting in the right AV ring (RAVR) tissue. The RAVR encircles the future tricuspid valve while the retro-aortic ring bundle (RARB) surrounds the future mitral valve. These myocardial ring structures must be further remodeled while AV junction maturation takes place. As mentioned, the only remaining conducting tract through the central fibrous body of the AV junction is the bundle of His in the adult. Due to lack of proper regression, remnants of these embryonic myocardial rings can become the substrate of ectopic conduction in the diseased heart.

In amniote hearts, the inferior and superior cushions, develop at the luminal side of the AVC. Subsequently, two additional endocardial cushions, known as the left and right lateral AV cushions, develop. The mesenchyme of these cushions is derived from the endocardium through a process termed endocardial-to-mesenchymal transformation (EMT) that is induced by the underlying myocardium. Meanwhile, the outer side of the AVC is covered by a layer of epithelium known as the epicardium, which forms the epicardial AV sulcus tissue (Wessels et al., 1992). Subsequently, fusion of the endocardial cushion-derived and epicardial sulcus tissues at the lower boundary of the AVC leads to the incorporation of the embryonic AV myocardium into the lower atria. Connective tissue of the AV sulcus then begins to penetrate the loose AVC myocardium causing the remainder of the specialized AV ring tissue to regress as insulation between the upper and lower chambers reach completion.

In both mice and humans, the small segment of the inner cell layer of the double-layered AVC that does not become separated from the ventricles is the only myocardial connection that persists into postnatal life. Beginning at E10, the inner cell layer extends deeply into the left ventricular cavity while remaining connected to the early trabeculae and the developing IVS. At first, the IVS appears as a crescent-shaped muscular tissue formed by the consolidation of newly formed trabeculae. The proximal extremity of the IVS is made up of several trabeculae attached to the distal rim of the AVC. From these observations, it is clear that the AVC is continuous with the developing trabeculae through a number of interconnections. It is through this trabecular network that the impulse from the maturing AVN is conducted before septation is complete. The work of Virágh and Challice in the mouse supports the idea that the His bundle and Purkinje fibers are derived from the ventricular myocardial cells of the primary ring and early trabeculae.

The forming AVN is continuous with the His bundle primordium. Studies extending back to the early 1900s suggest that the Purkinje fibers are derived *in situ* from the ventricular myocardium. It is important to note that there is no need for the fusion of the AVN and His bundle primordia since they are in continuity from the onset and appear simultaneously during development. In E11–E12 embryos, the AVN and His bundle primordia interconnect with proliferating trabeculae and are in continuity with the myocardium of the IVS. At this time, the IVS has developed significantly leaving the interventricular foramen as the only communication between the ventricles. At E13, the developing AVN enlarges quickly in the dorsal AV cushion region at the distal end of the primitive conducting tract while the bundle branches differentiate from the surrounding trabeculae. As differentiation progresses, connective tissue progressively insulates the His–Purkinje system which is nearing its adult size by E14 (Viragh and Challice, 1982).

1.3. Molecular markers of the CCS

As discussed, a wealth of information on the mature and developing CCS was achieved through anatomical and histological methods. The use of molecular markers has made it possible to discern myocytes of the CCS from those of the working myocardium. Antigens and transcripts of many molecules, such as connexins, contractile proteins, and cytoskeletal proteins, have been found in a variety of animals that neatly pick out the developing CCS. These studies have shed light on how these primordial populations become converted into the definitive conduction system

(Moorman et al., 1998). In general, these markers have enriched expression within cells of the CCS in comparison to myocardial cells, rather than having exclusive expression in conduction cells. They tend to have a high degree of variability in expression patterns and intensities between different organisms and between embryos and adults of the same organism (Table 1).

Overlapping with the glycogen-rich track of cells described by Virágh and Challice, the sulfated carbohydrate epitope GLN2 recognizes is also found within the junction between the primary myocardial ring and the RAVR (Lamers et al., 1991; Wessels et al., 1992, 1996). Although the molecule recognized by GLN2 has not been elucidated, the localized and dynamic expression pattern that coincides with CCS maturation seems to support its involvement in early conduction system formation. Similar expression is seen with the HNK1/Leu7 antigen in the rat heart (Ikeda et al., 1990). The HNK-1 antibody, which is specific to some neuronal and endocrine cells, was originally used to search for differentiation antigens in human lymphoid tissues (Nakagawa et al., 1993). It was found to be expressed along the endocardial surface of the heart tube and subsequently within sparsely distributed myocytes along the looped heart. HNK1/Leu7 expression appears within the earliest ventricular trabeculae as they coalesce to form the IVS. Reactivity is also prominent at the intersection between the primary ring and the right AV myocardial ring and around the sinus node primordia. Little is known about the cardiac-specific function of the protein recognized by the HNK-1 antiserum. In birds, mammals, and fish, HNK-1 reactive glycoproteins outside of the heart are ligands in cell surface interactions between neurons (Metcalfe et al., 1990) suggesting that the heart-specific molecule could take part in the cell adhesion and extracellular matrix interactions. It is important to note that GLN2 and HNK1/Leu7 antigens are not detectable in the mouse and thus are limited in their applicability as markers of the CCS.

Certain gap junction subunits are found in the CCS. These proteins are responsible for the intercellular transfer of the depolarizing action potential within and between the atrial and ventricular myocardial syncytia. Gap junctions are formed by membrane channels made up of protein subunits called connexins encoded by a large family of genes. Four connexins are found in the mouse heart and include $Cx40$, $Cx43$, $Cx45$, and $Cx30.2$. In late embryonic and adult mouse hearts, $Cx40$ specifically marks the coronary arteries, atria, the AVN, and VCS. In addition, Coppen et al. (1998) showed $Cx45$ expression in the peripheral VCS in rats and mice, as well as its localized expression in the AV rings, His bundle, and branches during fetal stages. In the adult heart, $Cx45$ can be found in the proximal AVN and along the His bundle and bundle branches. The His bundle and bundle branches exhibit staining for both $Cx45$ and $Cx40$ suggesting that compartmental expression of these gap junction connexins can be used to

TABLE 1 Molecular makers of the CCS

	Gene	Species	SAN	AVN	HB, BB	Purkinje fibers
Transcription factors	Gata4	Chick	8	8	8	4
	Hf1b	Mouse	4	4	4	4
	Hop	Mouse	8	4	4	4
	Irx1, Irx2	Mouse	8	?	4	?
	Irx3	Mouse	8	?	4	4
	Msx2 (Hox8)	Mouse/chick	8	4	4	8
	MyoD	Chick	8	8	8	4
	Nkx2–5	Mouse/chick	8	4	4	4
	Tbx2, Tbx3	Mouse	4	4	4	8
	Tbx5	Mouse	4	4	4	?
Gap junction	Cx30.2	Mouse	4	4	8	8
	Cx43	Mouse	8	8	BB	4
	Cx40	Mouse/chick	8	8	4	4
	Cx45	Mouse	4	4	4	4
Ion channels	HCN1	Mouse	4	4	?	?
	HCN4	Mouse	4	8	8	8
	TASK-1	Mouse/chick	8	8	4	4
Structural	α-MHC	Mouse/chick	8	4	4	4
	β-MHC	Mouse	4	4	4	4
	α- Smooth muscle actin	Mouse/chick	8	8	4	4
	Cardiac troponin I enhancer	Mouse	8	4	8	8
	Desmin	Mouse	8	8	4	4

A representative list of molecular markers found exclusively or preferentially within the developing CCS. Absence or presence within the SAN, AVN, His bundle (HB), bundle branches (BB), and Purkinje fiber primordia is indicated with "8" and "4" respectively. For references, please see text.

distinguish the AVN from the His bundle and branches. *Cx30.2* is a newly described cardiac connexin that is predominantly expressed in the slow-conducting regions of the CCS, especially within the SAN and AVN (Kreuzberg et al., 2005, 2006). It further delineates *Cx40-* and *Cx43*-negative cells in the SAN, while in the AVN its expression overlaps with that of *Cx40* but not *Cx43*. Interestingly, *Cx30.2* is important for maintenance of the slow-conducting properties of the cells in which it is expressed.

Recently, a two-pore domain potassium channel protein, *TASK-1*, was found in the developing avian and mouse VCS (Graham et al., 2006). *TASK-1* and chick *cTASK-1* are initially found throughout the myocardium of the early heart tube. As development proceeds, ventricular expression becomes confined to the trabeculated myocardium. *TASK-1* then becomes localized to the bundle of His, bundle branches, and Purkinje fibers of the mature conduction system. These observations suggest that ventricular conduction cells differentiate from heart tube precursors that retain *TASK-1* expression. *TASK-1* channels have been demonstrated to be involved in the generation and modulation of the resting membrane potential (Lesage and Lazdunski, 2000). Given that the entire myocardium of the early heart tube functions as the conduction system and the working myocardium (Kamino, 1991) before trabeculated myocardium later takes over as the conductive component, the concordant expression pattern of *TASK-1* suggests that it could play a role in active cellular repolarization of developing conduction system cells. Its persistent expression in the adult VCS and sensitivity to oxygen and pH suggest that it could affect ventricular conduction cell repolarization under hypoxia and/or acidosis and alkalosis.

The developing SAN is also marked by expression of the cyclic nucleotide channel *HCN4* (Garcia-Frigola et al., 2003). *HCN4* expression in the embryonic mouse initially delineates the sinus venosa. It is subsequently predominantly found in the coronary sinus and right common cardinal vein in late fetal stages. Its expression is subsequently refined to the SAN by birth.

Preferential or exclusive expression of a growing number of transcription factors in developing CCS components include *Nkx2–5*, *Tbx5*, *Hf1b*, *Tbx2*, *Tbx3*, *Tbx18*, *Msx2*, *Irx1*, *Irx2*, *Irx3* among others. Their expression patterns are summarized (Tables 1 and 2) and discussed in depth below.

1.4. Mouse models: Tracking the developing CCS

The fusion of transgene reporters, such as green fluorescent protein (GFP) and *LacZ*, under the control of CCS-specific regulatory elements has facilitated the imaging of the conduction system in developmental and adult stages (Table 2).

TABLE 2 Mouse models of the conduction system

	Mouse model	Exclusive CCS expression?		Precardiac mesoderm			Adult expression?	
myocardium		SAN	AVN	HB, BB	OFT	Purkinje	AV cushions	AV
Unknown	MC4/Engrailed-2-LacZ (CCS-LacZ) (Rentschler et al., 2001)	8	8	8	8	8	4	4
Enhancer	cGATA6-LacZ (Davis et al., 2001)	4	4	4	8	4	4	8
	Cardiac troponin I-LacZ (Di Lisi et al., 2000)	4	4	4	8	8	4	8
Targeted KO	Targeted Cx40-EGFP (Miquerol et al., 2004)	8	8	8	4	8	4	4
	Targeted MinK-LacZ (Kupershmidt et al., 1999)	4	4	8	4	8	4	4
Targeted transcription factors	Hf1b-LacZ (Nguyen-Tran et al., 2000)	8	8	8	4	8	4	4
	HOP-LacZ (Ismat et al., 2005)	8	8	8	8	4	8	4
	Irx3LacZ (Unpublished)	4	4	8	4	8	8	8

Some of the existing enhancers and endogenous promoters driving CCS-specific reporter expression are shown. *CCS-LacZ* represents a site-dependent integration transgenic line. Conduction system components marked by each reporter are indicated. OFT, outflow tract; SAN, sinuatrial node; AVN, atrioventricular node; HB, bundle of His; BB, bundle branches. Absence or presence of expression is indicated with "8" and "4" respectively.

For example, a 2.3-kb cardiac-specific chicken *GATA6* (*cGATA6*) enhancer has been used to express *LacZ* in the precardiac mesoderm and in the most lateral portions of the left and right heart fields (Davis et al., 2001). As these fields fuse at the midline to form the heart tube, *LacZ* positive cells become confined to the AVC myocardium. Strong expression is detected at the ventral and dorsal sides of the AVC that correspond to the developing inferior and superior AV cushions, respectively. Subsequently, the transgene is found in the AV junction, AVN, His bundle, and bundle branches. From these observations, the transcription factors that regulate the *cGATA6* enhancer appear to be involved in AV cushion and node formation.

Desmin, a member of the intermediate filament family (Lazarides, 1982), has been found in all myogenic lineages including skeletal, cardiac, and smooth muscle cells (Pennisi et al., 2002). A proximal promoter region containing E-box sites for muscle determination transcription factors and the CArG-box of the serum response element was fused to a β-galactosidase reporter. Expression was found in skeletal muscle but not in smooth muscle or the working myocardium of the heart (Li et al., 1993). Interestingly, the enhancer directed scattered reporter expression in cardiac myocytes at E8 that are later found in the AVCS. These observations suggest that a skeletal muscle-specific program could be active in cardiac myocytes that differentiate into CCS cells. Interestingly, *MyoD* has been shown to be expressed in *ET*-induced Purkinje fibers of the chick embryo (Takebayashi-Suzuki et al., 2001).

The *CCS-LacZ* line, which has a random insertion of an *Engrailed-2* promoter region driving the reporter, exhibits conduction system-specific expression that persists into adulthood (Rentschler et al., 2002; Myers and Fishman, 2003). The integration site-dependent expression delineates the full extent of the conduction system. The first detectable pattern is restricted to the dorsal wall of the AV junction at E8.5 corresponding to the AVN primordia. At E10.5, strong expression is detectable at the right AV junction and atop of the growing IVS. At E12.5–E13.5, staining is found in the SAN, SA ring, and developing AVN of the posterior AVC that is continuous with the AV rings. At E13.5, $LacZ^+$ cells represent a nearly mature CCS with an elaborate and extensive Purkinje network. At this stage, the overall pattern is broad and encompasses regions that require remodeling and regression. The widespread expression in the left atrial myocardium does not address the controversial issue of the existence of internodal tracts that are distinct from the remaining atrial myocardium. Nevertheless, stronger expression of *CCS-LacZ* in cells that form a path between the SAN and AVN does suggest the existence of such a pathway. Optical mapping of cardiac electrical activation patterns using a voltage-sensitive dye confirmed that *LacZ*-positive regions are indeed part of the conduction system.

Connexin 40 is a functional marker of the VCS. The $Cx40^{eGFP/+}$ line has been a useful tool for studies of His–Purkinje morphology, maturation, and single cell electrophysiology (Miquerol et al., 2004). When GFP is expressed by the $Cx40$ promoter, the AVN and the His–Purkinje system are nicely revealed. Anatomical and functional asymmetries of the His–Purkinje system were also demonstrated. There appears to be differences in the number of strands making up the bundle branches with 1 on the right and around 20 on the left. Furthermore, the density of fibers extending over the ventricular wall is low on the right but high on the left. The electrical activation patterns reflected this anatomical asymmetry. When the CCS-LacZ adult heart is stained and compared with that of the Cx40-eGFP line, an architecturally similar network of Purkinje fibers is revealed (Myers and Fishman, 2004). This demonstrates that both mouse lines can serve as reporters of the CCS and will prove to be invaluable in studies of transcriptional control. The Cx40-eGFP mouse will allow for detailed single cell electrophysiology characterization that will no doubt facilitate future functional dissections of the complex Purkinje network.

Mutations in the potassium channel protein gene, minK (IsK, KCNE1), are associated with the hereditary cardiac arrhythmia termed long QT syndrome (Towbin et al., 2001). The minK-LacZ knockin showed unexpected reporter expression in the SAN, AV junction, and bundle branches (Kupershmidt et al., 1999). The OFT and the interventricular junction were β-galactosidase positive. OFT expression expanded into the right AV junction at E9.5. At E14, $LacZ^{+}$ cells were seen in the right and left AV junction, in the RARB, and on top of the IVS. Low levels of expression were also found in the myocardium of the apex. In neonatal hearts, LacZ was found in the AV junction and in both the proximal and the distal parts of the CCS including the Purkinje network.

Hf1b encodes a transcription factor preferentially expressed in the CCS and ventricular myocytes (Nguyen-Tran et al., 2000). Homozygous mice survive to birth, but die due to sudden cardiac death. Expression of the reporter, which is similar to that of minK-LacZ, is found in the OFT and the developing and mature CCS. Higher levels of expression are found in the right AV junctional ring in comparison to the left. Expression is detected in the ventricular myocardium and especially in the IVS toward the apex of the heart.

Transgenic and targeted mouse models with altered expression of specific transcription factors have become valuable tools in elucidating features of CCS morphology and function. Not only are the results consistent with morphological and histological data, these models offer a more detailed picture of the distinct components of the developing CCS. Different transcription factors are associated with distinct tissue components of the CCS. Therefore, this suggests that each region is under the governance of various sets of transcriptional cues/hierarchies. Investigation of these mouse lines and of those that have just been developed are continuously reshaping our understanding of the CCS. These transcriptional elements can serve as powerful tools for the expression of *Cre* recombinase to precisely eliminate gene function in the CCS. Moreover, the reporters themselves are useful tools when addressing the effect of the loss of function of certain genes on CCS formation. Mouse model studies, along with other approaches, will ultimately disentangle the regulatory cues behind formation of the heartbeat.

2. DEVELOPMENTAL ORIGIN AND INDUCTION OF THE CONDUCTION SYSTEM

Normal mechanisms underlying CCS formation need to be known before conduction diseases can be understood and addressed. The following sections showcase and examine recent findings and controversies on the origin of the CCS and on transcriptional regulation of its patterning and function.

There has been much debate over the developmental history of conduction system cells. Given that initial antigenic markers used to delineate CCS components recognize neural proteins, a neural crest (NC) origin was suggested (Gorza et al., 1988). Recent cell lineage studies in chick and in mice with NC-driven *Cre* recombinases have aimed to support the NC controversy. Furthermore, data from avian studies suggests a possible role of epicardial-derived cells (EPDCs) in CCS formation (Poelmann and Gittenberger-de Groot, 1999). Despite these new findings, extracardiac cells do not appear to incorporate into the CCS and likely do not have a direct role in its formation. Interestingly, lineage-tracing experiments advocate for early specification of CCS precursors at the cardiac crescent stage prior to heart tube formation. Small clusters of bilaterally symmetrical progenitors that do not express *Mesp1* within the heart fields later give rise to nonexpressing cells in the VCS (Kitajima et al., 2006). Moreover, a *cGATA6*-driven *Cre* recombinase showed that cells of the AVC are descended from a clusters of progenitors in the cardiac crescent (Davis et al., 2001).

2.1. Myogenic origin of the CCS

Evidence for the myogenic origin of the VCS is particularly strong. Throughout heart development, multipotent myocytes differentiate into specific cellular phenotypes. The atrial and ventricular lineages begin diversification in the precardiac mesoderm before heart tube formation. Likewise, the VCS forms by recruitment of adjacent multipotent cardiac myocytes. The ventricular trabecular component, represented by a separate transcriptional and molecular domain, is distinct from the compact myocardium and appears to give rise to the entire VCS. Lineage-tracing studies in chick embryos have provided clear evidence that recruitment takes place at specific embryonic stages beginning with central conduction system elements and ending with the peripheral Purkinje network(Cheng et al., 1999; Gourdie et al., 2003). A similar case has been demonstrated with birth dating experiments in the mouse that shows recruitment of myocytes followed by cell cycle withdrawal (Sedmera et al., 2003).

One of the first insights was provided by lineage tracing in the embryonic chick heart. Injection of low amounts of retrovirus into the looped, tubular heart identified segments of clonally related myocytes as development was allowed to proceed (Mikawa et al., 1992). It was then realized that some of the clonal segments within the central bundles and the peripheral conduction fibers contained conduction cells as well as working myocytes (Gourdie et al., 1995; Cheng et al., 1999). These results unequivocally demonstrate a common origin for working and conduction myocardial cells in the chick ventricle.

While lineage studies have demonstrated the presence of conduction progenitors in the early heart tube, different models exist for the elaboration of the conduction system. The specification model proposes that distinct domains of the looped tubular heart contain precursor cells from which the full extent of the CCS proliferates and differentiates (Moorman et al., 1998). These studies, along with anatomical findings, reveal the requirement of patterning events directing the formation of specialized myocardial rings/domains during early heart formation. By contrast, studies in chick suggest that nonterminally differentiated cells are continuously recruited to the forming conduction system. Cheng et al. (1999) demonstrated continued accretion of newly quiescent myocytes to the central bundles up until the end of septation in the chick.

A well-characterized mechanistic pathway for the recruitment and induction of the chick VCS has been described (Gourdie et al., 2003). The conversion within the cardiomyocyte lineage to form His–Purkinje cells is mediated by binding of the shear-stress responsive cytokine, endothelial cell-derived endothelin-1 (*ET-1*), to its receptors located on the surface of cardiomyocytes. Active *ET-1* is produced via proteolytic cleavage of the big *ET* precursor by endothelin converting enzyme-1 (*ECE-1*). *ECE-1* is expressed predominantly in the endocardial and arterial endothelial cells while not present in capillaries or veins. This unique and restricted pattern allows functional *ET-1* to be available at the endothelia of subendocardial and periarterial regions in the embryonic chick heart. These two sites of VCS cell differentiation and recruitment give rise to proximal and distal conducting fibers. A series of experiments in which coronary arterial branching was inhibited or activated *in vivo* led to altered Purkinje fiber development, suggesting that branching events are necessary and sufficient for peripheral conduction system formation (Takebayashi-Suzuki et al., 2000, 2001). Although under a common induction mechanism by *ET-1* signaling, the central elements of the VCS differentiate earlier than the distally located Purkinje fibers. A linkage between these two systems, *in situ*, has been shown to occur at the ventricular apex prior to apex-to-base activation (Hurtado and Mikawa, presented at the 2005 Weinstein Cardiovascular Development Conference).

The periarterial VCS is a unique feature of avian models and is not present in mice and humans, which have a mainly subendocardial VCS. A similar inductive paracrine signaling mechanism could be necessary in the mouse whereby signals emanate from the subendocardial layer. Rather than the CCS developing in its entirety from proliferation of fate-restricted sets of progenitors, continuous induction, and accretion of myocytes could also take place in the mouse. In 1998, Hertig et al. (1999) found induction of trabeculae and an increase in acetylcholinesterase staining (transient Purkinje marker) after treatment of cultured mouse embryos with the endocardial-derived growth and differentiation factor, *neuregulin-1* (*NRG-1*). The lack of a change in proliferation rate suggested that NRG-1 could recruit cells to the inner trabeculae layer that are otherwise destined to join the compact layer. In ensuing work, exogenous treatment of E8.5–E10.5 mouse embryos with *NRG-1* induced an expansion of *CCS-LacZ*-expressing cells (Rentschler et al., 2002). To test whether endocardial cells have an inductive role, Pennisi et al. (2002) cocultured myocytes of the *CCS-LacZ* line with several other cell types including endocardial endothelial cells. Persistence of conduction cells marked by the reporter was observed in cultured E8.5–E9.5 *CCS-LacZ* tissue segments. However, when myocytes at this stage were dissociated and cultured as a monolayer, *LacZ* staining was significantly diminished. A restoration of strong reporter expression was achieved by coculturing endocardial cells only but not Madin-Darby Canine Kidney cells or embryonic fibroblasts. Dissociated cells from E13.5 exhibited low but detectable levels of reporter expression. Under coculturing conditions, there was a marked increase in reporter expression without an overall change in the cell number. Taken together, these findings suggest that endocardial cells could be involved in maintaining high levels of *CCS-LacZ* expression in cardiomyocytes once they have differentiated into conduction cells.

Another study demonstrated a possible conserved role for *ET-1* in the induction of the mouse VCS (Gassanov et al., 2004). *ET-1*, but interestingly not *NRG-1*, induced *ANF-eGFP* embryonic stem cells to become pacemaker-like cells. It appears that parallels exist between *NRG-1* and *ET-1* signaling in addition to their coexpression by endothelial cells. Both molecules increased *ANF* expression in neonatal rodent ventricular myocytes (Zhao et al., 1998). Moreover, *ET-1* was shown to directly increase *NRG-1* in endothelial cells, suggesting cross talk between these signaling pathways. Indeed, *NRG-1* and *ET-1* interplay is important during peripheral nervous system differentiation (Jessen and Mirsky, 2002). Interestingly, both molecules can only recruit cells to the conduction lineage during the postlooping time frame in chicks and mice.

Taken together, these studies show that paracrine signaling from the endothelium underlies VCS development and could account for further elaboration of the prespecified conduction components via recruitment of multipotent myocytes. In essence, both specification and recruitment mechanisms could be required to achieve formation of the complex Purkinje meshwork. Of course, the specification and recruitment models are not mutually exclusive in the sense that both processes are probably required to achieve the elaborate arborization of the VCS. It is likely that the prespecified cells seen as glycogen-rich tracts or marked rings do not provide the progenitor pools from which all of the specialized conduction tissues can differentiate. Rather, in light of the progressive recruitment chick data, the segment- or ring-like domains can be considered as an initial framework or scaffold upon which subsequent maturation of the VCS can take place. When considering VCS development, the characterization of those molecular cues for directing the formation of this initial framework and those that govern its subsequent elaboration are of great interest. Moreover, understanding how these mechanisms are integrated to form the mature VCS remains an exciting challenge for the field. Whether central VCS components can induce and recruit myocytes through paracrine signaling to become conduction cells remains to be demonstrated.

2.2. Early myogenic lineage from the cardiac crescent

Progenitor cells that contribute to the AVC have been traced to the external margins of the cardiogenic fields using mice doubly transgenic for chick *cGATA6-Cre* and the *ROSA26-LacZ* reporter (Davis et al., 2001). This study shows that the AVN and bundle primordia may not differentiate from myocardial cells at relatively late stages of development, but rather that cell fate is set at the onset of the cardiogenic program. This study confirms that cells of the two bands of enhancer activity in the cardiac crescent later give rise to progeny of the AVN and AV bundle. Interestingly, progeny cells were also seen in the atria and in two tracks along each ventricular wall. Essentially, a transcriptional program converging on the *cGATA6* enhancer element encodes an early cardiac crescent lineage of AVN and working myocyte precursors. The next challenge is to determine those endogenous transcription factors regulating this chick enhancer to better understand how the early specification program affects murine CCS-specific targets.

Mesp1 and *Mesp2* encode transcription factors that contain nearly identical basic helix-loop-helix (bHLH) motifs. *Mesp1* disruption leads to defects in heart tube formation and looping, and cardia bifida resulting from a failure of ventral fusion of the precardial mesoderm (Saga, 1998). Chimera analysis and double knockouts of both genes show that they are indispensable for development of the cardiac mesoderm. *Mesp1* appears to be the earliest molecular marker of the early mesoderm at the onset of gastrulation and contributes to the formation of the heart and vasculature. *Mesp1*-expressing cells migrate out from the primitive streak before incorporating into the head mesenchyme and the heart field. *Mesp1-Cre; ROSA26* analysis revealed the presence of *Mesp1* progenitors in all cardiovascular system cell types including the endothelium, endocardium, epicardium, and myocardium at E9.5 (Saga et al., 1999). By E13.5, cardiogenic cells are not entirely contributed by *Mesp1* progenitors suggesting that they are of mixed origin. Through a series of elegant lineage-tracing experiments, those areas that do not express *Mesp1* were shown to include the OFT cushions and ~20% of the cells within the His bundle and branches (Kitajima et al., 2006).

Using *P0-Cre* and the *CAG-CAT-Z* reporter, neural crest-derived cells (NCCs) were mapped to the *Mesp1*-nonexpressing OFT cushions but not the *Mesp1*-nonexpressing IVS at E13.5. In addition, no NCC-derived cells were seen around the VCS in this study. The lack of NCC lineage and *Mesp1*-expressing mesodermal lineage contributions to VCS suggests that cells of these regions are derived from a distinct progenitor population. To test whether the *Mesp1*-nonexpressing cells are in fact a part of the VCS, the authors generated a triple transgenic to include *Mesp1-Cre, R26R,* and *CCS-LacZ*. *Mesp1*-nonexpressing cells are likely part of the VCS as all cells in this region showed *LacZ* activity expressed by the *CCS-LacZ* allele. OFT cells did not stain for *LacZ* in the triple transgenic further supporting their origin from the NC lineage.

Since the triple transgenic reporters were both *LacZ*, the authors examined the extent of the *Mesp1*-nonexpressing cell contribution to the VCS. To test this, a *CAG-CAT-GFP* reporter was used along with *Mesp1-Cre* and *CCS-LacZ*. Triple staining for GFP, *LacZ*, and nuclei showed that not all GFP-negative cells overlapped with *LacZ*. After quantification of sections from three embryos, around 80% of cells along the IVS showed colocalization of *Mesp1* (*GFP*) and *LacZ* leaving the rest of the *CCS-LacZ*-positive cells as presumably not derived from *Mesp1*-expressing precursors. Taken together, these studies suggest that there exists two distinct populations of *Mesp1*-nonexpressing cells, one of which gives rise to some of the VCS while the other eventually contributes to the OFT along with NCCs that do not express *Mesp1*. The early emergence of a cell population lacking *Mesp1* supports the idea of the requirement of different subsets of regulatory networks in the development of each CCS component and that these transcriptional programs can be established as early as the cardiac crescent stage.

Although the intrinsic rhythm of the heart is subject to nervous system modulation, it is under SAN control. Many observations support the idea that the SAN, which begins to initiate contractions at the early heart tube stage, is a derivative of existing myocardium. Myocytes at the AVC preferentially express the low conductance $Cx45$ gap junction channel that is later found in the adult SAN and AVN (Coppen et al., 2003). Along with their early functional differentiation, classic morphological studies show concurrent development of morphologically distinct nodal and VCS myocytes. Indeed, recent lineage-based analysis has determined that the SAN derives from $Nkx2$–5-negative myocardial precursors, although it does express cardiac markers such as TnI. It appears that the sinus horns, which give rise to the SAN, are under the control of $Tbx18$ (Christoffels et al., 2006).

2.3. Extracardiac controversies: Does the cardiac neural crest contribute to the CCS?

The heart is composed of three major cardiac cell types including the myocardium, endocardium, and epicardium. Other cell lineages such as the NCCs and the EPDCs from the proepicardial organ have been suggested to contribute to the heart. However, a direct contribution of nonmyocardial cells to the CCS has been highly debated and remains controversial.

The NC is a migratory cell population that begins its journey from the dorsal neural tube. These cells form at all axial levels of the embryo and adopt various ectodermal and mesodermal fates as they arrive at their peripheral destinations (Hall, 1988). A subpopulation of these cells termed the cardiac NC is a major extracardiac source of mesenchyme to the OFT and contributes to neural cells that innervate the heart. NC migration along the branchial arches and into as far as the distal cardiac OFT has been shown by chimeric avian studies and by labeled rat embryo experiments (Sumida et al., 1989; Fukiishi and Morriss-Kay, 1992). A Cre/lox approach has identified migrating cells populating the aorticopulmonary septum and the conotruncal cushions prior to and during OFT septation (Jiang et al., 2000).

Early immunohistochemical experiments suggest that CCS cells may originate from the NC. Gorza et al. (1988) demonstrated that an antineurofilament antibody iC8 picked out rabbit myocytes near the AV junction and atrial wall close to the fourth branchial arch. CCS-associated staining by the NC surface marker antibody HNK1/Leu7 was evident. This antibody was also detected in the sinus node and in the autonomic innervations of the rat heart (Nakagawa et al., 1993). Common labeling by these epitopes in NC and CCS cells began the debate over whether NCCs can migrate into the heart and directly participate in forming CCS components.

Using replication-defective retroviral vectors containing a *LacZ* reporter, Poelmann and Gittenberger-de Groot demonstrated that a population of cardiac NCCs enter the heart through the venous pole at the dorsal meso-cardium of the chick embryo (Gittenberger-de Groot et al., 1998). Scattered *LacZ*$^+$ cells were found around the developing AVN, RARB, RAVR, bundle of His, bundle branches, and the AV cushions. The area between the AVN and the RARB along with the aorticopulmonary septum were suggested to receive its NCC contribution from another migrating population of cells through the arterial pole. NCC-derived cells did not stain for HNK1 or α-actin suggesting that they are not myocytes of the conduction rings. Small groups of cells scattered around these areas were proposed not to directly contribute to the conduction system since they were apoptotic. Given their appearance around these areas and the correlation with the timing of CCS maturation, these apoptotic cells were proposed to partici-pate in the last phase of CCS cell differentiation, possibly by releasing transforming growth factors. An alternative suggestion was that these progenitors could mediate the physical separation of CCS cells from the surrounding working myocardium. However, Cheng et al. (1999) demon-strated that central conduction system cells are unequivocally derived from myocytes of the looped tubular heart. In contrast, they did not detect neurogenic contributions to these areas by labeling the migrating NC. Although some *LacZ*$^+$ NC-derived cells were found near the CCS in this study, a number of other *LacZ*$^+$ cells were interspersed randomly in the working myocardium. Whereas one study (Poelmann and Gittenberger-de Groot, 1999) introduced a large amount of virus through a flush method, Cheng et al. used a smaller amount of virus that was microinjected directly into the foci from which cardiac NCCs emigrate. Lack of CCS-associated cells in this more sensitive and precise approach argues against a direct contribution by the cardiac NC to CCS formation. Again, a more likely role for NCCs is to innervate the nodes and bundles of the adult heart. The migratory behavior and pattern of the cardiac NCCs, as revealed by all studies presented, support their involvement in establishing autonomous control of the heart.

In the mouse, Poelmann et al. (2004) proposed that the NC-derived cells could have an inductive role on the CCS by virtue of their association with CCS components at E14.5. Using the *Wnt1-Cre; ROSA26* system, two populations of NCCs were observed in the mouse heart. In contrast to the avian study, the arterial pole population was found to contribute to the bundle branches whereas the venous pole population provided cells to the SAN and AVN. To show that cells marked by *Wnt1-Cre; ROSA26* were near conduction cells, 3D reconstruction and subsequent superim-position of these images onto similarly processed *CCS-LacZ* results were performed. Scattered *LacZ*$^+$ cells were found near the anterior cardinal vein opening that includes the SAN. However, the rather late arrival of these cells at E14.5 cannot account for the presence of a functional peacemaker as early as E8.5. At the meeting point of the migrating populations, an even smaller number of cells were found neighboring the future AVN. All NCCs disappear by E17.5 presumably through apoptosis, although this possibility was not addressed. Overall, the late arrival of the NC-derived cells and their subsequent disappearance suggest that they could potentially influence maturation of the conduction system. This association, however, cannot be interpreted as implying that CCS cells have an extracardiac origin in the NC.

Multipotent cardiac NC derivatives are not only found in the arterial valves at early developmental stages but also persist deeper within the semilunar and AV valve leaflets (Nakamura et al., 2006). This is surprising given that the AV cushions were shown to be mainly derived from cells originating from the endocardium (Kisanuki et al., 2001) and that no contributions beyond the arterial valves were initially observed (Jiang et al., 2000). Nakamura et al. further suggested that NCC progenitors contribute to the His bundle and bundle branches. Two NC-specific promoters driving Cre recombinase, Wnt1-Cre and P0-Cre, were analyzed before aorticopulmonary septation and during adult stages. By crossing both Cre lines with either ROSA26 or CAG-CAT-eGFP, comparable results were obtained with each combination. At E17.5, a time point where all NCCs disappear (Poelmann et al., 2000), immunofluorescence was performed to look for GFP$^+$ NC-derived cells and markers for immature neuroglia. Disperse and rather faint staining was observed near the AV junction, the IVS, and in what the authors suggested to be the posterior internodal tract. Not all GFP$^+$ cells overlapped with markers of undifferentiated glia around the IVS nor was an explanation given for their functional significance. The peculiar persistence of NC-derived cells at E17.5 around the IVS suggests that NC-derived cells could participate in CCS innervation. These findings are at odds with other studies in which P0-Cre and Wnt1-Cre excision resulted in no reporter expression within the vicinity of the IVS (Yamauchi et al., 1999; Jiang et al., 2000; Ismat et al., 2005; Kitajima et al., 2006).

2.4. Extracardiac controversies: The role of EPDCs

The proepicardial organ is an extension of the coelomic wall encompassing the mesocardium at the sinus venosus and the liver. It gives rise to the epicardial-derived cells (EPDCs) that undergo EMT and migrate into the heart to populate coronary vasculature and subendocardium. Vascular cell adhesion molecule (VCAM-1)-deficient mice, which do not develop epicardium, have thin and poorly organized myocardium leading to embryonic lethality (Dansky et al., 2001). Studies have shown that EPDCs in these two regions could have the capacity to induce Purkinje network differentiation (Gourdie et al., 1995; Gittenberger-de Groot et al., 1998). In the 1998 study, chick–quail chimeras were used to show EPDCs differentiating into mural cells lining the coronary vascular system and into intramyocardial and subendocardial fibroblasts. A close relationship was demonstrated between differentiating Purkinje fibers, as indicated by the EAP-300 antibody, with EPDCs within the subendocardial layer and fibroblasts of the coronary arteries. Exactly how EPDCs contribute to Purkinje fiber formation remains to be tested.

3. TRANSCRIPTIONAL REGULATION OF CONDUCTION SYSTEM PATTERNING AND FUNCTION

Much is known and debated in regards to the origin and induction of the conduction system while less is known about the mechanisms that govern its patterning and maturation. Recent studies have demonstrated a critical role for heart-specific transcription factors in CCS formation. Transcription factors can regulate the expression of other genes in a tissue-specific and quantitative manner and are thus major regulators of embryonic developmental processes. Many transcription factors that specifically regulate cardiac genes have been described, several of which are not only indispensable for early heart formation but are also involved in CCS patterning and maturation. Transcription factors from several families have been found in the mouse CCS and include the T-box, homeodomain, and zinc-finger factors. Their functional significance in the conduction system is summarized in Table 3 and Fig. 2.

3.1. T-box transcription factors

The naturally occurring short-tailed T-mutation in mice is caused by multiple genes present in the T/t complex on chromosome 17 (Papaioannou, 2001). The T-mutant, or *Brachyury*, was cloned in 1990. It represents the first T-box factor to be studied. This family is characterized by a highly conserved 180-amino acid domain that is required for DNA binding. Heart-specific T-box factors include *Tbx1*, *Tbx2*, *Tbx3*, *Tbx5*, *Tbx18*, and *Tbx20*. Interestingly, transcription factors indispensable for critical stages in early heart development, such as *Tbx5* also play distinct roles in the patterning and/or maintenance of the conduction system. The duality of their function initially obscured their phenotypic manifestations in the CCS.

In humans, *TBX5* haploinsufficiency causes the rare inherited disease of Holt–Oram syndrome that is characterized in part by conduction system defects such as progressive AV block, bundle-branch block, and sick sinus syndrome (Basson et al., 1997, 1999; Li et al., 1997). Interestingly, some patients have electrophysiological defects in the absence of

TABLE 3 Transcriptional control of the conduction system

Genotype	Phenotypes
Tbx5$^{+/-}$; minK-LacZ	AVN maturation; bundle branch patterning/maintenance
Nkx2-5^{neo}/+ ; minK-LacZ	AVN, Cx45 (AVN), and His–Purkinje maintenance
Nkx2-5$^{flox/flox}$; Mlc2v-Cre	AV nodal and ventricular lineage maturation/maintenance
Nkx2-5 overexpression (chick)	Purkinje maturation
Msx2$^{-/-}$	No cardiac phenotype
Msx2$^{-/-}$; Nkx2-5$^{+/-}$	Nkx2-5 haploinsufficient phenotype
Hop-LacZ	Maintenance of Cx40 expression in the His–Purkinje and CCS function
Irx5$^{-/-}$	Establishment of the cardiac repolarization gradient
Hf1b$^{LacZ/LacZ}$	VCS maturation, and function
Pax3-Cre; Hf1b$^{flox/flox}$	AVN innervation
Mlc2v-Cre; Hf1b$^{flox/flox}$	Organization of cellular Cx40 expression

A summary of recent studies and the resulting CCS-specific phenotypes. For references, please see main text.

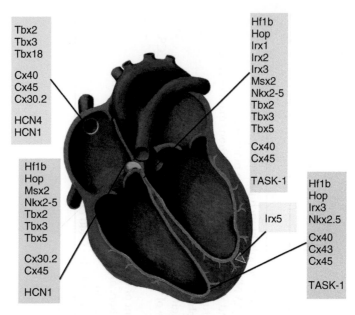

FIGURE 2   Molecular control of CCS development. Transcription factors, gap junction proteins, and ion channel subunits found in each of the conduction system components are shown as in Table 1.

structural abnormalities (Basson et al., 1994) arguing for a role of *Tbx5* in the conduction system that is separate from its role in cardiac septation. A mouse model of *Tbx5* haploinsufficiency recapitulated the electrophysiological defects seen in human patients and serves as a useful tool for investigating *Tbx5* dosage in the conduction system (Bruneau et al., 2001b). In this mouse model, atrial natriuretic factor (ANF) and *Cx40*, known to be expressed in the CCS, were discovered to be downstream targets of *Tbx5*. Recently, Moskowitz et al. (2004) demonstrated the preferential expression and requirement of *Tbx5* within the central conduction system. A role for this T-box factor in AVN maturation and patterning of the bundle branches was uncovered. *Tbx5* expression was found in the AV rings and the developing AVN, His bundle, and bundle branches in newborn mice. During normal development, the AV rings regress and slowly disappear leaving only the definitive AVN and bundle in the AV junction as indicated by *minK-LacZ*. However, reporter expression remained in the AV ring regions of *Tbx5* haploinsufficient adults suggesting a lack of CCS maturation. Electrophysiological studies show a lack of progressive PR shortening that is reflective of inadequate AVN maturation or a defect in its connection with the atria or the AV bundle. In addition to the hypoplastic AVCS, the right bundle branch appeared foreshortened. This was reflected by the prolongation of the QRS interval indicating abnormal ventricular depolarization and activation.

On the basis of their CCS-specific expression patterns and transcriptional repression activity, two other T-box genes, *Tbx2* and *Tbx3* could play an important role in the formation of the central CCS. Adult nodal myocytes are similar to embryonic cardiomyocytes in that they are small in size in comparison to those of the working myocardium and have poorly organized actin, myosin, and sarcoplasmic reticulum. Their embryonic or "primary" characteristics make them even more difficult to be traced during developmental stages. Although a universal molecular marker that is applicable at all stages is lacking, some species-specific genes have been found to be distinctively expressed within the SAN and AVN. *Tbx3* is expressed in the central conduction system, internodal region, and the AV junction (Hoogaars et al., 2004). At E8.5, it is found in the future IFT and AVC of the tubular heart. About one day later, it is expressed around the primary foramen within which the future conducting bundles would form. In older embryos, expression is detected in the AV and OFT valves. It was suggested through a series of *in vitro* studies that *Tbx3* permits the development of the nodal regions by repressing transcription of chamber-specific genes. One of these genes is *Cx40*, which is widespread during early stages but later becomes downregulated in the chamber myocardium, presumably by *Tbx3*. However, *Cx40* expression is later found to overlap with *Tbx3* in parts of the AV node and proximal bundle branches. This observation suggests that other factors are involved in promoting the critical transition between chamber versus AV nodal myocardium. *Tbx2* is also expressed in a similar pattern in the developing heart and has transcriptional repression properties that are identical to that of *Tbx3*. *Tbx2* null mice have defects of AVC patterning and of OFT morphogenesis (Harrelson et al., 2004). Their effects on patterning and differentiation of the nodal phenotype have not been investigated. Given that loss of either factor alone does not appear to affect CCS formation, there may be redundancy between *Tbx2* and *Tbx3* in the conduction system. Interestingly, *Tbx3* is regulated dose dependently by *Tbx5* (Mori et al., 2006), suggesting that a delicate balance between T-box genes is important for regulation of CCS form and/or function.

3.2. *Nkx2–5* is required for central conduction system formation and Purkinje maturation

Nkx2–5 encodes a homeodomain protein that includes N-terminal and C-terminal transcriptional activation domains in addition to the homeobox. *Nkx2–5* is homologous to the *Drosophila tinman* gene that is required for heart formation (Komuro and Izumo, 1993; Harvey et al., 2002).

Analysis of *Nkx2–5* expression in the chick, mouse, and humans show elevated levels in the forming CCS (Takebayashi-Suzuki et al., 2001; Thomas et al., 2001). *Nkx2–5* can act synergistically with other transcription factors to activate expression of cardiac-specific genes. *Nkx2–5, Gata4, Tbx5* (and other *T*-box factors), and serum response factor (*SRF*) function coordinately by binding to adjacent sites within the regulatory regions of cardiac target genes (Chen and Schwartz, 1996; Chen et al., 1996; Durocher et al., 1997; Bruneau et al., 2001b).

Nkx2–5 is initially localized within the cardiogenic mesoderm and is one of the earliest markers of the primary and secondary heart fields (Komuro and Izumo, 1993; Harvey et al., 2002). Functional studies show that the postgastrulation expression of *Nkx2–5* is required for maintenance and specification of the cardiomyogenic cell fate. In *Nkx2–5* null mice, heart development arrests around the looping stage resulting in a bulbous atrium and ventricle, a widened AVC, and a stenotic OFT (Harvey et al., 2002). The significance of transcriptional control of conduction system formation was not understood until the discovery of a human *Nkx2–5* kindred associated with AVN defects (Schott et al., 1998; Benson et al., 1999; Goldmuntz, 2001; Gutierrez-Roelens et al., 2002). Heterozygous mutations in *Nkx2–5* have been associated with AV conduction block and various defects in atrial, ventricular, and conotruncal septation and AV valve formation. The onset of the AV block phenotype emerges over the period of postnatal maturation of the individual suggesting a progressive disease of the CCS. There has been considerable interest in understanding how these mutations lead to the diverse cardiac phenotypes.

Around 30 *Nkx2–5* mutations have been described to date, of which 19 are associated with progressive AV block. Mutations causing reduced or absent DNA binding have been shown to be associated with the AV block phenotype. Overexpression of human *Nkx2–5* mutations that cannot bind DNA in transgenic mice recapitulates progressive AV block (Kasahara et al., 2001). Interestingly, recent studies show that a C-terminal truncation mutation with intact homeodomain is also associated with AV block. As shown by *in vitro* assays, this mutation turns out to be a hypomorphic allele that binds to DNA but cannot effectively activate downstream targets (Clark and Benson, presented at the 2005 Weinstein conference).

The *minK-LacZ* reporter was crossed with *Nkx2–5* haploinsufficient mice to examine the effect of lowered dosage of this transcription factor on the full extent of the conduction system (Jay et al., 2004). Interestingly, the CCS primordium at the inner curvature as indicated by *minK-LacZ* was completely missing in *Nkx2–5* null embryo at E9.5. This suggests that *Nkx2–5* is not only important during heart tube morphogenesis but its expression is also required for patterning of CCS precursors. An absence of the primordium as opposed to a requirement of *Nkx2–5* for *minK* expression appears to be the case since *minK* mRNA was preserved throughout the myocardium. Having established that normal *Nkx2–5* levels are crucial for patterning of the CCS primordium, the authors tested the effect of *Nkx2–5* haploinsufficiency on the adult conduction system. These mice form a smaller central conduction as indicated by diminished *minK-LacZ* expression. A dramatic decrease of the *Cx45*-expressing AVN region was observed along with an overall decrease in the number of cells coexpressing *Cx40* and *Cx45* in the His bundle and branches. As the peripheral conduction system begins to elaborate and mature, there is a significant decrease in the number of $LacZ^+$ and $Cx40^+$ Purkinje cells. Confocal microscopy quantification $Cx40^+$ plaques revealed a loss of approximately half of Purkinje cells while those that remained express normal levels of *Cx40*. Electrophysiological analysis revealed progressive AVN dysfunction upon atrial pacing in 7-week-old animals. These animals also exhibited progressive decrease of His bundle signal amplitude attributable to bundle branch thinning. Prolonged QRS intervals were observed and are accounted for by the low density of remaining conduction system cells at the Purkinje–myocyte junction. Consequently, a delay is produced as each terminal fiber cell is forced to depolarize a larger region of slow conducting myocardium. Taken together, these studies demonstrate a requirement for proper *Nkx2–5* dosage for patterning of the CCS primordium. Loss of approximately half of the dosage of this critical transcription factor in the heart leads to hypoplastic development of the AVN and His–Purkinje accompanied by a near absence of *Cx45* in the hypoplastic AVN. Thus, it is conceivable that *Nkx2–5* is required for the recruitment of myocytes into the conduction system or is crucial for their retention in the adult.

By generating a ventricular-specific knockout line, Pashmforoush et al. (2004) set out to address whether *Nkx2–5* participates in the specification of cardiac muscle lineages. Deletion of *Nkx2–5* at E10.5 by *MLC2v-Cre* led to the identification of an intrinsic role of this homeodomain transcription factor in the formation and maturation of the AV nodal and ventricular myocyte lineages. Fifty percent of adult *flox/flox;Cre⁺* mice exhibited first-degree AV block progressing to complete heart block. Electrophysiological studies showed clear defects in AVN and His bundle function. Pashmforoush et al., similar to Jay et al., also found that decreased *Nkx2–5* dosage led to a reduced number of cells in the conduction system. In the ventricular knockout study, histology revealed smaller central conduction system elements in adult mice along with their cellular degeneration as evidenced by cell drop out and fibrosis. A greater than 90% decrease of *Nkx2–5* in the ventricles led to myocardial noncompaction and hypertrabeculation. Pooled adult ventricular mRNA from *flox/flox* and *flox/flox;cre⁺* was used for microarray analysis showing upregulation and persistence of *BMP10*, *HCN-1*, the atrial-specific gene *sarcolipin* and downregulation of *HOP* and *minK*. Taken together, the dysregulation of downstream targets that include markers for atrial, conduction system, and trabeculae cell types argues for the requirement of *Nkx2–5* in cardiac lineage maturation and specification.

Interestingly, overexpression of a dominant negative *Nkx2–5* protein that cannot bind DNA but can undergo homodimerization during fetal and neonatal periods but not adult stages led to severe conduction defects (Wakimoto et al., 2002). Recent studies have shown that in addition to its early function in the establishing the cardiomyogenic cell fate, *Nkx2–5* plays an additional role in the formation and maturation of the VCS at a critical period between late fetal and neonatal stages. The spatiotemporal expression of elevated *Nkx2–5* in the chick CCS revealed a striking correlation with the timing and location of the recruitment of cells to the His bundle, bundle branches, and Purkinje system tissues (Thomas et al., 2001; Harris et al., 2006). In the mouse, postseptation restriction of upregulated *Nkx2–5* to the CCS was observed at E14.5. Disruption of the transient waves of preferential *Nkx2–5* expression in the chick CCS was performed with constitutive overexpression in the myocardium (Harris et al., 2006). This led to the dysregulation of markers of progressive stages of Purkinje development. The early Purkinje marker, *Cx40*, was shown to increase in the myocardium when *Nkx2–5* was artificially increased. Moreover, the terminal differentiation marker, slow tonic myosin heavy chain (*sMHC*), was blocked with constitutive overexpression. As a further confirmation that C-terminal truncation mutations seen in patients with progressive AV block are not dominant negative mutations, overexpression of this construct did not lead to disruption of Purkinje maturation. These results suggest that controlled endogenous upregulation of *Nkx2–5* preferentially in the Purkinje fibers is critical for their maturation. In light of the *ET-1/NRG-1* induction model, it is likely that the cell autonomous requirement for *Nkx2–5* transcriptional control and paracrine signaling mechanisms both play integral roles in Purkinje fiber recruitment and maturation.

3.3. *Msx2* does not mediate *Nkx2–5* function in the CCS

The mammalian *Msx* genes are closely related to *Drosophila msh* and include *Msx1*, *Msx2*, and *Msx3* (Davidson, 1995). Chick *Msx2* mRNA has been found in the developing conduction system at the inner curvature (Chan-Thomas et al., 1993). It is later found in the AV junction, AV rings, and the crest of the IVS. *Msx2* then becomes restricted to the developing His bundle and bundle branches (Thomas et al., 2001). A similar expression pattern is found in the developing central conduction system of the mouse. Despite its discovery as the first transcription factor in the CCS, *Msx2* knockout mice exhibit no cardiac defects (Satokata et al., 2000) suggesting that there could be functional redundancy among *Msx* genes or with other homeobox factors.

Msx2 is ectopically expressed throughout the myocardium in *Nkx2–5* homozygous null mice (Jay et al., 2004) suggesting that it could affect conduction system development through a transcriptional cascade starting with $Nkx2-5$. $Msx2$ has also been shown to be a regulator of BMP signaling in the AVC during cushion maturation and AV valve formation (Abdelwahid et al., 2001). It is expressed downstream of $Pax3$ which is a key regulator of cardiac NC development. Interestingly, the $Msx2$ null mutant rescued the OFT defects seen in the $Splotch$ mouse devoid of $Pax3$. To test whether the abnormal electrophysiology of $Nkx2-5^{+/-}$ mice could be mediated by $Msx2$, Jay et al. (2005) generated a double knockout model for both genes. This approach assumes that a lack of $Msx2$ would rescue $Nkx2-5$ haploinsufficiency. The authors showed that there were no significant functional or structural differences between $Msx2^{-/-}$ mice and wild-type littermate controls. When tested in the $Nkx2-5^{+/-}$ background, a complete lack of $Msx2$ led to PR interval prolongation, QRS widening, and decreased His-bundle amplitude. These phenotypic effects were attributable to a dramatic decrease in AVN size and Purkinje fiber number that are already a part of the $Nkx2-5^{+/-}$ phenotype (Jay et al., 2004). It is clear that although $Nkx2-5$ is upstream of $Msx2$ and restricts its expression in the heart, $Msx2$ does not mediate the conduction defects seen in the $Nkx2-5$ haploinsufficiency model.

3.4. HOP is required for CCS maturation and function

The homeodomain only protein Hop is the smallest known homeodomain protein which comprises just 73 amino acids (Chen et al., 2002; Shin et al., 2002). Its homeodomain appears to be divergent and cannot bind DNA. Instead, Hop has been shown to regulate transcription by binding to other transcriptional partners. At E7.75, Hop transcripts are found in the lateral wings of the cardiac crescent, then along the linear heart tube and later in the mesoderm dorsal to the heart. Expression in all four chambers is maintained until E14.5. $Hop^{-/-}$ mice have a partially penetrant phenotype with ~50% of embryos displaying poorly developed myocardium followed by lethality at midgestation.

In $Hop^{+/-}$ mice and those $Hop^{-/-}$ animals that survive to term, restricted $LacZ$ reporter expression from a $LacZ$ gene inserted at the Hop locus was detected in the adult CCS including the AVN, His bundle, bundle branches, and the Purkinje (Ismat et al., 2005). The staining is less restricted in newborn mice where there is more prominent expression in the endocardium and IVS. There was no change in adult proximal CCS cell volume between $Hop^{+/-}$ and $Hop^{-/-}$ mice indicating that a complete lack of Hop protein does not affect the establishment of CCS patterning or cell maintenance. This result is expected because by the time Hop begins to be restricted in newborns, proximal CCS elements have already started to mature.

Like Hop, Cx40 is expressed in the AVN, His–Purkinje system, and the atria. Homozygous $Cx40$ knockout mice have impaired sinoatrial conduction, slow atrial conduction, His bundle delay, unchanged ventricular conduction, and increase in the incidence of inducible atrial tachyarrhythmia

(Simon et al., 1998). *Cx40* is a target for *Nkx2–5* as indicated by its reduced levels within context of dominant *Nkx2–5* mutations. Given that *Nkx2–5* is upstream of HOP and that *Cx40* and *Hop* share similar adult expression patterns, Ismat et al. set out to test whether loss of *Hop* function in the surviving adult mice could lead to changes in *Cx40* expression. In the absence of *Hop*, *Cx40* expression was reduced in the atria and the AVCS with the distal conduction most severely affected. Although there was no change in CCS size, a reduction in number of cells expressing *Cx40* was found in adult $Hop^{-/-}$ mice. Specifically, *Cx40* was dramatically reduced below the proximal AVN and His bundle region. ECG and electrophysiological analysis of the $Hop^{-/-}$ mice indeed showed widened QRS complexes, longer QT intervals, and wider p-waves in comparison to wild-type littermates. A longer HV interval and normal AH interval indicates that conduction delay underlying the widened QRS complex on the surface ECGs is predominantly due to conduction defects below the AVN. These electrophysiological findings are consistent with HOP expression pattern and can be attributable to the observed decrease of Cx40-expressing cells below the AVN. Thus, *Hop* appears to be important for regulating and maintaining gap junction integrity in the CCS.

Interestingly, unlike *Nkx2–5* mutant mice that display predominantly proximal conduction defects, Hop mutant mice exhibit functional defects of the distal conduction system. Taken together, these studies show that similar to the duality that exists for *Nkx2–5* and *Tbx5*, *Hop* not only has a critical function during cardiogenesis but also functions downstream of *Nkx2–5* during CCS maturation and maintenance. Early in development, it modulates *Nkx2–5* function by inhibition of SRF, while later on in postnatal life HOP ensures that proper *Cx40* expression is achieved in conduction cells that require *Nkx2–5* to survive. Thus, *Hop* is likely to mediate some but not all of the downstream function of *Nkx2–5* in conduction tissues. The sensitivity of peripheral conduction system to loss of *Hop* and *Cx40* further highlights the concept that individual segments of the conduction system are regulated by different sets of transcriptional programs.

3.5. *Irx5* is required for ventricular conduction

The *Iroquois* genes encode transcription factors bearing a characteristic three-amino acid length extension (TALE) class homeobox and a highly conserved *Iro* motif that is unique to this family (Cavodeassi et al., 2001). They were discovered as important prepatterning molecules in *Drosophila* that have conserved functions during neural precursor specification in *Xenopus*. Six mammalian *Irx* genes are organized into two clusters of three orthologs. In mice, *Irx1*, *Irx2*, *Irx4* are found on chromosome 13, while *Irx3*, *Irx5*, *Irx6* reside on chromosome 8. These genes are found in the central nervous system, limbs, lungs, and skin with all six expressed in the heart.

Whether these genes function in the patterning and differentiation of the developing heart has been of great interest.

All six mammalian *Iroquois* genes show specific and overlapping patterns of expression during heart formation (Christoffels et al., 2000; Bruneau, 2002). In particular, *Irx1*, *Irx2*, and *Irx3* are found in the VCS (Christoffels et al., 2000). The expression of *Irx1*, *Irx2*, and *Irx3* appear to overlap in the His bundle, bundle branches, and Purkinje fiber primordium. Of these genes, a recent *Irx3-LacZ* knockout model shows that *Irx3* specifically marks the developing and mature VCS (our unpublished data). Loss of both alleles led to improper ventricular conduction suggesting that *Irx3* is important for VCS maturation and function. A requirement for *Irx4* and *Irx5* during heart development has been previously shown (Bao et al., 1999; Bruneau et al., 2001a; Lebel et al., 2003; Costantini et al., 2005).

Coordinated and synchronous contraction of the heart is not only dependent on the specialized CCS tissue, but is also a function of the electrical behavior of the working myocardium. In the mammalian heart, ventricular repolarization proceeds as a synchronous wave which progresses from base-to-apex and from epicardial to endocardial myocardium. For this sequence to occur, longer action potential durations (APDs) must occur in the endocardial myocytes than epicardial myocytes. Regional differences in the density of the fast component of the transient outward current, $I_{to,f}$, is required to achieve differential rates of repolarization in some mammalian species. It was shown that mice lacking *Irx5* failed to establish the cardiac repolarization gradient due to increased potassium channel subunit $Kv4.2$ expression in the endocardium (Costantini et al., 2005). Consequently, the major cardiac repolarization current, $I_{to,f}$, increased to cause susceptibility of these mice to inducible arrhythmias. These studies further demonstrated that *Irx5* is expressed in a gradient opposite of $Kv4.2$ expression, and regulates this inverse gradient by repressing this channel subunit in the endocardium. *Irx5* repression of $Kv4.2$ was shown to be dependent on the presence of the SET/MYND domain *mBOP* which recruits histone deacetylase activity to the target promoter, thus shifting the balance from a preferentially transcriptionally active complex to a more repressive state. As defects in the establishment or maintenance of the cardiac repolarization gradient are associated with lethal arrhythmias in human heart disease, it will be interesting to determine if *Irx5* is involved in disease-related responses in the myocardium.

3.6. Zinc-finger transcription factors

The GATA family of transcription factors includes six vertebrate members that possess a DNA-binding domain of either one or two zinc-fingers coupled to a region rich in basic residues. Their name is derived from the consensus recognition sequence of (A/T)GATA(A/G). *GATA4*, *GATA5*, and *GATA6* are expressed in the embryonic and adult mouse heart. *GATA4* is not expressed in the mammalian conduction system. However, its mRNA has been found to be in much higher levels in the Purkinje fiber than the working myocardium in the chick heart (Takebayashi-Suzuki et al., 2001). As *GATA4* can function in combination with *Nkx2–5*, this may indicate a similar requirement for high levels of *GATA4* for VCS development.

Hf1b is a zinc-finger transcription factor closely related to Sp-1 from *Xenopus*. Targeted disruption of *Hf1b* by *LacZ* reporter insertion revealed its expression in the working myocardium at E17 (Nguyen-Tran et al., 2000). Preferential expression is also found in the AVN and His–Purkinje system. Mice lacking *Hf1b* display normal cardiac structure and function but succumb to sudden cardiac death and conduction defects that include spontaneous ventricular tachycardia, atrial bradycardia, and AV block. Single cell electrophysiology analysis showed an increase in action potential heterogeneity and aberrant ventricular myocyte excitability. Mislocalization of connexins as well as defects of Purkinje fiber formation were evident. It was proposed that myocardial cells around the conduction fibers had a "confused" identity by expressing genes normally found in the conduction system. The *Hf1b* knockout initially suggests that it could be required in either specifying the transition between ventricular and conduction system cell lineages or for the physical separation and organization of the conducting fibers.

Support for the notion that NC-derived cells do not directly incorporate into the CCS but rather are important in its innervation and potentially maturation comes from a NC-specific knockout of the *Hf1b* transcription factor (St Amand et al., 2006). NC-specific *Hf1b* knockouts (*Pax3-Cre; Hf1bflox/flox*) exhibited a decrease in the neurotrophin receptor *TrkC* which is required for the survival of innervating neurons by interacting with neurotrophin *NT-1* (Story et al., 2000). The requirement of *Hf1b* in NCCs was proposed to be important for the extension of cardiac ganglia and their axon to innervate the AVN. Accordingly, antibody staining with the neuronal marker *β-III tubulin* revealed a substantial decrease of neurons around the AVN. Interestingly, the ventricular-specific deletion of *Hf1b* within the CCS led to *Cx40* disorganization but not conduction defects. This suggests that *Hf1b* expression within the ventricular myocyte lineage is not required for CCS formation or function.

4. CONCLUSIONS

Since the detailed anatomical discoveries made at the turn of the twentieth century, a tremendous amount of progress has been made toward understanding the regulatory mechanisms underlying conduction system development and function. Cells of the conduction axis have been shown to be descendents of the myocardial lineage and could be specified as early as the cardiac crescent stage. An increasing number of ion channels, gap junction proteins, and transcription factors regulating their expression are preferentially expressed in the developing conduction system. Their functional significance is beginning to be revealed by the generation of mouse models expressing conditional alleles, reporter genes, and lineage-specific *Cre* recombinases. Recent results from these approaches suggest that cells within conduction system components are heterologous in origin and harbor different sets of transcriptional regulatory networks required for patterning and maturation. Several transcription factors, including those of the T-box and homeodomain families, are not only indispensable for early stage cardiogenesis but also participate in conduction system formation. A current challenge of the field is to gain an even greater understanding of those inductive and regulatory cues that are important for CCS development. In particular, the precise origins of CCS components and the intrinsic regulators of CCS gene expression programs must be identified. Equipped with rapidly emerging mouse models and molecular markers, mechanisms driving CCS formation await their discovery before conduction disease can be fully understood and addressed.

ACKNOWLEDGMENTS

We would like to thank John Wylie for artwork, and Takashi Mikawa for critical review of the manuscript.

REFERENCES

Abdelwahid, E., Rice, D., Pelliniemi, L.J., Jokinen, E. 2001. Overlapping and differential localization of Bmp-2, Bmp-4, Msx-2 and apoptosis in the endocardial cushion and adjacent tissues of the developing mouse heart. Cell Tissue Res. 305, 67–78.

Anderson, R.H., Ho, S.Y., Smith, A., Becker, A.E. 1981. The internodal atrial myocardium. Anat. Rec. 201, 75–82.

Anderson, R.H., Christoffels, V.M., Moorman, A.F. 2004. Controversies concerning the anatomical definition of the conduction tissues. Anat. Rec. 280, 8–14.

Bao, Z.Z., Bruneau, B.G., Seidman, J.G., Seidman, C.E., Cepko, C.L. 1999. Regulation of-chamber-specific gene expression in the developing heart by Irx4. Science 283, 1161–1164.

Basson, C.T., Cowley, G.S., Solomon, S.D., Weissman, B., Poznanski, A.K., Traill, T.A., Seidman, J.G., Seidman, C.E. 1994. The clinical and genetic spectrum of the Holt-Oram syndrome (heart-hand syndrome). N. Engl. J. Med. 330, 885–891.

Basson, C.T., Bachinsky, D.R., Lin, R.C., Levi, T., Elkins, J.A., Soults, J., Grayzel, D., Kroumpouzou, E., Traill, T.A., Leblanc-Straceski, J., Renault, B., Kucherlapati, R., et al. 1997. Mutations in human TBX5 cause limb and cardiac malformation in Holt-Oram syndrome. Nat. Genet. 15, 30–35.

Basson, C.T., Huang, T., Lin, R.C., Bachinsky, D.R., Weremowicz, S., Vaglio, A., Bruzzone, R., Quadrelli, R., Lerone, M., Romeo, G., Silengo, M., Pereira, A., et al. 1999. Different TBX5 interactions in heart and limb defined by Holt-Oram syndrome mutations. Proc. Natl. Acad. Sci. USA 96, 2919–2924.

Benson, D.W., Silberbach, G.M., Kavanaugh-McHugh, A., Cottrill, C., Zhang, Y., Riggs, S., Smalls, O., Johnson, M.C., Watson, M.S., Seidman, J.G., Seidman, C.E., Plowden, J., et al. 1999. Mutations in the cardiac transcription factor NKX2.5 affect diverse cardiac developmental pathways. J. Clin. Investig. 104, 1567–1573.

Bruneau, B.G. 2002. Transcriptional regulation of vertebrate cardiac morphogenesis. Circ. Res. 90, 509–519.

Bruneau, B.G., Bao, Z.Z., Fatkin, D., Xavier-Neto, J., Georgakopoulos, D., Maguire, C.T., Berul, C.I., Kass, D.A., Kuroski-de Bold, M.L., de Bold, A.J., Conner, D.A., Rosenthal, N., et al. 2001a. Cardiomyopathy in Irx4-deficient mice is preceded by abnormal ventricular gene expression. Mol. Cell. Biol. 21, 1730–1736.

Bruneau, B.G., Nemer, G., Schmitt, J.P., Charron, F., Robitaille, L., Caron, S., Conner, D.A., Gessler, M., Nemer, M., Seidman, C.E., Seidman, J.G. 2001b. A murine model of Holt-Oram syndrome defines roles of the T-box transcription factor Tbx5 in cardiogenesis and disease. Cell 106, 709–721.

Cavodeassi, F., Modolell, J., Gomez-Skarmeta, J.L. 2001. The Iroquois family of genes: From body building to neural patterning. Development (Cambridge, England). 128, 2847–2855.

Chan-Thomas, P.S., Thompson, R.P., Robert, B., Yacoub, M.H., Barton, P.J. 1993. Expression of homeobox genes Msx-1 (Hox-7) and Msx-2 (Hox-8) during cardiac development in the chick. Dev. Dyn. 197, 203–216.

Chen, C.Y., Schwartz, R.J. 1996. Recruitment of the tinman homolog Nkx-2.5 by serum response factor activates cardiac alpha-actin gene transcription. Mol. Cell. Biol. 16, 6372–6384.

Chen, C.Y., Croissant, J., Majesky, M., Topouzis, S., McQuinn, T., Frankovsky, M.J., Schwartz, R.J. 1996. Activation of the cardiac alpha-actin promoter depends upon serum response factor, Tinman homologue, Nkx-2.5, and intact serum response elements. Dev. Genet. 19, 119–130.

Chen, F., Kook, H., Milewski, R., Gitler, A.D., Lu, M.M., Li, J., Nazarian, R., Schnepp, R., Jen, K., Biben, C., Runke, G., Mackay, J.P., et al. 2002. Hop is an unusual homeobox gene that modulates cardiac development. Cell 110, 713–723.

Cheng, G., Litchenberg, W.H., Cole, G.J., Mikawa, T., Thompson, R.P., Gourdie, R.G. 1999. Development of the cardiac conduction system involves recruitment within a multipotent cardiomyogenic lineage. Development (Cambridge, England) 126, 5041–5049.

Christoffels, V.M., Keijser, A.G., Houweling, A.C., Clout, D.E., Moorman, A.F. 2000. Patterning the embryonic heart: Identification of five mouse Iroquois homeobox genes in the developing heart. Dev. Biol. 224, 263–274.

Christoffels, V.M., Mommersteeg, M.T., Trowe, M.O., Prall, O.W., de Gier-de Vries, C., Soufan, A.T., Bussen, M., Schuster-Gossler, K., Harvey, R.P., Moorman, A.F., Kispert, A. 2006. Formation of the venous pole of the heart from an Nkx2-5-negative precursor population requires Tbx18. Circ. Res. 98, 1555–1563.

Coppen, S.R., Dupont, E., Rothery, S., Severs, N.J. 1998. Connexin45 expression is preferentially associated with the ventricular conduction system in mouse and rat heart. Circ. Res. 82, 232–243.

Coppen, S.R., Kaba, R.A., Halliday, D., Dupont, E., Skepper, J.N., Elneil, S., Severs, N.J. 2003. Comparison of connexin expression patterns in the developing mouse heart and human foetal heart. Mol. Cell. Biochem. 242, 121–127.

Costantini, D.L., Arruda, E.P., Agarwal, P., Kim, K.H., Zhu, Y., Zhu, W., Lebel, M., Cheng, C. W., Park, C.Y., Pierce, S.A., Guerchicoff, A., Pollevick, G.D., et al. 2005. The homeodomain transcription factor Irx5 establishes the mouse cardiac ventricular repolarization gradient. Cell 123, 347–358.

Dansky, H.M., Barlow, C.B., Lominska, C., Sikes, J.L., Kao, C., Weinsaft, J., Cybulsky, M.I., Smith, J.D. 2001. Adhesion of monocytes to arterial endothelium and initiation of atherosclerosis are critically dependent on vascular cell adhesion molecule-1 gene dosage. Arterioscler. Thromb. Vasc. Biol. 21, 1662–1667.

Davidson, D. 1995. The function and evolution of Msx genes: Pointers and paradoxes. Trends Genet. 11, 405–411.

Davis, D.L., Edwards, A.V., Juraszek, A.L., Phelps, A., Wessels, A., Burch, J.B. 2000. A GATA-6 gene heart-region-specific enhancer provides a novel means to mark and probe a discrete component of the mouse cardiac conduction system. Mech. Dev. 108, 105–119.

Di Lisi, R., Sandri, C., Franco, D., Ausoni, S., Moorman, A.F., Schiaffino, S. 2000. An atrioventricular canal domain defined by cardiac troponin I transgene expression in the embryonic myocardium. Anat. Embryol. 202, 95–101.

Durocher, D., Charron, F., Warren, R., Schwartz, R.J., Nemer, M. 1997. The cardiac transcription factors Nkx2-5 and GATA-4 are mutual cofactors. EMBO J. 16, 5687–5696.

Fleming, P. 1997. A Short History of Cardiology, Atlanta: Rodopi Publishing.

Fukiishi, Y., Morriss-Kay, G.M. 1992. Migration of cranial neural crest cells to the pharyngeal arches and heart in rat embryos. Cell Tissue Res. 268, 1–8.

Garcia-Frigola, C., Shi, Y., Evans, S.M. 2003. Expression of the hyperpolarization-activated cyclic nucleotide-gated cation channel HCN4 during mouse heart development. Gene Expr. Patterns 3, 777–783.

Gassanov, N., Er, F., Zagidullin, N., Hoppe, U.C. 2004. Endothelin induces differentiation of ANP-EGFP expressing embryonic stem cells towards a pacemaker phenotype. FASEB J. 18, 1710–1712.

Gittenberger-de Groot, A.C., Vrancken Peeters, M.P., Mentink, M.M., Gourdie, R.G., Poelmann, R.E. 1998. Epicardium-derived cells contribute a novel population to the myocardial wall and the atrioventricular cushions. Circ. Res. 82, 1043–1052.

Goldmuntz, E. 2001. The epidemiology and genetics of congenital heart disease. Clin. Perinatol. 28, 1–10.

Gorza, L., Schiaffino, S., Vitadello, M. 1988. Heart conduction system: A neural crest derivative? Brain Res. 457, 360–366.

Gourdie, R.G., Mima, T., Thompson, R.P., Mikawa, T. 1995. Terminal diversification of the myocyte lineage generates Purkinje fibers of the cardiac conduction system. Development (Cambridge, England) 121, 1423–1431.

Gourdie, R.G., Harris, B.S., Bond, J., Justus, C., Hewett, K.W., O'Brien, T.X., Thompson, R.P., Sedmera, D. 2003. Development of the cardiac pacemaking and conduction system. Birth Defects Res. C Embryo Today 69, 46–57.

Graham, V., Zhang, H., Willis, S., Creazzo, T.L. 2006. Expression of a two-pore domain K+ channel (TASK-1) in developing avian and mouse ventricular conduction systems. Dev. Dyn. 235, 143–151.

Gutierrez-Roelens, I., Sluysmans, T., Gewillig, M., Devriendt, K., Vikkula, M. 2002. Progressive AV-block and anomalous venous return among cardiac anomalies associated with two novel missense mutations in the CSX/NKX2-5 gene. Hum. Mutat. 20, 75–76.

Hall, B.K. 1988. Patterning of connective tissues in the head: Discussion report. Development (Cambridge, England) 103(Suppl.), 171–174.

Harrelson, Z., Kelly, R.G., Goldin, S.N., Gibson-Brown, J.J., Bollag, R.J., Silver, L.M., Papaioannou, V.E. 2004. Tbx2 is essential for patterning the atrioventricular canal and for morphogenesis of the outflow tract during heart development. Development (Cambridge, England) 131, 5041–5052.

Harris, B.S., Spruill, L., Edmonson, A.M., Rackley, M.S., Benson, D.W., O'Brien, T.X., Gourdie, R.G. 2006. Differentiation of cardiac Purkinje fibers requires precise spatiotemporal regulation of Nkx2-5 expression. Dev. Dyn. 235, 38–49.

Harvey, R.P., Lai, D., Elliott, D., Biben, C., Solloway, M., Prall, O., Stennard, F., Schindeler, A., Groves, N., Lavulo, L., Hyun, C., Yeoh, T., et al. 2002. Homeodomain factor Nkx2-5 in heart development and disease. Cold Spring Harb. Symp. Quant. Biol. 67, 107–114.

Heintzberger, C.F. 1974. The development of the sinu-atrial node in the mouse. Acta Morphol. Neerl.-Scand. 12, 317–330.

Hertig, M.C., Kubalak, W.S., Wang, Y., Chien, R.K. 1999. Synergistic roles of neuregulin-1 and insulin-like growth factor-I in activation of the phosphatidylinositol 3-kinase pathway and cardiac chamber morphogenesis. J. Biol. Chem. 274, 37362–37369.

Ho, S.Y. 2003. Clinical pathology of the cardiac conduction system. Novartis Found. Symp 250, 210–221; discussion 221–216, 276–219.

Hoogaars, W.M., Tessari, A., Moorman, A.F., de Boer, P.A., Hagoort, J., Soufan, A.T., Campione, M., Christoffels, V.M. 2004. The transcriptional repressor Tbx3 delineates the developing central conduction system of the heart. Cardiovasc. Res. 62, 489–499.

Ikeda, T., Iwasaki, K., Shimokawa, I., Sakai, H., Ito, H., Matsuo, T. 1990. Leu-7 immunoreactivity in human and rat embryonic hearts, with special reference to the development of the conduction tissue. Anat. Embryol. 182, 553–562.

Ismat, F.A., Zhang, M., Kook, H., Huang, B., Zhou, R., Ferrari, V.A., Epstein, J.A., Patel, V.V. 2005. Homeobox protein Hop functions in the adult cardiac conduction system. Circ. Res. 96, 898–903.

Jay, P.Y., Harris, B.S., Maguire, C.T., Buerger, A., Wakimoto, H., Tanaka, M., Kupershmidt, S., Roden, D.M., Schultheiss, T.M., O'Brien, T.X., Gourdie, R.G., Berul, C. I., et al. 2004. Nkx2-5 mutation causes anatomic hypoplasia of the cardiac conduction system. J. Clin. Invest. 113, 1130–1137.

Jay, P.Y., Maguire, C.T., Wakimoto, H., Izumo, S., Berul, C.I. 2005. Absence of Msx2 does not affect cardiac conduction or rescue conduction defects associated with Nkx2-5 mutation. J. Cardiovasc. Electrophysiol. 16, 82–85.

Jessen, K.R., Mirsky, R. 2002. Signals that determine Schwann cell identity. J. Anat. 200, 367–376.

Jiang, X., Rowitch, D.H., Soriano, P., McMahon, A.P., Sucov, H.M. 2000. Fate of the mammalian cardiac neural crest. Development (Cambridge, England) 127, 1607–1616.

Kamino, K. 1991. Optical approaches to ontogeny of electrical activity and related functional organization during early heart development. Physiol. Rev. 71, 53–91.

Kasahara, H., Wakimoto, H., Liu, M., Maguire, C.T., Converso, K.L., Shioi, T., Huang, W.Y., Manning, W.J., Paul, D., Lawitts, J., Berul, C.I., Izumo, S. 2001. Progressive atrioventricular conduction defects and heart failure in mice expressing a mutant Csx/Nkx2.5 homeoprotein. J. Clin. Invest. 108, 189–201.

Kisanuki, Y.Y., Hammer, R.E., Miyazaki, J., Williams, S.C., Richardson, J.A., Yanagisawa, M. 2001. Tie2-Cre transgenic mice: A new model for endothelial cell-lineage analysis *in vivo*. Dev. Biol. 230, 230–242.

Kitajima, S., Miyagawa-Tomita, S., Inoue, T., Kanno, J., Saga, Y. 2006. Mesp1-nonexpressing cells contribute to the ventricular cardiac conduction system. Dev. Dyn. 235, 395–402.

Komuro, I., Izumo, S. 1993. Csx: A murine homeobox-containing gene specifically expressed in the developing heart. Proc. Natl. Acad. Sci. USA 90, 8145–8149.

Kreuzberg, M.M., Sohl, G., Kim, J.S., Verselis, V.K., Willecke, K., Bukauskas, F.F. 2005. Functional properties of mouse connexin30.2 expressed in the conduction system of the heart. Circ. Res. 96, 1169–1177.

Kreuzberg, M.M., Schrickel, J.W., Ghanem, A., Kim, J.S., Degen, J., Janssen-Bienhold, U., Lewalter, T., Tiemann, K., Willecke, K. 2006. Connexin30.2 containing gap junction channels decelerate impulse propagation through the atrioventricular node. Proc. Natl. Acad. Sci. USA 103, 5959–5964.

Kupershmidt, S., Yang, T., Anderson, M.E., Wessels, A., Niswender, K.D., Magnuson, M.A., Roden, D.M. 1999. Replacement by homologous recombination of the minK gene with lacZ reveals restriction of minK expression to the mouse cardiac conduction system. Circ. Res. 84, 146–152.

Lamers, W.H., De Jong, F., De Groot, I.J., Moorman, A.F. 1991. The development of the avian conduction system, a review. Eur. J. Morphol. 29, 233–253.

Lazarides, E. 1982. Intermediate filaments: A chemically heterogeneous, developmentally regulated class of proteins. Annu. Rev. Biochem. 51, 219–250.

Lebel, M., Agarwal, P., Cheng, C.W., Kabir, M.G., Chan, T.Y., Thanabalasingham, V., Zhang, X., Cohen, D.R., Husain, M., Cheng, S.H., Bruneau, B.G., Hui, C.C. 2003. The Iroquois homeobox gene Irx2 is not essential for normal development of the heart and midbrain-hindbrain boundary in mice. Mol. Cell. Biol. 23, 8216–8225.

Lesage, F., Lazdunski, M. 2000. Molecular and functional properties of two-pore-domain potassium channels. Am. J. Physiol. Renal Physiol. 279, F793–F801.

Li, Z., Colucci, E., Babinet, C., Paulin, D. 1993. The human desmin gene: A specific regulatory programme in skeletal muscle both *in vitro* and in transgenic mice. Neuromuscul. Disord. 3, 423–427.

Li, Q.Y., Newbury-Ecob, R.A., Terrett, J.A., Wilson, D.I., Curtis, A.R., Yi, C.H., Gebuhr, T., Bullen, P.J., Robson, S.C., Strachan, T., Bonnet, D., Lyonnet, S., et al. 1997. Holt-Oram syndrome is caused by mutations in TBX5, a member of the Brachyury (T) gene family. Nat. Genet. 15, 21–29.

Luderitz, B. 2002. *History of the Disorders of Cardiac Rhythm,* New York: Blackwell Publishing.

Metcalfe, W.K., Myers, P.Z., Trevarrow, B., Bass, M.B., Kimmel, C.B. 1990. Primary neurons that express the L2/HNK-1 carbohydrate during early development in the zebrafish. Development (Cambridge, England) 110, 491–504.

Mikawa, T., Borisov, A., Brown, A.M., Fischman, D.A. 1992. Clonal analysis of cardiac morphogenesis in the chicken embryo using a replication-defective retrovirus: I. Formation of the ventricular myocardium. Dev. Dyn. 193, 11–23.

Miquerol, L., Meysen, S., Mangoni, M., Bois, P., van Rijen, H.V., Abran, P., Jongsma, H., Nargeot, J., Gros, D. 2004. Architectural and functional asymmetry of the His-Purkinje system of the murine heart. Cardiovasc. Res. 63, 77–86.

Moorman, A.F., de Jong, F., Denyn, M.M., Lamers, W.H. 1998. Development of the cardiac conduction system. Circ. Res. 82, 629–644.

Mori, A.D., Zhu, Y., Vahora, I., Nieman, B., Koshiba-Takeuchi, K., Davidson, L., Pizard, A., Seidman, J.G., Seidman, C.E., Chen, X.J., Henkelman, R.M., Bruneau, B.G. 2006. Tbx5-dependent rheostatic control of cardiac gene expression and morphogenesis. Dev. Biol. 297, 566–586.

Moskowitz, I.P., Pizard, A., Patel, V.V., Bruneau, B.G., Kim, J.B., Kupershmidt, S., Roden, D., Berul, C.I., Seidman, C.E., Seidman, J.G. 2004. The T-Box transcription factor Tbx5 is required for the patterning and maturation of the murine cardiac conduction system. Development (Cambridge, England) 131, 4107–4116.

Myers, D.C., Fishman, G.I. 2003. Molecular and functional maturation of the murine cardiac conduction system. Trends Cardiovasc. Med. 13, 289–295.

Myers, D.C., Fishman, G.I. 2004. Toward an understanding of the genetics of murine cardiac pacemaking and conduction system development. Anat. Rec. A Discov. Mol. Cell. Evol. Biol. 280, 1018–1021.

Nakagawa, M., Thompson, R.P., Terracio, L., Borg, T.K. 1993. Developmental anatomy of HNK-1 immunoreactivity in the embryonic rat heart: Co-distribution with early conduction tissue. Anat. Embryol. 187, 445–460.

Nakamura, T., Colbert, M.C., Robbins, J. 2006. Neural crest cells retain multipotential characteristics in the developing valves and label the cardiac conduction system. Circ. Res. 98, 1547–1554.

Nguyen-Tran, V.T., Kubalak, S.W., Minamisawa, S., Fiset, C., Wollert, K.C., Brown, A.B., Ruiz-Lozano, P., Barrere-Lemaire, S., Kondo, R., Norman, L.W., Gourdie, R.G.,

Rahme, M.M., et al. 2000. A novel genetic pathway for sudden cardiac death via defects in the transition between ventricular and conduction system cell lineages. Cell 102, 671–682.

Papaioannou, V.E. 2001. T-box genes in development: From hydra to humans. Int. Rev. Cytol. 207, 1–70.

Pashmforoush, M., Lu, J.T., Chen, H., Amand, T.S., Kondo, R., Pradervand, S., Evans, S.M., Clark, B., Feramisco, J.R., Giles, W., Ho, S.Y., Benson, D.W., et al. 2004. Nkx2-5 pathways and congenital heart disease; loss of ventricular myocyte lineage specification leads to progressive cardiomyopathy and complete heart block. Cell 117, 373–386.

Pennisi, D.J., Rentschler, S., Gourdie, R.G., Fishman, G.I., Mikawa, T. 2002. Induction and patterning of the cardiac conduction system. Int. J. Dev. Biol. 46, 765–775.

Poelmann, R.E., Gittenberger-de Groot, A.C. 1999. A subpopulation of apoptosis-prone cardiac neural crest cells targets to the venous pole: Multiple functions in heart development? Dev. Biol. 207, 271–286.

Poelmann, R.E., Molin, D., Wisse, L.J., Gittenberger-de Groot, A.C. 2000. Apoptosis in cardiac development. Cell Tissue Res. 301, 43–52.

Poelmann, R.E., Jongbloed, M.R., Molin, D.G., Fekkes, M.L., Wang, Z., Fishman, G.I., Doetschman, T., Azhar, M., Gittenberger-de Groot, A.C. 2004. The neural crest is contiguous with the cardiac conduction system in the mouse embryo: A role in induction? Anat. Embryol. 208, 389–393.

Rentschler, S., Vaidya, D.M., Tamaddon, H., Degenhardt, K., Sassoon, D., Morley, G.E., Jalife, J., Fishman, G.I. 2001. Visualization and functional characterization of the developing murine cardiac conduction system. Development(Cambridge, England) 128, 1785–1792.

Rentschler, S., Zander, J., Meyers, K., France, D., Levine, R., Porter, G., Rivkees, S.A., Morley, G.E., Fishman, G.I. 2002. Neuregulin-1 promotes formation of the murine cardiac conduction system. Proc. Natl. Acad. Sci. USA 99, 10464–10469.

Saga, Y. 1998. Genetic rescue of segmentation defect in MesP2-deficient mice by MesP1 gene replacement. Mech. Dev. 75, 53–66.

Saga, Y., Miyagawa-Tomita, S., Takagi, A., Kitajima, S., Miyazaki, J., Inoue, T. 1999. MesP1 is expressed in the heart precursor cells and required for the formation of a single heart tube. Development (Cambridge, England) 126, 3437–3447.

Satokata, I., Ma, L., Ohshima, H., Bei, M., Woo, I., Nishizawa, K., Maeda, T., Takano, Y., Uchiyama, M., Heaney, S., Peters, H., Tang, Z., et al. 2000. Msx2 deficiency in mice causes pleiotropic defects in bone growth and ectodermal organ formation. Nat. Genet. 24, 391–395.

Schott, J.J., Benson, D.W., Basson, C.T., Pease, W., Silberbach, G.M., Moak, J.P., Maron, B.J., Seidman, C.E., Seidman, J.G. 1998. Congenital heart disease caused by mutations in the transcription factor NKX2-5. Science 281, 108–111.

Sedmera, D., Reckova, M., DeAlmeida, A., Coppen, S.R., Kubalak, S.W., Gourdie, R.G., Thompson, R.P. 2003. Spatiotemporal pattern of commitment to slowed proliferation in the embryonic mouse heart indicates progressive differentiation of the cardiac conduction system. Anat. Rec. A Discov. Mol. Cell. Evol. Biol. 274, 773–777.

Shin, C.H., Liu, Z.P., Passier, R., Zhang, C.L., Wang, D.Z., Harris, T.M., Yamagishi, H., Richardson, J.A., Childs, G., Olson, E.N. 2002. Modulation of cardiac growth and development by HOP, an unusual homeodomain protein. Cell 110, 725–735.

Silverman, M.E., Upshaw, C.B., Jr. 2002. Walter Gaskell and the understanding of atrioventricular conduction and block. J. Am. Coll. Cardiol. 39, 1574–1580.

Silverman, M.E., Grove, D., Upshaw, C.B., Jr. 2006. Why does the heart beat? The discovery of the electrical system of the heart. Circulation 113, 2775–2781.

Simon, A.M., Goodenough, D.A., Paul, D.L. 1998. Mice lacking connexin40 have cardiac conduction abnormalities characteristic of atrioventricular block and bundle branch block. Curr. Biol. 8, 295–298.

St Amand, T.R., Lu, J.T., Zamora, M., Gu, Y., Stricker, J., Hoshijima, M., Epstein, J.A., Ross, J.J. Jr., Ruiz-Lozano, P., Chien, K.R. 2006. Distinct roles of HF-1b/Sp4 in ventricular and neural crest cells lineages affect cardiac conduction system development. Dev. Biol. 291, 208–217.

Story, G.M., Dicarlo, S.E., Rodenbaugh, D.W., Dluzen, D.E., Kucera, J., Maron, M.B., Walro, J. M. 2000. Inactivation of one copy of the mouse neurotrophin-3 gene induces cardiac sympathetic deficits. Physiol. Genomics 2, 129–136.

Sumida, H., Akimoto, N., Nakamura, H. 1989. Distribution of the neural crest cells in the heart of birds: A three dimensional analysis. Anat. Embryol. 180, 29–35.

Takebayashi-Suzuki, K., Yanagisawa, M., Gourdie, R.G., Kanzawa, N., Mikawa, T. 2000. *In vivo* induction of cardiac Purkinje fiber differentiation by coexpression of preproendothelin-1 and endothelin converting enzyme-1. Development (Cambridge, England) 127, 3523–3532.

Takebayashi-Suzuki, K., Pauliks, L.B., Eltsefon, Y., Mikawa, T. 2001. Purkinje fibers of the avian heart express a myogenic transcription factor program distinct from cardiac and skeletal muscle. Dev. Biol. 234, 390–401.

Thomas, P.S., Kasahara, H., Edmonson, A.M., Izumo, S., Yacoub, M.H., Barton, P.J., Gourdie, R.G. 2001. Elevated expression of Nkx-2.5 in developing myocardial conduction cells. Anat. Rec. 263, 307–313.

Towbin, J.A., Wang, Z., Li, H. 2001. Genotype and severity of long QT syndrome. Arch. Pathol. Lab. Med. 125, 116–121.

Van Mierop, L.H., Gessner, I.H. 1970. The morphologic development of the sinoatrial node in the mouse. Am. J. Cardiol. 25, 204–212.

Viragh, S., Challice, C.E. 1977a. The development of the conduction system in the mouse embryo heart. I. The first embryonic A-V conduction pathway. Dev. Biol. 56, 382–396.

Viragh, S., Challice, C.E. 1977b. The development of the conduction system in the mouse embryo heart. II. Histogenesis of the atrioventricular node and bundle. Dev. Biol. 56, 397–411.

Viragh, S., Challice, C.E. 1982. The development of the conduction system in the mouse embryo heart. Dev. Biol. 89, 25–40.

Viragh, S., Szabo, E., Challice, C.E. 1982. Glycogen-containing lysosomes and glycogen loss in the cardiocytes of embryonic and neonatal mice. Adv. Myocardiol. 3, 553–561.

Wakimoto, H., Kasahara, H., Maguire, C.T., Izumo, S., Berul, C.I. 2002. Developmentally modulated cardiac conduction failure in transgenic mice with fetal or postnatal overexpression of DNA nonbinding mutant Nkx2.5. J. Cardiovasc. Electrophysiol. 13, 682–688.

Wenink, A.C. 1976. Development of atrio-ventricular conduction pathways. Bull. Assoc. Anat. 60, 623–629.

Wessels, A., Vermeulen, J.L., Verbeek, F.J., Viragh, S., Kalman, F., Lamers, W.H., Moorman, A.F. 1992. Spatial distribution of "tissue-specific" antigens in the developing human heart and skeletal muscle. III. An immunohistochemical analysis of the distribution of the neural tissue antigen G1N2 in the embryonic heart; implications for the development of the atrioventricular conduction system. Anat. Rec. 232, 97–111.

Wessels, A., Markman, M.W., Vermeulen, J.L., Anderson, R.H., Moorman, A.F., Lamers, W.H. 1996. The development of the atrioventricular junction in the human heart. Circ. Res. 78, 110–117.

Yamauchi, Y., Abe, K., Mantani, A., Hitoshi, Y., Suzuki, M., Osuzu, F., Kuratani, S., Yamamura, K. 1999. A novel transgenic technique that allows specific marking of the neural crest cell lineage in mice. Dev. Biol. 212, 191–203.

Zhao, Y.Y., Sawyer, D.R., Baliga, R.R., Opel, D.J., Han, X., Marchionni, M.A., Kelly, R.A. 1998. Neuregulins promote survival and growth of cardiac myocytes. Persistence of ErbB2 and ErbB4 expression in neonatal and adult ventricular myocytes. J. Biol. Chem. 273, 10261–10269.

CHAPTER 11

Genetic Dissection of Hematopoiesis Using *Drosophila* as a Model System

Cory J. Evans,* Sergey A. Sinenko,* Lolitika Mandal,*
Julian A. Martinez-Agosto,‡ Volker Hartenstein,*,§
and Utpal Banerjee*,†,§

Contents

* Department of Molecular, Cell and Developmental Biology, University of California, Los Angeles, California
† Department of Biological Chemistry, University of California, Los Angeles, California
‡ Department of Pediatrics, Mattel Children's Hospital at UCLA, University of California, Los Angeles, California
§ Molecular Biology Institute, University of California, Los Angeles, California

Advances in Developmental Biology, Volume 18
ISSN 1574-3349, DOI: 10.1016/S1574-3349(07)18011-X

Abstract   Investigations into the developmental origins of blood cells have indicated that the genes and molecular pathways controlling hematopoiesis are highly conserved among metazoans. In this chapter we summarize the progress in understanding how the molecular mechanisms regulating *Drosophila* blood development compare with analogous processes in vertebrates. In both *Drosophila* and vertebrates, the ontogenetic origins of cardiovascular cells and blood cells are closely related. In *Drosophila*, there is *in vivo* evidence for the presence of hemangioblast-like cells. Furthermore, there are significant similarities between the molecular mechanisms regulating the development of the lymph gland, the *Drosophila* hematopoietic organ, and the formation of the mammalian AGM region. Other aspects are also shared, including the sequential maturation of progenitor cell types, the presence of multipotent or stem cell progenitors, and a requirement for a niche interaction to maintain these progenitors. During their development, *Drosophila* blood cells utilize an array of conserved signaling pathways and transcriptional regulators to mediate cell fate specification and differentiation. The power of *Drosophila* as a model system is well established and our understanding of hematopoiesis, in both normal and aberrant contexts, will surely illuminate similar mechanisms in vertebrate systems, including humans.

1. INTRODUCTION

The cardiovascular system bears close relationships, developmentally and evolutionarily, to the hematopoietic system. Blood and vascular cells arise from a common pool of mesodermal, aorta-gonad-mesonephros (AGM) precursors in mammalian embryos, or interstitial cell mass in fish embryos (Medvinsky and Dzierzak, 1996; Galloway and Zon, 2003). Shared signaling pathways and determinants of cell fate, such as the products of the zebrafish genes *cloche* and *scl*, play an essential role in both blood and vascular cells. Mesoderm-derived precursors giving rise to these two cell types have been called hemangioblasts and have been described in mouse, chick, zebrafish, and *Drosophila* (Choi et al., 1998; Mandal et al., 2004; Vogeli et al., 2006). Recent findings attest to a high degree of conservation of molecular mechanisms controlling blood vascular development in vertebrates and *Drosophila*, suggesting that both tissues already existed in the early triploblastic (bilaterian) animals from which both vertebrates and insects descended (Fig. 1). The close developmental relationship between vascular and blood cells becomes clearer on considering their evolutionary origin.

A comparative analysis of extant animal taxa indicates that the blood vascular system evolved as part of the coelom, the secondary body cavity that appears early in development as a cleft in the lateral mesoderm. The defining character of the coelom is its epithelial lining, called mesothelium, formed by mesodermally derived cells. Typically, the coelom is subdivided into several compartments (coeloms); in vertebrates, these are the peritoneal cavity, pleural cavity, and pericardial cavity. Blood vessels are originally formed in clefts (sinuses) left open in between the mesothelial walls of the coeloms (Nakao, 1974; Ruppert and Carle, 1983; Smith, 1986; Gardiner, 1992). These walls are also the sites of origin of blood progenitor cells (reviewed in Hartenstein and Mandal, 2006) and they also develop specialized cells that are responsible for excretion of waste products from the coelomic and vascular liquid (nephrocytes or podocytes).

A look at the structure and development of polychaetes, a group of annelids that may have retained many primitive bilaterian characteristics (Fig. 1), might illustrate the relationship between coelomic cavity and blood vascular system. In annelids, the coeloms filling the left and right half of each segment come in contact dorsally and ventrally with the intestine and with each other at segment boundaries. Where they come together, the mesothelia leave open spaces that link up to form a vascular system. Blood fluid is circulated through the body by a pumping action of the blood vessels. Thus, the mesothelial walls of all large blood vessels contain myofilaments by means of which they are able to contract. In other instances, only some strategically placed segments of the blood vessels are contractile and act as primitive "hearts." Nephrocytes develop as specialized mesothelial cells. Blood cells of polychaetes also segregate from specialized regions of the mesothelium, called "lymphoid tissue" or lymph glands, into the lumen of the coelom and the vasculature. In other words, cells forming the vascular lining and blood cells are generated from the mesothelium surrounding the coelomic cavity.

Vertebrates possess a blood vascular and excretory system that is highly evolved in comparison to that of polychaetes (schematically depicted in Fig. 1D). The heart and blood vessels of vertebrates are lined by endothelial cells, rather than mesothelium. Endothelial cells are not

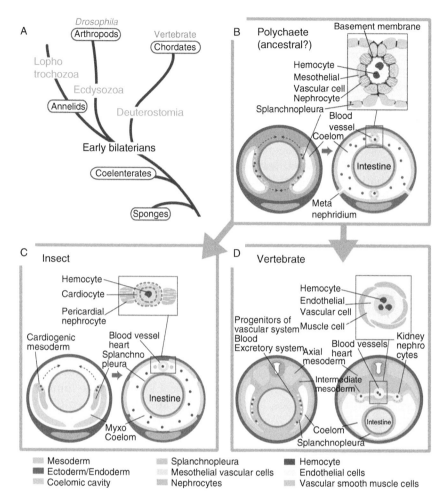

FIGURE 1 Blood vascular structure and development: an evolutionary perspective.
(A) Phylogeny of major animal taxa. The tree, at its base, highlights the two major
diploblast taxa, sponges and coelenterates. It then branches into the three main
clades of triploblastic animals, lophotrochozoa (represented by annelids), ecdysozoa
(represented by arthropods), and deuterostomia (represented by chordates). (B) Basal
coelomate invertebrate (e.g., polychaete annelid). Schematic cross sections of
embryo (bottom left) and adult (bottom right). Blood and vascular cells and nephrocytes
are derivatives of the mesothelial coelomic wall, in particular its inner leaf
(splanchnopleura). In the adult the coelom has expanded into the secondary body
cavities filling the interior of the animal. Mesothelia line the inner side of the body wall
and the outside of the intestine. Blood vessels, shown at higher magnification at the
top, are formed as clefts between the basal surfaces of the mesothelia. Precursors of
blood cells form specialized cell populations (lymph glands) in or close to the
mesothelium. (C) Derived invertebrate (e.g., insect). The embryonic mesoderm

contractile; the contraction of vertebrate blood vessels and the heart is mediated by a separate layer of muscle cells within the vessel. However, a close relationship between the mesothelium and the blood vascular system can still be observed in the vertebrate embryo. During early embryogenesis, the lateral mesoderm of vertebrates forms the mesothelium surrounding the coelom. The inner leaf of the mesothelium, called the splanchnopleura, gives rise to the progenitor cells that develop into endothelia and blood cells as well as the smooth muscle fibers surrounding the blood vessels. Progenitor cells of blood vessels and blood cells seem to be born in several ways from the splanchnopleura and migrate dorsomedially to assemble into blood vessels. Blood cell precursors, which ultimately populate the bone marrow and other blood-forming organs of the mature vertebrate animal, differentiate from the endothelial cells lining the dorsal aorta and the surrounding mesenchyme.

A close developmental relationship between blood and vascular cells is equally apparent in arthropods, including *Drosophila*. As in annelids, segmentally organized coeloms are formed in the embryo by the dorsal part of the mesoderm. However, these coeloms dissociate during later development. The vascular system consists of mesodermal cells that secondarily reaggregate and form an open dorsal blood vessel, or "heart." Rythmic contraction of the dorsal vessel generates a circulatory current through the intake of hemolymph at its posterior end and its expulsion anteriorly. The dorsal vessel is made up of a thin, contractile layer of myoepithelial cells that produce a basal lamina in the lumen, as well as at their outer surface, and in this way, structurally resembles the blood vessels of annelids. As described in more detail later, the same dorsolateral (cardiogenic) mesoderm from which the dorsal vessel descends also gives rise to the blood-forming tissue, the lymph gland.

(bottom left) transiently forms coeloms with an inner layer (splanchnopleura) and outer layer. Blood and vascular cells originate from the cardiogenic mesoderm located at the junction between the two layers. At later stages (right), vascular progenitors (cardioblasts) and blood precursors migrate dorsally, meet in the midline, and form the myoepithelial dorsal vessel (top), flanked by clusters of hematopoietic cells (lymph glands). (D) Vertebrate. The lateral plate mesoderm of the embryo (bottom left) cavitates, forming the mesothelial walls of the coelomic cavities. Progenitors of the vascular system and blood derive from the lateral plate. The vascular system is initially composed of endothelial cells (yellow). These cells recruit muscle cells (vascular smooth muscle cells, bright green) from the surrounding mesoderm. Hematopoietic precursors are closely associated with the endothelia of blood vessels and the mesenchyme surrounding these early blood vessels (AGM). Excretory nephrocytes derive from the intermediate mesoderm located adjacent to the lateral plate.

2. *DROSOPHILA* BLOOD CELL TYPES

The blood cells of *Drosophila*, called hemocytes, have been a subject of study for many decades. Three distinct blood cell types, called plasmatocytes, crystal cells, and lamellocytes (Fig. 2), can be found in circulation, with plasmatocytes representing greater than 90% of the hemocyte repertoire (Rizki, 1956, 1978). Most of the remaining hemocytes are crystal cells, with only a few or no lamellocytes present under normal growth conditions. However, specific stimuli, such as immune challenge by parasitic wasps, induce the differentiation of large numbers of these cells (Rizki and Rizki, 1992; Lanot et al., 2001; Sorrentino et al., 2002).

During development, hemocytes remove dead or dying cells by phagocytosis and secrete and remodel extracellular matrix components critical to morphogenesis (Fessler et al., 1994; Murray et al., 1995; Yasothornsrikul et al., 1997; Franc, 2002). The developmental function of hemocytes in this context is likely to be in tissue remodeling and in removing cells undergoing apoptosis; it is known that the development of the embryonic nervous system is severely disrupted in the absence of plasmatocyte function (Sears et al., 2003). Hemocytes are also key

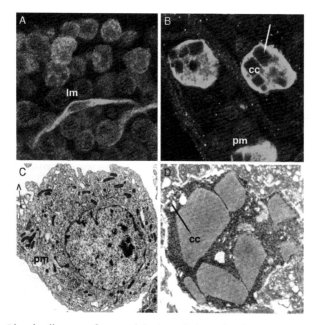

FIGURE 2   Blood cell types of *Drosophila*. (A and B) Confocal sections of the late larval lymph gland, showing differentiated plasmatocytes (pm), crystal cells (cc), and scattered lamellocytes (lm). (C and D) Electron microscopic section of plasmatocyte (C) and crystal cell (D). Crystalline inclusions are indicated by arrowheads.

mediators of innate immunity. They provide cellular immunity through the phagocytic clearance of microbial pathogens or the encapsulation of larger parasites. Flies compromised for plasmatocyte function exhibit increased susceptibility to microbial infection and a reduced capacity to potentiate humoral immunity (Elrod-Erickson et al., 2000; Matova and Anderson, 2006; Brennan et al., 2007; Pham et al., 2007). Additionally, plasmatocytes promote the humoral response by secreting antimicrobial peptides and cytokine-like molecules (Nappi, 1975; Samakovlis et al., 1990; Tzou et al., 2002; Agaisse et al., 2003; Hetru et al., 2003).

As macrophage-like cells, plasmatocytes utilize an array of specific receptors to identify targets and mediate their engulfment. Receptors recognizing apoptotic cells include Croquemort (Crq), a CD36 family member (Franc et al., 1996); PSR, a phosphatidyl serine receptor (Fadok et al., 2000); and Draper, a homologue of the *Caenorhabditis elegans* CED-1 phagocytosis receptor (Freeman et al., 2003; Manaka et al., 2004). Several receptors recognizing microbial pathogens are also known and include the scavenger receptor D-SR-CI and the peptidoglycan recognition protein PGRP-LC (Ramet et al., 2001, 2002; Royet et al., 2005). More recently, the receptors Eater, Nimrod, and Dscam have been found to be important for the phagocytosis of bacteria. Eater and Nimrod are related receptors that are characterized by the presence of multiple EGF-like motifs, called NIM repeats, whereas the extracellular domain of Dscam is composed of repeating immunoglobulin-like domains (Kocks et al., 2005; Kurucz et al., 2007). A subset of the Eater NIM repeats has been shown to directly bind bacteria and, collectively, is likely to mediate an array of intermolecular interactions (Kocks et al., 2005). Nimrod belongs to a family of 10 related proteins (grouped into three classes, NimA–NimC), 9 of which are expressed by hemocytes (Kurucz et al., 2007). Interestingly, the *nimB* class of genes appears to encode secreted molecules, suggesting that these proteins opsonize phagocytic targets in the free hemolymph. The ability to recognize and opsonize microbes for engulfment and degradation has also been proposed for Dscam, free forms of which have been found in the hemolymph (Watson et al., 2005). Subsets of the immunoglobulin-like repeats within the extracellular domain of Dscam are highly variable due to alternative splicing (Schmucker et al., 2000), and microarray analysis has revealed that greater than 18,000 alternate isoforms are expressed by hemocytes (Watson et al., 2005). Despite this variability, however, 80–90% of Dscam mRNAs are composed of only five isoforms, suggesting a blood-specific role for these particular Dscam types (Watson et al., 2005).

In contrast to the phagocytic role of plasmatocytes, the main function of crystal cells appears to be in the process of melanization. This darkening and hardening of tissue is due to the local deposition of melanin, which is generated by the oxidation of phenols to quinones mediated by the class of enzymes called phenoloxidase (PO). Activation of a serine protease cascade converts the zymogen prophenoloxidase (PPO) into its active PO form (Chosa et al., 1997; De Gregorio et al., 2002; Castillejo-Lopez and Hacker, 2005; Tang et al., 2006). It is thought that the large, cytoplasmic inclusions that crystal cells exhibit (and from which they derive their name) are composed primarily of PPO (Rizki and Rizki, 1980, 1985; Shrestha and Gateff, 1982; Lanot et al., 2001). *Drosophila* hemolymph exhibits PO activity that is absent in mutant lines that lack crystal cells, the primary source of this activity (Peeples et al., 1969; Rizki and Rizki, 1981; Rizki et al., 1985; Bidla et al., 2007). Genetic analysis has indicated that PO activity is delivered to the hemolymph through crystal cell rupture, occurring downstream of JNK signaling and the function of the TNF homologue Eiger (Bidla et al., 2007). Although PPO protein is distributed throughout the hemolymph, PO activity is tightly regulated and spatially restricted. For example, at wound sites, soft clots form via the coagulation of hemolymph and localized melanization causes hardening of the clot (Bidla et al., 2005).

Lamellocytes are large (15–40 μm across), flat, adherent cells that function during the encapsulation response, where a cellular barrier forms around foreign objects that cannot be removed by phagocytosis. Very few lamellocytes are found in normal larvae (Luo et al., 2002); however, many lamellocytes can differentiate in response to inductive signals that include parasitization, injection of foreign objects into the hemocoel, and sterile wounding (Nappi, 1975; Rizki and Rizki, 1992; Lanot et al., 2001; Markus et al., 2005). Subsequent to their differentiation, lamellocytes collaborate with plasmatocytes and crystal cells during the encapsulation and melanization process (Russo et al., 1996).

3. BLOOD CELL ORIGINS IN *DROSOPHILA*

In vertebrates, blood cell specification and differentiation occur in two general phases called primitive and definitive hematopoiesis, which differ with regard to developmental timing, location within the embryo, and the repertoire of cells generated in each phase (Cumano and Godin, 2007). In mammals, primitive hematopoiesis begins on the extraembryonic yolk sac and primarily gives rise to erythroid cells. At a later developmental stage, definitive hematopoiesis begins *de novo* in the embryo proper in an area called the AGM region (Dzierzak, 2005).

As in vertebrates, hematopoiesis in *Drosophila* also occurs in two waves (Fig. 3) that differ with regard to timing and location (Lebestky et al., 2000; Evans et al., 2003; Hartenstein, 2006). *Drosophila* hematopoiesis begins with the specification of blood cells in the head (procephalic) mesoderm. As these cells mature, they begin to migrate throughout

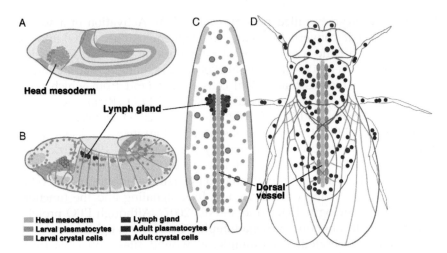

FIGURE 3   Hematopoietic origins of hemocytes in *Drosophila* development.
(A) Embryonic hematopoiesis occurs in a group of head mesoderm precursors. (B) The
lymph gland begins to develop in the dorsal aspect of the embryonic cardiogenic
mesoderm. (C) By the third larval instar, the lymph gland is fully formed and lies adjacent
to the dorsal vessel. (D) During adult stages, the lymph gland no longer exists as an intact
tissue and blood cells are seen in the circulating hemolymph.

the developing embryo (Tepass et al., 1994) and comprise the population
of cells in circulation during the subsequent larval period. At later stages
of embryogenesis, a second set of hematopoietic precursors is specified in
the cardiogenic mesoderm of the trunk region. These cells eventually
form a specialized organ called the lymph gland, which grows and supports the hematopoietic process throughout the larval stages (Rugendorff
et al., 1994; Jung et al., 2005). The lymph gland releases blood cells into
circulation at the onset of metamorphosis, leading to its disintegration
(Robertson, 1936; Holz et al., 2003). There is no evidence for *de novo*
hematopoiesis during the adult stage; however, this has not been analyzed rigorously. Transplantation experiments placing GFP-marked
blood progenitor cells into wild-type hosts have demonstrated that
hemocytes from both the embryonic head mesoderm and lymph gland
populate the adult (Holz et al., 2003).

4. EMBRYONIC HEMATOPOIESIS

4.1. Serpent: The central determinant of hemocyte fate

Approximately 800 blood cells differentiate from the head mesoderm
(Fig. 4) during the embryonic phase of hematopoiesis (Rugendorff et al.,
1994; Tepass et al., 1994). The vast majority of these cells become

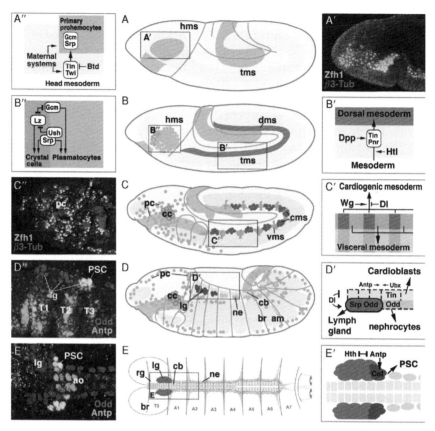

FIGURE 4 Early hematopoiesis in *Drosophila*. (A–D) Schematic lateral views of *Drosophila* embryos at stages 7 (A), 10 (B), 11 (C), and 12 (D). (E) Schematic dorsal view of late embryonic dorsal vessel and lymph gland. Panels B′ to E′ and A″ to B″ illustrate known essential molecular mechanisms acting at the corresponding stages. (A and A″) Early wave of hematopoiesis in the head mesoderm (hms). Maternal dorsoventral system specifies mesoderm by activating determinants such as Twist (Twi) and Tinman (Tin).
For specification of prohemocytes (ph) to occur, Tin is downregulated by head gap gene Buttonhead (Btd; SP1 transcription factor), which in turn is activated by the anterior maternal system. The anterior system is also required to positively regulate ph determinants Srp (GATA-1 to 3 homologue) and Gcm. (A′) Stage 7 hms (lateral view) labeled with anti-Zfh1 (mesoderm, green) and anti-β3 tubulin (red, larval hemocyte precursors). (B and B″) ph resulting from an early wave of hematopoiesis in the hms start spreading throughout the embryo. Expression of Gcm and Lz (vertebrate homologue: AML-1) in conjunction with Srp and its cofactor, Ush, specify two different blood cell lineages, plasmatocytes (pm) and crystal cells. (B′) Within the trunk mesoderm (trm), a dorsal domain is specified by means of the activity of the Htl (FGF homologue) signal. Dorsal mesoderm (dms) receives Dpp signal (BMP2/4 homologue) from the dorsal ecto-derm, which triggers expression of Srp and maintains expression of Tin. (C and C′) Initiation of second wave of hematopoiesis in dorsal trm. Among other signals, Wg (Wnt 1)

plasmatocytes that migrate throughout the embryo in response to specific cues (Tepass et al., 1994; Cho et al., 2002; Bruckner et al., 2004; Wood et al., 2006). Approximately 36 crystal cells are specified during this time and, unlike plasmatocytes, remain in the head region of the embryo until the larval stages (Milchanowski et al., 2004). Hemocytes differentiating from the embryonic head mesoderm are first distinguishable by embryonic stage 5 by the expression of Serpent (Srp), a member of the GATA family of zinc finger transcription factors (Rehorn et al., 1996; Lebestky et al., 2000; Patient and McGhee, 2002). Fate-mapping experiments using transplantation have shown that cells in the embryonic hemocyte anlagen are already committed to blood cell fate by this stage (Holz et al., 2003). Genetic evidence indicates that Srp expression in these cells is essential because mutations in *srp* abrogate their specification (Rehorn et al., 1996; Lebestky et al., 2000). The GATA family of transcriptional regulators is conserved from yeast to vertebrates and these factors mediate an array of developmental programs (Orkin et al., 1998; Lowry and Atchley, 2000; Cantor and Orkin, 2002; Maduro and Rothman, 2002; Patient and McGhee, 2002). Interestingly, much like Srp in *Drosophila*, murine GATA-1, GATA-2, and GATA-3 each have fundamental roles in promoting hematopoietic development (Pevny et al., 1991; Tsai et al., 1994; Ting et al., 1996; Shivdasani et al., 1997; Pai et al., 2003; Rodrigues et al., 2005). Furthermore, similar to the early function of Srp in *Drosophila* head mesoderm hematopoiesis, mammalian GATA proteins are one of the earliest known hematopoietic determinants (Fujiwara et al., 2004; Ling et al., 2004).

---

and Delta are required to specify, within the dms, segmental clusters of cells with cardiogenic fate (cardiogenic mesoderm: cms). The remainder of the dms develops as visceral mesoderm (vms). (C″) Confocal section of the head of a stage 11 embryo (lateral view), showing spreading pm. (D and D′) The cms splits up into three major lineages, cardioblasts (cb), pericardial nephrocytes (ne) and hemocyte progenitors (located in lymph gland, lg). Notch/Delta signaling is positively required for nephrocytes and lymph gland. Srp becomes restricted to the lymph gland; Tin to cb; Odd-skipped (Odd) marks lymph gland and nephrocytes. Input from the Hox complex (Antp, positively required; Ubx, negatively required) specifies lymph gland in the three thoracic segments, whereas abdominal segments produce nephrocytes. (D″) Confocal section of stage 12 embryo (lateral view) showing segmentally arranged clusters of cms (Odd-positive, red). The three clusters of the thoracic segments (T1–T3) represent the precursors of the lymph gland. The cluster of T3 expresses Antp (green) and develops as the posterior signaling center (PSC). (E and E′) Dorsal vessel and lymph gland of late *Drosophila* embryo, dorsal view. A1–A7, abdominal segments 1–7; br, brain; cb, cardioblasts; lg, lymph gland; ne, pericardial nephrocytes; rg, ring gland. (E″) Confocal section of late embryonic lymph gland and dorsal vessel (dorsal view) labeled with anti-Odd (red) and anti-Antp (green). ao is the aorta (anterior part of dorsal vessel).

Similar to vertebrate GATA factors, Srp contains two zinc finger domains, although alternative splicing also generates an isoform of Srp containing only the C-terminal finger (SrpC; Waltzer et al., 2002). Both isoforms are expressed in the embryo in identical patterns; however, *in vitro* and *in vivo* experiments indicate that they have differential regulatory activities. This is particularly true with regard to binding the Friend-of-GATA (FOG) homologue U-shaped (Ush; described further below), which, as in vertebrates, utilizes the GATA N-terminal zinc finger domain (Tevosian et al., 1999; Waltzer et al., 2002; Fossett et al., 2003).

The continued expression of Srp and the downregulation of the Nkx2.5 homologue Tinman (Tin; Azpiazu and Frasch, 1993; Bodmer, 1993) in progenitors are prerequisites for blood cell fate. Tin is a downstream target of the mesodermal determinant Twist (Castanon and Baylies, 2002) and, accordingly, is initially expressed throughout the entire mesoderm of the head and trunk. Subsequently, the SP1 homologue *Buttonhead* is expressed in a cylindrical domain of the embryonic head activating Srp and repressing Tin expression (Yin et al., 1997). Loss of *buttonhead* results in the continued presence of Tin in the entire head mesoderm and loss of Srp expression. This abolishes blood cell formation. A similar obligatory downregulation of Tin occurs at a later stage in blood cell progenitors of the lymph gland (Mandal et al., 2004; see below).

4.2. Srp interacts with Ush and Lozenge to control crystal cell development

As the name suggests, the FOG family of transcription factors interact with GATA factors in the regulation of gene expression. In the mouse, FOG-1 and GATA-1 are critical regulators of erythropoiesis and megakaryopoiesis (Tsang et al., 1997, 1998; Cantor et al., 2002; Chang et al., 2002; Katz et al., 2002) while FOG-2 and GATA-4 are important for heart morphogenesis (Svensson et al., 1999; Tevosian et al., 1999; Crispino et al., 2001). Likewise, the *Drosophila* FOG protein Ush has been shown to work with Pannier (Pnr, a GATA-4 homologue) to limit the production of cardioblasts and sensory bristles (Haenlin et al., 1997; Fossett et al., 2000). With regard to hematopoiesis, Ush appears to function with Srp to repress crystal cell fate. Mutations in *ush* cause an increase in the number of crystal cells while forced expression of Ush in hemocyte precursors reduces crystal cell production (Fossett et al., 2001, 2003). This repressive function of Ush on the crystal cell lineage can be enhanced if Ush is misexpressed along with Srp but not if it is misexpressed with SrpC, suggesting that the relative levels of full-length Srp and SrpC *in vivo* regulate Ush activity (Fossett et al., 2003). Consistent with this function, Ush is detectable in the head mesoderm by stage 8 of development and its expression is maintained in plasmatocytes but is eventually downregulated in crystal cells

(Fossett et al., 2001). Like all FOG proteins, Ush contains a PXDL motif that is capable of interacting with the transcriptional corepressor CtBP, which represents one possible mechanism by which Ush functions (Turner and Crossley, 1998; Holmes et al., 1999; Deconinck et al., 2000; Fossett et al., 2001).

In addition to Ush, Srp works in concert with the protein Lozenge (Lz) to control crystal cell differentiation. Lozenge belongs to the Runx family of transcription factors and exhibits significant similarity (71% identity within the Runt domain) to human AML-1/Runx1 (Daga et al., 1996), which is frequently the target of chromosomal translocations leading to acute myeloid leukemia (AML; Lutterbach and Hiebert, 2000; Speck and Gilliland, 2002). Runx1 forms a dimer with CBFβ, a protein homologous to the *Drosophila* proteins Brother and Big brother (Li and Gergen, 1999; Adya et al., 2000; Kaminker et al., 2001; Speck and Gilliland, 2002), and targeted disruption of either Runx1 or CBFβ in mice abolishes definitive hematopoiesis (Okuda et al., 1996; Sasaki et al., 1996; Wang et al., 1996a,b).

In the *Drosophila* embryo, *lz* transcripts can first be detected in crystal cell precursors that form bilaterally within the anterior Srp-expressing domain of the head mesoderm (Lebestky et al., 2000; Bataille et al., 2005). Null mutation of *lz* completely blocks the differentiation of the crystal cell lineage, with little or no effect on plasmatocytes (Lebestky et al., 2000). Although Lz is positively required for the specification of crystal cells, its misexpression throughout the entire embryonic mesoderm does not significantly alter the number of developing crystal cells (Fossett et al., 2003; Waltzer et al., 2003). If, however, Srp is misexpressed along with Lz, more crystal cells are induced to differentiate (Fossett et al., 2003; Waltzer et al., 2003). Thus, Srp and Lz appear to work together to promote crystal cell development while Ush represses this lineage (Fossett et al., 2003). Furthermore, *in vitro* data suggest that Lz and Srp proteins physically interact (Waltzer et al., 2003), similar to Runx1 and GATA-1 in mice (Elagib et al., 2003). Finally, evidence suggests that Lz works cooperatively with Srp to reinforce *lz* gene expression through a positive-feedback mechanism (Bataille et al., 2005).

4.3. Glial-cells-missing proteins promote the plasmatocyte lineage

The Glial-cells-missing (Gcm1 and Gcm2) proteins represent conserved transcription factors that function in *Drosophila* as determinants of glial and plasmatocyte cell fates (Hosoya et al., 1995; Jones et al., 1995; Vincent et al., 1996; Bernardoni et al., 1997; Lebestky et al., 2000; Kammerer and Giangrande, 2001; Alfonso and Jones, 2002). Several plasmatocyte-specific Gcm targets have been identified (Freeman et al., 2003), including the phagocytosis receptor Draper. The expression of Gcm in the *Drosophila* head mesoderm initially colocalizes with Srp throughout the hematopoie-tic primordium (Bataille et al., 2005). Subsequently, *gcm* expression is downregulated in an anterior subpopulation that then expresses *lz*. These *lz*-expressing cells represent crystal cell precursors, although ~40% of these cells will turn off *lz* expression and become plasmatocytes instead (Lebestky et al., 2000; Bataille et al., 2005). In addition to promoting the plasmatocyte lineage, the Gcm proteins also inhibit crystal cell development, in part, through the repression of *lz* expression in progenitor cells. The mechanism by which Gcm expression is downregulated in the anterior of the embryonic hemocyte primordium to allow for *lz* expression remains to be resolved.

4.4. PDGF/VEGF receptor signaling directs plasmatocyte migration and survival

In mammals, the vascular endothelial growth factor receptor (VEGFR) pathway controls the fate of vascular endothelial cells during angiogenesis and vasculogenesis (reviewed by Ferrara et al., 2003). VEGFR signaling also has a role in hematopoiesis where it is required for the maintenance and/or survival of hematopoietic stem cells (Gerber et al., 2002). Additionally, a large number of hematologic malignancies have been associated with increased or inappropriate VEGFR signaling (Gerber and Ferrara, 2003; Podar and Anderson, 2005; Hiramatsu et al., 2006). The *Drosophila* PDGF/VEGF receptor (PVR) is expressed in the embryonic head mesoderm by stage 8, but is not required for plasmatocyte specification or differentiation (Duchek et al., 2001; Heino et al., 2001; Cho et al., 2002; Sears et al., 2003). Instead, maturing plasmatocytes fail to properly disperse within the embryo when they lack PVR. Interestingly, each of the three PVR ligands (PVF1–3) is expressed along migratory routes normally followed by plasmatocytes, supporting the idea that directive cues are mediated by PVR signaling (Cho et al., 2002; Wood et al., 2006). Furthermore, misexpression of PVF2 was sufficient to redirect plasmatocyte migration in the embryo (Cho et al., 2002). The observed disruption in the migration of PVR-mutant plasmatocytes is also due, in part, to diminished survival of these cells. It has been shown that lack of PVR signaling causes a gradual reduction in the number of plasmatocytes through apoptosis (Bruckner et al., 2004). Interestingly, the plasmatocyte migration defects in PVR mutants can be significantly rescued by the misexpression of the baculovirus apoptotic inhibitor p35 (Bruckner et al., 2004), suggesting that PVR provides a cell-autonomous trophic survival signal to maturing plasmatocytes. Taken together, these results suggest that PVR signaling mediates both survival and migration, two distinct but intimately associated aspects of plasmatocyte development.

5. LARVAL HEMATOPOIESIS

5.1. Lymph gland specification: Hemangioblasts and similarities with the vertebrate AGM

In *Drosophila*, the lateral mesoderm of the embryo is subdivided several times prior to the specification of the lymph gland. Lateral mesoderm gives rise to dorsal mesoderm, which in turn yields the cardiogenic and visceral mesoderm (Fig. 4). Several signaling inputs then specify three general cell types within the cardiogenic mesoderm: (1) cardioblasts, which comprise the dorsal vessel; (2) the pericardial cells, which are proposed to have nephrocytic function; and (3) cells of the lymph gland anlagen, which will give rise to differentiated blood cells. In vertebrates, blood and vascular cells are thought to be derived from a common progenitor cell type called a hemangioblast (Murray, 1932; Choi et al., 1998). Analysis of the developmental fate of two-cell clones, derived from single, constitutively marked precursor cells within the *Drosophila* cardiogenic mesoderm, revealed that one sister cell often assumes lymph gland fate while the other may assume dorsal vessel cardioblast fate (Mandal et al., 2004). These data indicate that, as in vertebrates, the ontogeny of blood and vascular cell types is closely related in *Drosophila* and provided evidence for the existence of a cell type with properties similar to that of the hemangioblast.

Formation of the cardiogenic mesoderm requires the function of the homeobox protein Tin and the GATA factor Pnr. Accordingly, mutations in the corresponding genes not only prevent heart development (Azpiazu and Frasch, 1993; Bodmer, 1993; Klinedinst and Bodmer, 2003) but also prevent lymph gland specification (Mandal et al., 2004). Maintenance of Tin expression within the cardiogenic mesoderm requires the Decapentaplegic (Dpp, the *Drosophila* homologue of TGF-β) and Heartless (Htl, a *Drosophila* homologue of the FGF receptor) signaling pathways. Wingless (Wg) signaling is required within the cardiogenic mesoderm to promote all cell types, including the lymph gland. Notch signaling plays a dual role: early signaling (between 6 and 8 h of development) is required for restricting the formation of the cardiogenic mesoderm, whereas at later stages (between 8 and 10 h of development), signaling preferentially limits cardioblast development allowing the precursors to develop as lymph gland tissue (Mandal et al., 2004).

At stage 12 of embryogenesis, the ubiquitous expression of Tin and Pnr in the cardiogenic mesoderm becomes restricted to the cardioblasts. At the same stage, the zinc finger protein Odd-skipped (Odd) is expressed in cell clusters within segments T1–A6, flanking the dorsal vessel cardioblasts. The three thoracic Odd-positive clusters (T1–T3) come together to form lymph gland progenitors, whereas the abdominal clusters form pericardial cells (Mandal et al., 2004). Subsequently, the Odd-expressing lymph gland progenitors express Srp, converting them into true components of the hematopoietic system. As in the embryonic head mesoderm, Srp expression in the early lymph gland is primarily responsible for specifying hemocyte identity. In the absence of Srp, lymph gland progenitors express the pericardial cell marker Pericardin, suggesting that they have lost aspects of hemocyte identity or gained properties of nephrocytes. Additionally, misexpression of Srp throughout the cardiogenic mesoderm is sufficient to convert Odd-positive pericardial cells into lymph gland-like cells. The restriction of lymph gland specification to thoracic segments is due, in part, to a repressive function of the homeobox gene *ultrabithorax* (*ubx*), which is normally expressed in abdominal segments A2–A5. Loss of *ubx* causes an expansion of lymph gland progenitors into the abdominal segments, whereas misexpression of *ubx* converts thoracic lymph gland progenitors into pericardial cells (Mandal et al., 2004).

Many of the developmental principles operating during blood and dorsal vessel formation in *Drosophila* are similar to those regulating the progenitors of blood, aorta and renal systems in vertebrates. For example, the FGF, Dpp, and Wg signaling that function within the *Drosophila* cardiogenic mesoderm (Mandal et al., 2004) parallel the FGF (Nishikawa et al., 2001), BMP (Marshall et al., 2000), and Wnt (Orelio and Dzierzak, 2003) signaling that generates the mammalian AGM region. Likewise, GATA transcription factors play central roles in the establishment of vertebrate progenitor lineages during both primitive and definitive hematopoiesis as well as in the generation of the hemangioblast (Lugus et al., 2007), which is similar to the requirement of Pnr (along with Tin) and Srp in the *Drosophila* cardiogenic mesoderm and lymph gland, respectively. The expression of Tin, Pnr, and Srp within the thoracic region of the cardiogenic mesoderm is largely dependent on Notch signaling, which is similarly required in vertebrate blood and vascular development. In zebrafish, the Notch-signaling mutant *mind bomb* fails to specify definitive hematopoietic stem cells that are normally found associated with the ventral aortic wall. Furthermore, through the use of an inducible transgenic system, transient Notch activation was found to be sufficient to greatly expand the number of hematopoietic stem cells in the AGM (Burns et al., 2005).

In both vertebrates and *Drosophila*, many of the molecular mechanisms described above appear to feed into the expression of the conserved bHLH transcription factor Hand (McFadden et al., 2000; Yamagishi et al., 2001; Han and Olson, 2005; Han et al., 2006). The murine *hand1* and *hand2* genes are required for proper heart development (Srivastava et al., 1995, 1997) and disruption of the *hand* gene in zebrafish reduces cardiac cell number (Yelon et al., 2000). In addition to its expression in the cardiogenic region, mouse *hand1* is expressed in the developing AGM (Firulli et al., 1998). Interestingly, the *Drosophila hand* gene has been shown to be critical for both heart and lymph gland development and its expression completely overlaps with these cell types within the cardiogenic mesoderm (Han and Olson, 2005; Han et al., 2006). In cardioblasts and pericardial cells, Tin and Pnr mediate Hand expression whereas Srp mediates Hand expression in the lymph gland primordium. In *hand* mutant animals, ectopic cell death virtually ablates the developing lymph gland and causes a hyperplastic and disordered heart tube in late embryos and early larvae. This phenotype can be suppressed either by expressing an inhibitor of apoptosis or by expressing *Drosophila* or human Hand. This indicates that Hand functions, in part, to promote cell survival within the cardiogenic mesoderm and underscores the conserved function of these proteins.

5.2. The lymph gland contains three distinct cellular zones

By embryonic stage 16 the lymph gland primordium has coalesced from the three Odd-positive cell clusters in the thoracic cardiogenic mesoderm (Fig. 4). This occurs bilaterally, relative to the dorsal midline, with ~20 cells in each group of primitive hemocytes. These clusters correspond to what will be called the lymph gland primary lobes at later stages. Proliferation increases the number of cells to ~200 during the second larval instar, and two to three smaller, bilateral pairs of hemocyte clusters, called secondary lobes, form posteriorly (Fig. 5). The origin of these lobes is not well established, although during the later larval stages they are interspersed between pericardial nephrocytes suggesting the possibility that secondary lobes arise from cryptic precursors that alternate with pericardial cells flanking the embryonic dorsal vessel cardioblasts. By the third instar, the cell number in the primary lobes increases an additional tenfold, and in the latter half of this stage, it becomes possible to discern structurally distinct regions among these cells (Jung et al., 2005).

In the mid- to late-third instar, cells located in the periphery of the lymph gland primary lobe have a granular appearance, are loosely arranged, and exhibit a significant level of individualization. In contrast, cells in the medial regions of the lobe, near the dorsal vessel, are compactly arranged. Because of these differential characteristics, we have termed these regions of the primary lobe as the cortical zone (CZ) and the medullary zone (MZ), respectively, based on their relative positions (Jung et al., 2005). On comparison with the expression of known maturation markers, such as Hemolectin (Hml; Goto et al., 2003), Peroxidasin (Pxn; Nelson et al., 1994), and Lz (Lebestky et al., 2000), it was discovered that differentiating cells are restricted to the cortical zone. The medullary zone is devoid of mature hemocyte marker expression, suggesting that

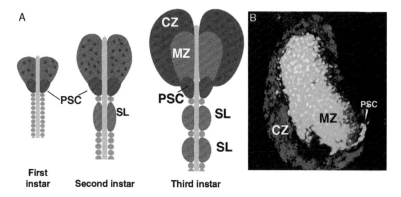

FIGURE 5   Development of the lymph gland during larval stages. (A) During the first larval instar, the lymph gland expands from about 30 to ~140 cells. Several of these precursors, shown in magenta, undergo cell division at this stage. The secondary lobes (SL) are not evident at this stage. The second larval instar lymph gland is composed of a homogeneous population of ~200 prohemocytes that proliferate (magenta), leading to a rapid expansion of its size. By this stage, the SL have started to develop in between pericardial cells (gray). By the third instar, the medullary zone (MZ) and cortical zone (CZ) have formed within the primary lobe. Progenitors located within the MZ cycle slowly and give rise to all hemocytes populating the CZ. Cells within the CZ divide rapidly (magenta). The SL are well developed. (B) Confocal section of a single lymph gland primary lobe showing all three zones: CZ (red, marked with Peroxidasin), MZ (green, marked with *ZCL2897*), and PSC (blue, marked with Hedgehog).

cells in this region are progenitor cells (prohemocytes) (Jung et al., 2005). The medullary zone population does express several distinct markers, including *Drosophila* E-cadherin (DE-cadherin or Shotgun), the reporters *domeless-gal4* (Bourbon et al., 2002) and *unpaired3-gal4* (Agaisse et al., 2003), and the GFP-trapline *ZCL2897*, among others.

Although not distinguishable by structure, a third zone of cells that we call the posterior signaling center (PSC) can be discerned in the lymph gland by gene expression and is always positioned at the posterior tip of the primary lobe. At the late third instar, the PSC in each lobe consists of ~50 cells that express *Antp-gal4*. Several additional PSC markers have been identified, including the Notch ligand Serrate (Ser) (Lebestky et al., 2000), the reporters *Dorothy-gal4* and *upd3-gal4* (Jung et al., 2005), and the transcription factors Collier (Crozatier et al., 2004) and Antp (Mandal et al., 2007). Cells of the PSC do not express any canonical cortical or medullary zone markers nor do they express receptors such as PVR or Domeless, which are found elsewhere in the lymph gland (Jung et al., 2005). These preliminary data suggested that the PSC cells are dedicated signaling cells.

Structural features similar to the cortical and medullary zones have not been observed in secondary lymph gland lobes. At the late third instar, secondary lobe cells rarely express maturation markers (Jung et al., 2005). Rather, the vast majority of secondary lobe cells express high levels of DE-cadherin and *dome-gal4*, suggesting that in this respect these cells are similar to medullary zone prohemocytes. In instances where maturation markers are seen in these lobes, regions containing them are devoid of DE-cadherin and *dome-gal4*, much like what occurs in the cortical zone of the primary lobes.

Analysis of cellular proliferation has revealed that the lymph gland exhibits distinct phases of growth as hematopoietic development proceeds. Proliferating cells, as assessed by BrdU incorporation, are evenly distributed throughout the lymph gland during the second instar, a stage in which the lymph gland is growing significantly in size. By the late third instar, however, most dividing cells (~90%) of the primary lobe are restricted to the cortical zone (Jung et al., 2005). This suggests that the rapidly cycling prohemocytes of the second instar lymph gland quiesce as they populate the medullary zone of the third instar. Furthermore, the process of differentiation from the prohemocyte stage appears to be coupled with the expansion of the mature cell population. The secondary lobes, which lack zonation, continue to proliferate in the third instar. In this respect, the cells of the secondary lobe are different from the quiescent medullary zone precursors.

The cells of the medullary zone should be considered to represent stem-like hematopoietic precursors. These cells do not express differentiation markers, give rise to all blood cell types, cycle slowly, and retain label in transient marking experiments (J. Marshall and Banerjee, unpublished data), and, as described later, their maintenance is dependent on signals from a hematopoietic niche.

5.3. Zone formation in the developing lymph gland

Formation of the medullary and cortical zones follows a stereotypic progression of marker expression as the lymph gland grows. By the second instar, the pan-hemocyte marker Hemese (Kurucz et al., 2003) is expressed throughout the lymph gland. Likewise, the expression of PVR is also apparent throughout the lymph gland, although it is excluded from the PSC. The reporter *dome-gal4* is expressed by a majority of the cells of the lymph gland, but is also excluded from the PSC. At the periphery of the lymph gland, a few cells can be found that have low or no *dome-gal4* expression and these cells correspond to maturing cells. Thus, the maturation process and the formation of the cortical zone at the lymph gland periphery begin in the mid-second instar (Jung et al., 2005). Peroxidasin and *hemolectin-gal4* are the earliest markers to be expressed and are soon followed by *Collagen-gal4* and P1-antigen (now known as Nimrod) (Yasothornsrikul et al., 1997; Asha et al., 2003;

Goto et al., 2003; Kurucz et al., 2007). A small subset of cells in the second instar lymph gland expresses neither maturation markers nor *dome-gal4* and is seen in the medial region of the lobe, close to the dorsal vessel. These cells are likely to be early precursors and have been considered to be pre-prohemocytes. Together these observations suggest a model in which maturing lymph gland hemocytes pass through a *dome-gal4*-positive prohemocyte stage. This concept is supported by lineage-tracing analysis that indicates that *dome-gal4*-expressing precursors indeed give rise to all the maturing cells of the lymph gland (Jung et al., 2005).

The cells of the PSC arise from the T3 segment and are distinguishable from the rest of the lymph gland cells as early as stage 11 of embryogenesis by virtue of their expression of the homeobox protein Antp (Fig. 4; Mandal et al., 2007). The transcription factor Col, which is required for PSC maintenance at later stages (Crozatier et al., 2004), is initially expressed throughout the embryonic lymph gland but is then refined to the PSC by stage 16. Epistasis experiments have demonstrated that Col expression in these cells is downstream of the function of Antp (Mandal et al., 2007). Examination of Homothorax (Hth), another homeobox protein and negative regulator of Antp, revealed that it is expressed throughout the embryonic lymph gland but becomes downregulated in the PSC, which becomes Antp-positive. The lymph gland primordium is largely missing in *hth* loss-of-function mutants, while *hth* misexpression throughout the mesoderm causes loss of PSC. Thus, PSC specification is mediated by downregulation of Hth and upregulation of Antp, while Col maintains PSC identity during subsequent stages.

5.4. Blood progenitors are maintained by interactions with a hematopoietic niche

Crozatier et al. (2004) noted that parasitization of *collier* (*col*) mutant larvae, where the lymph gland lacks the PSC, fails to induce lamellocyte differentiation. This suggested that the PSC provides some instructive signal to the hemocytes of the lymph gland to specify the lamellocyte lineage. Examination of various lymph gland markers in the *col* mutant background has subsequently revealed several interesting features. Consistent with a lack of PSC, expression of Antp is missing in the third instar, despite being expressed during embryogenesis (Mandal et al., 2007). At the third instar, medullary zone markers are also absent or severely reduced in *col* mutants (Fig. 6; Krzemien et al., 2007; Mandal et al., 2007) and BrdU incorporation is seen throughout the lobe (Mandal et al., 2007). Furthermore, cortical zone markers are abundantly expressed by hemocytes that mature throughout the lymph gland. Similar results are obtained using Antp mutants, which also lacks the PSC (Mandal et al., 2007). Together, these data indicate that the PSC functions to maintain medullary zone

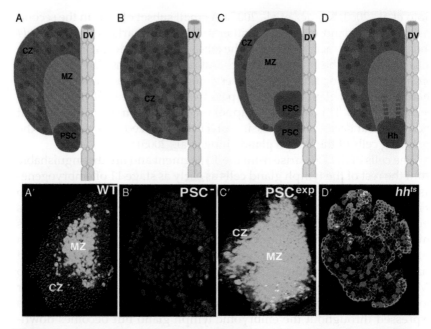

FIGURE 6 Maintenance of hematopoietic progenitors by a niche in the *Drosophila* lymph gland. Below each schematic representation is a representative confocal section showing the corresponding phenotype. (A and A′) Schematic representation of wild-type lymph gland zonation: cortical zone (CZ), medullary zone (MZ), and posterior signaling center (PSC). (B and B′) Absence of the PSC causes loss of MZ progenitors and leads to premature differentiation of hemocytes. (C and C′) Expansion of PSC size increases the population of MZ progenitors at the expense of differentiated hemocytes in the cortical zone. (D) Hedgehog signal initiated from the PSC maintains MZ progenitors. (D′) Lack of *hedgehog* function leads to loss of MZ progenitors and premature differentiation of hemocytes (red and green).

progenitors, and in the absence of the PSC, prohemocytes fail to be maintained and differentiate into mature cells. Consistent with this idea, experimentally increasing the size of the PSC causes an increase in the size of the medullary zone relative to the cortical zone (Mandal et al., 2007).

The interaction between the PSC and the medullary zone described above suggested that progenitor cells sense a signal emanating from the PSC, and is critical for their maintenance. Two signaling pathways, Hedgehog (Hh) (Mandal et al., 2007) and JAK/STAT (Krzemien et al., 2007), have been implicated in this process. The hh^{ts2}-mutant lymph gland phenotype is remarkably similar to that seen for *Antp* hypomorphic or *col* loss-of-function mutants, and Hh protein is expressed in the second and third instar PSC (Mandal et al., 2007). Additionally, the downstream components of the Hh pathway, the receptor Patched (Ptc) and activated

Cubitus interruptus (Ci, the *Drosophila* homologue of Gli), are preferentially expressed by medullary zone cells (Mandal et al., 2007), suggesting that these cells actively sense Hh. Interestingly, PSC cells extend numerous thin processes over many cell diameters into the adjacent medullary zone (Krzemien et al., 2007; Mandal et al., 2007). Taken together, these results indicate that the PSC functions as a hematopoietic niche that maintains, via Hh, the medullary zone population of multipotent blood cell progenitors within the lymph gland. Similar to the loss of Hh, mutation of *stat92E*, which encodes the *Drosophila* homologue of vertebrate STAT proteins, also causes a loss of medullary zone markers with increased differentiation of mature hemocytes (Krzemien et al., 2007). Therefore, JAK/STAT signaling is also important in the maintenance of the medullary zone progenitors (Krzemien et al., 2007).

5.5. Specification of cell fate

With regard to the hematopoietic development of specific lineages, several signaling and transcriptional components are known that control this process in the lymph gland. The Notch pathway has been shown to have an important role in the specification of the crystal cell lineage. Crystal cells in the lymph gland express Lz and PPO and both of these markers are missing in the *Notch* mutant background (Duvic et al., 2002; Lebestky et al., 2003). Additionally, misexpression of an activated form of Notch significantly increases the number of Lz-expressing cells within the lymph gland, particularly in secondary lobes where few such cells are normally found (Lebestky et al., 2003). Mutation and expression analyses indicate that the ligand Ser activates Notch in the lymph gland to promote crystal cell development. Ser is expressed during second and later larval stages in the PSC (Lebestky et al., 2003). A few other Ser-expressing cells can be found scattered within the primary lobes where most crystal cells develop (Lebestky et al., 2000; Crozatier et al., 2004). Cells of the PSC do not express Lz and do not require Notch signaling for their own development.

Consistent with its expression throughout the medullary and cortical zones, the PVR pathway has also been shown to play a role in lymph gland hematopoiesis. Clonal analysis has shown that cells lacking PVR fail to express the mature plasmatocyte marker Pxn, although PVR mutant cells are still able to differentiate into crystal cells (Jung et al., 2005). When and where PVR signaling occurs has not been established, although the ligand Pvf2 has been implicated in directing hemocyte proliferation (Munier et al., 2002).

As described previously, reporters for the JAK/STAT pathway receptor *domeless* and its ligand *upd3* are expressed in the lymph gland, suggesting that this signaling pathway influences hematopoietic development. Mutations that constitutively activate the *Drosophila* JAK kinase

Hopscotch (such as hop^{Tum-L} and hop^{T42}) cause hyperproliferation of lymph gland and circulating hemocytes, precocious differentiation of lamellocytes, and formation of melanotic tumors during larval stages (Harrison et al., 1995; Luo et al., 1995). Despite this strong effect, null mutations of *hop* have little or no effect on the number or differentiation of circulating or lymph gland plasmatocytes and crystal cells (Remillieux-Leschelle et al., 2002). In contrast, *hop* loss-of-function completely blocks lamellocyte differentiation when challenged by parasitization, indicating that JAK/STAT signaling is required for the differentiation of this lineage (Sorrentino et al., 2004; Krzemien et al., 2007).

The *Drosophila* Toll (Tl) signal transduction pathway controls embryonic dorsal/ventral patterning (Anderson et al., 1985) and is a critical regulator of humoral immunity during larval and adult stages (Leclerc and Reichhart, 2004; Lemaitre, 2004; Lemaitre and Hoffmann, 2007). Tl is a transmembrane receptor that controls the activity of the Rel-homology domain transcription factors Dorsal and Dif, which share similarity with mammalian NF-κB proteins (Hashimoto et al., 1988; Schneider et al., 1991; Rosetto et al., 1995; Govind, 1999; Silverman and Maniatis, 2001). Mutations that constitutively activate the Tl pathway cause blood cell phenotypes similar to those observed for the JAK/STAT pathway, including hemocyte hyperproliferation and precocious lamellocyte differentiation (Gerttula et al., 1988; Qiu et al., 1998). Furthermore, single-copy loss of *stat92E* suppresses Tl-activated lamellocyte differentiation (Remillieux-Leschelle et al., 2002), indicative of cross talk between these two pathways. Loss-of-function mutations in Tl, as well as downstream effectors, reduce the number of circulating hemocytes (Qiu et al., 1998; Sorrentino et al., 2004) but, unlike mutations in *hop* or *stat92E*, do not impair the lamellocyte differentiation response in lymph glands (Sorrentino et al., 2004).

6. MODELING HEMATOPOIETIC DISEASE IN *DROSOPHILA*

The conservation in many aspects of cell fate specification, proliferation, and differentiation of hematopoietic cells between *Drosophila* and mammals suggests that *Drosophila* may represent a useful system for modeling human hematopoietic disorders as has been reported for phenotypes in epithelia and solid tissues (Brumby and Richardson, 2005; Vidal and Cagan, 2006). In the last three decades, several genes have been identified in *Drosophila* that, when mutated, give aberrant hematopoietic phenotypes (Ghelelovitch, 1969; Gateff, 1994; Dearolf, 1998; Sinenko et al., 2004). Understanding how these genes function during normal blood development as well as how they cause proliferative disorders may prove to be useful in understanding how homologous genes function in vertebrates, including humans.

6.1. Transcription factors involved in select hematopoietic malignancies

The gene *acute myeloid leukemia-1* (*AML-1*, also known as *Runx1*) encodes a Runt-domain transcription factor critical for definitive hematopoiesis in mice and has multiple functions at later stages of hematopoietic development. This gene had been named *AML-1* because of the high frequency of chromosomal translocations into it that are associated with AML (Speck, 2001; Ito, 2004). Runx1 forms a heterodimer with CBFβ and enhances gene transcription by interacting with coactivators such as p300 and CREB-binding protein (CBP; Yamagata et al., 2005). AML-1 can also suppress gene transcription by interacting with corepressors such as Sin3A and TLE-1, which recruit histone deacetylase (HDAC) activity (Levanon et al., 1998; Reed-Inderbitzin et al., 2006). *Drosophila* encodes four Runt-domain proteins, including Runt (the founding member of this family; Gergen and Butler, 1988), Lz (Daga et al., 1996), and two other uncharacterized proteins, tentatively called RunxA and RunxB (Rennert et al., 2003; Boutros et al., 2004). Similar to AML-1, Lz cooperates with the CBFβ-like proteins Brother and Big brother to activate, and with the TLE-1 homologue Groucho to repress, gene expression (Canon and Banerjee, 2000; Kaminker et al., 2001). It has also been shown that Lz interacts with the GATA factor Srp and the FOG protein Ush to coordinate crystal cell specification and proliferation (Fossett et al., 2003; Waltzer et al., 2003).

Of the many chromosomal translocations that have been shown to cause AML, the one encoding the fusion protein AML-1-ETO has been the best studied (Hiebert et al., 2001; Speck and Gilliland, 2002; Peterson and Zhang, 2004). This protein functions in a dominant-negative fashion in mammals as well as when it is expressed from a transgene in *Drosophila* (Asou, 2003; Kurokawa and Hirai, 2003; Wildonger and Mann, 2005; Yamagata et al., 2005). Interestingly, expressing human AML-1-ETO protein specifically in *Drosophila* blood cells causes extensive proliferation of these cells (Fig. 7; S.A.S. and U.B., unpublished data). Thus, it is hoped that studying the function of AML-1-ETO in flies will aid in identifying components that are critical to the suppression of normal hematopoiesis and the onset of a proliferative response.

CBP has many roles in murine hematopoietic development, including interaction with GATA-1 to promote erythropoiesis (Bannister and Kouzarides, 1996; Letting et al., 2003) and with AML-1 to promote myelopoiesis (Kitabayashi et al., 1998). Mutations in CBP have also been associated with various forms of hematopoietic disease (Miller and Rubinstein, 1995; Petrij et al., 1995; Kung et al., 2000). Single-copy loss of *Drosophila CBP* (*dCBP*) enhances the melanotic tumor phenotype associated with mutations in *modulo*, which encodes a chromatin-associated factor first identified as a modifier of position effect variegation (Garzino et al., 1992; Bantignies et al., 2002).

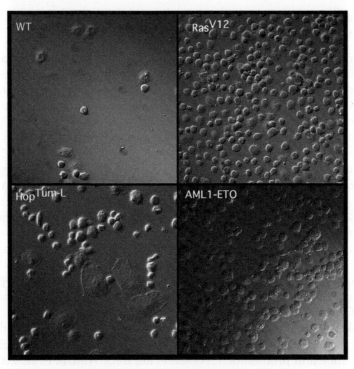

FIGURE 7 Abnormal differentiation and proliferation of hemocytes caused by expression of various oncogenic proteins: Ras^{V12}, Hop^{Tum-L}, and AML-1-ETO in circulating hemocytes of fly larva (Nomarski bright field microscopy images). The constitutively active form Ras^{V12} induces overproliferation of circulating cells in larvae. The oncogenic form, Hop^{Tum-L}, stimulates lamellocyte differentiation and increases in total hemocyte numbers. AML-1-ETO alters hemocyte differentiation inducing overproduction of all hemocyte lineages in circulation.

Chromatin remodeling factors, such as the SWI/SNF and NURF complexes, play prominent roles in hematopoietic development in mammals and *Drosophila* (Kowenz-Leutz and Leutz, 1999; Lee et al., 1999; Agalioti et al., 2000; Zhang et al., 2001; Ogawa et al., 2003). Two DNA-dependent ATPases, components of the SWI–SNF complex in *Drosophila*, have been studied in the hematopoietic system. The first, Brahma, is required to activate hemocyte proliferation while the second, Domino, represses the proliferation of blood cells (Ruhf et al., 2001; Remillieux-Leschelle et al., 2002). Similarly, the protein ISWI, a related DNA-dependent ATPase belonging to the NURF complex, as well as the NURF subunit NURF301 function as negative regulators of hemocyte proliferation (Badenhorst et al., 2002).

Members of the Polycomb (PcG) and trithorax (trxG) groups modify chromatin by maintaining established transcriptional programs (Lessard and Sauvageau, 2003). Mutation of the *Drosophila PcG* gene *multi sex combs* has been shown to cause hyperproliferation of both circulating and lymph gland hemocytes and to increase lamellocyte differentiation (Remillieux-Leschelle et al., 2002). Several mammalian *PcG* and *trxG* genes have been implicated in hematopoiesis including the well-characterized *Mixed-Lineage Leukemia (MLL)* gene (a homologue of *Drosophila trithorax*) that, like *AML-1*, is a frequent target of chromosomal translocations leading to leukemias in humans and mice (Ayton and Cleary, 2001; Muyrers-Chen et al., 2004). Additionally, targeted disruption of *MLL* in mice leads to defects in both primitive and definitive hematopoiesis (Hess et al., 1997; Yagi et al., 1998).

In vertebrates, *HOX* genes are important regulators of hematopoiesis and their expression patterns are controlled, in part, by PcG and trxG proteins. At least 22 of 39 *HOX* genes are known to be expressed in murine hematopoietic precursors (Grier et al., 2005) and misexpression of *HOXA5* is sufficient to cause expansion of myeloid progenitors while repressing erythroid differentiation (Crooks et al., 1999). As mentioned previously, the *Drosophila* homeobox gene *hth* is required for lymph gland specification and its expression is downregulated in the embryonic PSC that then begins to express the homeobox gene *Antp*. Similarly, targeted disruption of murine *meis1*, which encodes the homologue of Hth, causes a loss of definitive hematopoiesis and translocations in this gene are involved in leukemia (Abramovich et al., 2005; Azcoitia et al., 2005).

Abnormal hematopoiesis in fly larvae is often associated with development of melanotic tumors, which appear as melanized tumor-like aggregates of tissue, including hemocytes (Dearolf, 1998; Minakhina and Steward, 2006). For example, increasing the levels of the *Drosophila* homeobox protein Caudal through misexpression causes melanotic tumor formation (Hwang et al., 2002). This hallmark of blood abnormality has been used in several studies as a criterion to screen for hematopoietic mutants. Several genes have been identified in forward genetic screens, including those encoding the heterochromatin-associated protein, Su(var)205 (Minakhina and Steward, 2006); the importin superfamily protein, Pendulin (Kussel and Frasch, 1995; Torok et al., 1995); the nucleic acid-binding protein, Yantar (Sinenko et al., 2004); and the damaged DNA-binding protein, DDB1 (Takata et al., 2004a,b). Possible hematopoietic roles of these proteins in mammals remain to be addressed.

6.2. Signaling pathways involved in select hematopoietic malignancies

Gain-of-function mutations in RTK and Ras proteins are often involved in leukemia development. Oncogenic N-RAS and K-RAS mutations are present in ~25% and 15% of AML patients, respectively (Kelly and

Gilliland, 2002). Mutations activating the KIT and FLT3 receptors have been reported in 20–30% cases of AML and are particularly common in acute promyelocytic leukemia (APL) (Gilliland and Griffin, 2002; Steelman et al., 2004). Signal transduction through FLT3 activates several conserved pathways, including the RAS/MAP-kinase and the phosphoinositide-3-kinase (PI3-K)/Akt signaling cascades (Le and Shannon, 2002; Schmidt-Arras et al., 2004). As in mammals, Ras regulates proliferation and survival of *Drosophila* hematopoietic cells. Loss of the RTK receptor PVR or Ras negatively affects proper proliferation, differentiation, survival, and migration of hemocytes (Heino et al., 2001; Cho et al., 2002; Bruckner et al., 2004; Jung et al., 2005), while expression of activated Ras induces a dramatic increase in the number of circulating larval hemocytes (Asha et al., 2003). This expansion can be recapitulated by the misexpression of the PVR ligand PVF2 (Heino et al., 2001; Cho et al., 2002; Munier et al., 2002). Misexpression of an oncogenic form of Raf or the *Drosophila* EGF receptor can also cause a similar proliferative response (Sinenko and Mathey-Prevot, 2004; Zettervall et al., 2004). The function of tetraspanin proteins in mammalian hematopoiesis is not well understood; however, it has been shown they are able to suppress Ras-induced proliferation of *Drosophila* hemocytes (Sinenko and Mathey-Prevot, 2004).

The PI3-K signal transduction pathway integrates signals from multiple receptor tyrosine kinases and is crucial to many aspects of cell growth, survival, and apoptosis. Studies have also revealed that elevated PI3-K/Akt signaling is often associated with AML (Martelli et al., 2006; Witzig and Kaufmann, 2006). Key components of the pathway are the lipid kinase PI3-K, the small guanosine triphosphate-binding protein Rheb, and the protein kinases Akt and mTOR. PI3-K signaling activates S6 kinase, which phosphorylates ribosomal protein S6 and causes enhanced translation of mRNAs that encode components of the protein synthesis machinery (Lecomte et al., 1995). Interestingly, mutation of either the *Drosophila* S6 kinase protein or the ribosomal protein RpS6 causes overgrowth of lymph glands, abnormal blood cell differentiation, and melanotic tumor formation (Watson et al., 1992, 1994; Bryant et al., 1993; Stewart and Denell, 1993).

Chromosomal translocations within the JAK2 kinase and the transcription factor Tel are associated with leukemia (Lacronique et al., 1997; Peeters et al., 1997; Neubauer et al., 1998; Parganas et al., 1998), and mutations in JAK2 have been identified in various myeloproliferative disorders (Ferrajoli et al., 2006). Murine STAT5, which is most similar to *Drosophila* STAT92E, is required for Tel/JAK leukemic transformation (Schwaller et al., 2000), and increased activation of several different STAT homologues has been associated with various cancers of the blood (Bromberg, 2002). As described previously, mutations that constitutively activate the *Drosophila* JAK kinase Hopscotch (such as hop^{Tum-L})

cause leukemia-like hyperproliferation of both circulating and lymph gland hemocytes during larval stages and facilitate the formation of melanotic tumors (Harrison et al., 1995; Luo et al., 1995). The JAK/STAT and Ras signaling pathways appear to cooperate in the control of hemocyte proliferation and differentiation. The *Drosophila* gene *D-raf1* is a target of JAK/STAT signaling and has been shown to interact with Hop *in vitro* (Kwon et al., 2000). Furthermore, Raf is required for lamellocyte differentiation as well as the proliferation and survival of circulating hemocytes (Luo et al., 2002), which is similar to the function of murine Raf-1, which promotes B-cell development and facilitates hematopoietic cell survival (Iritani et al., 1997; Mikula et al., 2001). Finally, mutations in *brahma* suppress *hop*$^{Tum-L}$-mediated hyperproliferation whereas mutations in components of the NURF remodeling complex enhance tumor formation while specifically derepressing the expression of Hop target genes (Badenhorst et al., 2002; Remillieux-Leschelle et al., 2002).

The *Drosophila* Tl pathway regulates the transcriptional activity of Dorsal and Dif, proteins that share homology with mammalian NF-κB proteins. In mammals, NF-κB is a tightly regulated, positive mediator of T- and B-cell development, proliferation, and survival (Jimi and Ghosh, 2005). In *Drosophila*, several activating mutations have been identified that cause constitutive nuclear localization of Dorsal and Dif, which induces hyperproliferation, aberrant differentiation, and melanotic tumor formation among circulating and lymph gland hemocytes (Qiu et al., 1998; Sorrentino et al., 2004; Chiu et al., 2005; Huang et al., 2005). Given the conservation of these pathways, it is hoped that it may become possible to utilize *Drosophila* to understand the molecular mechanisms underlying hematopoietic disease caused by the aberrant regulation or function of NF-κB/Rel-domain factors.

7. CONCLUSIONS

In this chapter, we have described various aspects of hematopoietic development in *Drosophila*. Evidence suggests that there is a strong evolutionary link between the ontogenic relationships between cardiovascular and blood cell development. On the basis of functional criteria, it also appears that motile, phagocytic cells, such as *Drosophila* plasmatocytes and the macrophages and related cells of vertebrates, represent the earliest blood cell types. This phagocytic function is directly related to the conserved function of blood cells in innate immunity, for which *Drosophila* continues to be an excellent model. We have also described similarities between the developmental logic behind lymph gland development and the formation of the mammalian AGM region, and have described *in vivo* evidence that hemangioblast-like cells exist in *Drosophila*.

As in vertebrates, *Drosophila* blood development occurs sequentially as progenitors pass through various stages of maturation. Additionally, stem-like progenitor cells in *Drosophila* share many similarities with vertebrate hematopoietic stem cells, including niche-dependent maintenance. Finally, as *Drosophila* blood progenitors differentiate, they utilize an array of conserved signaling pathways and transcriptional regulators to mediate cell fate specification and differentiation. It is hoped that as our molecular understanding of *Drosophila* hematopoiesis becomes clearer, in both normal and aberrant contexts, it will illuminate similar mechanisms at work in vertebrate systems, including humans.

ACKNOWLEDGMENTS

We apologize for any pertinent work that has been inadvertently overlooked during the course of writing this chapter. Additionally, much of what is known about *Drosophila* hemocytes stems from the vast field of innate immunity. In an effort to limit this chapter to a developmental context, we have not focused on this work here. We thank Jamie Marshall for contributing the triple-stained confocal image in Fig. 5.

REFERENCES

Abramovich, C., Pineault, N., Ohta, H., Humphries, R.K. 2005. Hox genes: From leukemia to hematopoietic stem cell expansion. Ann. NY Acad. Sci. 1044, 109–116.

Adya, N., Castilla, L.H., Liu, P.P. 2000. Function of CBFbeta/Bro proteins. Semin. Cell Dev. Biol. 11, 361–368.

Agaisse, H., Petersen, U.M., Boutros, M., Mathey-Prevot, B., Perrimon, N. 2003. Signaling role of hemocytes in *Drosophila* JAK/STAT-dependent response to septic injury. Dev. Cell 5, 441–450.

Agalioti, T., Lomvardas, S., Parekh, B., Yie, J., Maniatis, T., Thanos, D. 2000. Ordered recruitment of chromatin modifying and general transcription factors to the IFN-beta promoter. Cell 103, 667–678.

Alfonso, T.B., Jones, B.W. 2002. gcm2 promotes glial cell differentiation and is required with glial cells missing for macrophage development in *Drosophila*. Dev. Biol. 248, 369–383.

Anderson, K.V., Bokla, L., Nusslein-Volhard, C. 1985. Establishment of dorsal-ventral polarity in the *Drosophila* embryo: The induction of polarity by the Toll gene product. Cell 42, 791–798.

Asha, H., Nagy, I., Kovacs, G., Stetson, D., Ando, I., Dearolf, C.R. 2003. Analysis of Ras-induced overproliferation in *Drosophila* hemocytes. Genetics 163, 203–215.

Asou, N. 2003. The role of a Runt domain transcription factor AML1/RUNX1 in leukemogenesis and its clinical implications. Crit. Rev. Oncol. Hematol. 45, 129–150.

Ayton, P.M., Cleary, M.L. 2001. Molecular mechanisms of leukemogenesis mediated by MLL fusion proteins. Oncogene 20, 5695–5707.

Azcoitia, V., Aracil, M., Martinez, A.C., Torres, M. 2005. The homeodomain protein Meis1 is essential for definitive hematopoiesis and vascular patterning in the mouse embryo. Dev. Biol. 280, 307–320.

Azpiazu, N., Frasch, M. 1993. Tinman and bagpipe: Two homeo box genes that determine cell fates in the dorsal mesoderm of *Drosophila*. Genes Dev. 7, 1325–1340.

Badenhorst, P., Voas, M., Rebay, I., Wu, C. 2002. Biological functions of the ISWI chromatin remodeling complex NURF. Genes Dev. 16, 3186–3198.

Bannister, A.J., Kouzarides, T. 1996. The CBP co-activator is a histone acetyltransferase. Nature 384, 641–643.

Bantignies, F., Goodman, R.H., Smolik, S.M. 2002. The interaction between the coactivator dCBP and Modulo, a chromatin-associated factor, affects segmentation and melanotic tumor formation in *Drosophila*. Proc. Natl. Acad. Sci. USA 99, 2895–2900.

Bataille, L., Auge, B., Ferjoux, G., Haenlin, M., Waltzer, L. 2005. Resolving embryonic blood cell fate choice in *Drosophila*: Interplay of GCM and RUNX factors. Development 132, 4635–4644.

Bernardoni, R., Vivancos, V., Giangrande, A. 1997. Glide/gcm is expressed and required in the scavenger cell lineage. Dev. Biol. 191, 118–130.

Bidla, G., Lindgren, M., Theopold, U., Dushay, M.S. 2005. Hemolymph coagulation and phenoloxidase in *Drosophila* larvae. Dev. Comp. Immunol. 29, 669–679.

Bidla, G., Dushay, M.S., Theopold, U. 2007. Crystal cell rupture after injury in *Drosophila* requires the JNK pathway, small GTPases and the TNF homolog Eiger. J. Cell Sci. 120, 1209–1215.

Bodmer, R. 1993. The gene tinman is required for specification of the heart and visceral muscles in *Drosophilu*. Development 118, 719–729.

Bourbon, H.M., Gonzy-Treboul, G., Peronnet, F., Alin, M.F., Ardourel, C., Benassayag, C., Cribbs, D., Deutsch, J., Ferrer, P., Haenlin, M., Lepesant, J.A., Noselli, S., et al. 2002. A P-insertion screen identifying novel X-linked essential genes in *Drosophila*. Mech. Dev. 110, 71–83.

Boutros, M., Kiger, A.A., Armknecht, S., Kerr, K., Hild, M., Koch, B., Haas, S.A., Paro, R., Perrimon, N. 2004. Genome-wide RNAi analysis of growth and viability in *Drosophila* cells. Science 303, 832–835.

Brennan, C.A., Delaney, J.R., Schneider, D.S., Anderson, K.V. 2007. Psidin is required in *Drosophila* blood cells for both phagocytic degradation and immune activation of the fat body. Curr. Biol. 17, 67–72.

Bromberg, J. 2002. Stat proteins and oncogenesis. J. Clin. Invest. 109, 1139–1142.

Bruckner, K., Kockel, L., Duchek, P., Luque, C.M., Rorth, P., Perrimon, N. 2004. The PDGF/VEGF receptor controls blood cell survival in *Drosophila*. Dev. Cell 7, 73–84.

Brumby, A.M., Richardson, H.E. 2005. Using *Drosophila melanogaster* to map human cancer pathways. Nat. Rev. Cancer 5, 626–639.

Bryant, P.J., Watson, K.L., Justice, R.W., Woods, D.F. 1993. Tumor suppressor genes encoding proteins required for cell interactions and signal transduction in *Drosophila*. Dev. Suppl., 239–249.

Burns, C.E., Traver, D., Mayhall, E., Shepard, J.L., Zon, L.I. 2005. Hematopoietic stem cell fate is established by the Notch-Runx pathway. Genes Dev. 19, 2331–2342.

Canon, J., Banerjee, U. 2000. Runt and Lozenge function in *Drosophila* development. Semin. Cell Dev. Biol. 11, 327–336.

Cantor, A.B., Orkin, S.H. 2002. Transcriptional regulation of erythropoiesis: An affair involving multiple partners. Oncogene 21, 3368–3376.

Cantor, A.B., Katz, S.G., Orkin, S.H. 2002. Distinct domains of the GATA-1 cofactor FOG-1 differentially influence erythroid versus megakaryocytic maturation. Mol. Cell. Biol. 22, 4268–4279.

Castanon, I., Baylies, M.K. 2002. A Twist in fate: Evolutionary comparison of Twist structure and function. Gene 287, 11–22.

Castillejo-Lopez, C., Hacker, U. 2005. The serine protease Sp7 is expressed in blood cells and regulates the melanization reaction in *Drosophila*. Biochem. Biophys. Res. Commun. 338, 1075–1082.

Chang, A.N., Cantor, A.B., Fujiwara, Y., Lodish, M.B., Droho, S., Crispino, J.D., Orkin, S.H. 2002. GATA-factor dependence of the multitype zinc-finger protein FOG-1 for its essential role in megakaryopoiesis. Proc. Natl. Acad. Sci. USA 99, 9237–9242.

Chiu, H., Ring, B.C., Sorrentino, R.P., Kalamarz, M., Garza, D., Govind, S. 2005. dUbc9 negatively regulates the Toll-NF-kappa B pathways in larval hematopoiesis and drosomycin activation in *Drosophila*. Dev. Biol. 288, 60–72.

Cho, N.K., Keyes, L., Johnson, E., Heller, J., Ryner, L., Karim, F., Krasnow, M.A. 2002. Developmental control of blood cell migration by the *Drosophila* VEGF pathway. Cell 108, 865–876.

Choi, K., Kennedy, M., Kazarov, A., Papadimitriou, J.C., Keller, G. 1998. A common precursor for hematopoietic and endothelial cells. Development 125, 725–732.

Chosa, N., Fukumitsu, T., Fujimoto, K., Ohnishi, E. 1997. Activation of prophenoloxidase A1 by an activating enzyme in *Drosophila melanogaster*. Insect Biochem. Mol. Biol. 27, 61–68.

Crispino, J.D., Lodish, M.B., Thurberg, B.L., Litovsky, S.H., Collins, T., Molkentin, J.D., Orkin, S.H. 2001. Proper coronary vascular development and heart morphogenesis depend on interaction of GATA-4 with FOG cofactors. Genes Dev. 15, 839–844.

Crooks, G.M., Fuller, J., Petersen, D., Izadi, P., Malik, P., Pattengale, P.K., Kohn, D.B., Gasson, J.C. 1999. Constitutive HOXA5 expression inhibits erythropoiesis and increases myelopoiesis from human hematopoietic progenitors. Blood 94, 519–528.

Crozatier, M., Ubeda, J.M., Vincent, A., Meister, M. 2004. Cellular immune response to parasitization in *Drosophila* requires the EBF orthologue collier. PLoS Biol. 2, E196.

Cumano, A., Godin, I. 2007. Ontogeny of the hematopoietic system. Annu. Rev. Immunol. 25, 745–785.

Daga, A., Karlovich, C.A., Dumstrei, K., Banerjee, U. 1996. Patterning of cells in the *Drosophila* eye by Lozenge, which shares homologous domains with AML1. Genes Dev. 10, 1194–1205.

De Gregorio, E., Han, S.J., Lee, W.J., Baek, M.J., Osaki, T., Kawabata, S., Lee, B.L., Iwanaga, S., Lemaitre, B., Brey, P.T. 2002. An immune-responsive Serpin regulates the melanization cascade in *Drosophila*. Dev. Cell 3, 581–592.

Dearolf, C.R. 1998. Fruit fly "leukemia." Biochim. Biophys. Acta 1377, M13–M23.

Deconinck, A.E., Mead, P.E., Tevosian, S.G., Crispino, J.D., Katz, S.G., Zon, L.I., Orkin, S.H. 2000. FOG acts as a repressor of red blood cell development in *Xenopus*. Development 127, 2031–2040.

Duchek, P., Somogyi, K., Jekely, G., Beccari, S., Rorth, P. 2001. Guidance of cell migration by the *Drosophila* PDGF/VEGF receptor. Cell 107, 17–26.

Duvic, B., Hoffmann, J.A., Meister, M., Royet, J. 2002. Notch signaling controls lineage specification during *Drosophila* larval hematopoiesis. Curr. Biol. 12, 1923–1927.

Dzierzak, E. 2005. The emergence of definitive hematopoietic stem cells in the mammal. Curr. Opin. Hematol. 12, 197–202.

Elagib, K.E., Racke, F.K., Mogass, M., Khetawat, R., Delehanty, L.L., Goldfarb, A.N. 2003. RUNX1 and GATA-1 coexpression and cooperation in megakaryocytic differentiation. Blood 101, 4333–4341.

Elrod-Erickson, M., Mishra, S., Schneider, D. 2000. Interactions between the cellular and humoral immune responses in *Drosophila*. Curr. Biol. 10, 781–784.

Evans, C.J., Hartenstein, V., Banerjee, U. 2003. Thicker than blood: Conserved mechanisms in *Drosophila* and vertebrate hematopoiesis. Dev. Cell 5, 673–690.

Fadok, V.A., Bratton, D.L., Rose, D.M., Pearson, A., Ezekewitz, R.A., Henson, P.M. 2000. A receptor for phosphatidylserine-specific clearance of apoptotic cells. Nature 405, 85–90.

Ferrajoli, A., Faderl, S., Ravandi, F., Estrov, Z. 2006. The JAK-STAT pathway: A therapeutic target in hematological malignancies. Curr. Cancer Drug Targets 6, 671–679.

Ferrara, N., Gerber, H.P., LeCouter, J. 2003. The biology of VEGF and its receptors. Nat. Med. 9, 669–676.

Fessler, L.I., Nelson, R.E., Fessler, J.H. 1994. *Drosophila* extracellular matrix. Meth. Enzymol. 245, 271–294.

Firulli, A.B., McFadden, D.G., Lin, Q., Srivastava, D., Olson, E.N. 1998. Heart and extra-embryonic mesodermal defects in mouse embryos lacking the bHLH transcription factor Hand1. Nat. Genet. 18, 266–270.

Fossett, N., Zhang, Q., Gajewski, K., Choi, C.Y., Kim, Y., Schulz, R.A. 2000. The multitype zinc-finger protein U-shaped functions in heart cell specification in the *Drosophila* embryo. Proc. Natl. Acad. Sci. USA 97, 7348–7353.

Fossett, N., Tevosian, S.G., Gajewski, K., Zhang, Q., Orkin, S.H., Schulz, R.A. 2001. The Friend of GATA proteins U-shaped, FOG-1, and FOG-2 function as negative regulators of blood, heart, and eye development in *Drosophila*. Proc. Natl. Acad. Sci. USA 98, 7342–7347.

Fossett, N., Hyman, K., Gajewski, K., Orkin, S.H., Schulz, R.A. 2003. Combinatorial interactions of serpent, lozenge, and U-shaped regulate crystal cell lineage commitment during *Drosophila* hematopoiesis. Proc. Natl. Acad. Sci. USA 100, 11451–11456.

Franc, N.C. 2002. Phagocytosis of apoptotic cells in mammals, *Caenorhabditis elegans* and *Drosophila melanogaster*: Molecular mechanisms and physiological consequences. Front Biosci. 7, d1298–d1313.

Franc, N.C., Dimarcq, J.L., Lagueux, M., Hoffmann, J., Ezekowitz, R.A. 1996. Croquemort, a novel *Drosophila* hemocyte/macrophage receptor that recognizes apoptotic cells. Immunity 4, 431–443.

Freeman, M.R., Delrow, J., Kim, J., Johnson, E., Doe, C.Q. 2003. Unwrapping glial biology: Gcm target genes regulating glial development, diversification, and function. Neuron 38, 567–580.

Fujiwara, Y., Chang, A.N., Williams, A.M., Orkin, S.H. 2004. Functional overlap of GATA-1 and GATA-2 in primitive hematopoietic development. Blood 103, 583–585.

Galloway, J.L., Zon, L.I. 2003. Ontogeny of hematopoiesis: Examining the emergence of hematopoietic cells in the vertebrate embryo. Curr. Top. Dev. Biol. 53, 139–158.

Gardiner, S.L. 1992. Polychaeta: General organization, integument, musculature, coelom, and vascular system. In: *Microscopic Anatomy of Invertebrates* (F.W. Harrison, S.L. Gardiner, Eds.), Vol. 7, New York: Wiley-Liss.

Garzino, V., Pereira, A., Laurenti, P., Graba, Y., Levis, R.W., Le Parco, Y., Pradel, J. 1992. Cell lineage-specific expression of modulo, a dose-dependent modifier of variegation in *Drosophila*. EMBO J. 11, 4471–4479.

Gateff, E. 1994. Tumor-suppressor genes, hematopoietic malignancies and other hematopoietic disorders of *Drosophila melanogaster*. Ann. NY Acad. Sci. 712, 260–279.

Gerber, H.P., Ferrara, N. 2003. The role of VEGF in normal and neoplastic hematopoiesis. J. Mol. Med. 81, 20–31.

Gerber, H.P., Malik, A.K., Solar, G.P., Sherman, D., Liang, X.H., Meng, G., Hong, K., Marsters, J.C., Ferrara, N. 2002. VEGF regulates haematopoietic stem cell survival by an internal autocrine loop mechanism. Nature 417, 954–958.

Gergen, J.P., Butler, B.A. 1988. Isolation of the *Drosophila segmentation* gene runt and analysis of its expression during embryogenesis. Genes Dev. 2, 1179–1193.

Gerttula, S., Jin, Y.S., Anderson, K.V. 1988. Zygotic expression and activity of the *Drosophila* Toll gene, a gene required maternally for embryonic dorsal-ventral pattern formation. Genetics 119, 123–133.

Ghelelovitch, S. 1969. Melanotic tumors in *Drosophila* melanogaster. Natl. Cancer Inst. Monogr. 31, 263–275.

Gilliland, D.G., Griffin, J.D. 2002. Role of FLT3 in leukemia. Curr. Opin. Hematol. 9, 274–281.

Goto, A., Kadowaki, T., Kitagawa, Y. 2003. *Drosophila* hemolectin gene is expressed in embryonic and larval hemocytes and its knock down causes bleeding defects. Dev. Biol. 264, 582–591.

Govind, S. 1999. Control of development and immunity by rel transcription factors in *Drosophila*. Oncogene 18, 6875–6887.

Grier, D.G., Thompson, A., Kwasniewska, A., McGonigle, G.J., Halliday, H.L., Lappin, T.R. 2005. The pathophysiology of HOX genes and their role in cancer. J. Pathol. 205, 154–171.

Haenlin, M., Cubadda, Y., Blondeau, F., Heitzler, P., Lutz, Y., Simpson, P., Ramain, P. 1997. Transcriptional activity of pannier is regulated negatively by heterodimerization of the GATA DNA-binding domain with a cofactor encoded by the U-shaped gene of *Drosophila*. Genes Dev. 11, 3096–3108.

Han, Z., Olson, E.N. 2005. Hand is a direct target of Tinman and GATA factors during *Drosophila* cardiogenesis and hematopoiesis. Development 132, 3525–3536.

Han, Z., Yi, P., Li, X., Olson, E.N. 2006. Hand, an evolutionarily conserved bHLH transcription factor required for *Drosophila* cardiogenesis and hematopoiesis. Development 133, 1175–1182.

Harrison, D.A., Binari, R., Nahreini, T.S., Gilman, M., Perrimon, N. 1995. Activation of a *Drosophila* Janus kinase (JAK) causes hematopoietic neoplasia and developmental defects. EMBO J. 14, 2857–2865.

Hartenstein, V. 2006. Blood cells and blood cell development in the animal kingdom. Annu. Rev. Cell Dev. Biol. 22, 677–712.

Hartenstein, V., Mandal, L. 2006. The blood/vascular system in a phylogenetic perspective. Bioessays 28, 1203–1210.

Hashimoto, C., Hudson, K.L., Anderson, K.V. 1988. The Toll gene of *Drosophila*, required for dorsal-ventral embryonic polarity, appears to encode a transmembrane protein. Cell 52, 269–279.

Heino, T.I., Karpanen, T., Wahlstrom, G., Pulkkinen, M., Eriksson, U., Alitalo, K., Roos, C. 2001. The *Drosophila* VEGF receptor homolog is expressed in hemocytes. Mech. Dev. 109, 69–77.

Hess, J.L., Yu, B.D., Li, B., Hanson, R., Korsmeyer, S.J. 1997. Defects in yolk sac hematopoiesis in Mll-null embryos. Blood 90, 1799–1806.

Hetru, C., Troxler, L., Hoffmann, J.A. 2003. *Drosophila melanogaster* antimicrobial defense. J. Infect. Dis. 187(Suppl. 2), S327–S334.

Hiebert, S.W., Lutterbach, B., Amann, J. 2001. Role of co-repressors in transcriptional repression mediated by the t(8;21), t(16;21), t(12;21), and inv(16) fusion proteins. Curr. Opin. Hematol. 8, 197–200.

Hiramatsu, A., Miwa, H., Shikami, M., Ikai, T., Tajima, E., Yamamoto, H., Imai, N., Hattori, A., Kyo, T., Watarai, M., Miura, K., Satoh, A., et al. 2006. Disease-specific expression of VEGF and its receptors in AML cells: Possible autocrine pathway of VEGF/type1 receptor of VEGF in t(15;17) AML and VEGF/type2 receptor of VEGF in t(8;21) AML. Leuk. Lymphoma 47, 89–95.

Holmes, M., Turner, J., Fox, A., Chisholm, O., Crossley, M., Chong, B. 1999. hFOG-2, a novel zinc finger protein, binds the co-repressor mCtBP2 and modulates GATA-mediated activation. J. Biol. Chem. 274, 23491–23498.

Holz, A., Bossinger, B., Strasser, T., Janning, W., Klapper, R. 2003. The two origins of hemocytes in *Drosophila*. Development 130, 4955–4962.

Hosoya, T., Takizawa, K., Nitta, K., Hotta, Y. 1995. Glial cells missing: A binary switch between neuronal and glial determination in *Drosophila*. Cell 82, 1025–1036.

Huang, L., Ohsako, S., Tanda, S. 2005. The lesswright mutation activates Rel-related proteins, leading to overproduction of larval hemocytes in *Drosophila melanogaster*. Dev. Biol. 280, 407–420.

Hwang, M.S., Kim, Y.S., Choi, N.H., Park, J.H., Oh, E.J., Kwon, E.J., Yamaguchi, M., Yoo, M.A. 2002. The caudal homeodomain protein activates *Drosophila* E2F gene expression. Nucleic Acids Res. 30, 5029–5035.

Iritani, B.M., Forbush, K.A., Farrar, M.A., Perlmutter, R.M. 1997. Control of B cell development by Ras-mediated activation of Raf. EMBO J. 16, 7019–7031.

Ito, Y. 2004. Oncogenic potential of the RUNX gene family: 'Overview.' Oncogene 23, 4198–4208.

Jimi, E., Ghosh, S. 2005. Role of nuclear factor-kappaB in the immune system and bone. Immunol. Rev. 208, 80–87.

Jones, B.W., Fetter, R.D., Tear, G., Goodman, C.S. 1995. Glial cells missing: A genetic switch that controls glial versus neuronal fate. Cell 82, 1013–1023.

Jung, S.H., Evans, C.J., Uemura, C., Banerjee, U. 2005. The *Drosophila* lymph gland as a developmental model of hematopoiesis. Development 132, 2521–2533.

Kaminker, J.S., Singh, R., Lebestky, T., Yan, H., Banerjee, U. 2001. Redundant function of Runt domain binding partners, Big brother and Brother, during *Drosophila* development. Development 128, 2639–2648.

Kammerer, M., Giangrande, A. 2001. Glide2, a second glial promoting factor in *Drosophila melanogaster*. EMBO J. 20, 4664–4673.

Katz, S.G., Cantor, A.B., Orkin, S.H. 2002. Interaction between FOG-1 and the corepressor C-terminal binding protein is dispensable for normal erythropoiesis *in vivo*. Mol. Cell. Biol. 22, 3121–3128.

Kelly, L.M., Gilliland, D.G. 2002. Genetics of myeloid leukemias. Annu. Rev. Genomics Hum. Genet. 3, 179–198.

Kitabayashi, I., Yokoyama, A., Shimizu, K., Ohki, M. 1998. Interaction and functional cooperation of the leukemia-associated factors AML1 and p300 in myeloid cell differentiation. EMBO J. 17, 2994–3004.

Klinedinst, S.L., Bodmer, R. 2003. Gata factor Pannier is required to establish competence for heart progenitor formation. Development 130, 3027–3038.

Kocks, C., Cho, J.H., Nehme, N., Ulvila, J., Pearson, A.M., Meister, M., Strom, C., Conto, S.L., Hetru, C., Stuart, L.M., Stehle, T., Hoffmann, J.A., et al. 2005. Eater, a transmembrane protein mediating phagocytosis of bacterial pathogens in *Drosophila*. Cell 123, 335–346.

Kowenz-Leutz, E., Leutz, A. 1999. A C/EBP beta isoform recruits the SWI/SNF complex to activate myeloid genes. Mol. Cell 4, 735–743.

Krzemien, J., Dubois, L., Makki, R., Meister, M., Vincent, A., Crozatier, M. 2007. Control of blood cell homeostasis in *Drosophila* larvae by the posterior signalling centre. Nature 446, 325–328.

Kung, A.L., Rebel, V.I., Bronson, R.T., Ch'ng, L.E., Sieff, C.A., Livingston, D.M., Yao, T.P. 2000. Gene dose-dependent control of hematopoiesis and hematologic tumor suppression by CBP. Genes Dev. 14, 272–277.

Kurokawa, M., Hirai, H. 2003. Role of AML1/Runx1 in the pathogenesis of hematological malignancies. Cancer Sci. 94, 841–846.

Kurucz, E., Zettervall, C.J., Sinka, R., Vilmos, P., Pivarcsi, A., Ekengren, S., Hegedus, Z., Ando, I., Hultmark, D. 2003. Hemese, a hemocyte-specific transmembrane protein, affects the cellular immune response in *Drosophila*. Proc. Natl. Acad. Sci. USA 100, 2622–2627.

Kurucz, E., Markus, R., Zsamboki, J., Folkl-Medzihradszky, K., Darula, Z., Vilmos, P., Udvardy, A., Krausz, I., Lukacsovich, T., Gateff, E., Zettervall, C.J., Hultmark, D., et al. 2007. Nimrod, a putative phagocytosis receptor with EGF repeats in *Drosophila* plasmatocytes. Curr. Biol. 17, 649–654.

Kussel, P., Frasch, M. 1995. Pendulin, a *Drosophila* protein with cell cycle-dependent nuclear localization, is required for normal cell proliferation. J. Cell Biol. 129, 1491–1507.

Kwon, E.J., Park, H.S., Kim, Y.S., Oh, E.J., Nishida, Y., Matsukage, A., Yoo, M.A., Yamaguchi, M. 2000. Transcriptional regulation of the *Drosophila* raf proto-oncogene by *Drosophila* STAT during development and in immune response. J. Biol. Chem. 26, 19824–19830.

Lacronique, V., Boureux, A., Valle, V.D., Poirel, H., Quang, C.T., Mauchauffe, M., Berthou, C., Lessard, M., Berger, R., Ghysdael, J., Bernard, O.A. 1997. A TEL-JAK2 fusion protein with constitutive kinase activity in human leukemia. Science 278, 1309–1312.

Lanot, R., Zachary, D., Holder, F., Meister, M. 2001. Postembryonic hematopoiesis in *Drosophila*. Dev. Biol. 230, 243–257.

Le, D.T., Shannon, K.M. 2002. Ras processing as a therapeutic target in hematologic malignancies. Curr. Opin. Hematol. 9, 308–315.

Lebestky, T., Chang, T., Hartenstein, V., Banerjee, U. 2000. Specification of *Drosophila* hematopoietic lineage by conserved transcription factors. Science 288, 146–149.

Lebestky, T., Jung, S.H., Banerjee, U. 2003. A Serrate-expressing signaling center controls *Drosophila* hematopoiesis. Genes Dev. 17, 348–353.

Leclerc, V., Reichhart, J.M. 2004. The immune response of *Drosophila melanogaster*. Immunol. Rev. 198, 59–71.

Lecomte, F., Champagne, B., Dasnoy, J.F., Szpirer, J., Szpirer, C. 1995. The mammalian RPS6 gene, homolog of the *Drosophila* air8 tumor suppressor gene: Is it an oncosuppressor gene? Somat. Cell Mol. Genet. 21, 443–450.

Lee, C.H., Murphy, M.R., Lee, J.S., Chung, J.H. 1999. Targeting a SWI/SNF-related chromatin remodeling complex to the beta-globin promoter in erythroid cells. Proc. Natl. Acad. Sci. USA 96, 12311–12315.

Lemaitre, B. 2004. The road to Toll. Nat. Rev. Immunol. 4, 521–527.

Lemaitre, B., Hoffmann, J. 2007. The host defense of *Drosophila melanogaster*. Annu. Rev. Immunol. 25, 697–743.

Lessard, J., Sauvageau, G. 2003. Polycomb group genes as epigenetic regulators of normal and leukemic hemopoiesis. Exp. Hematol. 31, 567–585.

Letting, D.L., Rakowski, C., Weiss, M.J., Blobel, G.A. 2003. Formation of a tissue-specific histone acetylation pattern by the hematopoietic transcription factor GATA-1. Mol. Cell. Biol. 23, 1334–1340.

Levanon, D., Goldstein, R.E., Bernstein, Y., Tang, H., Goldenberg, D., Stifani, S., Paroush, Z., Groner, Y. 1998. Transcriptional repression by AML1 and LEF-1 is mediated by the TLE/Groucho corepressors. Proc. Natl. Acad. Sci. USA 95, 11590–11595.

Li, L.H., Gergen, J.P. 1999. Differential interactions between Brother proteins and Runt domain proteins in the *Drosophila* embryo and eye. Development 126, 3313–3322.

Ling, K.W., Ottersbach, K., van Hamburg, J.P., Oziemlak, A., Tsai, F.Y., Orkin, S.H., Ploemacher, R., Hendriks, R.W., Dzierzak, E. 2004. GATA-2 plays two functionally distinct roles during the ontogeny of hematopoietic stem cells. J. Exp. Med. 200, 871–882.

Lowry, J.A., Atchley, W.R. 2000. Molecular evolution of the GATA family of transcription factors: Conservation within the DNA-binding domain. J. Mol. Evol. 50, 103–115.

Lugus, J.J., Chung, Y.S., Mills, J.C., Kim, S.I., Grass, J., Kyba, M., Doherty, J.M., Bresnick, E.H., Choi, K. 2007. GATA2 functions at multiple steps in hemangioblast development and differentiation. Development 134, 393–405.

Luo, H., Hanratty, W.P., Dearolf, C.R. 1995. An amino acid substitution in the *Drosophila* hopTum-l Jak kinase causes leukemia-like hematopoietic defects. EMBO J. 14, 1412–1420.

Luo, H., Rose, P.E., Roberts, T.M., Dearolf, C.R. 2002. The Hopscotch Jak kinase requires the Raf pathway to promote blood cell activation and differentiation in *Drosophila*. Mol. Genet. Genomics 267, 57–63.

Lutterbach, B., Hiebert, S.W. 2000. Role of the transcription factor AML-1 in acute leukemia and hematopoietic differentiation. Gene 245, 223–235.

Maduro, M.F., Rothman, J.H. 2002. Making worm guts: The gene regulatory network of the *Caenorhabditis elegans* endoderm. Dev. Biol. 246, 68–85.

Manaka, J., Kuraishi, T., Shiratsuchi, A., Nakai, Y., Higashida, H., Henson, P., Nakanishi, Y. 2004. Draper-mediated and phosphatidylserine-independent phagocytosis of apoptotic cells by *Drosophila* hemocytes/macrophages. J. Biol. Chem. 279, 48466–48476.

Mandal, L., Banerjee, U., Hartenstein, V. 2004. Evidence for a fruit fly hemangioblast and similarities between lymph-gland hematopoiesis in fruit fly and mammal aorta-gonadal-mesonephros mesoderm. Nat. Genet. 36, 1019–1023.

Mandal, L., Martinez-Agosto, J.A., Evans, C.J., Hartenstein, V., Banerjee, U. 2007. A Hedgehog- and Antennapedia-dependent niche maintains *Drosophila* haematopoietic precursors. Nature 446, 320–324.

Markus, R., Kurucz, E., Rus, F., Ando, I. 2005. Sterile wounding is a minimal and sufficient trigger for a cellular immune response in *Drosophila melanogaster*. Immunol. Lett. 101, 108–111.

Marshall, C.J., Kinnon, C., Thrasher, A.J. 2000. Polarized expression of bone morphogenetic protein-4 in the human aorta-gonad-mesonephros region. Blood 96, 1591–1593.

Martelli, A.M., Nyakern, M., Tabellini, G., Bortul, R., Tazzari, P.L., Evangelisti, C., Cocco, L. 2006. Phosphoinositide 3-kinase/Akt signaling pathway and its therapeutical implications for human acute myeloid leukemia. Leukemia 20, 911–928.

Matova, N., Anderson, K.V. 2006. Rel/NF-kappaB double mutants reveal that cellular immunity is central to *Drosophila* host defense. Proc. Natl. Acad. Sci. USA 103, 16424–16429.

McFadden, D.G., Charite, J., Richardson, J.A., Srivastava, D., Firulli, A.B., Olson, E.N. 2000. A GATA-dependent right ventricular enhancer controls dHAND transcription in the developing heart. Development 127, 5331–5341.

Medvinsky, A., Dzierzak, E. 1996. Definitive hematopoiesis is autonomously initiated by the AGM region. Cell 86, 897–906.

Mikula, M., Schreiber, M., Husak, Z., Kucerova, L., Ruth, J., Wieser, R., Zatloukal, K., Beug, H., Wagner, E.F., Baccarini, M. 2001. Embryonic lethality and fetal liver apoptosis in mice lacking the c-raf-1 gene. EMBO J. 20, 1952–1962.

Milchanowski, A.B., Henkenius, A.L., Narayanan, M., Hartenstein, V., Banerjee, U. 2004. Identification and characterization of genes involved in embryonic crystal cell formation during *Drosophila* hematopoiesis. Genetics 168, 325–339.

Miller, R.W., Rubinstein, J.H. 1995. Tumors in Rubinstein-Taybi syndrome. Am. J. Med. Genet. 56, 112–115.

Minakhina, S., Steward, R. 2006. Melanotic mutants in *Drosophila*: Pathways and phenotypes. Genetics 174, 253–263.

Munier, A.I., Doucet, D., Perrodou, E., Zachary, D., Meister, M., Hoffmann, J.A., Janeway, C.A., Jr., Lagueux, M. 2002. PVF2, a PDGF/VEGF-like growth factor, induces hemocyte proliferation in *Drosophila* larvae. EMBO Rep. 3, 1195–1200.

Murray, M.A., Fessler, L.I., Palka, J. 1995. Changing distributions of extracellular matrix components during early wing morphogenesis in *Drosophila*. Dev. Biol. 168, 150–165.

Murray, P.D.F. 1932. The development *in vitro* of the blood of the early chick embryo. Proc. R. Soc. Lond. B 11, 497–521.

Muyrers-Chen, I., Rozovskaia, T., Lee, N., Kersey, J.H., Nakamura, T., Canaani, E., Paro, R. 2004. Expression of leukemic MLL fusion proteins in *Drosophila* affects cell cycle control and chromosome morphology. Oncogene 23, 8639–8648.

Nakao, T. 1974. An electron microscopic study of the circulatory system in *Nereis japonica*. J. Morphol. 144, 217–236.

Nappi, A.J. 1975. Inhibition by parasites of melanotic tumour formation in *Drosophila melanogaster*. Nature 255, 402–404.

Nelson, R.E., Fessler, L.I., Takagi, Y., Blumberg, B., Keene, D.R., Olson, P.F., Parker, C.G., Fessler, J.H. 1994. Peroxidasin: A novel enzyme-matrix protein of *Drosophila* development. EMBO J. 13, 3438–3447.

Neubauer, H., Cumano, A., Muller, M., Wu, H., Huffstadt, U., Pfeffer, K. 1998. Jak2 deficiency defines an essential developmental checkpoint in definitive hematopoiesis. Cell 93, 397–409.

Nishikawa, M., Tahara, T., Hinohara, A., Miyajima, A., Nakahata, T., Shimosaka, A. 2001. Role of the microenvironment of the embryonic aorta-gonad-mesonephros region in hematopoiesis. Ann. NY Acad. Sci. 938, 109–116.

Ogawa, H., Ueda, T., Aoyama, T., Aronheim, A., Nagata, S., Fukunaga, R. 2003. A SWI2/ SNF2-type ATPase/helicase protein, mDomino, interacts with myeloid zinc finger protein 2A (MZF-2A) to regulate its transcriptional activity. Genes Cells 8, 325–339.

Okuda, T., van Deursen, J., Hiebert, S.W., Grosveld, G., Downing, J.R. 1996. AML1, the target of multiple chromosomal translocations in human leukemia, is essential for normal fetal liver hematopoiesis. Cell 84, 321–330.

Orelio, C., Dzierzak, E. 2003. Identification of 2 novel genes developmentally regulated in the mouse aorta-gonad-mesonephros region. Blood 101, 2246–2249.

Orkin, S.H., Shivdasani, R.A., Fujiwara, Y., McDevitt, M.A. 1998. Transcription factor GATA-1 in megakaryocyte development. Stem Cells 16(Suppl. 2), 79–83.

Pai, S.Y., Truitt, M.L., Ting, C.N., Leiden, J.M., Glimcher, L.H., Ho, I.C. 2003. Critical roles for transcription factor GATA-3 in thymocyte development. Immunity 19, 863–875.

Parganas, E., Wang, D., Stravopodis, D., Topham, D.J., Marine, J.C., Teglund, S., Vanin, E.F., Bodner, S., Colamonici, O.R., van Deursen, J.M., Grosveld, G., Ihle, J.N., et al. 1998. Jak2 is essential for signaling through a variety of cytokine receptors. Cell 93, 385–395.

Patient, R.K., McGhee, J.D. 2002. The GATA family (vertebrates and invertebrates). Curr. Opin. Genet. Dev. 12, 416–422.

Peeples, E.E., Geisler, A., Whitcraft, C.J., Oliver, C.P. 1969. Activity of phenol oxidases at the puparium formation stage in development of nineteen lozenge mutants of *Drosophila melanogaster*. Biochem. Genet. 3, 563–569.

Peeters, P., Raynaud, S.D., Cools, J., Wlodarska, I., Grosgeorge, J., Philip, P., Monpoux, F., Van Rompaey, L., Baens, M., Van den Berghe, H., Marynen, P. 1997. Fusion of TEL, the ETS-variant gene 6 (ETV6), to the receptor-associated kinase JAK2 as a result of t(9;12) in a lymphoid and t(9;15;12) in a myeloid leukemia. Blood 90, 2535–2540.

Peterson, L.F., Zhang, D.E. 2004. The 8;21 translocation in leukemogenesis. Oncogene 23, 4255–4262.

Petrij, F., Giles, R.H., Dauwerse, H.G., Saris, J.J., Hennekam, R.C., Masuno, M., Tommerup, N., van Ommen, G.J., Goodman, R.H., Peters, D.J., Breuning, M.H. 1995. Rubinstein-Taybi syndrome caused by mutations in the transcriptional co-activator CBP. Nature 376, 348–351.

Pevny, L., Simon, M.C., Robertson, E., Klein, W.H., Tsai, S.F., D'Agati, V., Orkin, S.H., Costantini, F. 1991. Erythroid differentiation in chimaeric mice blocked by a targeted mutation in the gene for transcription factor GATA-1. Nature 349, 257–260.

Pham, L.N., Dionne, M.S., Shirasu-Hiza, M., Schneider, D.S. 2007. A specific primed immune response in *Drosophila* is dependent on phagocytes. PLoS Pathog. 3, e26.

Podar, K., Anderson, K.C. 2005. The pathophysiologic role of VEGF in hematologic malignancies: Therapeutic implications. Blood 105, 1383–1395.

Qiu, P., Pan, P.C., Govind, S. 1998. A role for the *Drosophila* Toll/Cactus pathway in larval hematopoiesis. Development 125, 1909–1920.

Ramet, M., Pearson, A., Manfruelli, P., Li, X., Koziel, H., Gobel, V., Chung, E., Krieger, M., Ezekowitz, R.A. 2001. *Drosophila* scavenger receptor CI is a pattern recognition receptor for bacteria. Immunity 15, 1027–1038.

Ramet, M., Manfruelli, P., Pearson, A., Mathey-Prevot, B., Ezekowitz, R.A. 2002. Functional genomic analysis of phagocytosis and identification of a *Drosophila* receptor for *E. coli*. Nature 416, 644–648.

Reed-Inderbitzin, E., Moreno-Miralles, I., Vanden-Eynden, S.K., Xie, J., Lutterbach, B., Durst-Goodwin, K.L., Luce, K.S., Irvin, B.J., Cleary, M.L., Brandt, S.J., Hiebert, S.W. 2006. RUNX1 associates with histone deacetylases and SUV39H1 to repress transcription. Oncogene 25, 5777–5786.

Rehorn, K.P., Thelen, H., Michelson, A.M., Reuter, R. 1996. A molecular aspect of hematopoiesis and endoderm development common to vertebrates and *Drosophila*. Development 122, 4023–4031.

Remillieux-Leschelle, N., Santamaria, P., Randsholt, N.B. 2002. Regulation of larval hemato-poiesis in *Drosophila melanogaster*: A role for the multi sex combs gene. Genetics 162, 1259–1274.

Rennert, J., Coffman, J.A., Mushegian, A.R., Robertson, A.J. 2003. The evolution of Runx genes I. A comparative study of sequences from phylogenetically diverse model organisms. BMC Evol. Biol. 3, 4.

Rizki, T.M. 1956. Blood cells of *Drosophila* as related to metamorphosis. In: *Physiology of Insect Development* (F.L. Campbell, Ed.), Chicago: Chicago University Press, pp. 91–94.

Rizki, T.M. 1978. The circulatory system and associated cells and tissues. In: *The Genetics and Biology of Drosophila* (M. Ashburner, T.R.F. Wright, Eds.), Vol. 2b. London: Academic Press, pp. 397–452.

Rizki, T.M., Rizki, R.M. 1980. Properties of the larval hemocytes of *Drosophila melanogaster*. Eperientia 36, 1223–1226.

Rizki, T.M., Rizki, R.M. 1981. Alleles of lz as suppressors of the Bc-phene in *Drosophila melanogaster*. Genetics 97, S90.

Rizki, T.M., Rizki, R.M. 1985. Paracrystalline inclusions of *D. melanogaster* hemocytes have prophenoloxidases. Genetics 110, S98.

Rizki, T.M., Rizki, R.M. 1992. Lamellocyte differentiation in *Drosophila* larvae parasitized by Leptopilina. Dev. Comp. Immunol. 16, 103–110.

Rizki, T.M., Rizki, R.M., Bellotti, R.A. 1985. Genetics of a *Drosophila* phenoloxidase. Mol. Gen. Genet. 201, 7–13.

Robertson, C.W. 1936. The metamorphosis of *Drosophila melanogaster*, including an accurately timed account of the principle morphological changes. J. Morphol. 59, 351–399.

Rodrigues, N.P., Janzen, V., Forkert, R., Dombkowski, D.M., Boyd, A.S., Orkin, S.H., Enver, T., Vyas, P., Scadden, D.T. 2005. Haploinsufficiency of GATA-2 perturbs adult hematopoietic stem-cell homeostasis. Blood 106, 477–484.

Rosetto, M., Engstrom, Y., Baldari, C.T., Telford, J.L., Hultmark, D. 1995. Signals from the IL-1 receptor homolog, Toll, can activate an immune response in a *Drosophila* hemocyte cell line. Biochem. Biophys. Res. Commun. 209, 111–116.

Royet, J., Reichhart, J.M., Hoffmann, J.A. 2005. Sensing and signaling during infection in *Drosophila*. Curr. Opin. Immunol. 17, 11–17.

Rugendorff, A.E., Younossi-Hartenstein, A., Hartenstein, V. 1994. Embryonic origin and differentiation of the *Drosophila* heart. Roux's Arch. Dev. Biol. 203, 266–280.

Ruhf, M.L., Braun, A., Papoulas, O., Tamkun, J.W., Randsholt, N., Meister, M. 2001. The domino gene of *Drosophila* encodes novel members of the SWI2/SNF2 family of DNA-dependent ATPases, which contribute to the silencing of homeotic genes. Development 128, 1429–1441.

Ruppert, E.E., Carle, K.J. 1983. Morphology of metazoan circulatory systems. Zoomorphology 103, 193–208.

Russo, J., Dupas, S., Frey, F., Carton, Y., Brehelin, M. 1996. Insect immunity: Early events in the encapsulation process of parasitoid (*Leptopilina boulardi*) eggs in resistant and susceptible strains of *Drosophila*. Parasitology 112(Pt. 1), 135–142.

Samakovlis, C., Kimbrell, D.A., Kylsten, P., Engstrom, A., Hultmark, D. 1990. The immune response in *Drosophila*: Pattern of cecropin expression and biological activity. EMBO J. 9, 2969–2976.

Sasaki, K., Yagi, H., Bronson, R.T., Tominaga, K., Matsunashi, T., Deguchi, K., Tani, Y., Kishimoto, T., Komori, T. 1996. Absence of fetal liver hematopoiesis in mice deficient in transcriptional coactivator core binding factor beta. Proc. Natl. Acad. Sci. USA 93, 12359–12363.

Schmidt-Arras, D., Schwable, J., Bohmer, F.D., Serve, H. 2004. Flt3 receptor tyrosine kinase as a drug target in leukemia. Curr. Pharm. Des. 10, 1867–1883.

Schmucker, D., Clemens, J.C., Shu, H., Worby, C.A., Xiao, J., Muda, M., Dixon, J.E., Zipursky, S.L. 2000. *Drosophila* Dscam is an axon guidance receptor exhibiting extraordinary molecular diversity. Cell 101, 671–684.

Schneider, D.S., Hudson, K.L., Lin, T.Y., Anderson, K.V. 1991. Dominant and recessive mutations define functional domains of Toll, a transmembrane protein required for dorsal-ventral polarity in the *Drosophila* embryo. Genes Dev. 5, 797–807.

Schwaller, J., Parganas, E., Wang, D., Cain, D., Aster, J.C., Williams, I.R., Lee, C.K., Gerthner, R., Kitamura, T., Frantsve, J., Anastasiadou, E., Loh, M.L., et al. 2000. Stat5 is essential for the myelo- and lymphoproliferative disease induced by TEL/JAK2. Mol. Cell 6, 693–704.

Sears, H.C., Kennedy, C.J., Garrity, P.A. 2003. Macrophage-mediated corpse engulfment is required for normal *Drosophila* CNS morphogenesis. Development 130, 3557–3565.

Shivdasani, R.A., Fujiwara, Y., McDevitt, M.A., Orkin, S.H. 1997. A lineage-selective knockout establishes the critical role of transcription factor GATA-1 in megakaryocyte growth and platelet development. EMBO J. 16, 3965–3973.

Shrestha, R., Gateff, E. 1982. Ultrastructure and cytochemistry of the cell types in the larval hematopoietic organs and hemolymph of *Drosophila melanogaster*. Dev. Growth Differ. 24, 65–82.

Silverman, N., Maniatis, T. 2001. NF-kappaB signaling pathways in mammalian and insect innate immunity. Genes Dev. 15, 2321–2342.

Sinenko, S.A., Mathey-Prevot, B. 2004. Increased expression of *Drosophila* tetraspanin, Tsp68C, suppresses the abnormal proliferation of ytr-deficient and Ras/Raf-activated hemocytes. Oncogene 23, 9120–9128.

Sinenko, S.A., Kim, E.K., Wynn, R., Manfruelli, P., Ando, I., Wharton, K.A., Perrimon, N., Mathey-Prevot, B. 2004. Yantar, a conserved arginine-rich protein is involved in *Drosophila* hemocyte development. Dev. Biol. 273, 48–62.

Smith, P.R. 1986. Development of the blood vascular system in *Sabellaria cementarium* (Annelida, Polychaeta): An ultrastructural investigation. Zoomorphology 106, 67–74.

Sorrentino, R.P., Carton, Y., Govind, S. 2002. Cellular immune response to parasite infection in the *Drosophila* lymph gland is developmentally regulated. Dev. Biol. 243, 65–80.

Sorrentino, R.P., Melk, J.P., Govind, S. 2004. Genetic analysis of contributions of dorsal group and JAK-Stat92E pathway genes to larval hemocyte concentration and the egg encapsulation response in *Drosophila*. Genetics 166, 1343–1356.

Speck, N.A. 2001. Core binding factor and its role in normal hematopoietic development. Curr. Opin. Hematol. 8, 192–196.

Speck, N.A., Gilliland, D.G. 2002. Core-binding factors in haematopoiesis and leukaemia. Nat. Rev. Cancer 2, 502–513.

Srivastava, D., Cserjesi, P., Olson, E.N. 1995. A subclass of bHLH proteins required for cardiac morphogenesis. Science 270, 1995–1999.

Srivastava, D., Thomas, T., Lin, Q., Kirby, M.L., Brown, D., Olson, E.N. 1997. Regulation of cardiac mesodermal and neural crest development by the bHLH transcription factor, dHAND. Nat. Genet. 16, 154–160.

Steelman, L.S., Pohnert, S.C., Shelton, J.G., Franklin, R.A., Bertrand, F.E., McCubrey, J.A. 2004. JAK/STAT, Raf/MEK/ERK, PI3K/Akt and BCR-ABL in cell cycle progression and leukemogenesis. Leukemia 18, 189–218.

Stewart, M.J., Denell, R. 1993. Mutations in the *Drosophila* gene encoding ribosomal protein S6 cause tissue overgrowth. Mol. Cell. Biol. 13, 2524–2535.

Svensson, E.C., Tufts, R.L., Polk, C.E., Leiden, J.M. 1999. Molecular cloning of FOG-2: A modulator of transcription factor GATA-4 in cardiomyocytes. Proc. Natl. Acad. Sci. USA 96, 956–961.

Takata, K., Shimanouchi, K., Yamaguchi, M., Murakami, S., Ishikawa, G., Takeuchi, R., Kanai, Y., Ruike, T., Nakamura, R., Abe, Y., Sakaguchi, K. 2004a. Damaged DNA binding protein 1 in *Drosophila* defense reactions. Biochem. Biophys. Res. Commun. 323, 1024–1031.

Takata, K., Yoshida, H., Yamaguchi, M., Sakaguchi, K. 2004b. *Drosophila* damaged DNA-binding protein 1 is an essential factor for development. Genetics 168, 855–865.

Tang, H., Kambris, Z., Lemaitre, B., Hashimoto, C. 2006. Two proteases defining a melanization cascade in the immune system of *Drosophila*. J. Biol. Chem. 281, 28097–28104.

Tepass, U., Fessler, L.I., Aziz, A., Hartenstein, V. 1994. Embryonic origin of hemocytes and their relationship to cell death in *Drosophila*. Development 120, 1829–1837.

Tevosian, S.G., Deconinck, A.E., Cantor, A.B., Rieff, H.I., Fujiwara, Y., Corfas, G., Orkin, S.H. 1999. FOG-2: A novel GATA-family cofactor related to multitype zinc-finger proteins Friend of GATA-1 and U-shaped. Proc. Natl. Acad. Sci. USA 96, 950–955.

Ting, C.N., Olson, M.C., Barton, K.P., Leiden, J.M. 1996. Transcription factor GATA-3 is required for development of the T-cell lineage. Nature 384, 474–478.

Torok, I., Strand, D., Schmitt, R., Tick, G., Torok, T., Kiss, I., Mechler, B.M. 1995. The overgrown hematopoietic organs-31 tumor suppressor gene of *Drosophila* encodes an Importin-like protein accumulating in the nucleus at the onset of mitosis. J. Cell Biol. 129, 1473–1489.

Tsai, F.Y., Keller, G., Kuo, F.C., Weiss, M., Chen, J., Rosenblatt, M., Alt, F.W., Orkin, S.H. 1994. An early haematopoietic defect in mice lacking the transcription factor GATA-2. Nature 371, 221–226.

Tsang, A.P., Visvader, J.E., Turner, C.A., Fujiwara, Y., Yu, C., Weiss, M.J., Crossley, M., Orkin, S.H. 1997. FOG, a multitype zinc finger protein, acts as a cofactor for transcription factor GATA-1 in erythroid and megakaryocytic differentiation. Cell 90, 109–119.

Tsang, A.P., Fujiwara, Y., Hom, D.B., Orkin, S.H. 1998. Failure of megakaryopoiesis and arrested erythropoiesis in mice lacking the GATA-1 transcriptional cofactor FOG. Genes Dev. 12, 1176–1188.

Turner, J., Crossley, M. 1998. Cloning and characterization of mCtBP2, a co-repressor that associates with basic Kruppel-like factor and other mammalian transcriptional regulators. EMBO J. 17, 5129–5140.

Tzou, P., De Gregorio, E., Lemaitre, B. 2002. How *Drosophila* combats microbial infection: A model to study innate immunity and host-pathogen interactions. Curr. Opin. Microbiol. 5, 102–110.

Vidal, M., Cagan, R.L. 2006. *Drosophila* models for cancer research. Curr. Opin. Genet. Dev. 16, 10–16.

Vincent, S., Vonesch, J.L., Giangrande, A. 1996. Glide directs glial fate commitment and cell fate switch between neurones and glia. Development 122, 131–139.

Vogeli, K.M., Jin, S.W., Martin, G.R., Stainier, D.Y. 2006. A common progenitor for haematopoietic and endothelial lineages in the zebrafish gastrula. Nature 443, 337–339.

Waltzer, L., Bataille, L., Peyrefitte, S., Haenlin, M. 2002. Two isoforms of Serpent containing either one or two GATA zinc fingers have different roles in *Drosophila* haematopoiesis. EMBO J. 21, 5477–5486.

Waltzer, L., Ferjoux, G., Bataille, L., Haenlin, M. 2003. Cooperation between the GATA and RUNX factors Serpent and Lozenge during *Drosophila* hematopoiesis. EMBO J. 22, 6516–6525.

Wang, Q., Stacy, T., Binder, M., Marin-Padilla, M., Sharpe, A.H., Speck, N.A. 1996a. Disruption of the Cbfa2 gene causes necrosis and hemorrhaging in the central nervous system and blocks definitive hematopoiesis. Proc. Natl. Acad. Sci. USA 93, 3444–3449.

Wang, Q., Stacy, T., Miller, J.D., Lewis, A.F., Gu, T.L., Huang, X., Bushweller, J.H., Bories, J.C., Alt, F.W., Ryan, G., Liu, P.P., Wynshaw-Boris, A. 1996b. The CBFbeta subunit is essential for CBFalpha2 (AML1) function *in vivo*. Cell 87, 697–708.

Watson, F.L., Puttmann-Holgado, R., Thomas, F., Lamar, D.L., Hughes, M., Kondo, M., Rebel, V.I., Schmucker, D. 2005. Extensive diversity of Ig-superfamily proteins in the immune system of insects. Science 309, 1874–1878.

Watson, K.L., Konrad, K.D., Woods, D.F., Bryant, P.J. 1992. *Drosophila* homolog of the human S6 ribosomal protein is required for tumor suppression in the hematopoietic system. Proc. Natl. Acad. Sci. USA 89, 11302–11306.

Watson, K.L., Justice, R.W., Bryant, P.J. 1994. *Drosophila* in cancer research: The first fifty tumor suppressor genes. J. Cell Sci. Suppl. 18, 19–33.

Wildonger, J., Mann, R.S. 2005. The t(8;21) translocation converts AML1 into a constitutive transcriptional repressor. Development 132, 2263–2272.

Witzig, T.E., Kaufmann, S.H. 2006. Inhibition of the phosphatidylinositol 3-kinase/mammalian target of rapamycin pathway in hematologic malignancies. Curr. Treat Options Oncol. 7, 285–294.

Wood, W., Faria, C., Jacinto, A. 2006. Distinct mechanisms regulate hemocyte chemotaxis during development and wound healing in *Drosophila melanogaster*. J. Cell Biol. 173, 405–416.

Yagi, H., Deguchi, K., Aono, A., Tani, Y., Kishimoto, T., Komori, T. 1998. Growth disturbance in fetal liver hematopoiesis of Mll-mutant mice. Blood 92, 108–117.

Yamagata, T., Maki, K., Mitani, K. 2005. Runx1/AML1 in normal and abnormal hematopoiesis. Int. J. Hematol. 82, 1–8.

Yamagishi, H., Yamagishi, C., Nakagawa, O., Harvey, R.P., Olson, E.N., Srivastava, D. 2001. The combinatorial activities of Nkx2.5 and dHAND are essential for cardiac ventricle formation. Dev. Biol. 239, 190–203.

Yasothornsrikul, S., Davis, W.J., Cramer, G., Kimbrell, D.A., Dearolf, C.R. 1997. viking: Identification and characterization of a second type IV collagen in *Drosophila*. Gene 198, 17–25.

Yelon, D., Ticho, B., Halpern, M.E., Ruvinsky, I., Ho, R.K., Silver, L.M., Stainier, D.Y. 2000. The bHLH transcription factor hand2 plays parallel roles in zebrafish heart and pectoral fin development. Development 127, 2573–2582.

Yin, Z., Xu, X.L., Frasch, M. 1997. Regulation of the twist target gene tinman by modular cis-regulatory elements during early mesoderm development. Development 124, 4971–4982.

Zettervall, C.J., Anderl, I., Williams, M.J., Palmer, R., Kurucz, E., Ando, I., Hultmark, D. 2004. A directed screen for genes involved in *Drosophila* blood cell activation. Proc. Natl. Acad. Sci. USA 101, 14192–14197.

Zhang, W., Kadam, S., Emerson, B.M., Bieker, J.J. 2001. Site-specific acetylation by p300 or CREB binding protein regulates erythroid Kruppel-like factor transcriptional activity via its interaction with the SWI-SNF complex. Mol. Cell. Biol. 21, 2413–2422.

CHAPTER 12

Vascular Development
in the Zebrafish

Josette Ungos and **Brant M. Weinstein**

Contents

Abstract

The zebrafish has proven to be a powerful model organism for studying how the stereotypic and evolutionarily conserved vascular network is established during vertebrate development. The combination of genetic and experimental tools with high-resolution imaging has allowed rapid progress in our understanding of both the cellular and molecular mechanisms underlying the specification of endothelial progenitors as well as the differentiation and patterning of blood vessels into arterial-venous networks. Continued studies of vascular development in zebrafish promise to yield new insights into the developmental mechanisms regulating lymphangiogenesis and will undoubtedly reveal new information about the role of the endothelium in the patterning of various tissues. In this chapter, we discuss recent advances in vascular biology that have been achieved through use of the zebrafish.

Laboratory of Molecular Genetics, NICHD, NIH, Bethesda, Maryland

Advances in Developmental Biology, Volume 18
ISSN 1574-3349, DOI: 10.1016/S1574-3349(07)18012-1

1. INTRODUCTION

The vertebrate circulatory system is a highly organized network of arteries, capillaries, and veins that penetrates and supplies tissues and organs in the body with oxygen, nutrients, and cellular and humoral factors. It is the first functional organ system to develop during vertebrate embryogenesis, since embryonic growth and differentiation are critically dependent on the transport of oxygen, nutrients, and waste products through the early vasculature. The heart and blood vessels form a closed circulatory loop, with blood retained entirely within the vessels, except through leakage or hemorrhage. Arteries carry blood away from the heart to tissues, while veins generally return blood back to the heart. Blood vessels all have a similar histological structure. They are composed of two basic cell types, vascular endothelial cells (ECs) and vascular smooth muscle cells. The inner epithelial lining of blood vessels, adjacent to the lumen, is a thin, single-layered epithelium of ECs, while smooth muscle cells or pericytes surround the EC layer. In larger blood vessels, the inner endothelial lining is called the "intimal layer," and it is surrounded by a "medial layer" composed of multiple layers of vascular smooth muscle cells embedded in elastin-rich extracellular matrix. The medial layer is itself surrounded by an extracellular matrix-rich layer called the "adventitial layer." The muscular vascular smooth cell layer is highly developed in arteries. In larger blood vessels, the adventitia also contains nerves and nutrient capillaries that supply the muscular layer. In addition to the blood vessels of the circulatory system, vertebrates have an additional blind-ended vascular system responsible for draining fluids and macromolecules to maintain tissue homeostasis, the lymphatic system. Like blood vessels, lymphatic vessels are also lined by a single EC layer, but smaller lymphatic capillaries lack the surrounding smooth muscle or pericyte cells found in blood capillaries.

The zebrafish offers a number of advantageous features as a model organism for studying vascular development. Zebrafish embryos develop externally to the mother and are very small and optically clear, so noninvasive high-resolution light microscopic imaging methods can be used to visualize the finest details of developing blood vessels throughout the entire animal at every stage of its early development. Zebrafish are also amenable to forward-genetic analysis. Large numbers of adult fish can be maintained in a small space, and they have a short breeding cycle (every 1–2 weeks), have short generation time (3 months), and give large numbers of progeny (hundreds of eggs laid per clutch), all of which greatly facilitate large-scale genetic analysis. Furthermore, zebrafish embryos are small enough to receive sufficient oxygen by passive diffusion to survive and develop reasonably normally for a number of days in the complete absence of blood circulation. This makes it possible to screen for vascular mutants and accurately assess the vascular specificity of their effects.

A variety of experimental techniques and tools have been developed to exploit and amplify the intrinsic advantages. These include widely used tools and methods such as cell fate and lineage analysis, single-cell transplantation, microinjection of functional molecules, functional "knockdown" of specific genes by injection of morpholine antisense oligonucleotides, and transgenic expression of fluorescent tracers or functionally active genes. In addition to these more generally useful methods, a number of vascular-specific tools and resources have also been made available. High-resolution optical imaging of the vasculature can be accomplished via a number of different methods (Kamei et al., 2004). Confocal microangiography allows three-dimensional (3-D) reconstruction of the patent (open) vascularized spaces of living zebrafish embryos and larvae (Weinstein et al., 1995). Fluorescent dyes are microinjected into the circulation, optical "slices" of the fluorescent vasculature are collected using a confocal or multiphoton microscope, and the slices are then digitally reconstructed into 3-D reconstruction images (Fig. 1). Vascular imaging can also be accomplished via digital motion analysis, a method that uses successive frame subtraction of digitally collected light microscopic images, to generate "shifting vector" casts of the movement of circulating erythrocytes to delineate the functioning vasculature (Schwerte and Pelster, 2000). Transgenic lines have also been generated that express fluorescent proteins under the control of vascular-specific promoters (Motoike et al., 2000; Lawson and Weinstein, 2002; Cross et al., 2003). These lines permit very high-resolution dynamic imaging of vascular dynamics. They can be used to visualize vessels that have not yet

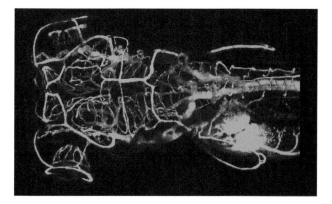

FIGURE 1   Microangiographic image of the cranial vasculature of a 4.5-day-old zebrafish embryo reveals vessels throughout the entire depth of the zebrafish head. Since the data collected are actually 3-D, complex microangiograpic data such as these are readily parsed to determine anatomical patterns (see Isogai et al., 2001, or access http://eclipse.nichd.nih. gov/nichd/Img/redirect.html, for wiring diagrams corresponding to this image).

begun to carry circulation, vascular endothelial cords that have not yet lumenized, and even single migratory angioblasts. Methods have been developed for very long-term multiphoton time-lapse imaging of developing vessels in these transgenics (Kamei and Weinstein, 2005). The capability to perform high-resolution fluorescence optical imaging of the entire vasculature of the zebrafish has also made it possible to assemble a highly detailed staged anatomical atlas of the developing vasculature of the embryonic and larval zebrafish, the most complete description of early vascular development available for any vertebrate (Isogai et al., 2001). This atlas has revealed striking conservation between the anatomical pattern of developing vessels in zebrafish and other vertebrates, including mammals, reinforcing the usefulness of the fish as a comparative model for vascular development.

The experimental accessibility and optical properties of the zebrafish are complemented by its robust genetic capabilities. Large-scale forward-genetic screens have been performed in the fish, and have already generated large numbers of different vascular mutants for study. Sequencing of the zebrafish genome is well under way, and reasonably dense genetic and radiation hybrid maps are available, making positional cloning of mutated genes increasingly straightforward. Methods for transgenesis have also become much more robust and efficient, although targeted integration is still not feasible. Together, the basic attributes of the zebrafish and the tools and resources available have made this an exceptional model for both high-resolution imaging and experimental analysis of developing blood vessels, and for large-scale genetic screening. Below, we describe some of the insights into vascular development that have resulted from zebrafish studies.

2. ENDOTHELIAL FATE SPECIFICATION AND VASCULOGENESIS

The earliest vertebrate blood vessels formed by *in situ* aggregation of individual angioblasts to form cellular cords and, then, lumenized endothelial tubes through a process termed vasculogenesis. In mammals, this process begins shortly after gastrulation with the formation of the blood islands in the yolk sac and angioblast precursors in the head mesenchyme (reviewed in Risau and Flamme, 1995; Rossant and Howard, 2002). In zebrafish, the first angioblasts arise from the lateral plate mesoderm (LPM), migrating to the trunk midline between the 10- and 15-somite stages where they coalesce to form the primary axial vessels of the trunk, the dorsal aorta and posterior cardinal vein (PCV). Early observations that primitive hematopoietic cells and developing endothelium arise in physical proximity to each other promoted the theory that these cell types share a common progenitor termed the "hemangioblast" (Sabin, 1917; Murray, 1932). The idea of a hemangioblast progenitor makes sense, given that the proper functioning of the circulatory system requires the parallel development of the blood and the vessels to transport the blood. A common progenitor would ensure that these different components would develop synchronously. Evidence for the existence of the hemangioblast came from *in vitro* studies showing that mouse embryoid bodies cultured in the presence of vascular endothelial growth factor (VEGF) give rise to individual cells capable of producing colonies composed of both hematopoietic and ECs, termed blast colony-forming cells (BL-CFC) (Choi et al., 1998). Further studies showed that cells isolated from the primitive streak of gastrulating mouse embryos could also produce progenitors displaying both hematopoietic and vascular potential, when cultured under conditions that support the growth of BL-CFCs (Huber et al., 2004). Although these results are compelling, they do not exclude the possibility that the BL-CFCs are actually multipotent cells whose full potential is not evident under the *in vitro* assay conditions used in these experiments.

The most definitive evidence to date for the existence of the hemangioblast comes from *in vivo* studies in zebrafish. Single-cell resolution fate maps of the marginal region of zebrafish early gastrula or blastula stage embryos were constructed by injecting caged fluorescein at the one-cell stage and using laser activation to uncage the dye in single blastula cells (Vogeli et al., 2006). They labeled superficial cells within the first five rows up from the margin, along the ventral side of early gastrula embryos, and found that 12.5% of the cells labeled at shield stage specifically gave rise to only endothelial and hematopoietic lineages. These cells were interspersed along the extent of the ventral mesoderm with cells that normally give rise to blood or to ECs alone. In studies performed on blastula stage embryos, they surveyed the entire margin of the embryo and found that only 2.9% of the cells labeled gave rise to both endothelial and blood cells. Many more cells gave rise to either ECs plus other nonblood cell types or blood cells plus other non-EC types, suggesting that most endothelial and blood cells do not arise from a common exclusive progenitor.

In the mouse, the VEGF-A receptor Flk1 (VEGFR2, KDR) is the earliest marker of angioblast precursors. Flk1-positive cells within the primitive streak migrate into the yolk sac to form the blood islands that give rise to both endothelial and hematopoietic progenitors. Flk1 also marks the angioblast precursors distributed throughout the head mesenchyme and other areas of the embryo. A targeted knockout of *Flk1* demonstrates a pivotal role for Flk1 in mouse hematopoiesis and vasculogenesis. Flk1-deficient embryos die between 8.5 and 9.5 dpc due to an early defect in hematopoietic and endothelial development. Yolk sac blood islands are absent, and organized blood vessels are not observed in the yolk sac or embryo at any stage of development (Shalaby et al., 1995). In zebrafish, there are two VEGFR2 (Flk1) orthologues, *kdra* and *kdrb*. Based on sequence, *kdrb* clusters together with human and mouse *kdr*, while *kdra* clusters with VEGFR-like homologues and appears more closely related to VEGFR1. However, based on loss-of-function studies, *kdra* appears to have evolved as the functional homologue of mammalian VEGFR2 (Covassin et al., 2006). Zebrafish *flk1* (*kdra*) mutants do not display the same early defects in vasculogenesis and hematopoiesis as seen in the mouse targeted knockout (Habeck et al., 2002; Covassin et al., 2006). Although *flk1/kdra* expression is significantly reduced, beginning at early somitogenesis stages, mutant embryos continue to express normal levels of other markers for hematopoiesis and vasculogenesis, like *scl, gata1, fli1*, and *vegf*. Microangiography in mutant embryos reveals that the large vessels formed before 1.5 dpf develop normally in the absence of functional *flk1*. Vascular defects only become apparent after 2 dpf in vessels that form via angiogenic sprouting from existing large vessels (Habeck et al., 2002). It is possible that other VEGFRs act earlier during vascular development or that there may be functional redundancy that compensates for the loss of *kdra*. Interestingly, following treatment of zebrafish embryos with the pan-VEGFR inhibitor SU5416, which blocks all three mammalian VEGFRs, the segmental arteries in the trunk are eliminated (Covassin et al., 2006). Similar results are obtained following morpholino-mediated knockdown of different combinations of VEGF ligands and VEGFRs (Covassin et al., 2006). These results suggest that loss of VEGF signaling may disrupt formation of the dorsal aorta and, therefore, may be affecting vasculogenesis. However, angioblasts appear to be specified normally since ECs are still found aggregated along the midline in the normal position of the axial vessels.

Along with Flk1, expression of the stem cell leukemia transcription factor (*Scl*) delineates the early precursors of the developing endothelial lineage (reviewed in Patterson et al., 2006). Loss-of-function studies in mouse suggest that while *Scl* is essential for hematopoiesis, it is only required later during vascular development (Robb et al., 1995). This is in contrast to studies in zebrafish, where overexpression of *scl* mRNA promotes the differentiation of mesodermal precursors toward endothelial and hematopoietic fates and is able to partially rescue blood and endothelial gene expression in *cloche* mutant embryos, suggesting a role for *scl* during hemangioblast specification (Gering et al., 1998; Liao et al., 1998). Loss-of-function studies have been carried out in zebrafish, using morpholino-mediated knockdown of Scl (Dooley et al., 2005; Patterson et al., 2005). In *scl* morphants, both primitive and definitive hematopoietic lineages are lost; however, expression of early hematopoietic genes is unaffected. Furthermore, although the dorsal aorta fails to form properly as evidenced by loss of numerous arterial-specific markers, angioblasts are still specified. These results suggest that zebrafish *scl* is unlikely to participate in specification of the hemangioblast; however, it may still be important for vasculogenesis. Results from morpholino-mediated knockdown of the LIM domain protein Lmo2 are very similar to those observed during loss of *scl* function, suggesting that Lmo2 and *scl* may act together during hematopoiesis and vascular development (Patterson et al., 2006).

Mutant and morpholino-mediated knockdown analysis in zebrafish more clearly points toward a role for E26 transformation-specific sequence (ETS) factors as regulators of vasculogenesis in zebrafish (Pham et al., 2006; Sumanas and Lin, 2006). ETS factors comprise a large family of transcriptional regulators sharing a highly conserved winged-helix-turn-helix DNA-binding domain. A number of different ETS family members have been shown to be expressed in ECs as well as putative hemangioblast cells (Pardanaud and Dieterlen-Lievre, 1993; Spyropoulos et al., 2000), and a number of *ex vivo* and *in vitro* studies (Wernert et al., 1999; Nakano et al., 2000) have pointed toward a role for ETS factors in angiogenesis. However, targeted gene knockdowns of individual ETS factors in mice have failed to establish a requirement for them during early vasculogenesis (reviewed in Bartel et al., 2000). A novel ETS1-related protein (Etsrp) that is specifically expressed in ECs was identified in zebrafish and shown to be required for proper vasculogenesis (Pham et al., 2006; Sumanas and Lin, 2006). Expression of *etsrp* appears in bilateral stripes within the lateral mesodermal domain in both the anterior and posterior regions of the embryo and persists later in the main axial, head and intersegmental vessels (ISVs). Loss of *etsrp* function produces defects in both vascular morphogenesis and angiogenesis. Sumanas and Lin (2006) reported almost complete loss of ECs in *etsrp* morphants. However, Pham et al. (2006) find using a null mutant of *etsrp* that although the primary vasculogenic vessels fail to undergo proper morphogenesis, there is no reduction in the initial numbers of ECs in the absence of Etsrp function. Instead, they find that a more complete vascular defect requires a simultaneous loss of multiple ETS factors. One possible explanation for this discrepancy may be the difference in transgenic reporters used to examine vessels in Etsrp-deficient animals (fli-EGFP in Pham et al. vs flk1-EGFP in Sumanas and Lin). The expression of *flk1*, but not *fli1*, is strongly reduced in Etsrp-deficient animals (Pham et al., 2006). The differing results may also reflect differences between the experiments such as developmental stage at which marker genes were analyzed.

In addition to *etsrp*, Pham et al. (2006) assayed the function of the other three identified zebrafish vascular-specific ETS factors, *fli1*, *fli1b*, and *ets1*. By injecting combinations of morpholinos targeting the four ETS genes, they uncovered synergy between ETS factors. Simultaneous injection of all four morpholinos resulted in near-complete loss of vascular marker expression (*flt4, vecdn, flk1, plxnD1, efnb2*), and a strong reduction in the number of fli-EGFP-positive ECs was observed, suggesting that loss of marker expression resulted from loss of expressing cells rather than simple downregulation of expression. Together, these results show that multiple ETS factors function cooperatively during establishment and maintenance of the zebrafish vasculature, and establish ETS factors as the earliest known regulators of a hierarchy of transcription factors governing endothelial specification. The failure to uncover such a role for ETS factors during early vascular development in mice is likely due to the redundant requirement of multiple factors within the vasculature, making it difficult to analyze the overall loss of ETS signaling. The ease of performing combinatorial loss-of-function experiments in zebrafish highlights the potential power of this model system for furthering our understanding of vascular development.

3. ARTERIAL-VENOUS SPECIFICATION

The vasculature is divided into two separate but interconnected networks of arterial and venous blood vessels. The arterial and venous networks are both morphologically and functionally very distinct from one another. Arteries transport oxygenated blood at high pressure from the heart, while veins serve as conduits for low-pressure blood return. The acquisition of structural differences between arteries and veins was historically attributed mainly to physiologic forces such as direction and pressure of blood flow. Although cellular differences between the ECs that line arteries and veins were observed long ago, until recently little was known about the complex genetic pathways responsible for these differences. The observation that *EphrinB2* (*efnb2*, in zebrafish), a member of a family of membrane-bound ligands, is expressed specifically in arterial, but not in venous, ECs before the onset of circulation in mouse embryos was the first recognition of a molecular distinction between arteries and veins (Wang et al., 1998). Conversely, *Ephb4*, a gene that encodes the putative receptor for ephrinB2, was found at much higher levels in veins than in arteries.

Recent studies in the zebrafish have made important contributions to our understanding of how and when the acquisition of arterial versus venous identity occurs. In zebrafish, angioblasts arise in the LPM and migrate to the midline where they form the dorsal aorta and PCV of the trunk. As in the mouse, these vessels display arterial- and venous-specific marker expression as soon as they begin to assemble at the trunk midline, well before the onset of circulation. In fact, cell-tracing experiments show that individual angioblasts in the LPM labeled with a fluorescent lineage tracer give rise to only arterial or venous ECs, not both cell types (Zhong et al., 2000, 2001). These results indicate that the arterial-venous fate decision begins to be made very early, before angioblasts migrate to the midline and coalesce into nascent vascular cords. They further suggest that this early fate choice is genetically programmed rather than determined by flow dynamics or local tissue needs. Additional studies in the zebrafish have defined a molecular pathway leading to the acquisition of arterial fate during embryogenesis. This pathway includes well-known signaling molecules including sonic hedgehog (Shh), Vegf, Notch, mitogen-activated protein kinase (MAPK), and phosphatidylinositol-3 kinase (PI3K), some of which had already been shown to play other roles in ECs. By using mRNA and morpholino microinjections in combination with mutants, transgenics, and drug-treated embryos, it has been possible to affect multiple signaling pathways simultaneously and uncover the genetic hierarchy responsible for arterial differentiation (Fig. 2).

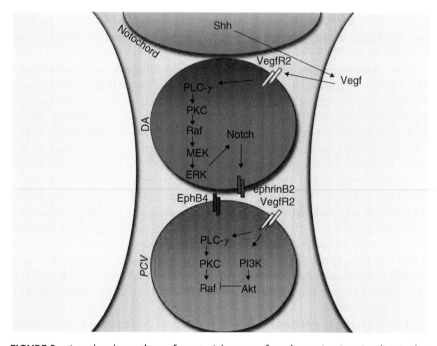

FIGURE 2   A molecular pathway for arterial-venous fate determination. Studies in the zebrafish have shown that Vegf acts downstream of Shh and upstream of the Notch pathway to determine arterial cell fate. Loss of Notch, Vegf, or Shh signaling results in loss of arterial identity, while exogenous activation or overexpression of these factors causes ectopic expression of arterial markers. The *plcg1* gene lies downstream of Vegf signaling, with ERK signals involved downstream of *plcg1*. PI3K signaling is also downstream of Vegf signaling. For detailed information on pathways in arterial-venous fate determination, see Lamont and Childs (2006).

Hedgehog signaling lies at the top of this genetic hierarchy. Shh is expressed by the notochord, a midline structure juxtaposed to the dorsal aorta during its assembly. Loss of Shh signaling through genetic mutations in hedgehog signaling genes (e.g., *sonic you, you too* mutants) or chemical inhibition (cyclopamine) results in a failure to properly form the dorsal aorta (Lawson et al., 2002). The ability of Shh to regulate dorsal aorta formation and arterial marker expression in particular is mediated by Vegf, since overexpression of *vegf* mRNA restores expression of arterial markers in the absence of Shh signaling. Like Shh, Vegf is required for arterial fate. *Vegf* is expressed by the somites (future trunk muscle) adjacent to both the notochord and dorsal aorta, and its expression in the somites is induced by and dependent on Shh signals from the notochord. Morpholino-mediated knockdown of Vegf results in a loss of arterial-specific marker expression from the dorsal aorta, while injection of *vegf* mRNA results in ectopic expression of arterial markers in normally venous vessels such as the PCV. Injection of *vegf* mRNA resulted in induction of arterial marker expression, even in mutants lacking hedgehog signaling, showing that Vegf acts downstream of hedgehog signaling (Lawson et al., 2002).

The Notch signaling pathway is also required for arterial fate determination, downstream of both hedgehog and VEGF signaling (Fig. 3). Interfering with Notch activity via microinjection of dominant-negative suppressor of hairless [Su(H)], a downstream mediator of Notch signaling, or use of Notch signaling-deficient *mindbomb* (*mib*) mutants (Itoh et al., 2003), results in loss of arterial-specific gene expression (Lawson et al., 2001). Conversely, activation of Notch signaling by ectopic expression of the Notch intracellular domain leads to suppression of venous-specific marker expression. Molecular epistasis experiments were used to show that Notch acts downstream of VEGF signaling. In wild-type animals, overexpression of *vegf* mRNA results in ectopic expression of arterial markers, as noted above; but in *mib* mutant embryos, Vegf does not induce artery marker expression. In contrast, ubiquitous or vascular-specific local ectopic activation of Notch is capable of inducing ectopic arterial marker expression in either the presence or absence (via morpholino knockdown) of VEGF signaling.

The signaling pathways downstream of Notch have also been explored in zebrafish studies. Using forward-genetic screening of transgenic zebrafish, a mutation was identified in the zebrafish homologue of phospholipase C γ-1 (*plcg1*), a known downstream effector of many receptor tyrosine kinases required for normal VEGF function (Lawson et al., 2003). Like *vegf* morphants, *plcg1* mutant embryos display specific defects in artery formation and show loss of artery-specific expression of markers. Overexpression of *vegf* mRNA normally results in upregulation of artery-specific markers, but injection of *vegf* mRNA into a *plcg1* mutant

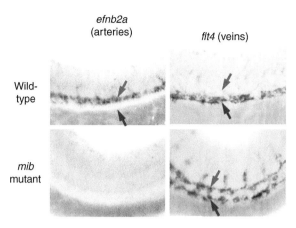

efnb2a
(arteries)

flt4 (veins)

Wild-
type

mib
mutant

FIGURE 3   Reduction in Notch signaling in zebrafish embryos perturbs arterial-venous identity. *In situ* hybridization of the trunk dorsal aorta (red arrows) and cardinal vein (blue arrows) in 25-somite stage Notch-deficient *mindbomb* (*mib*^{ta52b}) mutant and wild-type sibling zebrafish embryos. In wild-type animals, *ephrinB2a* (*efnb2a*) expression is apparent in the dorsal aorta but not the cardinal vein (upper left), but in Notch-deficient *mib*^{ta52b} mutant embryos, *efnb2a* expression is absent (lower left). In contrast, *flt4* expression is restricted to the cardinal vein in wild-type animals by the 25-somite stage (upper right), while in *mib*^{ta52b} mutant embryos, *flt4* expression persists within both the cardinal vein and the dorsal aorta (lower right). All panels show lateral views of the mid-trunk, dorsal up, and anterior to the left. Figure modified from Lawson et al. (2001).

background fails to elicit induction of arterial markers, suggesting that this gene acts downstream from *vegf*. Mutations in *gridlock* (*grl/hey2*), a member of the Hairy-related transcription factor family reported to act downstream of Notch, fail to establish trunk circulation due to defects in formation of the dorsal aorta (Weinstein et al., 1995; Zhong et al., 2000; Zhong et al., 2001). The expression of *grl* is restricted to arterial ECs, and overexpression of *grl* mRNA can suppress expression of venous markers. Injection of *grl* morpholinos results in partial to complete loss of dorsal aorta, with concomitant expansion of the PCV in a dosage-dependent fashion (Zhong et al., 2001), suggesting that *grl* is somehow involved in arterial/venous fate decisions. However, the exact role of *grl* downstream of Notch signaling during arterial specification remains unclear, given that *grl* expression is not downregulated in *mib* mutants or in embryos injected with a dominant-negative Su(H) construct (Lawson et al., 2001).

The ability to perform large-scale screening in the zebrafish has also been used for small-molecule screens to identify chemical modifiers of vascular development, including compounds that suppress the *grl* mutant phenotype and potentially act downstream in arterial-venous development (Peterson et al., 2004; Hong et al., 2006). Out of 5000 small molecules screened, two structurally related compounds were found that suppressed the *grl* phenotype, restoring circulation to the trunk and tail (Peterson et al., 2004). The compounds, GS4012 and GS3999, represent a novel class of molecules not previously implicated in vascular development. VEGF expression is increased in embryos exposed to GS4012, suggesting that GS4012 might compensate for the loss of *grl* function by upregulating VEGF, which is sufficient to bypass the requirement for *grl* (Peterson et al., 2004). An additional structurally distinct suppressor of the *grl* phenotype, GS4898, was identified in a separate chemical screen of 7000 compounds (Hong et al., 2006). GS4898 inhibits activation of Akt, a downstream effector of PI3K signaling, while promoting activation of the extracellular signal-related kinase 1 and 2 (ERK1/2). These results led Hong et al. to propose that PI3K, a known downstream mediator of VEGF signaling, promotes venous specification by inhibiting the arteriogenic phospholipase C gamma/extracellular regulated *kinase* (PLCg/ERK) branch of the VEGF pathway (Fig. 2). In support of this idea, known inhibitors of PI3K signaling suppress the vascular defects in *grl* mutants, while expanding the population of ECs that show labeling with antibodies specific for activated ERK. Conversely, inhibition of mitogen activated protein kinase kinase (MEK), an upstream activator of ERK, leads to loss of dorsal aorta and expansion of the cardinal vein. Hong et al. confirmed their results from chemical treatments by genetically inhibiting AKT signaling. Following coinjection of green fluorescent protein (GFP) and a dominant-negative AKT (dn-AKT) construct, GFP-positive cells were preferentially localized to the dorsal aorta, whereas GFP-positive cells were generally restricted to the vein when embryos were injected with a constitutively activated AKT (myr-AKT).

Together, the experimental and genetic tools available in the fish have made it possible to uncover genes involved in arterial-venous fate determination and assemble an ordered genetic pathway responsible for the arterial differentiation during embryogenesis. Experimental studies performed using other species have confirmed the involvement of many of the signaling pathways implicated in arterial-venous development in the zebrafish studies, notably VEGF and Notch signaling [for a detailed review of relevant zebrafish and nonzebrafish studies of arterial-venous differentiation, see Torres-Vazquez et al. (2003)]. Further work in the zebrafish will undoubtedly lead to the identification of additional players in this important vascular cell fate decision.

4. LYMPHATIC DEVELOPMENT

In addition to the blood (circulatory) vascular system with its two separate but interconnected networks of arterial and venous vessels, vertebrates possess an additional vascular network, the lymphatic vasculature.

The blood and lymphatic vascular systems serve complementary functions, but the lymphatic vasculature is structurally distinct from the blood vasculature. Like the blood vascular system, the lymphatic vasculature is a complex branched system of endothelium-lined vessels that ramifies through virtually every organ and tissue in the body (the brain being the one notable exception). Unlike the closed circulatory system, however, the lymphatic system is an open-ended, unidirectional system. Lymphatic vessels maintain fluid homeostasis by collecting fluid and macromolecules that leak from blood vessels into small lymphatic capillaries. These capillaries drain into larger lymphatic collecting vessels, eventually emptying lymph back into the blood circulation at several evolutionarily conserved drainage points. In addition to their role in fluid homeostasis, the lymphatic vessels are critical for white blood cell transport and immunity, fat absorption, and function as a key route for the spread of many types of metastatic tumors. Identification of molecular markers specific for lymphatic vessels and the finding that malignant tumors may directly promote lymphangiogenesis has spurred renewed interest in both normal and pathological development of the lymphatic vasculature (Saharinen et al., 2004; Oliver and Alitalo, 2005; Cueni and Detmar, 2006).

The embryonic origins of the ECs that make up the lymphatic system have been somewhat controversial. The most widely accepted model for early lymphatic development was first proposed by Florence Sabin (1902). Based on ink injection experiments, she proposed that the two primitive jugular lymph sacs originate from ECs that bud from large veins early during development. The peripheral lymphatic vessels then form by centrifugal sprouting from these primary lymph sacs. While studies have provided support for this model, evidence has also been obtained supporting an alternative model proposed by Huntington and McClure (1910) that lymphatic endothelium arises *de novo* from mesenchyme. Recent studies have begun to shed light on the molecular mechanisms responsible for development of the lymphatic system. In mammals, the expression of the *Prox1* transcription factor, initially restricted to a subpopulation of embryonic venous ECs, is required for differentiation of lymphatic endothelium and maintenance of lymphatic endothelial sprouting (Wigle and Oliver, 1999). As development proceeds, these Prox1-positive cells appear to bud from the cardinal vein, and migrate and coalesce to give rise to the lymphatic jugular sacs. This migration was shown to be critically dependent on the presence of VEGF-C, a member of the VEGF family of ligands. In mice deficient for VEGF-C, Prox1-expressing lymphatic ECs (LECs) form but fail to migrate toward the lymph sacs (Karkkainen et al., 2004). VEGF-C specifically binds to its high-affinity tyrosine kinase receptor VEGFR3 (see Jussila and Alitalo, 2002 for review). During mouse embryogenesis, the pattern of expression of VEGFR3 (Flt-4) coincides as well with Sabin's model of lymphatic development. VEGFR3 is first expressed in a subset of blood vascular EC and, subsequently, becomes restricted to lymphatic

EC (Kaipainen et al., 1995; Oh et al., 1997). Although these results support Sabin's original proposal that lymphatic vessels arise from veins, this model had not been directly tested *in vivo* and other evidence has suggested that lymphatic vessels can arise independently from mesenchyme (Ny et al., 2005; Wilting et al., 2006).

There has been conflicting data on whether zebrafish have a true lymphatic vascular system. Although a "secondary vessels" had been described in a number of different teleost fish species, it was reported that these vessels differed from authentic lymphatic vessels in a number of ways, notably in the presence of direct connections to the arterial blood vascular system (these are entirely absent from mammalian lymphatic vessels). However, two studies definitively established the existence of a *bona fide* lymphatic vascular system in the zebrafish (Kuchler et al., 2006; Yaniv et al., 2006). Histology, electron microscopy, and confocal imaging of $Tg(fli1:EGFP)^{y1}$ fish showed that zebrafish possess a third trunk axial vascular tube (in addition to the dorsal aorta and cardinal vein) that appears to correspond to the lymphatic thoracic duct, the major trunk lymphatic vessel found in mammals and other vertebrates. *Prox1* and *neuropilin2*, two genes that mark mammalian lymphatic endothelium, are both expressed in the zebrafish thoracic duct. Morpholino-mediated knockdown of the Vegf-C and Prox1 leads to loss of the thoracic duct, without significant effects on the formation of adjacent blood vessels. These results indicate that zebrafish lymphatics express genes known to mark mammalian lymphatics, and require the function of genes necessary for mammalian lymphatic development. Additional experiments demonstrated that zebrafish lymphatics also share functional characteristics of mammalian lymphatic vessels (Fig. 4). Injection of dyes into either the blood vascular system or the lymphatic vessels showed that they are separate networks of vessels, with each bearing remarkable similarity to the anatomical pattern of blood or lymphatic vessels described in other species. Importantly, the lymphatic vessels lacked any connections with arteries and connected only with the venous vasculature, later in development through a conserved drainage point, as in mammals and other vertebrates. One of the most important and conserved functional features of lymphatic vessels is their ability to take up and drain fluid and macromolecules from surrounding tissues. Dye injected subcutaneously into the tail of a developing zebrafish larva was rapidly taken up and transported through the lymphatic (but not the blood) vessels, demonstrating that they carry out this important basic function in a manner analogous to mammalian lymphatics.

Taking advantage of the optical clarity of the developing zebrafish, Yaniv et al. (2006) performed multiphoton time-lapse imaging of transgenic zebrafish to directly examine the embryonic origins of lymphatic endothelium. Using Tg (fli:nEGFP)^{y7} transgenic zebrafish in which EC

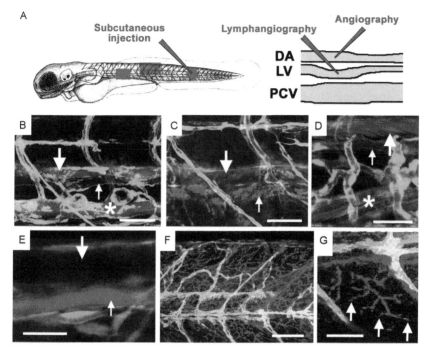

FIGURE 4   Functional characterization of zebrafish lymphatic vessels. (A) Methods used to image blood or lymphatic vessels. The red box on the fish drawing (modified from Kimmel et al., 1995) shows the approximate region of the trunk imaged in panels B, C, D, E, and G. The blue box on the drawing shows the approximate region of the trunk imaged in panel F. (B) Angiography of a 14-dpf *Tg(fli1:EGFP)^{y1}* zebrafish (green) injected with fluorescent microspheres (red), labeling dorsal aorta (large arrow) and PCV (asterisk) but not lymphatic thoracic duct (small arrow). (C) Lymphangiography of 3-week *Tg(fli1:EGFP)^{y1}* zebrafish (green) injected with fluorescent microspheres (red), labeling thoracic duct (small arrow) but not dorsal aorta (large arrow). (D) Time-averaged confocal image of a 7-dpf *Tg(fli1:EGFP)^{y1}* (green) and *Tg(gata1:dsRed)* (red) double transgenic animal, showing red fluorescence in the dorsal aorta (large arrow) and cardinal vein (asterisk) but not lymphatic thoracic duct (small arrow). (E–G) Confocal imaging of an 18-dpf *Tg(fli1:EGFP)^{y1}* zebrafish (green) injected subcutaneously with 2 Md rhodamine-dextran (red). (E) Subcutaneously injected rhodamine-dextran drains into the thoracic duct (small arrow) but does not label the adjacent dorsal aorta (large arrow). (F) Numerous rhodamine-dextran labeled vessels (red) are visible between the blood vessels (green). (G) Higher magnification image of blind-ended rhodamine-dextran labeled vessels. Scale bars = 20μm (E), 50μm (B,C,D,G), and 100μm (F). Figure modified from Yaniv et al. (2006).

nuclei were marked with GFP, they tracked the origins of the ECs populating the developing thoracic duct and showed that their precursors could be traced back to primitive embryonic veins, including the trunk parachordal vessels and the PCV that the parachordals themselves sprout from. These results provide direct *in vivo* evidence to support Sabin's century-old model (Sabin, 1902) proposing a venous origin for primitive lymphatic vessels. Although the alternative model that LECs develop from mesenchyme independent of veins (Huntington and McClure, 1910) may still apply to subpopulations of lymphatic vessels not examined in this study, it is clear that this model does not hold true for the zebrafish thoracic duct since every cell that was analyzed was traceable back to the parachordal vessels. Further *in vivo* imaging studies in zebrafish should help to determine how other vessels within the lymphatic network develop.

Together, these results establish the zebrafish as an important new model organism for genetic and experimental dissection of lymphatic development. Combinations of *in vivo* imaging, functional analyses, and the opportunity to perform genetic screening should allow for rapid progress in our understanding of lymphangiogenesis.

5. BLOOD VESSEL LUMEN FORMATION

In order to form a functional circulatory system, ECs that are lining blood vessels must develop a continuous luminal space. Although there are many possible ways to form tubes (Lubarsky and Krasnow, 2003), early observations hinted at a role for intracellular vacuoles during vascular lumen formation (Sabin, 1902; Downs, 2003). Cytoplasmic vesicles have been noted in ECs during angiogenesis *in vivo*, with the vesicles appearing to be reduced in number and increased in size as vessels develop. Similar observations were later made *in vitro* during some of the first EC culture experiments. Folkman and Haudenschild (1980) reported "longitudinal vacuoles" within ECs cultured in 3-D collagen or fibrin matrices. They also noted that these vacuoles appeared to traverse cell boundaries and make connections with neighboring cells. Since these initial observations, a number of studies (Davis et al., 2002) have verified and extended these findings, and led to a proposed model for lumen formation via intracellular vacuolation and intercellular fusion of endothelial vacuole (Fig. 5). Despite compelling *in vitro* data, however, there has been no evidence that this mechanism is used actually in endothelial lumen formation *in vivo*.

Kamei et al. (2006) took advantage of the optical clarity of the developing zebrafish and availability of methods for long-term time-lapse imaging (Kamei and Weinstein, 2005) to examine the mechanism of endothelial tube assembly *in vivo*. They used two-photon time-lapse imaging of *Tg(fli1:EGFP-cdc42wt)^{y48}* transgenic embryos to look at the dynamics of endothelial lumen formation *in vivo*. In these embryos, the cdc42wt-EGFP fusion protein preferentially localizes to pinocytic

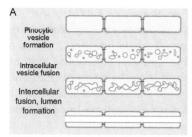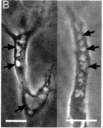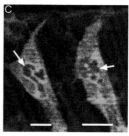

FIGURE 5   Endothelial tubes assemble from intracellular vacuoles *in vivo*. (A) A model for vascular lumen formation by intracellular and intercellular fusion of endothelial vacuoles. The diagram shows the mechanistic sequence of steps proposed to lead to intercellular lumen formation: intracellular vesicle formation, intracellular vesicle fusion, and finally intercellular merging of vacuolar compartments and lumen formation. (B and C) Visualization of vacuoles in endothelial cells *in vitro* and *in vivo*. (B) High-resolution light microscopic imaging of vacuoles (arrows) in EC in 3-D *in vitro* culture. (C) High-resolution multiphoton imaging of growing ISVs in *Tg(fli1:EGFP-cdc42wt)^{y48}* transgenic zebrafish reveals similar vacuolar structures (arrows). Images from Kamei et al. (2006); see this reference for time-lapse movies showing the dynamics of endothelial lumen assembly from vacuoles *in vivo*.

vacuolar structures as shown previously in cultured ECs (Bayless and Davis, 2002), while use of the fli1 promoter allows specific expression of the fusion protein within zebrafish ECs (Lawson and Weinstein, 2002). They focused on the simple, reiterated network of angiogenic trunk ISVs that form from three ECs that emerge from the dorsal aorta and extend as a chain along the intersomitic boundaries (Childs et al., 2002; Isogai et al., 2003). Time-lapse imaging reveals that *in vivo* vacuolar dynamics are very similar to what has been described in cultured ECs (Kamei et al., 2006). Vacuoles emerge, disappear, and coalesce to form larger vacuoles on a timescale of minutes, and eventually, the merging vacuoles enlarge to create a luminal space that occupies most of the cell (Fig. 5). In order to examine how vacuolar compartments within different ECs become linked to create a common endothelial tube, they injected nanometer-sized fluorescent beads (quantum dots or Qdots) into the dorsal aorta and examined the transfer of Qdots to adjacent developing ISVs. They observed serial transfer of the Qdots from the dorsal aorta, first, into the vacuolar lumens of the proximal ISV EC and, next, into the more distal ECs. These results demonstrate that the continuous luminal space in ISVs emerges from the fusion of preformed intracellular vacuolar compartments in neighboring ECs. Taken together, their data support the previously proposed model for vascular lumen formation by intracellular and intercellular fusion of endothelial vacuoles.

Whether a similar mechanism also applies to other vessel types, like the larger caliber vessels that form from larger aggregates of ECs, remains to be determined. Analysis of transverse sections of $Tg(flk1:EGFP)^{s843}$ embryos suggests that these vessels may form through cord hollowing (Jin et al., 2005). Angioblasts initially aggregate at the midline to form a cord-like structure without any detectable cell–cell junctions. Within hours, cell–cell junctions become apparent and the cells begin to elongate, and subsequently, a lumen appears with four to six angioblasts around its circumference. Parker et al. (2004) make similar observations but also note that, prior to morphological changes, substantial refinement of the tight junctions between angioblasts occurs. They further show that morpholino-mediated knockdown of EGF-like domain 7 (Egfl7), a putative secreted factor that is endothelial specific, results in maintenance of tight junctions and failure of angioblasts to separate. Although these results do not provide definitive evidence that the large axial vessels form tubes through a cord hollowing, this seems the most probable mechanism. It will be interesting to see whether a mechanism analogous to that observed by Kamei et al. (2006) in ISVs also applies to lumen formation within large vessels. In this case, exocytosis of vacuoles into a common space bounded by multiple ECs would contribute to the formation of a common lumenal space sealed off from the rest of the extracellular environment.

6. PATTERNING AND GUIDANCE

Although the vasculature is a highly complex and intricate network of vessels, its overall organization is remarkably stereotyped both between individuals of a given species and across species boundaries. Larger vessels form in defined positions and make predictable interconnections. Much of the recent progress that has been made toward elucidating the mechanisms underlying vessel patterning comes from studies of the developing trunk vasculature. The anatomy of the trunk vasculature is relatively simple, highly reproducible from animal to animal, and in its basic plan, is conserved across the vertebrate phyla, including zebrafish (Fig. 6A). All vertebrates possess longitudinal axial vessels (dorsal aorta and cardinal vein) that form via vasculogenesis, the coalescence of lateral mesoderm-derived angioblasts at the embryonic midline (Risau and Flamme, 1995). There is also a conserved network of secondary vessels, including the reiterated series of ISVs that form at the vertical myotomal boundaries between somites and the longitudinal parachordal vessels flanking the notochord. Multiphoton time-lapse imaging of $Tg(fli1:EGFP)^{y1}$ transgenic zebrafish embryos (Lawson and Weinstein, 2002) has revealed the detailed dynamics of assembly of the trunk angiogenic vascular network (Isogai et al., 2003). Angiogenic sprouts emerge from axial vessels in two spatially and temporally distinct steps (Fig. 6B). Bilateral sprouts emerge from the dorsal aorta close to intersomitic

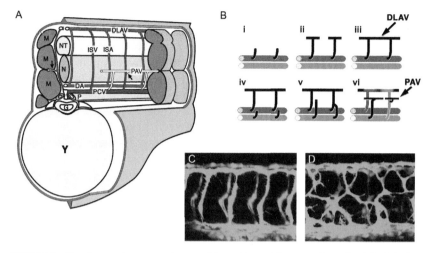

FIGURE 6   Trunk vascular network assembly and its guidance in the zebrafish. (A) The anatomy of the zebrafish trunk and its blood vessels by ~3 days postfertilization. At this stage, there is active flow through the dorsal aorta (DA), posterior cardinal vein (PCV), and most intersegmental arteries (ISA) and intersegmental vessels (ISV). The ISA and ISV are linked together dorsally via paired dorsal longitudinal anastomotic vessels (DLAV). All of these vessels are shown relative to adjacent tissues and structures in the mid-trunk including the gut (G), myotomes (M), notochord (N), neural tube (NT), left pronephric duct (P), and yolk mass (Y). In addition to the functioning vessels noted above, parachordal veins (PAV) run longitudinally to either side of N, along the horizontal myoseptum. Anterior is to the left and above the plane of the page, and dorsal is up. (B) Schematic diagram illustrating steps leading to assembly of the trunk angiogenic vascular network. For clarity, the diagram shows the vessels on only one side of the trunk. (B.i) Primary sprouts emerge bilaterally exclusively from the dorsal aorta (red). (B.ii) Primary sprouts grow dorsally, branching cranially and caudally at the level of the dorsal-lateral roof of the NT. (B.iii) Branches interconnect on either side of the trunk to form two DLAV. (B.iv) Secondary sprouts begin to emerge, exclusively from the PCV (blue). (B.v) Some secondary sprouts connect to the base of primary segments, while others do not. (B.vi) Primary segments with patent connections to secondary segments become intersegmental veins (blue), while primary segments that remain connected only to the dorsal aorta become ISA (red). Most of the secondary sprouts that do not connect to primary segments serve instead as ventral roots for the PAV. (C and D) Blood vessels in the mid-trunk of control morpholino (C) or plxnD1 morpholino (D) injected 48-hpf Fli1-EGFP transgenic embryos. In control morpholino-injected animals, ISVs extend along the boundaries between somites, avoiding the semaphorin-rich central regions (C). In animals deficient in plxnD1, intersegmental vessels sprout, branch, and grow without regard to somitic boundaries (D). Anterior is to the left, and dorsal up in all panels. Panels A and B modified from Isogai et al. (2003); panels C and D modified from Torres-Vazquez et al. (2004); see these references for further details.

boundaries and track along the intersomitic boundaries to generate a primary vascular network. This is followed by the emergence of vein-derived secondary sprouts that interconnect with the primary network to form a functional vasculature. Assembly of this network of vessels occurs in the absence of circulatory flow, supporting the idea that "hardwired" genetic cues are the primary force driving early vascular patterning. However, hemodynamic forces do appear to be important for refining the pattern of vessel interconnections between primary and secondary networks and for defining the final arterial or venous identity of vessels within the network.

Local interactions with neighboring tissues play an important role in directing vessel growth and patterning. In the trunk, the assembly of the dorsal aorta is dependent on the adjacent notochord, an important embry-onic axial patterning structure. The development of the dorsal aorta is disrupted in *floating head* (*flh*) and *no tail* (*ntl*) notochord mutants (Fouquet et al., 1997; Sumoy et al., 1997). Wild-type notochord transplanted into *flh* mutants can rescue assembly of *flk1*-expressing cells (Fouquet et al., 1997). As discussed above, hedgehog signaling is probably a critical notochord-derived cue for dorsal aorta assembly. Local cues also play an important role in proper development of the angiogenic vessels in the trunk, includ-ing the ISVs. ISVs track closely along the boundaries between the somites, and somite-derived cues are important for proper patterning of ISVs. ISV patterning is disrupted in zebrafish *fused somite* (*fss/tbx24*) and *beamter* (*bea/deltaC*) somite patterning mutants (Shaw et al., 2006). Where the somitic boundaries are ill defined in these mutants, ectopic branching and disorganized migration of the ISVs is observed. Relatively nonspe-cific, permissive cues such as proper assembly of extracellular matrix may be responsible for some of the patterning. Regions between somites are rich in laminins, extracellular matrix glycoproteins shown to be important for ECs *in vitro*. Indeed, morpholino-mediated knockdown of Laminins a1 and a4 together results in aberrant ISV patterning (Pollard et al., 2006). ISV sprouts emerge in the Laminin morphants, but their growth is retarded and they branch inappropriately away from the intersegmental boundaries.

Recent evidence has shown that specific repulsive and attractive cues from local tissues act to guide the patterning of blood vessel in the trunk as well as elsewhere in the developing animal. The role of these sorts of cues in guidance and patterning has been best characterized in the ner-vous system, for axon guidance. There are obvious parallels between the vertebrate vascular system and the nervous system. Both systems are highly branched networks that extend across and innervate every tissue in the animal. Both are composed of largely separate efferent and afferent networks (e.g., motor and sensory nerves in the nervous system, arteries and veins in the vascular system). In many cases, both systems are not only similar in anatomical structure and form but also track along the same paths. Given the frequent close physical association between vessels and nerves and the similar patterning of both systems, it is not surprising that many of the molecules that have been characterized as guidance cues for neuronal patterning have also recently been implicated in development of the vascular system. Four ligand–receptor pairs known as classical neural guidance cues have now been implicated in vascular patterning. These signaling families include ephrins and their Eph receptors, semaphorins and their plexin and neuropilin receptors, netrins and their DCC/neogenin and Unc5 receptors, and Slit ligands and their roundabout (Robo) receptors. All of these signaling molecules are expressed in the developing zebrafish and may play roles in guidance and patterning of different sets of vessels.

Exclusion of trunk vessels from the somites may be mediated in part by Eph–ephrin signaling. Eph receptor tyrosine kinases and their membrane-bound ephrin ligands are found both in blood vessels (where they play a role in arterial-venous patterning, as noted above) and in the tissues surrounding them. Loss of somite-expressed EphrinB2 in mice results in ISV branching across somite boundaries (Adams et al., 1999). Similarly, injection of ephrinB ligand or *EphB4* mRNA into *Xenopus* embryos results in aberrant growth of vessels into somitic tissue (Helbling et al., 2000). These studies suggest that ephrinB ligands expressed in the somites may restrict growth of adjacent EphB4-expressing intersomitic vessels in the trunk, preventing vascular sprouts from entering the somites (reviewed in Weinstein, 2005). Although the role of Eph–ephrin signaling in vessel patterning this has not yet been explored in the zebrafish, experimental studies in the fish have shown that semaphorin–plexin signaling plays a critical role in repulsive guidance of the ISV, restricting the growth of these vessels to defined corridors along the intersegmental boundaries (Torres-Vazquez et al., 2004).

Semaphorins are part of a diverse family of membrane-associated and -secreted ligands characterized as short-range inhibitory cues that direct axons away from inappropriate regions or steer them toward specific corridors. These ligands signal through multimeric receptor complexes composed of Plexin receptors and Neuropilin coreceptors (reviewed in Weinstein, 2005; Suchting et al., 2006). Although most Plexin receptors are expressed within the nervous system, zebrafish and mice both possess a novel endothelial-specific Plexin receptor, PlexinD1 (PlxnD1) (Gitler et al., 2004; Torres-Vazquez et al., 2004; Gu et al., 2005). Loss of *plxnD1* function in zebrafish causes ISVs to sprout and branch randomly across somites (Torres-Vazquez et al., 2004; Fig. 6C and D). The Semaphorin3a1 and Semaphorin3a2 (*sema3a1* and *sema3a2*) ligands are expressed in the central portions of the adjacent somites, but not along intersomitic boundaries where the ISVs grow. Morpholino-mediated knockdown of either *sema3a1* or *sema3a2* results in ISV patterning defects similar to (albeit milder than) those seen in *plxnD1* mutants. Overexpression of *sema3a2* within somitic tissue inhibits the growth of ISVs in wild-type animals but not in *plxnD1* mutants or morphants, demonstrating that semaphorin signaling within the vasculature is *PlxnD1* dependent. Although neuropilins are obligatory coreceptors for most plexin signaling within the nervous system, their role in *PlxnD1*-dependent vascular guidance is not clear. In mice, one report has reported that PlxnD1 signals together with Neuropilin (Gitler et al., 2004), but another suggests that PlxnD1 signaling is Neuropilin independent (Gu et al., 2005). There are four neuropilin genes expressed in the zebrafish (only two are present in mammals), and it has been suggested that at least some of these may be involved in vascular development, although a role for these genes in plxnD1-dependent signaling has not been uncovered (Lee et al., 2002; Martyn and Schulte-Merker, 2004; Yu et al., 2004).

Netrins are another family of molecules that play a role in axonal guidance in the nervous system, through both repulsive and attractive signaling. They have also been implicated in vascular guidance, although somewhat conflicting results have been obtained by two different groups on the vascular role of netrin signaling (Lu et al., 2004; Park et al., 2004; Wilson et al., 2006). Lu et al. (2004) find that loss of the *unc5b* gene in mice leads to ectopic capillary branching, suggesting that netrin signaling may normally mediate repulsive guidance of vessels. Zebrafish *netrin1a* (*ntn1a*) is restricted to the ventral neural keel and the horizontal myoseptum, the dividing line between the dorsal and ventral compartments of the somites. Lu *et al.* (2004) find that ISV sprouting and pathfinding is normal in *ntn1a* or *unc5b* morphants until the ISVs reach the horizontal myoseptum where *ntn1a* is normally expressed, at which point the vessels begin branching inappropriately. Interestingly, the parachordal veins (PAV) normally form along the horizontal myoseptum later in development, suggesting that netrin signaling may be responsible for keeping primary ISVs from branching prematurely into the horizontal myoseptum. In contrast, Wilson et al. (2006) mainly observed loss of PAV in their *ntn1a*–MO-injected embryos, and these results, together with *in vitro* and *in vivo* data from mouse (Park et al., 2004; Wilson et al., 2006), led them to propose a proangiogenic role for netrins. Given the dual nature of netrin activity in nervous system guidance and patterning, it is possible that both activities are also present in the vasculature, but further studies will be needed to resolve this.

Another neural guidance cue, Slit–Robo signaling, has also been suggested to play a role in vascular guidance, although the results to date are not clear. There are four Slit receptors known in mammals, and expression of the most divergent receptor, Robo4, is endothelial specific. Although Slit inhibits the migration of Robo4-expressing cells *in vitro*, *robo4* knockout mice do not have substantial vascular defects and none of the reported *slit* knockouts show vascular phenotypes (reviewed in Weinstein, 2005). In zebrafish, *robo4* is expressed in neural tissue, in addition to its vascular expression in the dorsal aorta, the PCV, and the ISVs. Either morpholino-mediated knockdown or overexpression of zebrafish *robo4* causes disruptions in the timing of ISV sprouting and results mainly in loss of ISV, although in both cases, these vascular defects are accompanied by severe general morphological defects and possible overall developmental delay (Bedell et al., 2005). Three Slit homologues have been identified in zebrafish (Yeo et al., 2001; Hutson et al., 2003), but none show spatial and temporal expression complementary to *robo4* expression in the developing ISV.

In developing axons, guidance cues are received and interpreted by specialized structures called growth cones at the leading edge of the growing axon. It has been proposed that the cells situated at the tip of growing vessel sprouts represent specialized cells (termed "tip cells") that play an analogous role. These cells are characterized by the presence of multiple dynamic filopodia that sample their surroundings and either retract in response to repulsive cues or extend further in response to attractive signals. Recent studies in zebrafish provide evidence for specialization among ECs within vascular sprouts and show that that Notch signaling is required for determination of tip cell identity (Leslie, J. et al., 2007; Siekmann, A. and Lawson, N. 2007). Loss of Notch signaling results in excessive angiogenesis in the trunk vasculature. By analyzing segmental artery formation in embryos transgenic for a nuclear localized EGFP [$Tg(fli1:nEGFP)^{y7}$], Siekmann and Lawson (2007) find that ECs exhibit stereotypical patterns of migration and proliferation. Loss of Notch signaling leads to recruitment of additional cells to the initial sprout, and excessive numbers of cells within the sprout undergo cell division, resulting in an excess of cells within the sprout. Moreover, the supernumerary sprout cells display migratory behavior more typical of the tip cell, with most cells migrating dorsally. The Notch ligand Delta4 (*Delta-like 4, dll4*) seems to be critical for mediating Notch signaling in zebrafish vessels, since loss of Dll4 function produces the same phenotype (Leslie J. et al., 2007; Siekmann, A. and Lawson, N. 2007). Taken together, these results suggest that Notch signaling limits the number of endothelial tip cells to prevent inappropriate branching and sprouting of vessels.

It is clear that studies of guidance and patterning in the nervous system have and will continue to provide a tremendous boost in our understanding of the molecular mechanisms underlying vascular patterning. Although many of the molecular players involved in vascular guidance have now been elucidated, in most cases, their roles are still poorly defined. With the genetic tools available in zebrafish, a larger repertoire of guidance molecules involved in vascular network formation will undoubtedly be revealed. The intrinsic advantages of zebrafish, combined with an ever-increasing arsenal of tools for imaging and perturbing vessel development, will also allow for significant refinement of our understanding of the mechanisms underlying vascular morphogenesis at the cellular level.

7. BLOOD VESSELS AND ORGANOGENESIS

Historically, blood vessels have been thought of as merely pipelines in the body allowing efficient transport of oxygen and vital nutrients. However, the physical proximity and coordinated development of endothelium with many other developing tissues hints at a potential role for reciprocal signaling between the endothelium and its surrounding tissues. Although it is well established that other tissues provide signals to ECs that are necessary for their proper development, only recently has the endothelium been recognized as a source of developmental signals for neighboring tissues. Endothelial-derived signals have now been implicated in development of a number of organs, some of which are briefly reviewed below.

Endothelial signals are important for the development of the pancreas. Removal of the endothelial precursors of the dorsal aorta in *Xenopus* embryos leads to loss of pancreatic gene expression in underlying endoderm, indicating that ECs are required for pancreatic differentiation. *In vitro* experiments combining mouse dorsal endoderm with aortic endothelium further show that endothelium is sufficient to induce pancreatic differentiation as measured by insulin expression (Lammert et al., 2001). Likewise, overexpression of VEGF-A in transgenic mice using a pancreatic promoter demonstrates that ECs are capable of instructing foregut cells to differentiate into pancreatic cells. Endothelial-derived signals have also been implicated in hepatic development. Loss of VEGFR2 function in embryos or in cultured liver explants leads to arrested liver growth and differentiation, suggesting that endothelial signals are necessary for normal liver morphogenesis (Matsumoto et al., 2001). Similarly, disruption of vessel formation in liver bud explants from wild-type embryos using the NK4 angiogenesis inhibitor selectively impairs the growth of liver epithelium, supporting the hypothesis that endothelial-derived signals are continuously required to promote early hepatic morphogenesis. Surprisingly, both pancreatic and hepatic development appear to proceed normally in zebrafish *cloche* mutants that are deficient in ECs (Field et al., 2003a,b). The possible explanations for this apparent lack of conservation between zebrafish and other vertebrates during pancreatic development are unclear at present. The molecular nature of the *cloche* mutation has not yet been determined, and it is possible that endothelial progenitors capable of endodermal signaling are still present in *cloche* mutants. There may, however, be underlying differences in the way these tissues differentiate in fish and mammals. Further work will be needed to examine the role of endothelium in endoderm specialization in the zebrafish.

Endothelial-derived signals are required for normal morphogenesis of the kidney in zebrafish (Majumdar and Drummond, 1999; Serluca et al., 2002). In *cloche* mutants, although differentiation of podocytes proceeds normally, assembly of podocytes into a glomerulus is disrupted, suggesting that ECs are required for glomerular morphogenesis. Serluca et al. (2002) show that blood flow is required for glomerulus formation by analyzing zebrafish embryos either with mutations disrupting cardiac function or by treating embryos with 2,3-butanedione monoxime (BDM) which reversibly stops the heart. Furthermore, they show that matrix metalloproteinase-2 (MMP-2), whose expression in renal ECs depends on blood flow, is required for glomerulogenesis. Taken together, these results suggest that blood flow regulates glomerulogenesis by inducing MMP-2 in renal cells. This expression of MMP-2 in turn may allow podocytes to assemble into glomeruli, possibly degrading extracellular matrix components.

The zebrafish counterpart to the adrenal cortex, interrenal tissue, also appears to require signals from ECs for normal morphogenesis (Liu and Guo, 2006). During normal development, the interrenal tissue develops in close association with the developing axial vessels. The early primordia arise as bilateral clusters of cells that migrate toward the midline. In *cloche* mutant embryos, midline convergence of interrenal primordia is disrupted. However, normal migration of interrenal primorida occurs in zebrafish mutants where blood flow is disrupted (*silent heart*) and in mutants with defective arteriogenesis (*gridlock*). These results suggest that the signals normally required for interrenal morphogenesis may arise from earlier interactions of interrenal primordia with angioblasts. This hypothesis could be tested through mosaic analysis of wild-type cells in a *cloche* mutant background. If interrenal morphogenesis depends on signals from angioblasts, transplanted wild-type cells might be capable of rescuing interrenal midline convergence in the absence of *cloche* function.

Finally, work has also revealed a role for endothelium during development of the thyroid (Alt et al., 2006). In zebrafish embryos where pharyngeal vessel development is disrupted, thyroid follicles are found lateral to their normal position along the ventral aorta, suggesting that ECs are important for localizing thyroid tissue to the midline. Disruption of blood flow in *tnnt2* morphants (thin-filament contractile protein cardiac troponin T) does not affect proper patterning of the thyroid, and ectopic ECs are sufficient to mislocalize thyroid tissue. These results suggest that endothelium may normally influence thyroid morphogenesis by providing attractive cues. From analysis of mouse mutants, they also find that thyroid tissue develops along ectopically positioned carotid arteries, suggesting that, as in fish, arteries may define the position of the thyroid.

From these studies, it is clear that endothelial-derived signals play an important role in patterning of several different organ systems. Further research will likely uncover roles for the endothelium in the development of additional tissues and identify more molecular players. Development of more transgenic lines specific for different organ primordia, combined with the genetic and experimental tools available in zebrafish, should allow for rapid progress in our understanding of the interdependent development of blood vessels and organs.

8. CONCLUSIONS

The zebrafish has proven itself a superb model organism for experimental and genetic analysis of vascular development. Studies in the fish have contributed significantly in recent years to advances in our understanding of how ECs are specified, how they assemble into vessels and acquire differentiated arterial-venous identity, and how the anatomical patterning of developing vascular networks is achieved. The powerful genetic tools of the fish, combined with high-resolution imaging, have made possible detailed description of many dynamic aspects of vascular development *in vivo*, and have allowed us to begin to define the underlying mechanisms both at the molecular and cellular levels. Ongoing forward-genetic screens (Jin et al., 2007), as well as chemical (Peterson et al., 2004; Sumanas et al., 2005), reverse genetic (Draper et al., 2004; Wienholds and Plasterk, 2004; Sood et al., 2006), and microarray screens (Chan et al., 2002; Qian et al., 2005; Sumanas et al., 2005; Covassin et al., 2006), will continue to expand the known repertoire of molecular players and enhance our overall picture of vascular development at the molecular level. These screens will also generate invaluable new tools (i.e., mutants, markers) for teasing apart the complex regulatory networks and cellular mechanisms that drive development of the vasculature.

Although zebrafish arguably represents the best vertebrate system for analyzing the combined functions of multiple factors simultaneously, redundant functions of many genes during development and the frequent recycling of signaling pathways for use in multiple different developmental processes pose certain technical challenges for *in vivo* functional analyses. With regard to this problem, zebrafish are currently limited by a lack of available tools for allowing specific spatial, temporal, or conditional expression or knockdown of genes within the zebrafish vasculature and within the surrounding tissues. However, improved technologies for identifying promoter elements (Fisher et al., 2006) and more efficient methods for transgenesis hold the promise that some of the technical challenges of multifactoral analyses may be overcome. With its many genetic and experimental advantages, zebrafish will undoubtedly be swimming at the forefront of the next big wave of discoveries in vascular development.

ACKNOWLEDGMENT

This work was supported by the intramural program on the National Institutes of Health.

REFERENCES

Adams, R.H., Wilkinson, G.A., Weiss, C., Diella, F., Gale, N.W., Deutsch, U., Risau, W., Klein, R. 1999. Roles of ephrinB ligands and EphB receptors in cardiovascular development: Demarcation of arterial/venous domains, vascular morphogenesis, and sprouting angiogenesis. Genes Dev. 13, 295–306.

Alt, B., Elsalini, O.A., Schrumpf, P., Haufs, N., Lawson, N.D., Schwabe, G.C., Mundlos, S., Gruters, A., Krude, H., Rohr, K.B. 2006. Arteries define the position of the thyroid gland during its developmental relocalisation. Development 133, 3797–3804.

Bartel, F.O., Higuchi, T., Spyropoulos, D.D. 2000. Mouse models in the study of the Ets family of transcription factors. Oncogene 19, 6443–6454.

Bayless, K.J., Davis, G.E. 2002. The Cdc42 and Rac1 GTPases are required for capillary lumen formation in three-dimensional extracellular matrices. J. Cell Sci. 115, 1123–1136.

Bedell, V.M., Yeo, S.Y., Park, K.W., Chung, J., Seth, P., Shivalingappa, V., Zhao, J., Obara, T., Sukhatme, V.P., Drummond, I.A., Li, D.Y., Ramchandran, R., et al. 2005. Roundabout4 is essential for angiogenesis *in vivo*. Proc. Natl. Acad. Sci. USA 102, 6373–6378.

Chan, J., Bayliss, P.E., Wood, J.M., Roberts, T.M. 2002. Dissection of angiogenic signaling in zebrafish using a chemical genetic approach. Cancer Cell 1, 257–267.

Childs, S., Chen, J.N., Garrity, D.M., Fishman, M.C. 2002. Patterning of angiogenesis in the zebrafish embryo. Development 129, 973–982.

Choi, K., Kennedy, M., Kazarov, A., Papadimitriou, J.C., Keller, G. 1998. A common precursor for hematopoietic and endothelial cells. Development 125, 725–732.

Cleaver, O., Krieg, P.A. 1999. Molecular mechanisms of vascular development. In: *Heart Development* (R. P. Harvey, N. Rosenthal, Eds.), San Diego: Academic Press, pp. 221–252.

Covassin, L.D., Villefranc, J.A., Kacergis, M.C., Weinstein, B.M., Lawson, N.D. 2006. Distinct genetic interactions between multiple Vegf receptors are required for development of different blood vessel types in zebrafish. Proc. Natl. Acad. Sci. USA 103, 6554–6559.

Cross, L.M., Cook, M.A., Lin, S., Chen, J.N., Rubinstein, A.L. 2003. Rapid analysis of angiogenesis drugs in a live fluorescent zebrafish assay. Arterioscler. Thromb. Vasc. Biol. 23, 911–912.

Cueni, L.N., Detmar, M. 2006. New insights into the molecular control of the lymphatic vascular system and its role in disease. J. Invest. Dermatol. 126, 2167–2177.

Davis, G.E., Bayless, K.J., Mavila, A. 2002. Molecular basis of endothelial cell morphogenesis in three-dimensional extracellular matrices. Anat. Rec. 268, 252–275.

Dooley, K.A., Davidson, A.J., Zon, L.I. 2005. Zebrafish scl functions independently in hematopoietic and endothelial development. Dev. Biol. 277, 522–536.

Downs, K.M. 2003. Florence Sabin and the mechanism of blood vessel lumenization during vasculogenesis. Microcirculation 10, 5–25.

Draper, B.W., McCallum, C.M., Stout, J.L., Slade, A.J., Moens, C.B. 2004. A high-throughput method for identifying N-ethyl-N-nitrosourea (ENU)-induced point mutations in zebrafish. Methods Cell Biol. 77, 91–112.

Field, H.A., Dong, P.D., Beis, D., Stainier, D.Y. 2003a. Formation of the digestive system in zebrafish. II. Pancreas morphogenesis. Dev. Biol. 261, 197–208.

Field, H.A., Ober, E.A., Roeser, T., Stainier, D.Y. 2003b. Formation of the digestive system in zebrafish. I. Liver morphogenesis. Dev. Biol. 253, 279–290.

Fisher, S., Grice, E.A., Vinton, R.M., Bessling, S.L., McCallion, A.S. 2006. Conservation of RET regulatory function from human to zebrafish without sequence similarity. Science 312, 276–279.

Folkman, J., Haudenschild, C. 1980. Angiogenesis in vitro. Nature 288, 551–556.

Fouquet, B., Weinstein, B.M., Serluca, F.C., Fishman, M.C. 1997. Vessel patterning in the embryo of the zebrafish: Guidance by notochord. Dev. Biol. 183, 37–48.

Gering, M., Rodaway, A.R., Gottgens, B., Patient, R.K., Green, A.R. 1998. The SCL gene specifies haemangioblast development from early mesoderm. EMBO J. 17, 4029–4045.

Gitler, A.D., Lu, M.M., Epstein, J.A. 2004. PlexinD1 and semaphorin signaling are required in endothelial cells for cardiovascular development. Dev. Cell 7, 107–116.

Gu, C., Yoshida, Y., Livet, J., Reimert, D.V., Mann, F., Merte, J., Henderson, C.E., Jessell, T.M., Kolodkin, A.L., Ginty, D.D. 2005. Semaphorin 3E and plexin-D1 control vascular pattern independently of neuropilins. Science 307, 265–268.

Habeck, H., Odenthal, J., Walderich, B., Maischein, H., Schulte-Merker, S. 2002. Analysis of a zebrafish VEGF receptor mutant reveals specific disruption of angiogenesis. Curr. Biol. 12, 1405–1412.

Helbling, P.M., Saulnier, D.M., Brandli, A.W. 2000. The receptor tyrosine kinase EphB4 and ephrin-B ligands restrict angiogenic growth of embryonic veins in Xenopus laevis. Development 127, 269–278.

Hong, C.C., Peterson, Q.P., Hong, J.Y., Peterson, R.T. 2006. Artery/vein specification is governed by opposing phosphatidylinositol-3 kinase and MAP kinase/ERK signaling. Curr. Biol. 16, 1366–1372.

Huber, T.L., Kouskoff, V., Fehling, H.J., Palis, J., Keller, G. 2004. Haemangioblast commitment is initiated in the primitive streak of the mouse embryo. Nature 432, 625–630.

Huntington, G., McClure, C. 1910. The anatomy and development of the jugular lymph sac in the domestic cat (Felis domestica). Am. J. Anat. 10, 177–311.

Hutson, L.D., Jurynec, M.J., Yeo, S.Y., Okamoto, H., Chien, C.B. 2003. Two divergent slit1 genes in zebrafish. Dev. Dyn. 228, 358–369.

Isogai, S., Horiguchi, M., Weinstein, B.M. 2001. The vascular anatomy of the developing zebrafish: An atlas of embryonic and early larval development. Dev. Biol. 230, 278–301.

Isogai, S., Lawson, N.D., Torrealday, S., Horiguchi, M., Weinstein, B.M. 2003. Angiogenic network formation in the developing vertebrate trunk. Development 130, 5281–5290.

Itoh, M., Kim, C.H., Palardy, G., Oda, T., Jiang, Y.J., Maust, D., Yeo, S.Y., Lorick, K., Wright, G.J., Ariza-McNaughton, L., Weissman, A.M., Lewis, J., et al. 2003. Mind bomb is a ubiquitin ligase that is essential for efficient activation of Notch signaling by Delta. Dev. Cell 4, 67–82.

Jin, S.W., Beis, D., Mitchell, T., Chen, J.N., Stainier, D.Y. 2005. Cellular and molecular analyses of vascular tube and lumen formation in zebrafish. Development 132, 5199–5209.

Jin, S.W., Herzog, W., Santoro, M.M., Mitchell, T.S., Frantsve, J., Jungblut, B., Beis, D., Scott, I.C., D'Amico, L.A., Ober, E.A., Verkade, H., Field, H.A., et al. 2007. A transgene-assisted genetic screen identifies essential regulators of vascular development in vertebrate embryos. Dev. Biol. 307, 29–42.

Jussila, L., Alitalo, K. 2002. Vascular growth factors and lymphangiogenesis. Physiol. Rev. 82, 673–700.

Kaipainen, A., Korhonen, J., Mustonen, T., van Hinsbergh, V.W., Fang, G.H., Dumont, D., Breitman, M., Alitalo, K. 1995. Expression of the fms-like tyrosine kinase 4 gene becomes restricted to lymphatic endothelium during development. Proc. Natl. Acad. Sci. USA 92, 3566–3570.

Kamei, M., Weinstein, B.M. 2005. Long-term time-lapse fluorescence imaging of developing zebrafish. Zebrafish 2, 113–123.

Kamei, M., Isogai, S., Weinstein, B.M. 2004. Imaging blood vessels in the zebrafish. Methods Cell Biol. 76, 51–74.

Kamei, M., Saunders, W.B., Bayless, K.J., Dye, L., Davis, G.E., Weinstein, B.M. 2006. Endothelial tubes assemble from intracellular vacuoles *in vivo*. Nature 442, 453–456.

Karkkainen, M.J., Haiko, P., Sainio, K., Partanen, J., Taipale, J., Petrova, T.V., Jeltsch, M., Jackson, D.G., Talikka, M., Rauvala, H., Betsholtz, C., Alitalo, K., et al. 2004. Vascular endothelial growth factor C is required for sprouting of the first lymphatic vessels from embryonic veins. Nat. Immunol. 5, 74–80.

Kimmel, C.B., Ballard, W.W., Kimmel, S.R., Ullmann, B., Schilling, T.F. 1995. Stages of embryonic development of the zebrafish. Dev. Dyn. 203, 253–310.

Kuchler, A.M., Gjini, E., Peterson-Maduro, J., Cancilla, B., Wolburg, H., Schulte-Merker, S. 2006. Development of the zebrafish lymphatic system requires VEGFC signaling. Curr. Biol. 16, 1244–1248.

Lammert, E., Cleaver, O., Melton, D. 2001. Induction of pancreatic differentiation by signals from blood vessels. Science 294, 564–567.

Lamont, R.E., Childs, S. 2006. MAPping out arteries and veins. Sci STKE 2006, pe39.

Lawson, N.D., Weinstein, B.M. 2002. *In vivo* imaging of embryonic vascular development using transgenic zebrafish. Dev. Biol. 248, 307–318.

Lawson, N.D., Scheer, N., Pham, V.N., Kim, C.H., Chitnis, A.B., Campos-Ortega, J.A., Weinstein, B.M. 2001. Notch signaling is required for arterial-venous differentiation during embryonic vascular development. Development 128, 3675–3683.

Lawson, N.D., Vogel, A.M., Weinstein, B.M. 2002. Sonic hedgehog and vascular endothelial growth factor act upstream of the Notch pathway during arterial endothelial differentiation. Dev. Cell 3, 127–136.

Lawson, N.D., Mugford, J.W., Diamond, B.A., Weinstein, B.M. 2003. Phospholipase C gamma-1 is required downstream of vascular endothelial growth factor during arterial development. Genes Dev. 17, 1346–1351.

Lee, P., Goishi, K., Davidson, A.J., Mannix, R., Zon, L., Klagsbrun, M. 2002. Neuropilin-1 is required for vascular development and is a mediator of VEGF-dependent angiogenesis in zebrafish. Proc. Natl. Acad. Sci. USA 99, 10470–10475.

Leslie, J.D., Ariza-McNaughton, L., Bermange, A.L., McAdow, R., Johnson, S.L., Lewis, J. 2007. Endothelial signalling by the Notch ligand Delta-like 4 restricts angiogenesis. Development 134, 839–844.

Liao, E.C., Paw, B.H., Oates, A.C., Pratt, S.J., Postlethwait, J.H., Zon, L.I. 1998. SCL/Tal-1 transcription factor acts downstream of cloche to specify hematopoietic and vascular progenitors in zebrafish. Genes Dev. 12, 621–626.

Liu, Y.W., Guo, L. 2006. Endothelium is required for the promotion of interrenal morphogenetic movement during early zebrafish development. Dev. Biol. 297, 44–58.

Lu, X., Le Noble, F., Yuan, L., Jiang, Q., De Lafarge, B., Sugiyama, D., Breant, C., Claes, F., De Smet, F., Thomas, J.L., et al. 2004. The netrin receptor UNC5B mediates guidance events controlling morphogenesis of the vascular system. Nature 432, 179–186.

Lubarsky, B., Krasnow, M.A. 2003. Tube morphogenesis: Making and shaping biological tubes. Cell 112, 19–28.

Majumdar, A., Drummond, I.A. 1999. Podocyte differentiation in the absence of endothelial cells as revealed in the zebrafish avascular mutant, cloche. Dev. Genet. 24, 220–229.

Martyn, U., Schulte-Merker, S. 2004. Zebrafish neuropilins are differentially expressed and interact with vascular endothelial growth factor during embryonic vascular development. Dev. Dyn. 231, 33–42.

Matsumoto, K., Yoshitomi, H., Rossant, J., Zaret, K.S. 2001. Liver organogenesis promoted by endothelial cells prior to vascular function. Science 294, 559–563.

Motoike, T., Loughna, S., Perens, E., Roman, B.L., Liao, W., Chau, T.C., Richardson, C.D., Kawate, T., Kuno, J., Weinstein, B.M., Stainier, D.Y., Sato, T.N., et al. 2000. Universal GFP reporter for the study of vascular development. Genesis 28, 75–81.

Murray, P.D.F. 1932. The development in vitro of the blood of the early chick embryo. Proc. R. Soc. Lond. B 11, 497–521.

Nakano, T., Abe, M., Tanaka, K., Shineha, R., Satomi, S., Sato, Y. 2000. Angiogenesis inhibition by transdominant mutant Ets-1. J. Cell. Physiol. 184, 255–262.

Ny, A., Koch, M., Schneider, M., Neven, E., Tong, R.T., Maity, S., Fischer, C., Plaisance, S., Lambrechts, D., Heligon, C., Terclavers, S., Ciesiolka, M., et al. 2005. A genetic Xenopus laevis tadpole model to study lymphangiogenesis. Nat. Med. 11, 998–1004.

Oh, S.J., Jeltsch, M.M., Birkenhager, R., McCarthy, J.E., Weich, H.A., Christ, B., Alitalo, K., Wilting, J. 1997. VEGF and VEGF-C: Specific induction of angiogenesis and lymphangiogenesis in the differentiated avian chorioallantoic membrane. Dev. Biol. 188, 96–109.

Oliver, G., Alitalo, K. 2005. The lymphatic vasculature: Recent progress and paradigms. Annu. Rev. Cell Dev. Biol. 21, 457–483.

Pardanaud, L., Dieterlen-Lievre, F. 1993. Expression of C-ETS1 in early chick embryo mesoderm: Relationship to the hemangioblastic lineage. Cell Adhes. Commun. 1, 151–160.

Park, K.W., Crouse, D., Lee, M., Karnik, S.K., Sorensen, L.K., Murphy, K.J., Kuo, C.J., Li, D.Y. 2004. The axonal attractant Netrin-1 is an angiogenic factor. Proc. Natl. Acad. Sci. USA 101, 16210–16215.

Parker, L.H., Schmidt, M., Jin, S.W., Gray, A.M., Beis, D., Pham, T., Frantz, G., Palmieri, S., Hillan, K., Stainier, D.Y., De Sauvoge, F.J., Ye, W., et al. 2004. The endothelial-cell-derived secreted factor Egfl7 regulates vascular tube formation. Nature 428, 754–758.

Patterson, L.J., Gering, M., Patient, R. 2005. Scl is required for dorsal aorta as well as blood formation in zebrafish embryos. Blood 105, 3502–3511.

Patterson, L.J., Gering, M., Eckfeldt, C.E., Green, A.R., Verfaillie, C.M., Ekker, S.C., Patient, R. 2006. The transcription factors, Scl and Lmo2, act together during development of the haemangioblast in zebrafish. Blood 109, 2389–2398.

Peterson, R.T., Shaw, S.Y., Peterson, T.A., Milan, D.J., Zhong, T.P., Schreiber, S.L., MacRae, C. A., Fishman, M.C. 2004. Chemical suppression of a genetic mutation in a zebrafish model of aortic coarctation. Nat. Biotechnol. 22, 595–599.

Pham, V.N., Lawson, N.D., Mugford, J.W., Dye, L., Castranova, D., Lo, B., Weinstein, B.M. 2006. Combinatorial function of ETS transcription factors in the developing vasculature. Dev. Biol. 303, 772–783.

Pollard, S.M., Parsons, M.J., Kamei, M., Kettleborough, R.N., Thomas, K.A., Pham, V.N., Bae, M.K., Scott, A., Weinstein, B.M., Stemple, D.L. 2006. Essential and overlapping roles for laminin alpha chains in notochord and blood vessel formation. Dev. Biol. 289, 64–76.

Qian, F., Zhen, F., Ong, C., Jin, S.W., Meng Soo, H., Stainier, D.Y., Lin, S., Peng, J., Wen, Z. 2005. Microarray analysis of zebrafish cloche mutant using amplified cDNA and identification of potential downstream target genes. Dev. Dyn. 233, 1163–1172.

Risau, W., Flamme, I. 1995. Vasculogenesis. Annu. Rev. Cell Dev. Biol. 11, 73–91.

Robb, L., Lyons, I., Li, R., Hartley, L., Kontgen, F., Harvey, R.P., Metcalf, D., Begley, C.G. 1995. Absence of yolk sac hematopoiesis from mice with a targeted disruption of the scl gene. Proc. Natl. Acad. Sci. USA 92, 7075–7079.

Rossant, J., Howard, L. 2002. Signaling pathways in vascular development. Annu. Rev. Cell. Dev. Biol. 18, 541–573.

Sabin, F. 1902. On the origin of the lymphatic system from the veins, and the development of the lymph hearts and thoracic duct in the pig. Am. J. Anat. 1, 367–389.

Sabin, F. 1917. Preliminary note on the differentiation of angioblasts and the method by which they produce blood-vessels, blood-plasma and red blood-cells as seen in the living chick. Anat. Rec. 13, 199–204.

Saharinen, P., Tammela, T., Karkkainen, M.J., Alitalo, K. 2004. Lymphatic vasculature: Development, molecular regulation and role in tumor metastasis and inflammation. Trends Immunol. 25, 387–395.

Schwerte, T., Pelster, B. 2000. Digital motion analysis as a tool for analysing the shape and performance of the circulatory system in transparent animals. J. Exp. Biol. 203, 1659–1669.

Serluca, F.C., Drummond, I.A., Fishman, M.C. 2002. Endothelial signaling in kidney morphogenesis: A role for hemodynamic forces. Curr. Biol. 12, 492–497.

Shalaby, F., Rossant, J., Yamaguchi, T.P., Gertsenstein, M., Wu, X.F., Breitman, M.L., Schuh, A.C. 1995. Failure of blood-island formation and vasculogenesis in Flk-1-deficient mice. Nature 376, 62–66.

Shaw, K.M., Castranova, D.A., Pham, V.N., Kamei, M., Kidd, K.R., Lo, B.D., Torres-Vasquez, J., Ruby, A., Weinstein, B.M. 2006. Fused-somites-like mutants exhibit defects in trunk vessel patterning. Dev. Dyn. 235, 1753–1760.

Siekmann, A.F., Lawson, N.D. 2007. Notch signalling limits angiogenic cell behaviour in developing zebrafish arteries. Nature 445, 781–784.

Sood, R., English, M.A., Jones, M., Mullikin, J., Wang, D.M., Anderson, M., Wu, D., Chandrasekharappa, S.C., Yu, J., Zhang, J., Liu, P.P. 2006. Methods for reverse genetic screening in zebrafish by resequencing and TILLING. Methods 39, 220–227.

Spyropoulos, D.D., Pharr, P.N., Lavenburg, K.R., Jackers, P., Papas, T.S., Ogawa, M., Watson, D.K. 2000. Hemorrhage, impaired hematopoiesis, and lethality in mouse embryos carrying a targeted disruption of the Fli1 transcription factor. Mol. Cell. Biol. 20, 5643–5652.

Suchting, S., Bicknell, R., Eichmann, A. 2006. Neuronal clues to vascular guidance. Exp. Cell Res. 312, 668–675.

Sumanas, S., Lin, S. 2006. Ets1-related protein is a key regulator of vasculogenesis in zebrafish. PLoS Biol. 4, e10.

Sumanas, S., Jorniak, T., Lin, S. 2005. Identification of novel vascular endothelial-specific genes by the microarray analysis of the zebrafish cloche mutants. Blood 106, 534–541.

Sumoy, L., Keasey, J.B., Dittman, T.D., Kimelman, D. 1997. A role for notochord in axial vascular development revealed by analysis of phenotype and the expression of VEGR-2 in zebrafish flh and ntl mutant embryos. Mech. Dev. 63, 15–27.

Torres-Vazquez, J., Kamei, M., Weinstein, B.M. 2003. Molecular distinction between arteries and veins. Cell Tissue Res. 314, 43–59.

Torres-Vazquez, J., Gitler, A.D., Fraser, S.D., Berk, J.D., Van, N.P., Fishman, M.C., Childs, S., Epstein, J.A., Weinstein, B.M. 2004. Semaphorin-plexin signaling guides patterning of the developing vasculature. Dev. Cell 7, 117–123.

Vogeli, K.M., Jin, S.W., Martin, G.R., Stainier, D.Y. 2006. A common progenitor for haematopoietic and endothelial lineages in the zebrafish gastrula. Nature 443, 337–339.

Wang, H.U., Chen, Z.F., Anderson, D.J. 1998. Molecular distinction and angiogenic interaction between embryonic arteries and veins revealed by ephrin-B2 and its receptor Eph-B4. Cell 93, 741–753.

Weinstein, B.M. 2005. Vessels and nerves: Marching to the same tune. Cell 120, 299–302.

Weinstein, B.M., Stemple, D.L., Driever, W., Fishman, M.C. 1995. Gridlock, a localized heritable vascular patterning defect in the zebrafish. Nat. Med. 1, 1143–1147.

Wernert, N., Stanjek, A., Kiriakidis, S., Hugel, A., Jha, H.C., Mazitschek, R., Giannis, A. 1999. Inhibition of angiogenesis *in vivo* by Ets-1 antisense oligonucleotides—inhibition of Ets-1 transcription factor expression by the antibiotic fumagillin. Angew. Chem. Int. Ed. Engl. 38, 3228–3231.

Wienholds, E., Plasterk, R.H. 2004. Target-selected gene inactivation in zebrafish. Methods Cell Biol. 77, 69–90.

Wigle, J.T., Oliver, G. 1999. Prox1 function is required for the development of the murine lymphatic system. Cell 98, 769–778.

Wilson, B.D., Ii, M., Park, K.W., Suli, A., Sorensen, L.K., Larrieu-Lahargue, F., Urness, L.D., Suh, W., Asai, J., Kock, G.A., Thorne, T., Silver, M., et al. 2006. Netrins promote developmental and therapeutic angiogenesis. Science 313, 640–644.

Wilting, J., Aref, Y., Huang, R., Tomarev, S.I., Schweigerer, L., Christ, B., Valasek, P., Papoutsi, M. 2006. Dual origin of avian lymphatics. Dev. Biol. 292, 165–173.

Yaniv, K., Isogai, S., Castranova, D., Dye, L., Hitomi, J., Weinstein, B.M. 2006. Live imaging of lymphatic development in the zebrafish. Nat. Med. 12, 711–716.

Yeo, S.Y., Little, M.H., Yamada, T., Miyashita, T., Halloran, M.C., Kuwada, J.Y., Huh, T.L., Okamoto, H. 2001. Overexpression of a slit homologue impairs convergent extension of the mesoderm and causes cyclopia in embryonic zebrafish. Dev. Biol. 230, 1–17.

Yu, H.H., Houart, C., Moens, C.B. 2004. Cloning and embryonic expression of zebrafish neuropilin genes. Gene Expr. Patterns 4, 371–378.

Zhong, T.P., Rosenberg, M., Mohideen, M.A., Weinstein, B., Fishman, M.C. 2000. Gridlock, an HLH gene required for assembly of the aorta in zebrafish. Science 287, 1820–1824.

Zhong, T.P., Childs, S., Leu, J.P., Fishman, M.C. 2001. Gridlock signalling pathway fashions the first embryonic artery. Nature 414, 216–220.

CHAPTER 13

Development and Function of the Epicardium

Jörg Männer* and **Pilar Ruiz-Lozano**[†]

Contents			

*Department of Anatomy and Embryology, Georg-August-University of Göttingen, Göttingen 37075, Germany
†Burnham Institute for Medical Research, La Jolla, California

Advances in Developmental Biology, Volume 18
ISSN 1574-3349, DOI: 10.1016/S1574-3349(07)18013-3

Abstract   The epicardium is the outermost layer of the heart and it is formed by mesothelial cells that derive from a transient structure of precursor cells (the proepicardium). Proepicardial cells migrate over the postlooped heart, followed by migration of committed endothelial and smooth muscle precursors from the proepicardium through the subepicardial matrix where the coronary arteries develop. Epicardial cells undergo epithelial–mesenchymal transition to become coronary vascular smooth muscle, perivascular fibroblasts, and intermyocardial fibroblasts. The origin of coronary endothelial cells is still under debate. Here, we provide an overview of the current knowledge on epicardial development with special emphasis on the cellular processes and genetic networks that regulate coronary arteriogenesis and myocardial growth.

1. INTRODUCTION

The function of the mature vertebrate heart not only depends on the integrity of its force-producing elements (myocardial cells) but also on structural elements such as the cardiac connective tissue (interstitium), the cardiac pacemaking and conduction system, the coronary blood and lymph vessels, and nerves. A well-known feature of the heart is the three-layered structure of its free walls. These tissue layers consist of the epicardium, the myocardium, and the endocardium, starting from the outside. The myocardium is the musculature of the heart that contracts to propel the blood through the arteries. The epi- and endocardium are the outer and inner "skins" of the heart that are in contact with the pericardial fluid and the blood, respectively. Both "skins" consist of an epithelial layer (epicardial mesothelium and endocardial endothelium) and an underlying layer of connective tissue (subepicardium and subendocardium) that harbor several of the above-mentioned nonmyocardial structures. The subepicardium, for example, holds the main stems of the coronary arteries and veins, the main stems of the cardiac lymph vessels, and some nervous elements.

At the time point of its first peristaltoid contractions, the embryonic heart of vertebrates is a relatively simple tubular structure that is composed of only two different cell types. These are the primitive myocardial cells which form the outer epithelial wall of the heart tube and the endothelial cells that form the inner endocardial wall of the heart tube. Both cell layers are connected with each other by a thick layer of a cell-free extracellular matrix called the cardiac jelly. The early embryonic heart thus lacks the epicardium as well as other well-known structural elements of the mature heart such as the cardiac interstitium, the coronary vasculature, and the cardiac nervous system.

For more than a century, embryological research on the heart has preferentially focused on the origin and development of its myo- and endocardial components. Questions on the origin and development of the epicardium did not receive much recognition since it was generally thought that the epicardium and the subepicardial connective tissue were derived from the outermost layer of the primitive myocardium (Kölliker, 1879; Mollier, 1906; Streeter, 1945; De Haan, 1965). During the past decade, however, the epicardium has come into the focus of contemporary embryology. This was inspired by new embryological and genetic data that did not only change the traditional view on its origin but additionally showed that the embryonic epicardium fulfills several important functions in cardiac development.

2. ORIGIN OF THE EPICARDIUM

2.1. The epicardium derives from a primarily extracardiac progenitor cell population, the proepicardium

According to the traditional textbook knowledge, the epicardial mesothelium (epicardium proper) and subepicardial connective tissue both were thought to derive from the outermost layer of the primitive myocardium of the early embryonic heart tube (Kölliker, 1879; Mollier, 1906; Streeter, 1945; De Haan, 1965). The primitive myocardium, therefore, was called the myo-epi-cardium (Mollier, 1906). Research in the past three decades has challenged this view. Doubts about the validity of the traditional view first arose from electron microscopic data showing that the primitive myocardium of chick embryos was composed of myoblasts (Manasek, 1968). This finding suggested that the epicardium either must evolve from a process of dedifferentiation of superficial myocardioblasts or must derive from a hitherto unknown source (Manasek, 1968). Subsequent studies showed that the epicardium derives from an accumulation of villous or vesicular protrusions of the pericardial mesothelium that is located near the venous pole of the embryonic heart loop (Viragh and Challice, 1973; Shimada and Ho, 1980; Komiyama et al., 1987a; Hiruma and Hirakow, 1989; Männer, 1992; Viragh et al., 1993; Männer et al., 2001). These structures are nowadays called the proepicardium (PE) (Viragh et al., 1993; Kálmán et al., 1995). Interestingly, the accumulation of PE villi was not a new discovery as this structure was well known among many renowned embryologists of the nineteenth century (Lieberkuhn, 1876; Kölliker, 1879; His, 1881; Remak, 1885a,b; Lockwood, 1888; Born, 1889). For unknown reasons, however, these researchers failed to identify its developmental significance. As far as we know, it was the relatively unknown polish anatomist Kurkiewicz (1909) who first identified these villi as the source for the epicardium of the chick embryo heart (Romanoff, 1960).

Since its rediscovery in the 1970s, the PE villi or vesicles have been identified as the source of the embryonic epicardium in representatives of almost all the major vertebrate groups.

2.2. The PE arises from bilaterally paired PE anlagen

Initial morphological data suggested that the PE was a single midline structure (Kurkiewicz, 1909; Viragh and Challice, 1981; Komiyama et al., 1987a; Männer, 1992, 1993, 1999; Viragh et al., 1993). This view was first challenged by data from dogfish embryos, which showed that the PE of this species appears in the form of bilaterally paired clusters of mesothelial protrusions (Munoz-Chapuli et al., 1997). The existence of primarily paired PE anlagen was also suggested in avian and mammalian embryos (Kuhn and Liebherr, 1988; Männer et al., 2001; Schlueter et al., 2006). Exact documentation of the paired nature of the PE anlagen and of their developmental fate, however, was unavailable for any of the frequently used vertebrate model organisms such as the zebrafish, *Xenopus*, chick, and mouse. In a study on mouse and chick embryos, we have documented that both species principally have bilaterally paired PE anlagen (Schulte et al., 2007).

The developmental fates of the two PE anlagen, however, differ remarkably between species (Figs. 1–3). Mouse embryos display a bilaterally symmetric pattern of PE development in which the left and right PE anlagen appear simultaneously on E8.5. The two PE anlagen grow in a bilaterally symmetric fashion and move toward the embryonic midline where they merge to form a single PE on E9.25 until a mature PE forms on E9.5. In contrast, chick embryos display a bilaterally asymmetric pattern of PE development that is characterized by a right-sided dominance. The right PE anlage appears earlier than the left one [Hamburger & Hamilton (HH)-stage 14 vs HH-stage 15], and only the right PE anlage undergoes rapid growth and acquires the full PE phenotype on HH-stage 16. The left PE anlage remains in a rudimentary state, shows a high rate of apoptosis, and disappears on HH-stage 18/19. Experimental data suggest

FIGURE 1   Scanning electron micrographs showing ventral views of the PE of chick embryos shortly before (A; HH-stage 16) and after the attachment of proepicardial villi to the dorsal surface of the heart (B; HH-stage 17/18). Note that in chick embryos only the right PE (rPE) establishes firm contact to the heart and delivers the primitive epicardium (epicardial border marked by arrowheads) and further PE-derived cells. The left PE (lPE) remains in a rudimentary state and does not contribute to the formation of the epicardium. Abbreviations: H = heart; L = liver.

FIGURE 2   Scanning electron micrographs showing the surface morphology of the mature PE in chick embryos (A; HH-stage 16) and in mouse embryos (B; E9.5).

that the bilaterally asymmetric growth of the PE anlagen in chick embryos might be linked to the left–right signaling pathways that are known to control the bilaterally asymmetric morphogenesis of several internal organs (Schulte et al., 2007).

Descriptive data from dogfish (Munoz-Chapuli et al., 1997) and *axolotl* embryos (Fransen and Lemanski, 1990) have suggested that the bilaterally asymmetric pattern of PE development is not confined to avian species.

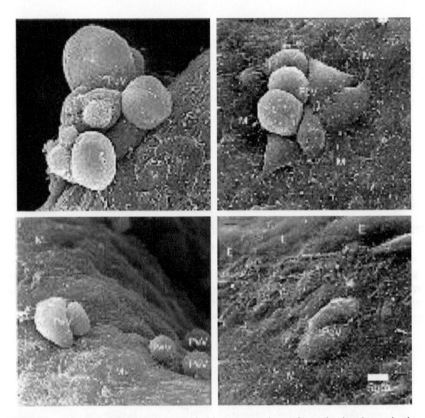

FIGURE 3   Scanning electron micrographs showing PE-derived vesicles (PeV) attached to the naked myocardial surface of the heart from an E9.75 mouse embryo. Abbreviations: E = epicardial cells; M = myocardial surface.

We have therefore analyzed the formation of the PE in *Xenopus* embryos (J.M. in preparation). In this species, normally developed embryos also display a right-sided formation of the PE. *Xenopus* embryos, however, in which the heart tube undergoes abnormal looping toward the left side show left-sided formation of the PE, supporting the above-mentioned hypothesis that asymmetric growth of the PE might be linked to the left–right signaling pathways. At the present time, we do not know the functional relevance of bilaterally symmetric or asymmetric pattern of PE development. Future studies on zebrafish and *Xenopus* might help to answer this question.

2.3. Induction of the PE

The tissues and factors involved in the induction and control of PE development have not been identified for a long time (Männer et al., 2001). Recent studies have started to address this topic. It has been found that bone morphogenetic protein (BMP) and fibroblast growth factor (FGF) signaling play important regulatory roles in the commitment of pericardial mesoderm to the PE identity in chick embryos (Kruithof et al., 2006; Schlueter et al., 2006).

Organs and tissues with PE-inducing or -repressing activity are expected to lie in proximity to the PE anlagen. This is the case for the liver anlage, which forms close to the PE in all species studied so far. Up to now, however, no experimental data have been published that could support or disprove this idea. Knockout mice have been generated that do not form a liver (Lee et al., 2005). Unfortunately, no information has been provided about PE development in these embryos.

In chick embryos, the right PE anlage forms opposite to the ventricular bend of the heart, whereas the left PE anlage forms opposite to the yolk sac (Schulte et al., 2007). This suggests that the asymmetric growth of the two PE anlagen of chick embryos (see above) might be the consequence of its different topographical relationship to tissues with PE-inducing (ventricular myocardium) or PE-repressing (yolk sac mesoderm) activity. To test this hypothesis, we analyzed the development of the chicken PE in two sets of experiments: first, in organ cultures in which the two PE anlagen were physically isolated from the suspected PE inducer and repressor, and, second, in whole embryo cultures in which the normal topographical relationship of the two PE anlagen had been inverted by the mechanical induction of inversion of cardiac looping and embryonic body rotation. The fact that the development of the two PE anlagen proceeds in the normal asymmetric pattern in both sets of our experiments excludes the possibility that the ventricular myocardium and the yolk sac mesoderm have a strong influence on PE development. These studies suggest that the bilaterally asymmetric development of the chicken PE is not the consequence of side-specific differences in the topographical relationships of the two PE anlagen to neighboring tissues (Schulte et al., 2007). Since the development of the chicken PE proceeds in an asymmetric pattern, candidate inducers as well as repressors might be identified by corresponding asymmetric expression patterns in the wall of the sinus venosus.

2.4. The transfer of PE cells to the developing heart is accomplished by three different mechanisms

Three different transfer mechanisms have been identified for PE cells to reach the naked myocardial surface. The first mechanism was discovered in mouse, tupaia, and dogfish embryos, where it represents the predominant mechanism for PE cell transfer (Komiyama et al., 1987; Kuhn and Liebherr, 1988; Munoz-Chapuli et al., 1997). Solid or vesicular aggregates of PE cells are released into the fluid of the pericardial cavity where they float passively, and on making contact with the heart, PE cell aggregates attach to its naked myocardial surface. The attached cell aggregates flatten into isolated patches of epicardium, which subsequently coalesce to form a coherent cell sheet. Due to its position opposite to the PE, the dorsal wall of the ventricular bend is the first portion of the cardiac wall that becomes covered by PE-derived epicardium.

The second mechanism was described in amphibian (Fransen and Lemanski, 1990) and avian embryos (Männer, 1992, 1993; Vrancken Peeters et al., 1995; Nahirney et al., 2003). Here, a secondary tissue bridge is established between the PE and the dorsal wall of the ventricular bend. PE-derived mesothelial cells cross the free pericardial cavity by moving along this tissue bridge toward the developing heart. Formation of the primitive epicardium starts at the point of attachment of the secondary tissue bridge to the heart, from where the epicardial mesothelium spreads over the naked myocardial surface in a continuous epithelial sheet. The formation of the secondary tissue bridge has been studied in detail in chick embryos (Männer, 1992; Nahirney et al., 2003). This bridge is formed by the direct attachment of the PE villous protrusions to the dorsal surface of the rhythmically contracting heart (Männer, 1992). This mechanically difficult process is mediated by the formation of an extra-cellular matrix bridge (Nahirney et al., 2003). PE cell movement along the secondary tissue bridge seems to be the predominant transfer mechanism in birds and reptiles (Männer, 1992, 1993; Männer et al., 2001), although there is also evidence for a small contribution of free-floating PE vesicles to the establishment of the primitive epicardium (Vrancken Peeters et al., 1995; Männer, 1999; Sejima et al., 2001).

The third mechanism of PE cell transfer to the heart does not involve PE-derived cells crossing the free pericardial cavity. Rather, cells travel along the continuous walls of the sinus venosus and heart. The movement of PE-derived cells along this pathway has frequently been neglected since it is usually masked by the dominance of one of the two above-mentioned mechanisms. This third mechanism becomes apparent, how-ever, when the predominant transfer mechanism is experimentally blocked (Männer, 1993). In this case, the formation of the epicardium is severely delayed since PE-derived cells can reach the heart exclusively by this alternative pathway (Männer, 1993).

Why PE-derived cells bridge the free pericardial cavity in some spe-cies, predominantly in the form of free-floating cell aggregates (dogfish, mice, tupaia), remains an open question; while in other species, this is accomplished by movement along a secondary tissue bridge (chicks, quails, and some reptiles). Fransen and Lemanski (1990) suggested that the use of different cell transfer mechanisms might be explained by species-specific differences in the development of the pericardial cavity. They hypothesize that an effective cell transfer via free-floating cell aggregates requires that the developing heart and the pericardial cavity form an enclosed system. Such a situation is found in dogfish and mammalian embryos. In chick and quail embryos, however, the pericardial cavity normally communicates with the extraembryonic coelom via a large opening (Steding and Klemeyer, 1969; Männer et al., 1995) unable to contain free-floating cell aggregates.

2.5. The epicardium of the intrapericardial portion of the great arterial trunks does not derive from the PE

Initial descriptive studies could not clarify whether the entire epicardium derives from the PE. This question was resolved by fate-mapping studies using quail–chick chimera technology. These studies revealed that the PE provides only the epicardium of the avian heart but not the epicardium of the intrapericardial portions of the great arterial trunks (Männer, 1999; Perez-Pomares et al., 2001; Perez-Pomares et al., 2003). The latter population of epicardial cells seems to derive from the pericardial mesothelium surrounding the arterial end of the embryonic heart loop (Männer, 1999; Perez-Pomares et al., 2003), and it has been found that its biological properties [e.g., morphology, gene expression profile, and ability to undergo epithelial–to–mesenchymal transition (EMT)] differ in several respects from the PE-derived epicardium (Perez-Pomares et al., 2003). It is still unclear whether this finding has any functional relevance for the developing heart. The fact, however, that the border between PE-derived and non-PE-derived epicardia corresponds to the border between the proximal and distal portions of the outflow tract of the embryonic heart suggests that the disappearance of the myocardium in the distal outflow tract might be linked to its covering with non-PE-derived epicardium (Männer, 1999; Perez-Pomares et al., 2003).

2.6. Evidence for the presence of a third population of epicardial precursor cells in lower vertebrates

The PE and the pericardial mesothelium at the arterial end of the embryonic heart might not be the only sources for epicardial cells. The hearts of some fishes, amphibia, and reptiles, for example, have a tissue bridge that connects the cardiac apex with the ventral pericardial wall. This structure is called the apical ligament (Grant and Regnier, 1926) or the gubernaculum cordis (Fritsch, 1869; Foxon, 1950; MacKinnon and Heatwole, 1981). Greil (1903) and Hochstetter (1906, 1908) report that this tissue bridge is a secondary structure formed by the fusion of pericardial villi with the apex of the embryonic heart. These findings suggest that, in some lower vertebrates, there might be a third population of epicardial precursor cells at the pericardial wall opposite to the apex of the embryonic heart.

Currently, we have no exact information about the biology of this cell population. There are good reasons to clarify the possible contributions of this cell population to the heart. First, it is possible that the third population of epicardial precursor cells might also provide myocardial precursor cells to the heart. This idea would explain the observations that chicken PE-derived cells can differentiate into myocardial cells in *in vitro* cultures (Kruithof et al., 2006; Schlueter et al., 2006) and that the hearts of some lower vertebrates show remarkable capacities for myocardial regeneration. Second, the mature gubernaculum cordis contains an artery and vein that connect the cardiac blood vessels with the left-sided internal "mammary" (anterior epigastric) vessels (MacKinnon and Heatwole, 1981). This suggests that the "third population" might make important contributions to the development of the coronary vascular system whose further definition might improve our understanding of coronary vessel development.

3. CONTRIBUTION OF THE EPICARDIUM TO THE CORONARY VASCULATURE AND CARDIAC GROWTH

3.1. EMT of the PE-derived cells produces mesenchymal cells that colonize the cardiac wall

Despite interspecific differences in the transfer of PE cells to the heart, the process of epicardial covering of the outer myocardium remains conserved in all the species analyzed (reviewed in Männer et al., 2001). The dorsal surface of the primitive ventricle opposite to the PE vesicles is the first area of the heart covered by epicardial cells. Subsequently, epicardial cells cover the dorsal surfaces of the atria and outflow tract, and the lateral and ventral surfaces of the ventricles. In the primitive epicardium, the epicardial mesothelial cells are directly attached to the myocardium. Shortly after that, a space filled with extracellular matrix forms between the primitive epicardium and the myocardium. The formation of this subepicardial space is probably due to a combination of local downregulation of cell adhesion molecules and enhanced secretion of extracellular matrix proteins, including fibronectin and collagens I, IV, V, and VI (Tidball, 1992; Hurle et al., 1994; Kalman et al., 1995; Hur et al., 1999), proteoglycans, laminin (Kalman et al., 1995), GP68 (Morita et al., 1998), vitronectin, fibrillin-2, elastin (Bouchey et al., 1996), tenascin-X (Burch et al., 1995), and flectin (Tsuda et al., 1998). The source of the extracellular matrix proteins is unclear.

The subepicardial space is soon populated by mesenchymal cells, which originate from epicardial cells undergoing an EMT (Viragh et al., 1993; Munoz-Chapuli et al., 1994; Markwald et al., 1996; Munoz-Chapuli and

Hamlett, 1996; Perez-Pomares et al., 1997; Dettman et al., 1998; Perez-Pomares et al., 1998) and from PE-derived mesenchymal cells reaching the heart either via free-floating PE vesicles (Van den Eijnde et al., 1995) or secondary tissue bridges (cardiac ligaments). Following the colonization of the subepicardial space, epicardium- and PE-derived mesenchymal cells invade the myocardium and subendocardial space (Gittenberger-de Groot et al., 1998; Männer, 1999).

EMT is a critical process that occurs during gastrulation and organogenesis as a means to distribute cells from epithelia (Hay and Zuk, 1995). As mentioned earlier, in the heart, epicardial EMT is an essential process in normal cardiac development at least in part because the epicardium provides precursor cells to the coronary vascular plexus forming in the subepicardial space (Mikawa and Fischman, 1992; Mikawa and Gourdie, 1996; Männer et al., 2001). Mutations that alter epicardial EMT impair coronary vessel formation (Morabito et al., 2002). PE explants and epicardial cells extracted from explanted hearts have become an invaluable experimental tool to assess the factors that regulate epicardial EMT. In particular, explanted tissues have served to identify positive and negative factors that regulate epicardial EMT, including serum response factor (SRF), fibroblast growth factors (FGFs), platelet derived growth factors (PDGFs), vascular endothelial growth factor (VEGF), transforming growth factor-β (TGF-β), integrins, and retinoids.

With the use of PE cells from chicken embryos, SRF has been identified as an inductor of epicardial EMT (Landerholm et al., 1999). Explanted PE cells upregulated the expression of SRF and a subset of cells acquired a motile phenotype. Together with SRF induction and mesenchymal transformation, smooth muscle markers were detected. Remarkably, different dominant-negative SRF constructs prevented the appearance of smooth muscle markers without blocking mesenchymal transformation, thus defining a stepwise differentiation of coronary smooth muscle cells from PE cells that requires SRF (Landerholm et al., 1999). A subsequent study by the same group further demonstrates that PDGF-BB acts upstream of the SRF-dependent induction of smooth muscle markers in PE cells. Inhibition of rhoA signaling blocks PDGF-BB-induced EMT and inhibits coronary smooth muscle differentiation, both *in vitro* and *in vivo* (Lu et al., 2001). Epicardial-specific mutation of both SRF and PDGF-BB will ultimately demonstrate their physiological requirement in cardiac development.

Furthermore, experiments using explanted epicardial tissues have investigated the ability of known activators of EMT in the cardiac cushion mesenchyme to regulate epicardial EMT, including FGFs and TGF-β (Morabito et al., 2001; Compton et al., 2006; Olivey et al., 2006). Based on these experiments, FGFs, VEGFs, and EGF stimulate epicardial EMT (Morabito et al., 2001; Morabito et al., 2002; Dettman et al., 2003; Lavine et al., 2005; Merki et al., 2005).

In contrast, the role of TGF-β is less clear. Currently, two models of TGF-β action in epicardial cells have been described. In the first one, TGF-β-1 stimulates EMT weakly, while TGF-β-2 and -3 do not stimulate EMT. The FGF-2-mediated induction of EMT is strongly inhibited by addition of TGF-β-1, -2, or -3 to the culture. TGF-β-3 does not block EMT *per se* (since expression of vimentin, a mesenchymal marker, appears normal) but appears to inhibit EMT by blocking epithelial cell dissociation and invasion of the extracellular matrix. The authors propose a model in which myocardially derived FGF-1, -2, or -7 promote epicardial EMT, while TGF-β-1, -2, or -3 in the myocardium restrains EMT (Morabito et al., 2001). These studies suggest that epicardial EMT is repressed by TGF-β, in contrast to EMT in the endocardium, where TGF-β serves as an EMT inductor. In the second model, TGF-β-1 and -2 are activators of epicardial EMT and their function is dependent on the type I TGF-β receptors ALK2 (Olivey et al., 2006) and ALK5 (Compton et al., 2006).

Interestingly, the balance between positive and negative EMT regulators is an essential factor for the proper formation of the coronary vasculature, as excessive EMT may result in failure of coronary vessel development. This has been suggested by mechanistic studies of α4-integrin (Dettman et al., 2003) and *α4-integrin* mouse mutants, which failed to form the epicardium (Yang et al., 1995). In quail–chick chimeras, transplanted epicardial cells with knockdown expression of α4-integrin were particularly invasive and did not contribute to the smooth musculature of the developing coronary vessels, suggesting that α4-integrin is required to restrain epicardial–mesenchymal transition. Preventing excessive EMT seems, therefore, essential for correct targeting of epicardium-derived cells (EPDCs) to the smooth musculature of the developing coronary vasculature.

3.2. PE- and epicardium-derived mesenchymal cells differentiate into several cell types that form important structural components of the heart, such as the cardiac interstitium and the coronary blood vessels

Fate-mapping studies have shown that the PE- and epicardium-derived mesenchyme delivers nearly all cellular elements of the cardiac connective tissue (fibroblasts) and coronary vasculature, but [in contrast to *in vitro* studies (Kruithof et al., 2006; Schlueter et al., 2006)] does not contribute a substantial number of myocardioblasts to the developing heart (Mikawa and Fischman, 1992; Mikawa and Gourdie, 1996; Gittenberger-de Groot et al., 1998; Männer, 1999; Merki et al., 2005).

Several seminal articles have described the use of retroviral β-galactosidase-expressing vectors (retroviral-LacZ) to analyze the cell fate of clonal PE cells (Mikawa and Fischman, 1992; Mikawa and Gourdie, 1996; Dettman et al., 1998; Gourdie et al., 2000) in chicken embryos. Retroviral-LacZ injection of PE cells initially identified two different cell types in the coronary vasculature that originated in the PE: a spirally arranged spindle-shaped type cell oriented transverse to the vessel, identified as smooth muscle cells; and elongated, flattened cells oriented longitudinal to the vessel that were identified as endothelial cells. Furthermore, these experiments also detected LacZ-positive clones in the interstitial connective tissue of the myocardium. Although no additional marker was analyzed, this third population of EPDCs very likely constitutes intramyocardial fibroblasts. It is clear from this study that vascular smooth muscle, perivascular fibroblasts, and coronary endothelial cells all derive from independent precursors when EPDCs migrate into the heart (Mikawa and Fischman, 1992). A subsequent study concludes that prior to the migration of PE cells, the coronary smooth muscle linage is already established (Mikawa and Gourdie, 1996).

Less clear is the origin of coronary endothelial cells. On one hand, experiments performed in quail–chick chimeras support an epicardial origin for the endothelial and smooth muscle cells of coronary vasculature (Männer, 1999; Perez-Pomares et al., 2002), and quail PE explants showed colocalization of cytokeratin (mesothelial marker) and QH1 (endothelial marker) after 24 hours of culture (Perez-Pomares et al., 2002). On the other hand, work with quail–chick chimeras showed that EMT is an essential step in coronary morphogenesis *in vivo* (Dettman et al., 1998), and chick epicardial cells labeled *in ovo* with DiI invaded the subepicardial matrix and myocardial wall, and became coronary vascular smooth muscle, perivascular fibroblasts, and intramyocardial fibroblasts (Dettman et al., 1998). No staining of the coronary endothelium, however, was evident in the latter set of experiments (Dettman et al., 1998). Differences in the labeling strategy of PE cells are of great difficulty in the study of the precise origin of endothelial coronary precursor cells. A recently developed trans-genic mouse model that targets the PE (GATA-5-Cre) has served as a linage tracing tool in mammals (Merki et al., 2005). Smooth muscle and cardiac connective tissue were positively identified as epicardial deriva-tives, whilst no staining was detected in the coronary endothelium (Merki et al., 2005). Whether this is a promoter effect of the GATA-5 system or reflects an alternate origin of the mammalian coronary endothelium requires further investigation.

In culture, however, it seems clear that epicardium-derived vascular progenitor cells (EPDCs) display bipotency. EPDCs can either form endo-thelial cells, in response to a combination of myocardial VEGF and basic FGF (bFGF) signaling, or differentiate into smooth muscle cells on expo-sure to PDGF and TGF-β (Tomanek et al., 1999; Perez-Pomares et al., 2002; Guadix et al., 2006; Kruithof et al., 2006).

3.3. Epicardial/myocardial cross talk directs epicardial development and controls growth of the outer compact layer of the myocardium

Several observations suggest the existence of a dynamic cross talk between EPDCs and the developing myocardium. Surgical ablation of the PE results in thin myocardium, defects of the coronary vasculature, and abnormal tissue bridges between the ventricles and the pericardial wall (Männer et al., 2005). The formation of the coronary vasculature coincides with increasing muscle mass, suggesting myocardial signals that induce increased vascularization of the heart, perhaps mediated by hypoxia (Mancini et al., 1991). However, the existence of a hypoxia-mediated mechanism has yet to be demonstrated. In addition, the conserved distribution of the coronary vessels suggests a strict regulation of the cardiac vascular patterning and the existence of signals in the myocardium that direct vascular growth in a three-dimensional manner. Recent reports have demonstrated that ablation of some myocardial genes has a tremendous impact on the development of the coronary vasculature. Disruption of the GATA cofactor FOG-2 (friend of GATA-2) in mice results in embryonic death at midgestation characterized by thin ventricular myocardium and absent coronary vasculature, despite formation of an intact epicardial layer expressing the epicardial-specific genes *WT-1* and *epicardin* (Tevosian et al., 2000). Markers for the coronary vasculature, FLK-1 and ICAM-2, are not detected in *FOG-2* mutant mice, indicating either a failure to activate their expression or initiate EMT in epicardial cells (Tevosian et al., 2000). Furthermore, a single amino acid replacement in GATA-4 that impairs its physical interaction with FOG-2 also results in lethality around E12.5, exhibiting features in common with *FOG-2* mutant embryos (Crispino et al., 2001). Remarkably, transgenic re-expression of FOG-2 in cardiomyocytes using the αMHC promoter rescues the vascular phenotype, thus demonstrating that myocardial FOG-2 function is required for the induction of the formation of the coronary vasculature (Tevosian et al., 2000). These findings establish that the GATA-4-FOG-2 interaction in the myocardium is a critical step in the development of epicardium-derived tissue.

Independent studies have recently underscored the importance of a second myocardial system required for epicardial development. Thymosinβ4 (Tβ4) is a G-actin monomer-binding protein implicated in the reorganization of the actin cytoskeleton and required for cell migration. Increasing interest in the role of Tβ4 in the heart has been triggered by the discovery that Tβ4 promotes myocardial and endothelial cell migration in the embryonic heart. Tβ4 also promotes survival of embryonic and postnatal cardiomyocytes in culture and on myocardial infarction (Bock-Marquette et al., 2004). These effects are, at least partially, due to activation of the survival kinase AKT, downstream of Tβ4 (Bock-Marquette et al., 2004). In addition, exogenous Tβ4 enhances vascular sprouting in the coronary artery ring angiogenesis assay. Tβ4 induces an increase in cell–matrix attachment, proliferation, and endothelial tube formation (Grant et al., 1999). These findings suggest that Tβ4 may be a new therapeutic target in the setting of acute myocardial damage (Bock-Marquette et al., 2004). Although Tβ4 is mainly expressed in the vasculature (Dathe and Brand-Saberi, 2004), cardiac myocyte-conditional knockdown *Tβ4* mouse mutants display a thin noncompacted myocardium and partially detached epicardium mottled with surface endothelial nodules and abnormal vascularization, suggesting that Tβ4 plays a role in regulating coronary development (Smart et al., 2006). The mechanism by which Tβ4 regulates the development of the epicardium and epicardium-derived tissues has yet to be defined. Whether the FOG-2 and Tβ4 pathways are interdependent pathways remains to be determined.

If myocardial mutations affect the developing epicardium, alterations in epicardial formation have always had a dramatic effect on myocardial growth. Several pathways are known to be critical in signaling myocardial growth from the epicardium. Cardiac-muscle defects have been reported in several mutant mice in which epicardium development or function was affected. For instance, α4-integrin is a subunit of a cell surface receptor that mediates cell–extracellular matrix and cell–cell adhesion by interacting with fibronectin and vascular cell adhesion molecule (VCAM-1). α4-integrin is produced in the PE and epicardium, whereas its cognate ligand VCAM-1 is secreted by the myocardium. Mutations of both α4-integrin and VCAM-1 were both embryonic and lethal, showing two sets of defects: failure of fusion of the allantois with the chorion during placentation and defects in the development of the heart. The latter included abnormal epicardium and coronary vessels and thin myocardium (Kwee et al., 1995; Yang et al., 1995). As mentioned earlier, a critical role for α4-integrin in restraining epicardial EMT has been demonstrated (Dettman et al., 2003).

Further evidence for a putative role of the epicardium in myocardial formation comes from the mutation of erythropoietin (EPO). Targeted mutation of EPO and its receptor (EPOR) result in embryonic lethality with thinning of the ventricular myocardium coupled to defects in ventricular septum, linked to a reduction in cell proliferation that appears to be specific to the heart. EPOR expression is endocardium and epicardium specific, yet it is the myocardium that shows a consistent reduction in cell number, suggesting that EPO triggers cardiomyocyte proliferation (Wu et al., 1999). The mechanism by which EPO induces myocardial proliferation is not clear. The fact, however, that EPOR is not expressed in myocytes implicates the existence of an epicardium-derived myocardial growth factor, downstream of EPO activation.

A thin compact myocardium associated with impaired epicardial development (Moore et al., 1998, 1999) was also observed in the mutation of the Wilms' tumor-1 (WT-1) gene, where malformation leads to embryonic death at midgestation. *WT-1* mutant mouse embryos lack epicardium in the ventricles, atria, and aorta. No subepicardial mesenchymal cell formation occurred in areas without epicardium. Even in the areas where epicardium was formed, subepicardial mesenchymal cells were either not detected or much reduced.

Detachment of the primitive epicardium (Sucov et al., 1994) and defect in myocyte proliferation (Kastner et al., 1994) were also detected in the retinoid receptor *RXRα* mouse mutants. Despite the dramatic myocardial growth defect in *RXRα* embryos, myocardium-specific *RXRα* mutants do not display an abnormal myocardial phenotype, suggesting a non-cell-autonomous effect in *RXRα* mutation on cardiomyocytes (Chen et al., 1998).

It has recently been determined that the epicardium is the most prominent tissue to transduce retinoid signal to the myocardium. Blockade of either retinoic acid (RA) or EPO signaling from the epicardium inhibits cardiac myocyte proliferation and survival (Stuckmann et al., 2003). The blockade of cardiac myocyte proliferation following administration of an RA antagonist can be rescued by exogenous EPO. Conversely, blockade following administration of anti-EPOR antisera can be rescued by exogenous RA, suggesting that RA and EPO are parallel signals that support myocardial growth (Stuckmann et al., 2003). These authors postulate that the mechanism of RA and EPO on cardiac growth is mediated through the induction of another soluble factor(s) in the epicardium that directly regulate(s) proliferation of cardiac myocytes (Stuckmann et al., 2003). In addition, epicardial conditional *RXRα* mouse mutants display hypoplastic myocardium (Merki et al., 2005), further supporting that the epicardium induces myocardial growth.

Therefore, it seems that epicardial dysmorphogenesis is consistently associated with defective myocardial growth, suggesting that EPDCs have a signaling function, perhaps independent of their role in coronary vasculogenesis. Epicardium-conditioned media induce myocardial growth in culture on treatment with RA, further supporting this hypothesis (Chen et al., 2002).

3.4. The epicardium as a niche of stem cells

The potential of EPDCs to differentiate into a variety of different cell types, together with the observation that some EPDCs remain in a more or less undifferentiated state, made several authors postulate that EPDCs could be considered as the "ultimate cardiac stem cell" (Wessels and

Perez-Pomares, 2004). Two recent studies on phylogenetically distant species such as the zebrafish and humans are supportive of this hypothesis.

The zebrafish heart displays a unique capacity for myocardial regeneration that, subsequent to resection of the ventricular apex, occurs through two coordinated stages. It starts with the formation of a blastema that harbors a population of progenitor cells expressing early myocardial genes. These progenitor cells proliferate and differentiate into mature myocardial cells. Subsequently, epicardial tissue surrounding both atria and ventricles expands, providing a new epithelial covering over the injured myocardium. A subpopulation of these epicardial cells undergoes EMT to provide vascular progenitor cells for vascularization and to support the regenerating musculature. Importantly, blockade of FGF signaling through of a dominant-negative FGFR impairs epicardial EMT and myocardial regeneration (Lepilina et al., 2006).

In an attempt to study the cell biology of EPDCs, several groups have developed and characterized cell lines derived from mesothelial cell (Chen et al., 2002; Wada et al., 2003). Rat mesothelial cells retain many characteristics of the epicardial epithelium, including the formation of a polarized epithelium, the expression of epicardial genes, and the ability to produce mesenchyme in response to specific growth factors (Wada et al., 2003). A major breakthrough toward the identification of the epicardium as a source for cardiac stem cells came only with the establishment of primary epicardial cell cultures from adult human hearts (van Tuyn et al., 2006). EPDCs can be purified from atrial tissue obtained in human biopsies that spontaneously undergo EMT under certain culture conditions (van Tuyn et al., 2006). The cell fate of adult human EPDCs can be modulated by addition of various factors to the media. On experimental induction of the transcription factor myocardin or after treatment with TGF-β-1 or BMP-2, EPDCs obtain characteristics of smooth muscle cells. Adult EPDCs can also undergo osteogenesis but fail to form adipocytes or endothelial cells *in vitro*. Therefore, cultured epicardial cells from human adults recapitulate at least part of the differentiation potential of embryonic epicardial cells. Whether adult EPDCs constitute a valuable therapeutic substrate remains to be seen, but this data certainly suggests the existence of an autologous source of cells with tremendous clinical potential.

4. GENETICS OF EPICARDIAL DEVELOPMENT AND CORONARY ARTERIOGENESIS

The combination of the phenotypic analysis of diverse mutant mice with *in vitro* experiments blocking or enhancing activation of selected growth factors has recently expanded our knowledge on the pathways that regulate the differentiation and cell fate of the PE and EPDCs. In addition to the mutations that affect epicardial/myocardial signaling, altered expression of several other factors affect epicardial polarity, the capacity of epicardial-to-mesenchyme transition, and the ability of EDPCs to fully differentiate into their derivatives.

4.1. Polarity and cell adhesion

Among the earliest alterations that adversely affect the formation of the epicardium and coronary vasculature are alterations of the ability to form PE vesicles, as recently shown by mutations in PAR3. PAR3 is a mammalian homologue of *Caenorhabditis elegans* polarity proteins. PAR3 forms a conserved protein complex with PAR6 and αPKC, and the polarized distribution of the PAR3-PAR6-αPKC complex is crucial for establishing epithelial cell polarity by regulating junctional structures (Joberty et al., 2000; Suzuki et al., 2001). Targeted disruption of the mouse *Par3* gene results in midgestational embryonic lethality with defective formation of the epicardium (Hirose et al., 2006). The PE of PAR3-deficient mice fails to form PE cysts and its mesothelial cells show defects in the establishment of a basoapical axis, thus demonstrating that polarity of the PE mesothelium plays a crucial role in mammalian cardiac morphogenesis (Hirose et al., 2006).

4.2. Transcription factors

We mentioned earlier that the mutation of the genes VCAM and α4-integrin in the mouse affect epicardial EMT (Dettman et al., 2003) and result in incomplete development of the epicardium and coronary vasculature leading to cardiac hemorrhage (Kwee et al., 1995; Yang et al., 1995). Interestingly, the transcription factor WT-1 suppressor, which is necessary for normal development of the epicardium (Moore et al., 1999), activates transcription of the α4-integrin gene (Kirschner et al., 2006).

WT-1 knockout mice die at E13.5 of heart failure probably due to severe defects in the epicardial layer and impaired formation of subepicardial mesenchymal cells. A LacZ reporter gene inserted into a YAC construct demonstrates that WT-1 is expressed in the early PE, the epicardium, and the subepicardial mesenchymal cells. Following these studies, the expression of the *WT-1* gene has been extensively used as a marker for epicardial derivatives in early stages. Unfortunately for linage-tracing purposes, the WT-1-LacZ reporter expression is downregulated in later stages of epicardial development (Moore et al., 1999).

Mouse embryos lacking the RA receptor RXRα properly undergo the early steps of heart development, but then fail to initiate expansion of cardiomyocytes that normally form the compact zone of the ventricular

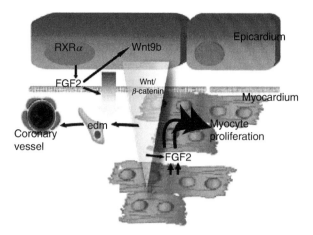

FIGURE 4 Working model for molecular mechanism for RXRα control of coronary formation and myocardial growth: RXRα activates the expression of epicardial FGF-2. Epicardial FGF-2 stimulates epithelial–mesenchymal transition of epithelial epicardial cells and induces epicardial Wnt expression. Epicardial Wnt stabilizes myocardial β-catenin which induces myocardial synthesis of FGF-2. In turn, FGF signaling induces myocardial proliferation.

chamber wall. RXRα mutant embryos have a hypoplastic ventricular chamber and die in midgestation from cardiac insufficiency. Despite the fact that a hypoplastic myocardium is the most prominent characteristic of the RXRα mutant embryos, a myocardium-restricted RXRα mutation does not perturb heart morphogenesis (Chen et al., 1998). Additional studies showed that the pattern of response to RA in the mouse embryo coincided with the progression of epicardium formation, suggesting a key role for epicardium-derived tissues in directing RA response in the developing heart (Xavier-Neto et al., 1999, 2001). A series of tissue-restricted mutations of the *RXRα* gene in the cardiac neural crest, endothelial, and epicardial linages revealed that RXRα signaling in the epicardium is essential for proper cardiac morphogenesis (Merki et al., 2005). The same work identifies a retinoid-dependent Wnt signaling pathway that cooperates with FGFs (Fig. 4) in epicardial EMT (Merki et al., 2005).

4.3. Secreted factors

FGFs FGF-9, FGF-16, and FGF-20 are expressed in the endocardium and epicardium, and RA can induce epicardial expression of FGF-9. Studies using FGFR-1 and FGFR-2 knockout mice have shown that endocardial- and epicardial-derived FGF signals regulate myocardial proliferation during midgestation heart development (Lavine et al., 2005). Furthermore, FGF signals promote coronary growth indirectly by signaling to the cardiomyoblast through redundant function of FGFR-1 and FGFR-2. Myocardial FGF signaling triggers a wave of hedgehog (HH) activation that is essential for VEGF-A, VEGF-B, VEGF-C, and angiopoietin-2 (ANG-2) expression. Activation of HH signaling was sufficient to promote coronary growth and rescue coronary defects due in *FGFR* mutant mice (Lavine et al., 2006).

5. CONCLUSION

Despite numerous questions concerning the formation of the coronary vasculature and the mechanisms of epicardial–myocardial interaction, the recent development of tools to target epicardial derivatives will be instrumental to address these issues. The generation of epicardial-specific mutant mice will very likely provide insight into the genetics of epicardial development and coronary formation.

REFERENCES

Bock-Marquette, I., Saxena, A., White, M.D., Dimaio, J.M., Srivastava, D. 2004. Thymosin beta4 activates integrin-linked kinase and promotes cardiac cell migration, survival and cardiac repair. Nature 432, 466–472.

Born, G. 1889. Beiträge zur Entwicklungsgeschichte des Säugetierherzens. Arch. Mikrosk. Anat. 33, 284–378.

Bouchey, D., Drake, C.J., Wunsch, A.M., Little, C.D. 1996. Distribution of connective tissue proteins during development and neovascularization of the epicardium. Cardiovasc. Res. 31(Spec. No.), E104–E115.

Burch, G.H., Bedolli, M.A., McDonough, S., Rosenthal, S.M., Bristow, J. 1995. Embryonic expression of tenascin-X suggests a role in limb, muscle, and heart development. Dev. Dyn. 203, 491–504.

Chen, J., Kubalak, S.W., Chien, K.R. 1998. Ventricular muscle-restricted targeting of the RXRalpha gene reveals a non-cell-autonomous requirement in cardiac chamber morphogenesis. Development 125, 1943–1949.

Chen, T.H., Chang, T.C., Kang, J.O., Choudhary, B., Makita, T., Tran, C.M., Burch, J.B., Eid, H., Sucov, H.M. 2002. Epicardial induction of fetal cardiomyocyte proliferation via a retinoic acid-inducible trophic factor. Dev. Biol. 250, 198–207.

Compton, L.A., Potash, D.A., Mundell, N.A., Barnett, J.V. 2006. Transforming growth factor-beta induces loss of epithelial character and smooth muscle cell differentiation in epicardial cells. Dev. Dyn. 235, 82–93.

Crispino, J.D., Lodish, M.B., Thurberg, B.L., Litovsky, S.H., Collins, T., Molkentin, J.D., Orkin, S.H. 2001. Proper coronary vascular development and heart morphogenesis depend on interaction of GATA-4 with FOG cofactors. Genes Dev. 15, 839–844.

Dathe, V., Brand-Saberi, B. 2004. Expression of thymosin beta4 during chick development. Anat. Embryol. (Berl.) 208, 27–32.

De Haan, R.L. 1965. *Morphogenesis of the Vertebrate Heart*. New York: Holt Reinhardt and Winston.

Dettman, R.W., Denetclaw, W.Jr., Ordahl, C.P., Bristow, J. 1998. Common epicardial origin of coronary vascular smooth muscle, perivascular fibroblasts, and intermyocardial fibroblasts in the avian heart. Dev. Biol. 193, 169–181.

Dettman, R.W., Pae, S.H., Morabito, C., Bristow, J. 2003. Inhibition of alpha4-integrin stimulates epicardial-mesenchymal transformation and alters migration and cell fate of epicardially derived mesenchyme. Dev. Biol. 257, 315–328.

Foxon, G.E. 1950. A description of the coronary arteries in dipnoan fishes and some remarks on their importance from the evolutionary standpoint. J. Anat. 84, 121–131.

Fransen, M.E., Lemanski, L.F. 1990. Epicardial development in the axolotl, Ambystoma mexicanum. Anat. Rec. 226, 228–236.

Fritsch, G. 1869. Zur vergleichenden Anatomie der Amphibienherzen. Arch. Anat. Physiol. Jg 1869, 654–758.

Gittenberger-de Groot, A.C., Vrancken Peeters, M.P., Mentink, M.M., Gourdie, R.G., Poelmann, R.E. 1998. Epicardium-derived cells contribute a novel population to the myocardial wall and the atrioventricular cushions. Circ. Res. 82, 1043–1052.

Gourdie, R.G., Cheng, G., Thompson, R.P., Mikawa, T. 2000. Retroviral cell lineage analysis in the developing chick heart. Methods Mol. Biol. 135, 297–304.

Grant, D.S., Rose, W., Yaen, C., Goldstein, A., Martinez, J., Kleinman, H. 1999. Thymosin beta4 enhances endothelial cell differentiation and angiogenesis. Angiogenesis 3, 125–135.

Grant, R.T., Regnier, M. 1926. The comparative anatomy of cardiac coronary vessels. Heart 13, 285–317.

Greil, A. 1903. Beitrage zur vergleichenden Anatomie und Entwicklungsgeschichte des Herzens und des Truncus arteriosus der Wirbelthiere. Gegenbaurs Morphol. JB 31, 123–310.

Guadix, J.A., Carmona, R., Munoz-Chapuli, R., Perez-Pomares, J.M. 2006. In vivo and in vitro analysis of the vasculogenic potential of avian proepicardial and epicardial cells. Dev. Dyn. 235, 1014–1026.

Hay, E.D., Zuk, A. 1995. Transformations between epithelium and mesenchyme: Normal, pathological, and experimentally induced. Am. J. Kidney Dis. 26, 678–690.

Hirose, T., Karasawa, M., Sugitani, Y., Fujisawa, M., Akimoto, K., Ohno, S., Noda, T. 2006. PAR3 is essential for cyst-mediated epicardial development by establishing apical cortical domains. Development 133, 1389–1398.

Hiruma, T., Hirakow, R. 1989. Epicardial formation in embryonic chick heart: Computer-aided reconstruction, scanning, and transmission electron microscopic studies. Am. J. Anat. 184, 129–138.

His, W. 1881. Mittheilungen zur Embryologie der Säugethiere und des Menschen. Arch. Anat. Entwickl. Gesch. Jg 1881, 1303–1329.

Hochstetter, F. 1906. Beitrage zur Anatomie und Entwicklungsgeschichte des Blutgefässsystems der krokodile. In: Reise in Ostarfrika (A. Voeltzkow, Ed.), Vol. 4, Wissenschaftliche Ergebnisse.

Hochstetter, F. 1908. Diskussionsbeitrag. Verh. Anat. Ges. 22, 180.

Hur, H., Kim, Y.J., Noh, C.I., Seo, J.W., Kim, M.H. 1999. Molecular genetic analysis of the DiGeorge syndrome among Korean patients with congenital heart disease. Mol. Cells 9, 72–77.

Hurle, J.M., Kitten, G.T., Sakai, L.Y., Volpin, D., Solursh, M. 1994. Elastic extracellular matrix of the embryonic chick heart: An immunohistological study using laser confocal microscopy. Dev. Dyn. 200, 321–332.

Joberty, G., Petersen, C., Gao, L., Macara, I.G. 2000. The cell-polarity protein Par6 links Par3 and atypical protein kinase C to Cdc42. Nat. Cell Biol. 2, 531–539.

Kalman, F., Viragh, S., Modis, L. 1995. Cell surface glycoconjugates and the extracellular matrix of the developing mouse embryo epicardium. Anat. Embryol. (Berl.) 191, 451–464.

Kastner, P., Grondona, J.M., Mark, M., Gansmuller, A., LeMeur, M., Decimo, D., Vonesch, J. L., Dolle, P., Chambon, P. 1994. Genetic analysis of RXR alpha developmental function: Convergence of RXR and RAR signaling pathways in heart and eye morphogenesis. Cell 78, 987–1003.

Kirschner, K.M., Wagner, N., Wagner, K.D., Wellmann, S., Scholz, H. 2006. The Wilms' tumor suppressor WT1 promotes cell adhesion through transcriptional activation of the alpha 4integrin gene. J. Biol. Chem. 281(42), 31930–31939.

Kölliker, A. 1879. Entwicklungsgeschichte des Menschen und der Thiere. Leipzig, Germany: Engleman.

Komiyama, M., Ito, K., Shimada, Y. 1987. Origin and development of the epicardium in the mouse embryo. Anat. Embryol. (Berl.) 176, 183–189.

Kruithof, B.P., van Wijk, B., Somi, S., Kruithof-de Julio, M., Perez Pomares, J.M., Weesie, F., Wessels, A., Moorman, A.F., van den Hoff, M.J. 2006. BMP and FGF regulate the differentiation of multipotential pericardial mesoderm into the myocardial or epicardial lineage. Dev. Biol. 295(2), 507–522.

Kuhn, H.J., Liebherr, G. 1988. The early development of the epicardium in tupaia belangeri. Anat. Embryol. (Berl.) 177, 225–234.

Kurkiewicz, T. 1909. O histogenezie miesna sercowego zwierzat kregowych-Zur Histogenese des Herzmuskels der Wirbeltiere. Bull. Int. Acad. Sci. Cracovie. 148–191.

Kwee, L., Baldwin, H.S., Shen, H.M., Stewart, C.L., Buck, C., Buck, C.A., Labow, M.A. 1995. Defective development of the embryonic and extraembryonic circulatory systems in vascular cell adhesion molecule (VCAM-1) deficient mice. Development 121, 489–503.

Landerholm, T.E., Dong, X.R., Lu, J., Belaguli, N.S., Schwartz, R.J., Majesky, M.W. 1999. A role for serum response factor in coronary smooth muscle differentiation from proepicardial cells. Development 126, 2053–2062.

Lavine, K.J., Yu, K., White, A.C., Zhang, X., Smith, C., Partanen, J., Ornitz, D.M. 2005. Endocardial and epicardial derived FGF signals regulate myocardial proliferation and differentiation in vivo. Dev. Cell 8, 85–95.

Lavine, K.J., White, A.C., Park, C., Smith, C.S., Choi, K., Long, F., Hui, C.C., Ornitz, D.M. 2006. Fibroblast growth factor signals regulate a wave of hedgehog activation that is essential for coronary vascular development. Genes Dev. 20, 1651–1666.

Lee, C.S., Friedman, J.R., Fulmer, J.T., Kaestner, K.H. 2005. The initiation of liver development is dependent on Foxa transcription factors. Nature 435, 944–947.

Lepilina, A., Coon, A.N., Kikuchi, K., Holdway, J.E., Roberts, R.W., Burns, C.G., Poss, K.D. 2006. A dynamic epicardial injury response supports progenitor cell activity during zebrafish heart regeneration. Cell 127, 607–619.

Lieberkuhn, N. 1876. Über die Allantois und die Nieren von Säugethierembryonen. Sitzungsberichte d Ges z Beförderung der Gesamten Naturwiss z Marburg 1, 1–11.

Lockwood, C.B. 1888. The early development of the pericardium, diaphragm, and great veins. Phil. Trans. R. Soc. Lond. Seri. B 179, 365–384.

Lu, J., Landerholm, T.E., Wei, J.S., Dong, X.R., Wu, S.P., Liu, X., Nagata, K., Inagaki, M., Majesky, M.W. 2001. Coronary smooth muscle differentiation from proepicardial cells requires rhoA-mediated actin reorganization and p160 rho-kinase activity. Dev. Biol. 240, 404–418.

MacKinnon, M.R., Heatwole, H. 1981. Comparative cardiac anatomy of the reptilia. IV. The coronary arterial circulation. J. Morphol. 170, 1–27.

Manasek, F.J. 1968. Embryonic development of the heart. I. A light and electron microscopic study of myocardial development in the early chick embryo. J. Morphol. 125, 329–365.

Mancini, L., Bertossi, M., Ribatti, D., Bartoli, F., Nico, B., Lozupone, E., Roncali, L. 1991. The effects of long-term hypoxia on epicardium and myocardium in developing chick embryo hearts. Int. J. Microcirc. Clin. Exp. 10, 359–371.

Männer, J. 1992. The development of pericardial villi in the chick embryo. Anat. Embryol. (Berl.) 186, 379–385.

Männer, J. 1993. Experimental study on the formation of the epicardium in chick embryos. Anat. Embryol. (Berl.) 187, 281–289.

Männer, J. 1999. Does the subepicardial mesenchyme contribute myocardioblasts to the myocardium of the chick embryo heart? A quail-chick chimera study tracing the fate of the epicardial primordium. Anat. Rec. 255, 212–226.

Männer, J., Seidl, W., Steding, G. 1995. The role of extracardiac factors in normal and abnormal development of the chick embryo heart: Cranial flexure and ventral thoracic wall. Anat. Embryol. (Berl.) 191, 61–72.

Männer, J., Perez-Pomares, J.M., Macias, D., Munoz-Chapuli, R. 2001. The origin, formation and developmental significance of the epicardium: A review. Cells Tissues Organs 169, 89–103.

Männer, J., Schlueter, J., Brand, T. 2005. Experimental analyses of the function of the proepicardium using a new microsurgical procedure to induce loss-of-proepicardial-function in chick embryos. Dev. Dyn. 233, 1454–1463.

Markwald, R., Eisenberg, C., Eisenberg, L., Trusk, T., Sugi, Y. 1996. Epithelial-mesenchymal transformations in early avian heart development. Acta Anat. (Basel) 156, 173–186.

Merki, E., Zamora, M., Raya, A., Kawakami, Y., Wang, J., Zhang, X., Burch, J., Kubalak, S.W., Kaliman, P., Belmonte, J.C., Chien, K.R., Ruiz-Lozano, P. 2005. From the cover: Epicardial retinoid X receptor {alpha} is required for myocardial growth and coronary artery formation. Proc. Natl. Acad. Sci. USA 102, 18455–18460.

Mikawa, T., Fischman, D.A. 1992. Retroviral analysis of cardiac morphogenesis: Discontinuous formation of coronary vessels. Proc. Natl. Acad. Sci. USA 89, 9504–9508.

Mikawa, T., Gourdie, R.G. 1996. Pericardial mesoderm generates a population of coronary smooth muscle cells migrating into the heart along with ingrowth of the epicardial organ. Dev. Biol. 174, 221–232.

Mollier, S. 1906. Die erste Anlage des Herzens bei den Wirbeltieren. In: *Handbuch der Vergleichenden und Experimentellen Entwicklungslehre der Wirbeltiere.* Sect. 1(O. Hertwig, Ed.), Vol. 1, Jena: Gustav Fischer, pp. 1026–1051.

Moore, A.W., Schedl, A., McInnes, L., Doyle, M., Hecksher-Sorensen, J., Hastie, N.D. 1998. YAC transgenic analysis reveals Wilms' tumour 1 gene activity in the proliferating coelomic epithelium, developing diaphragm and limb. Mech. Dev. 79, 169–184.

Moore, A.W., McInnes, L., Kreidberg, J., Hastie, N.D., Schedl, A. 1999. YAC complementation shows a requirement for Wt1 in the development of epicardium, adrenal gland and throughout nephrogenesis. Development 126, 1845–1857.

Morabito, C.J., Dettman, R.W., Kattan, J., Collier, J.M., Bristow, J. 2001. Positive and negative regulation of epicardial-mesenchymal transformation during avian heart development. Dev. Biol. 234, 204–215.

Morabito, C.J., Kattan, J., Bristow, J. 2002. Mechanisms of embryonic coronary artery development. Curr. Opin. Cardiol. 17, 235–241.

Morita, T., Shinozawa, T., Nakamura, M., Awaya, A., Sato, N., Ishiwata, I., Kami, K. 1998. Expressions of a 68kDa-glycoprotein (GP68) and laminin in the mesodermal tissue of the developing mouse embryo. Okajimas Folia Anat. Jpn. 75, 185–195.

Munoz-Chapuli, R., Hamlett, W.C. 1996. Epilogue: Comparative cardiovascular biology of lower vertebrates. J. Exp. Zool. 275, 249–251.

Munoz-Chapuli, R., Macias, D., Ramos, C., de Andres, V., Gallego, A., Navarro, P. 1994. Cardiac development in the dogfish (*Scyliorhinus canicula*): A model for the study of vertebrate cardiogenesis. Cardioscience 5, 245–253.

Munoz-Chapuli, R., Macias, D., Ramos, C., Fernandez, B., Sans-Coma, V. 1997. Development of the epicardium in the dogfish. Acta Zool. 78, 39–46.

Nahirney, P.C., Mikawa, T., Fischman, D.A. 2003. Evidence for an extracellular matrix bridge guiding proepicardial cell migration to the myocardium of chick embryos. Dev. Dyn. 227, 511–523.

Olivey, H.E., Mundell, N.A., Austin, A.F., Barnett, J.V. 2006. Transforming growth factor-beta stimulates epithelial-mesenchymal transformation in the proepicardium. Dev. Dyn. 235, 50–59.

Perez-Pomares, J.M., Macias, D., Garcia-Garrido, L., Munoz-Chapuli, R. 1997. Contribution of the primitive epicardium to the subepicardial mesenchyme in hamster and chick embryos. Dev. Dyn. 210, 96–105.

Perez-Pomares, J.M., Macias, D., Garcia-Garrido, L., Munoz-Chapuli, R. 1998. Immunolocalization of the vascular endothelial growth factor receptor-2 in the subepicardial mesenchyme of hamster embryos: Identification of the coronary vessel precursors. Histochem. J. 30, 627–634.

Perez-Pomares, J.M., Munoz-Chapuli, R., Wessels, A. 2001. The contribution of the proepicardum to avian cardiovascular development. Int. J. Dev. Biol. 45, S155–S156.

Perez-Pomares, J.M., Carmona, R., Gonzalez-Iriarte, M., Atencia, G., Wessels, A., Munoz-Chapuli, R. 2002. Origin of coronary endothelial cells from epicardial mesothelium in avian embryos. Int. J. Dev. Biol. 46, 1005–1013.

Perez-Pomares, J.M., Phelps, A., Sedmerova, M., Wessels, A. 2003. Epicardial-like cells on the distal arterial end of the cardiac outflow tract do not derive from the proepicardium but are derivatives of the cephalic pericardium. Dev. Dyn. 227, 56–68.

Remak, R. 1885a. Über die Entwicklung des Hühnchens im Ei. Arch. Anat. Physiol. Wiss Med. Jg 1843, 478–484.

Remak, R. 1885b. Untersuchungen über die Entwicklung der Wirbelthiere. Berlin: Reimer.

Romanoff, A.L. 1960. The Avian Embryo. Structural and Functional Development. New York: Macmillan Co.

Schlueter, J., Männer, J., Brand, T. 2006. BMP is an important regulator of proepicardial identity in the chick embryo. Dev. Biol. 295, 546–558.

Schulte, I., Schlueter, J., Abu-Issa, R., Brand, T., Männer, J. 2007. Morphological and molecular left-right asymmetries in development of the proepicardium: A comparative analysis on mouse and chick embryos. Dev. Dyn. 236, 684–695.

Sejima, H., Isokawa, K., Shimizu, O., Morikawa, T., Ootsu, H., Numata, K., Fukai, M., Kubota, S., Toda, Y. 2001. Possible participation of isolated epicardial cell clusters in the formation of chick embryonic epicardium. J. Oral. Sci. 43, 109–116.

Shimada, Y., Ho, E. 1980. Scanning electron microscopy of the embryonic chick heart: Formation of the epicardium and surface structure of the four heterotypic cells that contribute to the embryonic heart. In: Etiology and Morphogenesis of Congenital Heart Disease (R. Van Praagh, A. Takao, Eds.), New York: Futura, pp. 63–80.

Smart, N., Risebro, C.A., Melville, A.A., Moses, K., Schwartz, R.J., Chien, K.R., Riley, P.R. 2006. Thymosin beta4 induces adult epicardial progenitor mobilization and neovascularization. Nature 445(7124), 177–182.

Steding, G., Klemeyer, L. 1969. [The development of the pericardial fold in the chick embryo]. Z. Anat. Entwicklungsgesch. 129, 223–233.

Streeter, G.L. 1945. Developmental horizons in human embryos. description of age group XIII, embryos about 4 or 5 milimeters long, and age group XIV, period of indentation of the lens vesicle. Contrib. Embryol. 31, 29–64.

Stuckmann, I., Evans, S., Lassar, A.B. 2003. Erythropoietin and retinoic acid, secreted from the epicardium, are required for cardiac myocyte proliferation. Dev. Biol. 255, 334–349.

Sucov, H.M., Dyson, E., Gumeringer, C.L., Price, J., Chien, K.R., Evans, R.M. 1994. RXR alpha mutant mice establish a genetic basis for vitamin A signaling in heart morphogenesis. Genes Dev. 8, 1007–1018.

Suzuki, A., Yamanaka, T., Hirose, T., Manabe, N., Mizuno, K., Shimizu, M., Akimoto, K., Izumi, Y., Ohnishi, T., Ohno, S. 2001. Atypical protein kinase C is involved in the evolutionarily conserved par protein complex and plays a critical role in establishing epithelia-specific junctional structures. J. Cell Biol. 152, 1183–1196.

Tevosian, S.G., Deconinck, A.E., Tanaka, M., Schinke, M., Litovsky, S.H., Izumo, S., Fujiwara, Y., Orkin, S.H. 2000. FOG-2, a cofactor for GATA transcription factors, is essential for heart morphogenesis and development of coronary vessels from epicardium. Cell 101, 729–739.

Tidball, J.G. 1992. Distribution of collagens and fibronectin in the subepicardium during avian cardiac development. Anat. Embryol. (Berl.) 185, 155–162.

Tomanek, R.J., Ratajska, A., Kitten, G.T., Yue, X., Sandra, A. 1999. Vascular endothelial growth factor expression coincides with coronary vasculogenesis and angiogenesis. Dev. Dyn. 215, 54–61.

Tsuda, T., Majumder, K., Linask, K.K. 1998. Differential expression of flectin in the extracellular matrix and left- right asymmetry in mouse embryonic heart during looping stages. Dev. Genet. 23, 203–214.

Van den Eijnde, S.M., Wenink, A.C., Vermeij-Keers, C. 1995. Origin of subepicardial cells in rat embryos. Anat. Rec. 242, 96–102.

van Tuyn, J., Atsma, D.E., Winter, E.M., van der Velde-van Dijke, I., Pijnappels, D.A., Bax, N. A., Knaan-Shanzer, S., Gittenberger-de Groot, A.C., Poelmann, R.E., van der Laarse, A., van der Wall, E.E., Schalij, M.J., et al. 2006. Epicardial cells of human adults can undergo an epithelial-to-mesenchymal transition and obtain characteristics of smooth muscle cells *in vitro*. Stem Cells 25, 271–278.

Viragh, S., Challice, C.E. 1973. Origin and differentiation of cardiac muscle cells in the mouse. J. Ultrastruct. Res. 42, 1–24.

Viragh, S., Challice, C.E. 1981. The origin of the epicardium and the embryonic myocardial circulation in the mouse. Anat. Rec. 201, 157–168.

Viragh, S., Gittenberger-de Groot, A.C., Poelmann, R.E., Kalman, F. 1993. Early development of quail heart epicardium and associated vascular and glandular structures. Anat. Embryol. (Berl.) 188, 381–393.

Vrancken Peeters, M.P., Mentink, M.M., Poelmann, R.E., Gittenberger-de Groot, A.C. 1995. Cytokeratins as a marker for epicardial formation in the quail embryo. Anat. Embryol. (Berl.) 191, 503–508.

Wada, A.M., Smith, T.K., Osler, M.E., Reese, D.E., Bader, D.M. 2003. Epicardial/mesothelial cell line retains vasculogenic potential of embryonic epicardium. Circ. Res. 92(5), 525–531.

Wessels, A., Perez-Pomares, J.M. 2004. The epicardium and epicardially derived cells (EPDCs) as cardiac stem cells. Anat. Rec. A Discov. Mol. Cell. Evol. Biol. 276, 43–57.

Wu, H., Lee, S.H., Gao, J., Liu, X., Iruela-Arispe, M.L. 1999. Inactivation of erythropoietin leads to defects in cardiac morphogenesis. Development 126, 3597–3605.

Xavier-Neto, J., Neville, C.M., Shapiro, M.D., Houghton, L., Wang, G.F., Nikovits, W., Stockdale, F.E., Rosenthal, N. 1999. A retinoic acid-inducible transgenic marker of sinoatrial development in the mouse heart. Development 126, 2677–2687.

Xavier-Neto, J., Rosenthal, N., Silva, F.A., Matos, T.G., Hochgreb, T., Linhares, V.L. 2001. Retinoid signaling and cardiac anteroposterior segmentation. Genesis 31, 97–104.

Yang, J.T., Rayburn, H., Hynes, R.O. 1995. Cell adhesion events mediated by alpha 4 integrins are essential in placental and cardiac development. Development 121, 549–560.

CHAPTER 14

Genetics of Transcription Factor Mutations

Vijaya Ramachandran and D. Woodrow Benson

Contents

Abstract

Congenital heart disease (CHD) is an important form of cardiovascular disease in the young, which is characterized by malformations of the cardiovascular system discovered at or near the time of birth. Mutations in genes coding for transcription factors have been associated with CHD demonstrating that transcription factors play a critical role in many aspects of cardiac development. In the postgenomic era, a variety of approaches including family-based

Department of Pediatrics, Division of Cardiology, MLC 7042, University of Cincinnati, Cincinnati Children's Hospital Medical Center, Cincinnati, Ohio

Advances in Developmental Biology, Volume 18
ISSN 1574-3349, DOI: 10.1016/S1574-3349(07)18014-5

linkage studies, cytogenetic studies, association studies, and mutation analysis of candidate genes are available for use in human
genetic studies. Such approaches have been used to identify
seven genes encoding transcription factors (*FOG2, GATA4, NKX2.5,
TBX1, TBX5, TFAP2B,* and *ZIC3*) implicating mutations in transcription
factors as a significant cause of CHD. Identification of transcription factor mutations has been useful to the clinician for diagnosis, disease classification, genetic counseling, and population
screening. At the same time, human transcription factor mutations
and their corresponding phenotypes have informed developmental
biologists and biochemists of transcription factor function. In this
chapter, we describe the genetic approaches used to identify the
causative genes for CHD and assess the genotype–phenotype correlations of transcription factor mutations.

1. INTRODUCTION

Transcription factors play a critical role in various developmental processes,
initiating expression of different genes in specific tissues and also mediating
responses to environmental and hormonal stimulation. Mutations in the
genes coding for transcription factors disrupt the transcriptional machinery
and have been associated with a variety of disease phenotypes, mostly
confined to the tissues expressing the mutated gene. Such mutations are
an important cause of congenital heart disease (CHD). In this chapter,
we describe the genetic approaches used to identify the causative genes
for CHD and assess the genotype–phenotype correlations of transcription
factor mutations identified to date.

2. WHAT IS CHD?

CHD is a form of cardiovascular disease in the young, which is characterized by malformations of the cardiovascular system discovered at or near
the time of birth. Cardiac malformations are an important component of
CHD and constitute a major portion of clinically significant birth defects
with a definition dependent estimated incidence of 4–50 per 1000 live
births. For example, it has been estimated that 4–10 live born infants per
1000 have a cardiac malformation, 40% of which are diagnosed in the first
year of life (Benson, 2002; Hoffman and Kaplan, 2002; Hinton et al., 2005).
However, bicuspid aortic valve (BAV), the most common cardiac malformation, is usually excluded from this estimate. BAV is associated with
considerable morbidity and mortality in affected individuals, and by itself
occurs in 10–20 per 1000 of the population (Ward, 2000; Cripe et al., 2004).
When isolated aneurysm of the atrial septum and persistent left superior

vena cava, for each occuring in 5–10 per 1000 (Benson et al., 1998), is taken into account, the incidence of cardiac malformations approaches 50 per 1000 (~5%) live births. Although it was believed that the etiology of CHD was multifactorial, the recent advances in molecular genetics suggest that single-gene defects are more common than previously thought, and the same malformation may be caused by mutant genes at different loci (genetic heterogeneity) (Benson, 2002; Gelb, 2004). Molecular genetics is revolutionizing our understanding of CHD by providing unparalleled insights into the pathogenesis based on precise knowledge of genes, the encoded proteins, and their phenotypic consequences.

3. HUMAN GENETIC STUDY DESIGNS

Before undertaking a genetic study, it is essential to establish statistical evidence for a genetic component of the disease. All "familial" diseases are not genetic; familial disease can be due to shared environmental factors. Hence, the cause for familial clustering has to be established. Several methodologies, for example, familial aggregation, twin studies, adoption studies, familial recurrent risk ratio, heritability, and segregation analysis have been utilized for this purpose (Ashley-Koch, 2006).

3.1. General considerations

Once the genetic component of a disease has been established, designing a study to identify the causative genes is an elaborate process requiring a stepwise approach. Phenotype definition is the first step in a genetic study, and hence, the role of experienced clinicians is vital to success. It is important to obtain the family history in all patients, identify individuals and families with the trait of interest (characterize phenotype), and construct a pedigree. This information will help in establishing the mode of inheritance and will direct the method of statistical analysis. If the mode of inheritance is not clear, a nonparametric statistical analysis can be performed. Genetic studies of CHD are challenging because major factors affecting phenotype, including heterogeneity (at the allelic or locus level), incomplete penetrance, and variable expressivity, are observed frequently (Hinton et al., 2005; Clark et al., 2006).

Once the phenotype is defined, a decision as to whether a linkage study or an association study is appropriate can be made. If large multiplex families are available, then linkage study is ideal. Additionally, the choice of markers, either microsatellites or single nucleotide polymorphisms (SNPs), has to be decided. Both decisions have an effect on the sample size necessary to achieve linkage. For example, BAV has a prevalence of 1–2%, and familial clustering has long been known

(Cripe et al., 2004). Nearly equal proportions of single and multiple affected families with BAV have been identified (Martin et al., 2007). Comparison of power analyses for linkage versus association methods indicated that a fivefold increase in the sample size would be required for association analysis to have the same power as for linkage analysis to detect a significant effect ($p \leq 0.001$). In such a situation, a family-based linkage analysis proved successful in identifying a locus on chromosome 18q (Martin et al., 2007). Since BAV is a complex trait, a family-based association study using SNPs is ideal for narrowing the region on 18q. Such a study design is robust and will identify the BAV-causing genes because the various approaches are complementary and utilize all families whether they contain single or multiple affected individuals, as well as prior evidence (linkage) that this region is implicated in trait etiology.

3.2. Specific study designs

Four basic human genetic approaches have been utilized to identify genes causing CHD: family-based linkage studies, cytogenetic studies, population-based or family-based association studies, and mutation analysis of candidate genes. Candidate genes may be identified as positional candidates or biological candidates (Fig. 1). Positional candidate genes are considered candidates because they are encoded at a chromosomal locus that is linked to disease. Biological candidates are usually identified based on biological or developmental insight that comes from studies in model systems (Fig. 2).

3.2.1. Family-based linkage studies

Linkage analysis is used to identify a disease locus on a chromosome by demonstrating that the causative gene is in proximity to a genetic marker on the same chromosome. Linkage analysis relies on "informative families"; affected as well as unaffected family members contribute information to a linkage study. The genetic transmission of disease in families can be Mendelian or complex. In Mendelian inheritance, the disease or trait transmission conforms to the Laws of Mendel; patterns of Mendelian inheritance include autosomal dominant, autosomal recessive, X-linked dominant, X-linked recessive, and Y-linked. In complex inheritance, the disease or trait is influenced by two or more loci with possible environmental influences and may exhibit multiple modes of inheritance. If the inheritance is Mendelian, a parametric analysis is performed using estimates of the penetrance of the disease gene, phenocopy rate, and allele frequency. If the inheritance is complex, both parametric and nonparametric analyses are performed. The odds for and against linkage are calculated and the logarithmic odds are referred to as LOD score. A LOD score greater than 3 provides strong evidence for linkage and a LOD lesser than -2 excludes

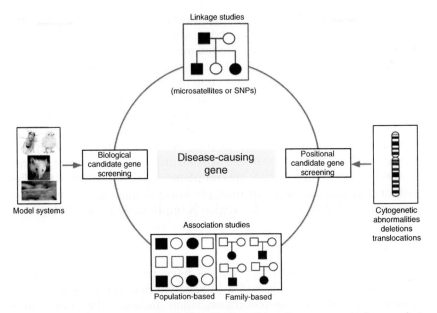

FIGURE 1 Four basic genetic approaches for identifying disease genes. Linkage analysis is a traditional method that utilizes families with multiple affected individuals. Association studies use unrelated cases and controls or families; when families are used, the unaffected members serve as the controls. Genotyping is performed using either microsatellites or SNPs. Cytogenetic abnormalities involving breaks, usually translocations or deletions, provide clues about the location of causative genes. Candidate genes are prioritized based on biological importance and screened for mutations.

linkage. Linkage provides an approximate location (locus) of the disease gene within the genome, but the number of genes in these regions varies from being relatively gene-poor to hundreds of genes. For example, the linkage region on 18q harboring BAV-causing genes is ~10 cM but has only 29 genes, which is a relatively gene-poor region (Martin et al., 2007). After the identification of a locus, fine mapping is used to further narrow the locus to reduce the size of the linkage region and the number of genes. Then the positional candidate genes are prioritized and sequenced to discover disease-causing alterations (Fig. 2). Linkage analysis has proven successful in identifying CHD-causing transcription factor mutations, including *GATA4, NKX2.5, TBX5, TFAP2B,* and *ZIC3,* all of which have become important tools for the study of normal heart development and the definition of CHD pathogenesis.

3.2.2. Cytogenetic studies

Cytogenetic studies have been useful in elucidating the genetic cause of many syndromes that are associated with CHD, but they have also lead to identification of genes causing nonsyndromic CHD (Curran et al., 1993;

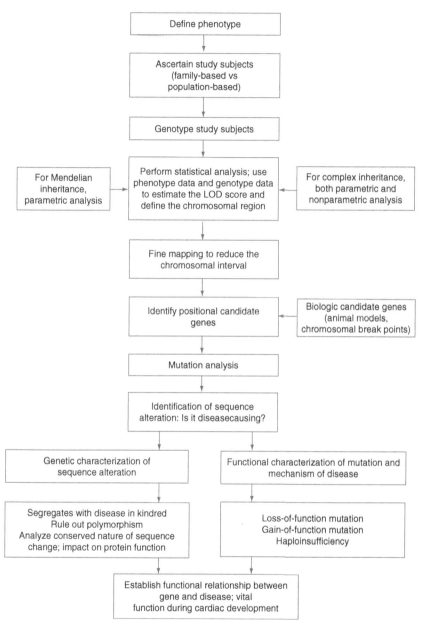

FIGURE 2 Flowchart illustrating the steps involved in traditional linkage or association studies.

Pehlivan et al., 1999; Muncke et al., 2003). Chromosomal abnormalities such as translocations and deletions provide clues for localizing disease-causing genes as an alternative to linkage analysis (Fig. 1). Chromosomal break points can cause a loss-of-function phenotype: sometimes, the breakage may occur at a point where it separates the gene from the regulatory sequences or it may disrupt the coding sequence; either scenario may cause disease due to haploinsufficiency. For example, the elastin gene on chromosome 7 was identified at a translocation break point in patients with familial supravalvular aortic stenosis (SVAS) (Curran et al., 1993). A similar approach was used to identify the *PROSIT240 (TRAP)* gene in patients with transposition of great arteries (Muncke et al., 2003). Deletion mapping identified a locus on 3p25 where *CRELD1* is encoded; mutations in *CRELD1* (AVSD2, OMIM 606217) gene were subsequently identified as a cause for atrioventricular septal defect (AVSD) (Green et al., 2000; Robinson et al., 2003).

3.2.3. Candidate gene analysis
In this approach, mutation analysis is performed directly on candidate genes. Candidate genes may be positional (identified in a linkage analysis or cytogenetic study) or biological (identified based on results of analysis in a developmental system or pathological tissues). The availability of human genome databases (Table 1) has facilitated the *"in silico"* search for candidate genes by providing information on both detailed chromosomal location and gene expression in different embryonic and adult tissues. In applying this technique to identify genes causing CHD, a specific gene is identified as a strong candidate based on its involvement in cardiac development (expressed in the right place at the right time). Clues for involvement in cardiac development may also be surmised during the

TABLE 1   List of genome browsers

Name	URL
National Center for Biotechnology Information (NCBI)	www.ncbi.nlm.nih.gov/ mapview/map_search.cg? taxid+9606
Ensembl Genome Browser	www.ensembl.org/ Homo_sapiens/
University of California at Santa Cruz (UCSC) Genome Browser	www.genome.cse.ucsc.edu/
GeneCards	www.genecards.org
HapMap Project	www.hapmap.org
ENCODE Project	www.genome.gov

course of phenotype characterization in gene deletion studies in animal models, for example, identification of *FOG2/ZFPM2* gene as a cause of CHD (Pizzuti et al., 2003).

3.2.4. Association studies

A more recent technology in gene identification is association studies, which are well suited for genetic studies of complex traits (Botstein and Risch, 2003; Edwards et al., 2005; Klein et al., 2005; Duerr et al., 2006). It has been hypothesized that complex or polygenic traits are caused by the interaction of genes and environment. Hence, a single genetic or environmental variant may not be sufficient to cause disease and the absence of the variant does not prevent development of disease. Many genetic studies of cardiovascular disease in the young have observed that reduced penetrance and variable expressivity result in limited genotype–phenotype correlation, that is the genotype does not completely predict the phenotype. In this context, many forms of CHD may be considered to be complex traits. For example, BAV and hypoplastic left heart syndrome (HLHS) may be complex traits where multiple genes interact resulting in increased risk of disease susceptibility. It is difficult to map genes for complex disease by traditional linkage analysis. An alternate approach to map the susceptibility genes of complex traits is by performing association studies, which assess a significantly increased or decreased frequency of a marker allele, genotype, or haplotype with a disease phenotype, than would be expected by chance if there were no association between markers and phenotype. Association studies may be family-based, but most often they are population-based and use a case-control design. In population-based case-control studies, the frequency of haplotypes and genotypes are compared and differences in the frequencies between the groups are tested for statistical significance. In family-based association studies, the statistical analysis is performed using the unaffected family members as internal controls.

Association studies are also performed to narrow known linkage regions facilitating identification of causative genes. Based on technical considerations, the current trend is to use SNPs for genotyping. Since SNPs have only two alleles, they are relatively inexpensive to genotype, and they require less DNA template. SNPs also have fewer genotyping errors and a low inconsistency rate when compared to microsatellites (Hinrichs and Suarez, 2005). SNP-based case-control studies require a large sample size, which is a limitation for many forms of CHD. A benefit of using microsatellites in family-based association studies [transmission disequilibrium test (TDT)] is that they require a smaller sample size. Moreover, validation of the results of the association studies is also important. For instance, replicating the results in a second dataset is advantageous in addition to providing evidence by functional studies and defining interactions. Theoretical analyses predict that association studies have greater potential than linkage studies in identifying the genetic components of common complex diseases and will require large-scale testing to identify the genes causing the disease. Accordingly, it has been argued that linkage analysis that has been used successfully to find major genes has limited power to detect genes of modest effect, but association studies that utilize candidate genes has far greater power, even if one needs to test every gene in the genome (Risch and Merikangas, 1996). However, allelic heterogeneity (multiple variants within a gene) substantially reduces the power of association-based methods. For instance, many genes that have been implicated in CHD exhibit different mutations in different families (Hinton et al., 2005). Population stratification, which can result in spurious association, is another issue in population-based association studies; it arises due to the different ethnic background of the cases and controls (Mayeux, 2005). It has been shown that allele frequencies differ in major racial/ethnic groups owing to historical isolation (Dean et al., 1994; Stephens et al., 2001). Therefore, the results of the association studies must be confirmed by replication studies in a similar second independent dataset.

With the completion of the International HapMap Project (Table 1), SNPs are also used in whole genome association studies (WGA). WGA is a powerful genome screen that may genotype up to 500,000 SNPs distributed across the genome with high fidelity and comparatively low cost using a population-based or a family-based approach. The International HapMap Project, Phase I, genotyped 1 million SNPs spaced at 5-kb density across the genome in 269 samples from 4 different populations. This data is freely available for use and facilitates WGA studies. The Phase II HapMap targets to genotype an additional 4.6 million SNPs in HapMap samples and also genotype additional members of population and other populations in ENCyclopedia Of DNA Elements (ENCODE) regions (www.genome.gov). Estimates suggest that with 500,000 SNPs, about 70–75% of the common variation in the genome will be identified (Martin, 2006). The extent to which common variation is linked to common versus rare disease remains to be determined.

Linkage and association approaches can also be combined. Many forms of CHD are identified as single affected individuals in families; such families are ideal for performing association studies. For example, ~50% of BAV families have a single affected individual; hence, a combination of traditional linkage studies and association studies is an ideal way to make use of such multiplex and single affected families to identify causative genes. The most common method to perform family-based association studies is the TDT proposed by Spielman et al. (1993) as a robust test for association due to two loci being tightly linked. In performing TDT analysis, usually families with single affected individuals are used as trios, that is probands are cases and their parents are controls. This test examines the transmission of a particular allele from heterozygous parents to their offspring. If the parents are not available, the unaffected sibs may be used as controls (Sib-TDT). TDT analysis has also been applied for narrowing the regions identified by linkage analysis. For example, Qui and associates narrowed the susceptibility gene of ventricular septal defect (VSD) to 3.56 cM on 12q13 by performing TDT analysis using microsatellite markers in trios (proband and parents from each families) nuclear families and found the candidate gene, *GLI* (glioma-associated oncogene homologue), a member of the GLI-Kruppel family, to be associated with VSD (Qiu et al., 2006).

4. CARDIAC TRANSCRIPTION FACTORS IDENTIFIED BY GENETIC STUDIES

In the past decade, human genetic studies, using techniques outlined above, have identified an important role for transcription factor mutations in human CHD. Results of genetic studies identifying mutations in cardiac transcription factor genes, which include *FOG2, GATA4, NKX2.5, TBX1, TBX5, TFAP2B*, and *ZIC3*, will be reviewed.

4.1. *FOG2*

FOG2/ZFPM2 is a multitype zinc finger transcription factor that modulates the activity of the cardiac transcription factor *GATA4* (Svensson et al., 2000). The similarity of complex cardiac malformations in *FOG2*-null mice to human tetralogy of Fallot (TOF) led to the analysis of *FOG2* as a candidate gene in 47 unrelated people with this cardiac phenotype (Pizzuti et al., 2003). Approximately 4% (2 of 47) of these unrelated patients with TOF showed a heterozygous missense mutation (*S657G, E30G*) within the *FOG2* gene, indicating its contribution in at least a subset of isolated cases of TOF (Pizzuti et al., 2003). To date, familial transmission of CHD associated with *FOG2* mutations has not been analyzed.

4.2. *GATA4*

A role for *GATA4* in CHD was initially suggested by Pehlivan et al. (1999) when GATA4 was identified in an interstitial deletion on chromosome 8p23.1 in four children with CHD. Subsequently, linkage analysis in large kindred segregating septal defects, atrial septal defects (ASD), AVSDs, VSDs, and pulmonary valve thickening identified a heterozygous *GATA4* missense mutation, G296S, which segregated with disease in the kindred but was not found in unrelated, unaffected control subjects

(Garg et al., 2003). A second mutation, E359del, was identified in a small kindred (Garg et al., 2003). Seven additional heterozygous *GATA4* mutations were subsequently reported (Okubo et al., 2004; Hirayama-Yamada et al., 2005; Nemer et al., 2006; Rajagopal et al., 2007). Studies to date suggest that the pathogenesis of human *GATA4* mutations is the result of loss of function (haploinsufficiency). *Gata4* is expressed in cardiomyocytes and their mesodermal precursors as well as in the endocardium and epicardium. *Gata4* binds to and regulates expression of a number of genes expressed in myocardium (Charron and Nemer, 1999; Molkentin, 2000). Homozygous mutant mice embryos lacking *Gata4* had arrested development and died due to a defect in the visceral endoderm that caused aberrant ventral morphogenesis and cardiac bifida (Kuo et al., 1997; Molkentin et al., 1997; Pu et al., 2004). Using conditional knockout approaches, it has been shown that myocardial expression of *Gata4* is necessary for myocardial growth and for morphogenesis of the right ventricle (Zeisberg et al., 2005), while expression of *Gata4* in endocardium and endocardial cushions is necessary for normal valve formation (Rivera-Feliciano et al., 2006). A recent study showed that a constitutive mutation of one copy of *GATA4* results in ASD, VSD, AVSD, right ventricular hypoplasia, and cardiomyopathy, consistent with disruption of *GATA4* function in both myocardial and endocardial compartments (Rajagopal et al., 2007). Such studies have provided new insight into the pathogenesis of human CHD and emphasize that normal cardiac development and function depend on maintaining precise levels of *GATA4* activity.

4.3. *NKX2.5*

Using linkage analysis in a kindred with autosomal dominant transmission of nonsyndromic CHD, a locus was mapped to chromosome 5q35 where *NKX2.5*, a homeobox transcription factor, is encoded (Schott et al., 1998); analysis of *NKX2.5* sequence in affected individuals identified a heterozygous mutation. To date, ~30 missense, nonsense, and frameshift mutations in the *NKX2.5* gene have been reported (Schott et al., 1998; Benson et al., 1999; Goldmuntz et al., 2001; Gutierrez-Roelens et al., 2002; Ikeda et al., 2002; Watanabe et al., 2002; Elliott et al., 2003; McElhinney et al., 2003; Kasahara and Benson, 2004; Konig et al., 2006; Rifai et al., 2007). Studies to date indicate that (1) 4% of CHD are due to heterozygous *NKX2.5* mutations, (2) AV block and ASD are the most common (highly penetrant) phenotypes, (3) *NKX2.5* mutations may cause AV block without associated CHD, and (4) *NKX2.5* mutations can result in varied CHD, including VSD, TOF, and tricuspid valve anomalies ranging from Ebstein anomaly of the tricuspid valve to tricuspid valve atresia (Benson et al., 1999; Kasahara and Benson, 2004).

The early embryonic lethality of homozygous deletion of Nkx2.5 in mouse has been a limitation to studies of cardiogenesis (Lyons et al., 1995). However, genotype–phenotype correlations have begun to emerge. For example, the principal determinant of the human AV block phenotype, and to a lesser extent the ASD phenotype, is the total dose of NKX2.5 capable of binding to DNA. This observation is supported by studies in mouse models that demonstrate that tight dosing of Nkx2.5 wild-type protein is required for normal development of the AV conduction system (Jay et al., 2004; Pashmforoush et al., 2004). In addition, nonhomeodomain NKX2.5 missense mutations (with normal DNA binding) are associated with varied CHD but not conduction abnormalities (Goldmuntz et al., 2001; McElhinney et al., 2003). The basis for the varied CHD is unknown, but may be due to dominant negative effects, modifying alleles or environmental factors.

4.4. TBX1

Del22q11 syndrome (aka DiGeorge syndrome, velocardiofacial syndrome, conotruncal anomaly face syndrome) has been associated with deletions of chromosomal region 22q11.2. TBX1, a member of the T-box family of transcription factors, is located on chromosome 22q11 in the DiGeorge chromosomal region (DGCR) (Chieffo et al., 1997). Mice with a heterozygous deletion of a region homologous to the DGCR (mouse chromosome 16) exhibit many of the aortic–arch, thymic, and parathyroid defects of del22q11.2 syndrome (Lindsay et al., 2001; Merscher et al., 2001). The generation of $Tbx1^{-/-}$ mice identified an essential role for Tbx1 during normal pharyngeal and cardiac development and making it an attractive biological candidate gene for the cardiovascular manifestations of del22q11.2 syndrome. These observations were used to support TBX1 as a candidate gene for conotruncal defects in patients without a 22q11.2 deletion. However, despite evaluation of large cohorts of deletion-negative patients with conotruncal defects, only three TBX1 mutations have been identified, and these account for <1% of conotruncal malformations in this population (Gong et al., 2001; Yagi et al., 2003; Stoller and Epstein, 2005). Similar findings have been reported for psychiatric illness, an important part of the del22q11 syndrome (Paylor et al., 2006).

4.5. TBX5

Holt-Oram syndrome (HOS) is an autosomal dominant disorder characterized by anomalies of the upper extremities and CHD. The skeletal abnormalities affect the upper limbs exclusively, predominantly involving the radial ray; limb involvement is always bilateral but often asymmetric. The thumb is the most commonly affected structure and can be triphalangeal, hypoplastic, or completely absent. The commonly observed cardiac manifestations in HOS patients include secundum ASD, VSD, and AVSD. Linkage studies in families segregating HOS identified a locus on chromosome 12q; *TBX5* was identified during positional cloning at this locus (Basson et al., 1997; Li et al., 1997). Heterozygous *TBX5* mutations were also detected in HOS patients (Basson et al., 1997; Li et al., 1997).

Mutations of *TBX5* causing HOS include missense and nonsense mutations, deletions, splice site mutations, and a translocation disrupting the *TBX5* gene (Basson et al., 1997; Li et al., 1997; Basson et al., 1999; Brassington et al., 2003; Fan et al., 2003; Gruenauer-Kloevekorn and Froster, 2003; Gruenauer-Kloevekorn et al., 2005; Heinritz et al., 2005). To date, ~38 mutations have been reported, and the majority of these mutations result in truncated proteins, suggesting that HOS pathogenesis is due to *TBX5* haploinsufficiency. Many aspects of the HOS phenotype are recapitulated in a mouse model of *Tbx5* haploinsufficiency (Bruneau et al., 2001).

4.6. TFAP2B

Char syndrome is an autosomal dominant disorder characterized by patent ductus arteriosus, facial dysmorphism, and an abnormal fifth digit of the hand. Satoda and associates initially identified a locus on chromosome 6p12-p21 for Char syndrome by traditional linkage analysis in two families (Satoda et al., 1999). Using a positional candidate approach, two missense mutations were found in the *TFAP2B* gene (Satoda et al., 2000). *TFAP2B* is a member of the transcription factor activating enhancer binding protein-2-β family of transcription factors. Subsequent studies in an additional eight families identified four novel mutations (Zhao et al., 2001). To date, eight *TFAP2B* mutations (six missense and two splice-site) have been identified to be associated with Char syndrome (Mani et al., 2005). It has been suggested that the Char syndrome phenotype results from derangement of neural crest cell derivatives; functional studies to date suggest haploinsufficiency as the basis for pathogenesis of *TFAP2B* mutations (Satoda et al., 2000; Zhao et al., 2001; Mani et al., 2005). Mice with heterozygous disruptions of *Tfap2b* appear normal; homozygous mice develop polycystic kidney disease due to excessive apoptosis of renal epithelial cells, and die in the neonatal period (Moser et al., 1997).

4.7. ZIC3

Heterotaxy (situs ambiguus) is characterized by various congenital anomalies that include situs abnormalities and complex cardiac malformations, which are due to the disruption of the normal left–right asymmetric development. The congenital anomalies vary widely and involve both lateral and midline structures. The cardiac malformations include septal defects (ASD, VSD, and AVSD), venous anomalies (both systemic and pulmonary), transposition of the great arteries, double outlet right ventricle, and obstruction of right and left outflow tracts. Despite the predominance of sporadic cases, autosomal and sex-linked inheritance has been observed in families with situs abnormalities (Carmi et al., 1992). Using linkage analysis, Casey et al. (1993) identified a locus for X-linked heterotaxy, HTX1, in a family with nine affected males in two generations exhibiting variable expressivity of heterotaxy and X-linked recessive inheritance. Efforts to confirm the HTX1 locus in additional unrelated males with sporadic and familial heterotaxy led to the identification of a deletion interval spanning 600–1100 kb lying within the 1.3-Mb region in a sporadic case (Ferrero et al., 1997). Positional cloning of the genes in this deletion interval led to the characterization of ZIC3 gene and identification of loss-of-function mutations causing heterotaxy demonstrated that ZIC3 is required for normal left–right axis development in humans (Gebbia et al., 1997). To date, 11 mutations have been identified in sporadic and familial heterotaxy (Ferrero et al., 1997; Gebbia et al., 1997; Megarbane et al., 2000; Purandare et al., 2002; Ware et al., 2004). Predicted effects of mutations on protein structure suggest haploinsufficiency as the basis for pathogenesis of Zic3 mutation, but it has been difficult to establish a genotype–phenotype relationship (Ware et al., 2004). A mouse model of Zic3 haploinsufficiency replicates many elements of the human phenotype and should serve as a useful model for defining pathogenesis of Zic3 mutations (Purandare et al., 2002; Ware et al., 2006).

5. CONCLUSIONS

Transcription factor mutations are a significant cause of CHD. Studies to date indicate that transcription factor mutations demonstrate variable expression, that is varied phenotypes even from the same mutation, which suggests that other factors including coregulatory elements and epigenetic effects also play a role in the disease manifestations. Identification of transcription factor mutations is useful to the clinician for diagnosis, disease classification, genetic counseling, and population screening. At the same time, human transcription factor mutations have informed developmental biologists and biochemists of transcription factor function. Continued use of the complementary methods of human genetics, development, and biochemistry will provide a more thorough understanding of the molecular function of these transcription factors in cardiac development and disease. Such understanding promises lead to improved risk assessment, diagnosis, and novel therapeutic strategies for these important clinical problems.

ACKNOWLEDGMENTS

This work was supported by grants from National Institutes of Health HL069712 (D.W.B.) and HL074728 (V.R. and D.W.B.).

REFERENCES

Ashley-Koch, A.E. 2006. Determining genetic component of a disease. In: *Genetic Analysis of Complex Disease* (J. L. Hanines, M. A. Pericak-Vance, Eds.), Hoboken: Wiley-Liss, pp. 91–115.

Basson, C.T., Bachinsky, D.R., Lin, R.C., Levi, T., Elkins, J.A., Soults, J., Grayzel, D., Kroumpouzou, E., Traill, T.A., Leblanc-Straceski, J., Renault, B., Kucherlapati, R., et al. 1997. Mutations in human TBX5 [corrected] cause limb and cardiac malformation in Holt-Oram syndrome. Nat. Genet. 15, 30–35.

Basson, C.T., Huang, T., Lin, R.C., Bachinsky, D.R., Weremowicz, S., Vaglio, A., Bruzzone, R., Quadrelli, R., Lerone, M., Romeo, G., Silengo, M., Pereira, A., et al. 1999. Different TBX5 interactions in heart and limb defined by Holt-Oram syndrome mutations. Proc. Natl. Acad. Sci. USA 96, 2919–2924.

Benson, D.W. 2002. The genetics of congenital heart disease: A point in the revolution. Cardiol. Clin. 20, 385–394.

Benson, D.W., Sharkey, A., Fatkin, D., Lang, P., Basson, C.T., Mcdonough, B., Strauss, A.W., Seidman, J.G., Seidman, C.E. 1998. Reduced penetrance, variable expressivity, and genetic heterogeneity of familial atrial septal defects. Circulation 97, 2043–2048.

Benson, D.W., Silberbach, G.M., Kavanaugh-Mchugh, A., Cottrill, C., Zhang, Y., Riggs, S., Smalls, O., Johnson, M.C., Watson, M.S., Seidman, J.G., Seidman, C.E., Plowden, J., et al. 1999. Mutations in the cardiac transcription factor NKX2.5 affect diverse cardiac developmental pathways. J. Clin. Invest. 104, 1567–1573.

Botstein, D., Risch, N. 2003. Discovering genotypes underlying human phenotypes: Past successes for mendelian disease, future approaches for complex disease. Nat. Genet. 33(Suppl.), 228–237.

Brassington, A.M., Sung, S.S., Toydemir, R.M., Le, T., Roeder, A.D., Rutherford, A.E., Whitby, F.G., Jorde, L.B., Bamshad, M.J. 2003. Expressivity of Holt-Oram syndrome is not predicted by TBX5 genotype. Am. J. Hum. Genet. 73, 74–85.

Bruneau, B.G., Nemer, G., Schmitt, J.P., Charron, F., Robitaille, L., Caron, S., Conner, D.A., Gessler, M., Nemer, M., Seidman, C.E., Seidman, J.G. 2001. A murine model of Holt-Oram syndrome defines roles of the T-box transcription factor Tbx5 in cardiogenesis and disease. Cell 106, 709–721.

Carmi, R., Boughman, J.A., Rosenbaum, K.R. 1992. Human situs determination is probably controlled by several different genes. Am. J. Med. Genet. 44, 246–249.

Casey, B., Devoto, M., Jones, K.L., Ballabio, A. 1993. Mapping a gene for familial situs abnormalities to human chromosome Xq24-q27.1. Nat. Genet. 5, 403–407.

Charron, F., Nemer, M. 1999. GATA transcription factors and cardiac development. Semin. Cell Dev. Biol. 10, 85–91.

Chieffo, C., Garvey, N., Gong, W., Roe, B., Zhang, G., Silver, L., Emanuel, B.S., Budarf, M.L. 1997. Isolation and characterization of a gene from the DiGeorge chromosomal region homologous to the mouse Tbx1 gene. Genomics 43, 267–277.

Clark, K.L., Yutzey, K.E., Benson, D.W. 2006. Transcription factors and congenital heart defects. Annu. Rev. Physiol. 68, 97–121.

Cripe, L., Andelfinger, G., Martin, L.J., Shooner, K., Benson, D.W. 2004. Bicuspid aortic valve is heritable. J. Am. Coll. Cardiol. 44, 138–143.

Curran, M.E., Atkinson, D.L., Ewart, A.K., Morris, C.A., Leppert, M.F., Keating, M.T. 1993. The elastin gene is disrupted by a translocation associated with supravalvular aortic stenosis. Cell 73, 159–168.

Dean, M., Stephens, J.C., Winkler, C., Lomb, D.A., Ramsburg, M., Boaze, R., Stewart, C., Charbonneau, L., Goldman, D., Albaugh, B.J., Goedert, J.J., Beasley, R.P., et al. 1994. Polymorphic admixture typing in human ethnic populations. Am. J. Hum. Genet. 55, 788–808.

Duerr, R.H., Taylor, K.D., Brant, S.R., Rioux, J.D., Silverberg, M.S., Daly, M.J., Steinhart, A.H., Abraham, C., Regueiro, M., Griffiths, A., Dassopoulos, T., Bitton, A., et al. 2006. A genome-wide association study identifies IL23R as an inflammatory bowel disease gene. Science 314, 1461–1463.

Edwards, A.O., Ritter, R.3rd, Abel, K.J., Manning, A., Panhuysen, C., Farrer, L.A. 2005. Complement factor H polymorphism and age-related macular degeneration. Science 308, 421–424.

Elliott, D.A., Kirk, E.P., Yeoh, T., Chandar, S., Mckenzie, F., Taylor, P., Grossfeld, P., Fatkin, D., Jones, O., Hayes, P., Feneley, M., Harvey, R.P. 2003. Cardiac homeobox gene NKX2-5 mutations and congenital heart disease: Associations with atrial septal defect and hypoplastic left heart syndrome. J. Am. Coll. Cardiol. 41, 2072–2076.

Fan, C., Liu, M., Wang, Q. 2003. Functional analysis of TBX5 missense mutations associated with Holt-Oram syndrome. J. Biol. Chem. 278, 8780–8785.

Ferrero, G.B., Gebbia, M., Pilia, G., Witte, D., Peier, A., Hopkin, R.J., Craigen, W.J., Shaffer, L. G., Schlessinger, D., Ballabio, A., Casey, B. 1997. A submicroscopic deletion in Xq26 associated with familial situs ambiguus. Am. J. Hum. Genet. 61, 395–401.

Garg, V., Kathiriya, I.S., Barnes, R., Schluterman, M.K., King, I.N., Butler, C.A., Rothrock, C. R., Eapen, R.S., Hirayama-Yamada, K., Joo, K., Matsuoka, R., Cohen, J.C., et al. 2003. GATA4 mutations cause human congenital heart defects and reveal an interaction with TBX5. Nature 424, 443–447.

Gebbia, M., Ferrero, G.B., Pilia, G., Bassi, M.T., Aylsworth, A., Penman-Splitt, M., Bird, L.M., Bamforth, J.S., Burn, J., Schlessinger, D., Nelson, D.L., Casey, B. 1997. X-linked situs abnormalities result from mutations in ZIC3. Nat. Genet. 17, 305–308.

Gelb, B.D. 2004. Genetic basis of congenital heart disease. Curr. Opin. Cardiol. 19, 110–115.

Goldmuntz, E., Geiger, E., Benson, D.W. 2001. NKX2.5 mutations in patients with tetralogy of Fallot. Circulation 104, 2565–2568.

Gong, W., Gottlieb, S., Collins, J., Blescia, A., Dietz, H., Goldmuntz, E., Mcdonald-Mcginn, D.M., Zackai, E.H., Emanuel, B.S., Driscoll, D.A., Budarf, M.L. 2001. Mutation analysis of TBX1 in non-deleted patients with features of DGS/VCFS or isolated cardiovascular defects. J. Med. Genet. 38, E45.

Green, E.K., Priestley, M.D., Waters, J., Maliszewska, C., Latif, F., Maher, E.R. 2000. Detailed mapping of a congenital heart disease gene in chromosome 3p25. J. Med. Genet. 37, 581–587.

Gruenauer-Kloevekorn, C., Froster, U.G. 2003. Holt-Oram syndrome: A new mutation in the TBX5 gene in two unrelated families. Ann. Genet. 46, 19–23.

Gruenauer-Kloevekorn, C., Reichel, M.B., Duncker, G.I., Froster, U.G. 2005. Molecular genetic and ocular findings in patients with Holt-Oram syndrome. Ophthalmic Genet. 26, 1–8.

Gutierrez-Roelens, I., Sluysmans, T., Gewillig, M., Devriendt, K., Vikkula, M. 2002. Progressive AV-block and anomalous venous return among cardiac anomalies associated with two novel missense mutations in the CSX/NKX2-5 gene. Hum. Mutat. 20, 75–76.

Heinritz, W., Moschik, A., Kujat, A., Spranger, S., Heilbronner, H., Demuth, S., Bier, A., Tihanyi, M., Mundlos, S., Gruenauer-Kloevekorn, C., Froster, U.G. 2005. Identification of new mutations in the TBX5 gene in patients with Holt-Oram syndrome. Heart 91, 383–384.

Hinrichs, A.L., Suarez, B.K. 2005. Genotyping errors, pedigree errors, and missing data. Genet Epidemiol. 29(Suppl. 1), S120–S124.

Hinton, R.B.Jr., Yutzey, K.E., Benson, D.W. 2005. Congenital heart disease: Genetic causes and developmental insights. Prog. Ped. Cardiol. 20, 101–111.

Hirayama-Yamada, K., Kamisago, M., Akimoto, K., Aotsuka, H., Nakamura, Y., Tomita, H., Furutani, M., Imamura, S., Takao, A., Nakazawa, M., Matsuoka, R. 2005. Phenotypes with GATA4 or NKX2.5 mutations in familial atrial septal defect. Am. J. Med. Genet. A 135, 47–52.

Hoffman, J.I., Kaplan, S. 2002. The incidence of congenital heart disease. J. Am. Coll. Cardiol. 39, 1890–1900.

Ikeda, Y., Hiroi, Y., Hosoda, T., Utsunomiya, T., Matsuo, S., Ito, T., Inoue, J., Sumiyoshi, T., Takano, H., Nagai, R., Komuro, I. 2002. Novel point mutation in the cardiac transcription factor CSX/NKX2.5 associated with congenital heart disease. Circ. J. 66, 561–563.

Jay, P.Y., Harris, B.S., Maguire, C.T., Buerger, A., Wakimoto, H., Tanaka, M., Kupershmidt, S., Roden, D.M., Schultheiss, T.M., O'brien, T.X., Gourdie, R.G., Berul, C.I., et al. 2004. Nkx2-5 mutation causes anatomic hypoplasia of the cardiac conduction system. J. Clin. Invest. 113, 1130–1137.

Kasahara, H., Benson, D.W. 2004. Biochemical analyses of eight NKX2.5 homeodomain missense mutations causing atrioventricular block and cardiac anomalies. Cardiovasc. Res. 64, 40–51.

Klein, R.J., Zeiss, C., Chew, E.Y., Tsai, J.Y., Sackler, R.S., Haynes, C., Henning, A.K., Sangiovanni, J.P., Mane, S.M., Mayne, S.T., Bracken, M.B., Ferris, F.L., et al. 2005. Complement factor H polymorphism in age-related macular degeneration. Science 308, 385–389.

Konig, K., Will, J.C., Berger, F., Muller, D., Benson, D.W. 2006. Familial congenital heart disease, progressive atrioventricular block and the cardiac homeobox transcription factor gene NKX2.5: Identification of a novel mutation. Clin. Res. Cardiol. 95, 499–503.

Kuo, C.T., Morrisey, E.E., Anandappa, R., Sigrist, K., Lu, M.M., Parmacek, M.S., Soudais, C., Leiden, J.M. 1997. GATA4 transcription factor is required for ventral morphogenesis and heart tube formation. Genes Dev. 11, 1048–1060.

Li, Q.Y., Newbury-Ecob, R.A., Terrett, J.A., Wilson, D.I., Curtis, A.R., Yi, C.H., Gebuhr, T., Bullen, P.J., Robson, S.C., Strachan, T., Bonnet, D., Lyonnet, S., et al. 1997. Holt-Oram syndrome is caused by mutations in TBX5, a member of the Brachyury (T) gene family. Nat. Genet. 15, 21–29.

Lindsay, E.A., Vitelli, F., Su, H., Morishima, M., Huynh, T., Pramparo, T., Jurecic, V., Ogunrinu, G., Sutherland, H.F., Scambler, P.J., Bradley, A., Baldini, A. 2001. Tbx1 haploinsufficieny in the DiGeorge syndrome region causes aortic arch defects in mice. Nature 410, 97–101.

Lyons, I., Parsons, L.M., Hartley, L., Li, R., Andrews, J.E., Robb, L., Harvey, R.P. 1995. Myogenic and morphogenetic defects in the heart tubes of murine embryos lacking the homeo box gene Nkx2-5. Genes Dev. 9, 1654–1666.

Mani, A., Radhakrishnan, J., Farhi, A., Carew, K.S., Warnes, C.A., Nelson-Williams, C., Day, R.W., Pober, B., State, M.W., Lifton, R.P. 2005. Syndromic patent ductus arteriosus: Evidence for haploinsufficient TFAP2B mutations and identification of a linked sleep disorder. Proc. Natl. Acad. Sci. USA 102, 2975–2979.

Martin, E.R. 2006. Linkage disequilibrium and association analysis. In: *Genetic Analysis of Complex Disease* (J. L. Hanines, M. A. Pericak-Vance, Eds.), Hoboken: Wiley-Liss, pp. 329–353.

Martin, L.J., Ramachandran, V., Cripe, L.H., Hinton, R.B., Andelfinger, G., Tabangin, M., Shooner, K., Keddache, M., Benson, D.W. 2007. Evidence in favor of linkage to human chromosomal regions 18q, 5q and 13q for bicuspid aortic valve and associated cardiovascular malformations. Hum. Genet. 121(2), 275–284. Jan 4, Epub ahead of print.

Mayeux, R. 2005. Mapping the new frontier: Complex genetic disorders. J. Clin. Invest. 115, 1404–1407.

McElhinney, D.B., Geiger, E., Blinder, J., Benson, D.W., Goldmuntz, E. 2003. NKX2.5 mutations in patients with congenital heart disease. J. Am. Coll. Cardiol. 42, 1650–1655.

Megarbane, A., Salem, N., Stephan, E., Ashoush, R., Lenoir, D., Delague, V., Kassab, R., Loiselet, J., Bouvagnet, P. 2000. X-linked transposition of the great arteries and incomplete penetrance among males with a nonsense mutation in ZIC3. Eur. J. Hum. Genet. 8, 704–708.

Merscher, S., Funke, B., Epstein, J.A., Heyer, J., Puech, A., Lu, M.M., Xavier, R.J., Demay, M. B., Russell, R.G., Factor, S., Tokooya, K., Jore, B.S., et al. 2001. TBX1 is responsible for cardiovascular defects in velo-cardio-facial/DiGeorge syndrome. Cell 104, 619–629.

Molkentin, J.D. 2000. The zinc finger-containing transcription factors GATA-4, -5, and -6. Ubiquitously expressed regulators of tissue-specific gene expression. J. Biol. Chem. 275, 38949–38952.

Molkentin, J.D., Lin, Q., Duncan, S.A., Olson, E.N. 1997. Requirement of the transcription factor GATA4 for heart tube formation and ventral morphogenesis. Genes Dev. 11, 1061–1072.

Moser, M., Pscherer, A., Roth, C., Becker, J., Mucher, G., Zerres, K., Dixkens, C., Weis, J., Guay-Woodford, L., Buettner, R., Fassler, R. 1997. Enhanced apoptotic cell death of renal epithelial cells in mice lacking transcription factor AP-2beta. Genes Dev. 11, 1938–1948.

Muncke, N., Jung, C., Rudiger, H., Ulmer, H., Roeth, R., Hubert, A., Goldmuntz, E., Driscoll, D., Goodship, J., Schon, K., Rappold, G. 2003. Missense mutations and gene interruption in PROSIT240, a novel TRAP240-like gene, in patients with congenital heart defect (transposition of the great arteries). Circulation 108, 2843–2850.

Nemer, G., Fadlalah, F., Usta, J., Nemer, M., Dbaibo, G., Obeid, M., Bitar, F. 2006. A novel mutation in the GATA4 gene in patients with Tetralogy of Fallot. Hum. Mutat. 27, 293–294.

Okubo, A., Miyoshi, O., Baba, K., Takagi, M., Tsukamoto, K., Kinoshita, A., Yoshiura, K., Kishino, T., Ohta, T., Niikawa, N., Matsumoto, N. 2004. A novel GATA4 mutation completely segregated with atrial septal defect in a large Japanese family. J. Med. Genet. 41, e97.

Pashmforoush, M., Lu, J.T., Chen, H., Amand, T.S., Kondo, R., Pradervand, S., Evans, S.M., Clark, B., Feramisco, J.R., Giles, W., Ho, S.Y., Benson, D.W., et al. 2004. Nkx2-5 pathways and congenital heart disease; loss of ventricular myocyte lineage specification leads to progressive cardiomyopathy and complete heart block. Cell 117, 373–386.

Paylor, R., Glaser, B., Mupo, A., Ataliotis, P., Spencer, C., Sobotka, A., Sparks, C., Choi, C.H., Oghalai, J., Curran, S., Murphy, K.C., Monks, S., et al. 2006. Tbx1 haploinsufficiency is linked to behavioral disorders in mice and humans: Implications for 22q11 deletion syndrome. Proc. Natl. Acad. Sci. USA 103, 7729 7734.

Pehlivan, T., Pober, B.R., Brueckner, M., Garrett, S., Slaugh, R., Van Rheeden, R., Wilson, D. B., Watson, M.S., Hing, A.V. 1999. GATA4 haploinsufficiency in patients with interstitial deletion of chromosome region 8p23.1 and congenital heart disease. Am. J. Med. Genet. 83, 201–206.

Pizzuti, A., Sarkozy, A., Newton, A.L., Conti, E., Flex, E., Digilio, M.C., Amati, F., Gianni, D., Tandoi, C., Marino, B., Crossley, M., Dallapiccola, B. 2003. Mutations of ZFPM2/FOG2 gene in sporadic cases of tetralogy of Fallot. Hum. Mutat. 22, 372–377.

Pu, W.T., Ishiwata, T., Juraszek, A.L., Ma, Q., Izumo, S. 2004. GATA4 is a dosage-sensitive regulator of cardiac morphogenesis. Dev. Biol. 275, 235–244.

Purandare, S.M., Ware, S.M., Kwan, K.M., Gebbia, M., Bassi, M.T., Deng, J.M., Vogel, H., Behringer, R.R., Belmont, J.W., Casey, B. 2002. A complex syndrome of left-right axis, central nervous system and axial skeleton defects in Zic3 mutant mice. Development 129, 2293–2302.

Qiu, G.R., Gong, L.G., He, G., Xu, X.Y., Xin, N., Sun, G.F., Yuan, Y.H., Sun, K.L. 2006. Association of the GLI gene with ventricular septal defect after the susceptibility gene being narrowed to 3.56 cM in 12q13. Chin. Med. J. (Engl.) 119, 267–274.

Rajagopal, S.K., Ma, Q., Obler, D., Shen, J., Manichaikul, A., Tomita-Mitchel, A., Boardman, K., Briggs, C., Garg, V., Srivastava, D., Goldmuntz, E., Broman, K.W., et al. 2007. Spectrum of heart disease associated with murine and human GATA4 mutation. J. Mol. Cell. Cardiol. In press.

Rifai, L., Maazouzi, W., Sefiani, A. 2007. Novel point mutation in the NKX2-5 gene in a Moroccan family with atrioventricular conduction disturbance and an atrial septal defect in the oval fossa. Cardiol. Young 17(1), 107–109.

Risch, N., Merikangas, K. 1996. The future of genetic studies of complex human diseases. Science 273, 1516–1517.

Rivera-Feliciano, J., Lee, K.H., Kong, S.W., Rajagopal, S., Ma, Q., Springer, Z., Izumo, S., Tabin, C.J., Pu, W.T. 2006. Development of heart valves requires Gata4 expression in endothelial-derived cells. Development 133, 3607–3618.

Robinson, S.W., Morris, C.D., Goldmuntz, E., Reller, M.D., Jones, M.A., Steiner, R.D., Maslen, C.L. 2003. Missense mutations in CRELD1 are associated with cardiac atrioventricular septal defects. Am. J. Hum. Genet. 72, 1047–1052.

Satoda, M., Pierpont, M.E., Diaz, G.A., Bornemeier, R.A., Gelb, B.D. 1999. Char syndrome, an inherited disorder with patent ductus arteriosus, maps to chromosome 6p12-p21. Circulation 99, 3036–3042.

Satoda, M., Zhao, F., Diaz, G.A., Burn, J., Goodship, J., Davidson, H.R., Pierpont, M.E., Gelb, B.D. 2000. Mutations in TFAP2B cause Char syndrome, a familial form of patent ductus arteriosus. Nat. Genet. 25, 42–46.

Schott, J.J., Benson, D.W., Basson, C.T., Pease, W., Silberbach, G.M., Moak, J.P., Maron, B.J., Seidman, C.E., Seidman, J.G. 1998. Congenital heart disease caused by mutations in the transcription factor NKX2-5. Science 281, 108–111.

Spielman, R.S., Mcginnis, R.E., Ewens, W.J. 1993. Transmission test for linkage disequilibrium: The insulin gene region and insulin-dependent diabetes mellitus (IDDM). Am. J. Hum. Genet. 52, 506–516.

Stephens, J.C., Schneider, J.A., Tanguay, D.A., Choi, J., Acharya, T., Stanley, S.E., Jiang, R., Messer, C.J., Chew, A., Han, J.H., Duan, J., Carr, J.L., et al. 2001. Haplotype variation and linkage disequilibrium in 313 human genes. Science 293, 489–493.

Stoller, J.Z., Epstein, J.A. 2005. Identification of a novel nuclear localization signal in Tbx1 that is deleted in DiGeorge syndrome patients harboring the 1223delC mutation. Hum. Mol. Genet. 14, 885–892.

Svensson, E.C., Huggins, G.S., Dardik, F.B., Polk, C.E., Leiden, J.M. 2000. A functionally conserved N-terminal domain of the friend of GATA-2 (FOG-2) protein represses GATA4-dependent transcription. J. Biol. Chem. 275, 20762–20769.

Ward, C. 2000. Clinical significance of the bicuspid aortic valve. Heart 83, 81–85.

Ware, S.M., Peng, J., Zhu, L., Fernbach, S., Colicos, S., Casey, B., Towbin, J., Belmont, J.W. 2004. Identification and functional analysis of ZIC3 mutations in heterotaxy and related congenital heart defects. Am. J. Hum. Genet. 74, 93–105.

Ware, S.M., Harutyunyan, K.G., Belmont, J.W. 2006. Heart defects in X-linked heterotaxy: Evidence for a genetic interaction of Zic3 with the nodal signaling pathway. Dev. Dyn. 235, 1631–1637.

Watanabe, Y., Benson, D.W., Yano, S., Akagi, T., Yoshino, M., Murray, J.C. 2002. Two novel frameshift mutations in NKX2.5 result in novel features including visceral inversus and sinus venosus type ASD. J. Med. Genet. 39, 807–811.

Yagi, H., Furutani, Y., Hamada, H., Sasaki, T., Asakawa, S., Minoshima, S., Ichida, F., Joo, K., Kimura, M., Imamura, S., Kamatani, N., Momma, K., et al. 2003. Role of TBX1 in human del22q11.2 syndrome. Lancet 362, 1366–1373.

Zeisberg, E.M., Ma, Q., Juraszek, A.L., Moses, K., Schwartz, R.J., Izumo, S., Pu, W.T. 2005. Morphogenesis of the right ventricle requires myocardial expression of Gata4. J. Clin. Invest. 115, 1522–1531.

Zhao, F., Weismann, C.G., Satoda, M., Pierpont, M.E., Sweeney, E., Thompson, E.M., Gelb, B. D. 2001. Novel TFAP2B mutations that cause Char syndrome provide a genotype-phenotype correlation. Am. J. Hum. Genet. 69, 695–703.

CHAPTER 15

Human Genetics of Congenital Heart Disease

Jeffrey A. Towbin

Contents

Pediatric Cardiology, Texas Children's Hospital, Baylor College of Medicine, Houston, Texas

Advances in Developmental Biology, Volume 18
ISSN 1574-3349, DOI: 10.1016/S1574-3349(07)18015-7

Abstract   Human cardiovascular abnormalities occur with an incidence of approximately 1 in 100 live births, representing an estimated 25% of all congenital malformations. These cardiovascular defects are a major cause of morbidity and mortality, and these disorders are a leading cause of death in the first year of life. Hence, the costs related to congenital cardiovascular disease and the tragic consequences to families are immense. Despite the importance of these disorders, as well as the relative frequency of recurrence in family members, relatively little was known regarding the etiologies and mechanisms causing these diseases until the past 10–15 years.

In this chapter, we will explore the current understanding of the development and genetic causes of congenital cardiovascular disease. Knowledge has been gained over the past decade or more using molecular genetics and animal modeling studies and these advances in the understanding of valve disorders, septal defects, and diseases of the great vessels will be outlined.

1. INTRODUCTION

Human cardiovascular abnormalities occur with an incidence of $\sim$1 per 100 live births, representing an estimated 25% of all congenital malformations (Ransom and Srivastava, 2007). These cardiovascular disorders are believed to be the leading cause of death in the first year of life. Despite the significant numbers of affected individuals, the poor outcomes, and the impressively high costs accrued in the care of those with heart disease, particularly congenital heart disease (CHD), relatively little is known regarding etiology. In addition to genetic causes of disease, other known etiologies include environmental factors and teratogens. In the case of genetic-based causes of CHD, 5–8% are found to have gross chromosomal defects such as trisomy 21, trisomy 18, trisomy 13, and Turner syndrome. In the remainder, single gene mutations or contiguous-gene syndromes are typically sought as the etiology (Ransom and Srivastava, 2007). However, in 90% of all cases of CHD, no known etiology can be discerned to date.

Why so little information in the current "genetic era"? Cardiac development is a complex integration of various pathways, structures, cell types, gene networks, and regulatory components and this complexity has been difficult to overcome (Clark et al., 2006). However, the molecular basis of these complex disorders is beginning to unravel and in this chapter the current understanding of the human genetics of CHD will be discussed.

2. HEART DEVELOPMENT

Not unexpectedly, cardiac development has been found to require a complex set of interactions, genes, and regulatory forces, as well as internal and external modifiers (Clark et al., 2006; Sandler et al., 2006). Studies in model organisms from invertebrate and vertebrate species have revealed an evolutionarily conserved program of heart development, which occurs as the first organ to develop in the human, being initiated by specific signaling molecules and mediated by tissue-specific transcription factors. The current understanding of this development is described in great detail in other chapters but briefly is as follows: cells forming the heart originate from the lateral plate mesoderm in embryos and migrate toward the midline to form two crescent-shaped primordia that subsequently fuse to form the beating heart tube (Clark et al., 2006; Sandler et al., 2006; Ransom and Srivastava, 2007). The subsequent patterning of this tube into the atria and ventricles is accompanied with a differential gene expression profile responsible for the morphogenesis of the heart that ultimately becomes the normal four-chambered human heart. These integrated, timed steps include cardiac looping, septation, valve formation, and compaction (Srivastava, 2006). In addition to the structural formation of the four-chambered heart, the development of the conduction system from the pacemaker cells to the Purkinje fibers is required. Further, cardiac vascularization requires migration of the cells from the neural crest and their differentiation into smooth muscle cells and endothelial cells that make up the vascular components of the great vessels and veins (Hutson and Kirby, 2007). Coronary arteries, key vascular components necessary to nourish the myocardium, are formed from precursor cells migrating from the epicardium (Olson, 2006).

Any errors that occur during these developmental stages, whether it is during the cell-commitment stage all the way to valve formation or cardiac compaction stage, can have a serious impact on cardiac morphogenesis and function, leading to a variety of forms of CHD, some mild, some severe, and some lethal (Clark et al., 2006; Ransom and Srivastava, 2007). A variety of factors affecting development of the heart have been described using model systems. In Drosophila, the mutant known as heartless has permitted the identification of a fibroblast growth factor receptor (FGFR) as essential for the normal migration of precardiac cells. In this mutant, the heart is not formed as the precardiac cells do not migrate to the embryonic midline. Similarly, mouse mutants in which inactivation of genes encoding bone morphogenetic protein 2 (BMP-2), BMP-4, fibronectin, or the transcription factor GATA-4 has occurred develop a phenotype resembling the Drosophila heartless mutant, suggesting early developmental roles for these genes (Olson, 2006; Srivastava, 2006; Hutson and Kirby, 2007).

Errors in cardiac looping may also lead to lethal or phenotypically severe outcomes as well. In normal cases, left–right asymmetry to the heart occurs. In ~1 per 7000 live human births, inversion of this normal asymmetry occurs and is known as situs inversus. This developmental error may result in serious physiological and structural abnormalities of many organ systems, including the heart. In particular, disruption of cardiac septation is common in these disorders. Mutations in the zinc-finger transcription factor ZIC3 was the first gene to be found in humans to be involved in these disorders called heterotaxy syndrome (Belmont et al., 2004).

Myocyte differentiation into two subtypes, atrial and ventricular, will form the different chambers of the mature heart. Members of the basic helix-loop-helix family of transcription factors have been implicated in the regulation of cell fate specification and differentiation in different organisms. In the heart, the subfamilies of genes expressed include the Hand proteins (dHand and eHand) and the Hairy proteins (Hey 1–3) (Firulli et al., 2000; Fischer et al., 2005). Asymmetric expression of the Hand proteins has been well described, with dHand expressed primarily in the right ventricle and eHand in the left ventricle (LV), implying roles in chamber specification and function (Firulli et al., 2000).

In the case of cardiac valve formation, two molecular pathways are currently known to play a role in governing the epithelial–mesenchymal transformation; the endocardial cells undergo to form mature valves. These pathways involve the calcineurin/NFATc pathway whereby phosphorylation of the transcription factor NFATc (nuclear factor for activated T cells) is essential for the transformation, and the Ras proto-oncogene pathways essential for the proliferation of mesenchymal cells. Disturbance in the pathways lead to valvular disorders (Niessen and Karsan, 2007).

In recent years, gene mutations have been reported to cause a variety of human conditions, including CHD. In this chapter, the cardiac diseases in which gene mutations are known to play a significant role will be described.

3. NATURAL HISTORY OF CONGENITAL HEART DISEASE

Successful open-heart surgery to correct septal defects in humans was ushered in by the pioneering work of C. Walton Lillehei (Lillehei et al., 1955), who closed a ventricular septal defect (VSD) in a 4-year-old boy, using the boy's father for crosscirculation, on April 20, 1954. The boy is now a middle-aged man. Since that time, because of increasingly sophisticated diagnostic precision, better understanding of the altered hemodynamics, and development of surgical techniques, the natural history of CHD has changed considerably. In the early days of open-heart surgery, unoperated CHD among adults was not uncommon, whereas CHD is now routinely corrected by surgery in infancy and childhood. The pediatric cardiologist sees fetuses with CHD using sophisticated imaging modalities, as well as newborns, infants, and children with various forms of corrected or unoperated heart anomalies. Other pediatric specialists are also involved in the care of these patients and their families and should note that, in a portion of patients, the cardiac disease will be inherited and, therefore, good family history and screening exams (including echocardiography) may be warranted. In addition, in the current era, the adult cardiologist (and other adult specialists) with an interest in CHD in adults, sees mostly patients with CHD (e.g., Eisenmenger's complex) who are yet inoperable or those with CHD that has been surgically corrected and for whom the cardiologist is responsible for the long-term follow-up. In fact, in 2007, there are more adults with CHD than children with CHD and therefore physicians caring for all ages of patients must have a working understanding of CHD. Also included in this category are patients who are destined to develop cardiac abnormalities, such as those with Marfan syndrome, in whom the heart disease may not become overt until adulthood, and those with conditions that easily escape the noncardiologist physician, such as atrial septal defect and congenital bicuspid aortic valve.

There are many ways to classify CHD. Perhaps the most "scientific" classification would be in terms of abnormal embryologic development. Unfortunately, such a classification would be of limited clinical categories, such as whether or not there is cyanosis, whether pulmonary hypertension is present, whether cardiac valves are involved, which ventricle is more involved, and whether there are disturbances in the heart rhythm. Sophisticated echocardiographic and other noninvasive imaging techniques can now delineate the anatomy and often the physiologic disturbance or disturbances, and cardiac catheterization can be used to more precisely quantify the physiologic disturbances and to identify abnormalities that are easily visualized by echocardiography, such as coronary anomalies, unusual insertions of anomalous pulmonary veins, and pulmonary arteriovenous fistulas. The electrocardiogram (ECG) can be a very useful adjunct in clinical diagnosis because some forms of CHD, such as atrial septal defect, tricuspid atresia, and the endocardial cushion defects, have electrocardiographic markers. In the era ahead, cardiologists, geneticists, generalists, and other specialists will also need to understand the developmental and genetic basis of these disorders. Athough only a small percentage of cardiac disorders are currently understood at the level of genetics, development, and mechanism of disease, that knowledge will be upon us soon. In this chapter, the current knowledge will be summarized on the basis of the clinical disorder.

4. VALVAR ABNORMALITIES

When describing valvar abnormalities, it is typical to describe these disorders to include stenosis (blockage to flow due to narrowed passageways), regurgitation (backflow of blood), and atresia (lack of development of the valve).

4.1. Aortic stenosis

Congenital obstruction of the outflow tract of the LV is commonly at the level of the aortic valve itself and results in aortic stenosis either due to a small or thickened valve or a congenital bicuspid or unicuspid valve (Fenoglio et al., 1977; Braverman et al., 2005). The incidence of aortic valve disease is four to five times higher in males than in females (Roberts, 1970).

4.1.1. Valvar aortic stenosis

Aortic stenosis typically develops because of a small or thickened valve or a bicuspid (or unicuspid) aortic valve (Peckham et al., 1964; Johnson, 1971). This is sometimes seen in association with coarctation of the aorta, or other left-sided abnormalities such as mitral stenosis and this combination of abnormalities is known as Shone's syndrome (Clinical Path Conference, 1968). In general, the small or thick aortic valve can result in hemodynamic abnormalities and symptoms early in life, including in the fetus and newborn, while the bicuspid aortic valve causes little or no hemodynamic disturbance for the first several decades of life and, in fact, may never cause any hemodynamic disturbance at all. Progressive aortic regurgitation may develop, especially in young adulthood. It appears that most valves develop progressive thickening and fibrosis after the third or fourth decade of life and, as time goes on, increasing calcification and progressively severe aortic stenosis. The presence of mild aortic stenosis may be associated with progressively severe aortic regurgitation when inadequate coaptation outweighs inadequate opening of the valve (Clinical Path Conference, 1968). Bacterial endocarditis may also transform the pathophysiologic disturbance from predominant stenosis to predominant regurgitation.

The course of the disease, once stenosis begins, tends to be relentless but has a highly variable timetable. In general, the progression to severe aortic stenosis may take decades. After significant aortic stenosis develops, significant risk may be associated. The classic ominous symptoms are syncope, angina, and heart failure. Sudden death is a threat once even mild symptoms are noted (Keane et al., 1993; Pellikka et al., 2005). However, evaluation for surgery should be undertaken if there has been any subjective change in exercise tolerance or general sense of well-being. Most patients with pre-dominant aortic stenosis have well-preserved left ventricular function if exertional syncope is the primary manifestation. Approximately 5% of patients with a bicuspid aortic valve have associated cystic medial disease of the aorta, which can become the basis of an aortic dissection. Coarctation of the aorta is sometimes associated with a bicuspid aortic valve (it is far more common for coarctation of the aorta to be associated with a bicuspid aortic valve, rather than the reverse), which has its own inherent natural history and complications (Clinical Path Conference, 1968).

4.1.2. Genetics of aortic stenosis

Lewin and colleagues were the first to demonstrate clear evidence of herita-bility of left heart obstructive lesions, with aortic stenosis and bicuspid aortic valves prominently displayed within the families studied (Lewin et al., 2004; McBride et al., 2005b). A total of 113 probands with a nonsyn-dromic form of left ventricular outflow tract obstructive disease including aortic stenosis, bicuspid aortic valve, aortic coarctation, hypoplastic left heart syndrome, and mitral valve disease were studied by echocardio-graphic screening along with 282 asymptomatic first-degree relatives. In ~17% of first-degree relatives, left heart anomies were identified with a relative risk of bicuspid aortic valve being 5.05 in these relatives. The heritability of aortic valve disease was further supported by Cripe et al. (2004) who evaluated 50 probands with bicuspid aortic valve along with 259 relatives. In this case, the authors demonstrated a prevalence of 24% for bicuspid aortic valves and 31% for all left heart-associated disease.

The first gene for aortic valve disease was identified by Garg et al. when they defined a nonsense mutation in the NOTCH1 gene, along with a frameshift mutation in an unrelated family (Garg et al., 2005; Garg, 2006). The authors suggested that haploinsufficiency was responsible for the development of aortic valve malformations and calcification and that the NOTCH signaling pathway (Niessen and Karsan, 2007) was in fact linked to a molecular pathway for aortic calcification, repressing activation of Runx2, a transcription factor critical for osteoclast fate. The authors suggested that NOTCH1 mutations cause an early developmental defect in the aortic valve and a later derepression of calcium deposition resulting in progressive aortic valve disease. Support for the concept that NOTCH1 is a key player in the development of aortic valve disease was provided by Mohamed et al. (2006) who identified mutations in both sporadic and familial cases of bicuspid aortic valve.

4.2. Supravalvar aortic stenosis

Supravalvar aortic stenosis typically occurs in two settings (Roberts, 1973; Kim et al., 1999; Stamm et al., 2001; Brooke et al., 2003). Williams syndrome is a chromosome 7q11 microdeletion syndrome in which mutations of the elastin gene have been described. Often, there is hypercalcemia in infancy (Kim et al., 1999; Brooke et al., 2003; Tassabehji, 2003). Affected people have characteristic facies and characteristic personality traits and limits in learning. As infants, there is often pulmonary artery branch stenosis, which regresses with time. At the time of diagnosis in infancy, there may be no real supravalvar aortic obstruction, but over time, the aortic obstruction tends to be progressive. There is also another group of patients with supravalvar aortic stenosis which is familial. These individuals appear to have autosomal dominant transmission and do not have the findings of hypercalcemia in infancy or behavioral abnormalities. However, this autosomal dominant form of supravalve aortic stenosis (SVAS) is also believed to occur due to elastin mutations (Stamm et al., 2001). Clearly, those individuals with Williams syndrome have the other significant abnormalities due to disrupted function of other genes as well, thereby differentiating Williams syndrome and SVAS as two distinct disorders. The cardiovascular findings are the same in both disorders, however. On physical examination, there is a harsh ejection murmur at the right upper sternal border that transmits prominently into the carotids. Additionally, there is a peculiar physical examination finding, caused by the Coanda effect. The Coanda effect causes the jet of blood through the supravalvar narrowing to track along the greater curve of the aorta and into the innominate artery and subsequently into the right subclavian artery. This creates a higher systolic blood pressure in the right arm than in the left arm or the legs. While the pathophysiology seems similar to aortic valve stenosis, there are some important differences. The coronary artery origins are proximal to the obstruction. Since coronary flow is most prominent in diastole, coronary blood flow may be compromised in severe obstruction because blood flow back from the aorta into the coronary arteries is compromised. Also, coronary ostial stenosis is often present, causing even more coronary compromise. As the obstruction tends to be progressive, treatment typically consists of surgical repair once there is more than mild stenosis.

4.3. Mitral valve disease

4.3.1. Myxomatous mitral valvular dystrophy

Myxomatous valvular dystrophy most commonly affects the mitral valve, causing mitral valve prolapse (MVP), stenosis, and/or regurgitation (Kyndt et al., 2007). It affects ~3% of the population and has been demonstrated to be inherited as autosomal and X-linked disorder. Kyndt and colleagues recently identified mutations in filamin A, an X-linked gene located on Xq28 in a large multigenerational family, as well as three smaller families. The penetrance in men was complete and reduced in carrier females, and the clinical phenotype was highly variable (Kyndt et al., 2007).

4.3.2. Mitral valve prolapse (MVP)

MVP is a common abnormality in which the valve leaflets prolapse into the left atrium most commonly due to redundant valve tissue. It may occur alone or associated with other disorders, particularly aortic dilation with or without defined connective tissue disorder such as Marfan syndrome (Freed et al., 2002; Jonkaitiene et al., 2005; Hepnar et al., 2007). In some subjects with MVP, significant mitral regurgitation also occurs. In other situations, MVP occurs alone and is inherited in an autosomal dominant or X-linked fashion. To date, four loci have been identified including three autosomal dominant loci located at chromosomal positions 11p15.4, 16p11.2-p12.1, and 13q31.3-q32.1 and a lone X-linked locus at Xq28 (Disse et al., 1999; Towbin, 1999; Freed et al., 2003; Nesta et al., 2005; Kyndt et al., 2007; Levine and Slaugenhaupt, 2007). In the latter case, the gene filamin A has been identified. There has been speculation that the genetic "final common pathway" will include connective tissue-encoding genes (Towbin, 1999).

4.4. Tricuspid stenosis and atresia

Congenital tricuspid stenosis is extremely rare (Keefe et al., 1970; Geron et al., 1972). Echocardiography provides a definitive diagnosis by demonstrating a stenotic tricuspid valve with an enlarged right atrium (RA). Cardiac catheterization is usually unnecessary unless the coexistence of other anomalies cannot be entirely excluded by echocardiography.

Tricuspid atresia (Epstein, 2001) is unusual but is compatible with survival to adulthood, providing that there is a coexisting VSD (Patel et al., 1987; Hager et al., 2005). This anomaly never occurs alone because live birth obligates patency of the foramen ovale or, more commonly, a secundum type of atrial septal defect. In patients who survive to adulthood, coexisting pulmonary or subpulmonary stenosis is virtually always present. About one-fourth of the patients also will have transposition of the great arteries, with the aorta arising from a hypoplastic right ventricle. Patients with tricusipd atresia are invariably cyanotic because there is mixing at both the atrial and at the ventricular level. The degree of disability largely depends on the degree of arterial unsaturation (Epstein, 2001). This, in turn, depends on the magnitude of pulmonary blood flow. Patients with a large VSD and mild pulmonary stenosis will have the most pulmonary blood flow and be the least cyanotic initially. Over time, though, there will be high pulmonary blood flow and pulmonary vascular disease will develop. If pulmonary vascular disease develops, standard surgical treatments (e.g., single ventricle palliations such as the Fontan procedure) will be impossible.

Clinically, there is always marked cyanosis with clubbing. The characteristic murmur of pulmonary stenosis can be heard. There is some degree of cardiomegaly. On chest X-ray study, the enlarged heart is somewhat bottle shaped, with normal or somewhat decreased pulmonary vasculature. There may be associated skeletal anomalies, such as pectus excavatum and kyphoscoliosis (Epstein, 2001).

The echocardiogram demonstrates an absence of the inflow portion of the right ventricle. There is an atrial septal defect with obligate right-to-left shunting, a functionally single ventricle communicating with a rudimentary ventricle via a VSD, and often pulmonary stenosis. There may be transposition of the great arteries.

4.4.1. Genetics of tricuspid atresia

No single gene mutations have been identified to date for this form of CHD in humans but evidence for mutations in the gene FOG2 (Friend of GATA) in mice has been described (Svensson et al., 2000). In FOG2 mutant mice, a phenotype resembling tricuspid atresia with ASD and VSD has been noted. Further work screening for mutations in this pathway is required to define whether the human equivalent results from disruption in this gene and the associated pathway of genes, but till date has been negative (Sarkozy et al., 2005).

4.5. Pulmonary stenosis

Pulmonary stenosis can be valvar, supravalvar, or subvalvar. Although pulmonary stenosis may occur as an isolated anomaly, it is often associated with an interatrial or interventricular communication or with great artery transposition.

4.5.1. Pulmonary valvar stenosis

Pulmonary valvar stenosis typically is caused by commissural fusion and is associated with a dome-shaped valve. Alternatively, the valve may be dysplastic (Latson and Prioto, 2001; Almeda et al., 2003). The consequent obstruction to right ventricular outflow results in right ventricular hypertrophy and, when severe, an increase in the right ventricular filling pressure. Patients with isolated mild pulmonary stenosis usually do well (Mahoney, 1993; Waller et al., 1995; Sommer et al., 2000), whereas those with moderate (peak-to-peak systolic gradient of 50–80 mmHg) or severe stenosis (systolic gradient >80 mmHg) are usually considered for catheter or surgical intervention (Kern and Bach, 1998; Block and Bonhoeffer, 2005).

Multiple peripheral pulmonary stenoses can also occur. These are almost always associated with one of several clinical syndromes, including the following:

1. Williams syndrome (Kim et al., 1999; Stamm et al., 2001; Brooke et al., 2003)
2. Noonan syndrome (Noonan and O'Connor, 1996; Marino et al., 1999)
3. Congenital rubella syndrome (Nora and Nora, 1978)
4. Alagille syndrome (Kamath et al., 2004)

In patients who have an associated patent foramen ovale, right-to-left shunting at rest may occur during atrial systole (during which the right atrial pressure exceeds that of the LA) or during exercise or tachycardia, when the pressure in the RA may exceed that of the LA throughout the cardiac cycle, resulting in arterial desaturation. These elevated right atrial pressures result from the decreased compliance of the hypertrophied RV. The Valsalva maneuver at rest may cause a transient increase in the right atrial pressure relative to that of the LA during its release.

Mild pulmonary stenosis (systolic gradient <50 mmHg) is compatible with a normal existence, the chief risk being infective endocarditis (Krabill et al., 1985). In increasingly severe degrees of pulmonary stenosis, the degree of right ventricular hypertrophy increases and may result in early and significant right-to-left shunting at the atrial level if there is an atrial communication, or right ventricular failure. This is now an extremely rare eventuality when adequate medical facilities are available and early intervention can be instituted.

4.5.2. Treatment of pulmonary stenosis

Treatment of isolated pulmonary valvar stenosis formerly entailed surgical valvuloplasty, but increasingly, the treatment of choice is percutaneous transvenous balloon valvuloplasty (Kan et al., 1982; Gibbs, 2000; Almeda et al., 2003; Garty et al., 2005). Treatment with balloon valvuloplasty is highly successful and can be repeated if the dilatation is inadequate or restenosis occurs (Peppine et al., 1982; Fawzy et al., 1988; Ing et al., 1995). Long-term follow-up of patients with pulmonary stenosis shows that the probability of survival is similar to that of the general population after 25 years (95.7%) whether medically or, when indicated, surgically treated. Survival is somewhat shorter when associated cardiomegaly is present (Roos-Hesselink et al., 2006).

4.6. Subpulmonary stenosis

Subpulmonary stenosis is relatively uncommon as an isolated lesion but is the more common type of pulmonary stenosis associated with a VSD (tetralogy of Fallot) (Neches et al., 1998). Pulmonary stenosis can be either membranous or infundibular and may be clinically indistinguishable from pulmonary valvar stenosis. However, pulmonary valvar stenosis is associated with poststenotic dilatation of the main pulmonary artery, which is readily seen on chest X-ray study, and the echocardiogram is definitive. When the condition is severe, the treatment is surgical, with either resection of the obstruction or roofing of the outflow tract by patching (Murphy et al., 1993; Pokorski, 2000).

4.6.1. Genetics of pulmonary artery defects

The most common associated genetic defects associated with abnormalities of the pulmonary valve and pulmonary artery is the 22q11 microdeletion syndrome associated with DiGeorge syndrome and Velo-cardio-facial syndrome (VCFS) (Goldmuntz, 2005; Driscoll, 2006). In this group of disorders, mutations in multiple genes have been identified but TBX1 is the only gene thus far associated with the cardiovascular component of these syndromes (Zweier et al., 2007). Mouse models in which Tbx1 is disrupted and cardiovascular disease occurs was initially described by Baldini and colleagues and subsequently confirmed by others (Vitelli et al., 2002; Baldini, 2005; Paylor and Lindsay, 2006). In Alagille's syndrome, mutations in the gene Jagged-1 have been identified (Spinner et al., 2001). This same gene has also been found to cause nonsyndromic pulmonic stenosis, as well as tetralogy of Fallot (Krantz et al., 1999; Eldadah et al., 2001; Raas-Rothschild et al., 2002). Jagged-1 is a ligand to the Notch receptor, which has been implicated in different aspects of organogenesis in Drosophila and mammals (Li et al., 1997a). Another important discovery was the identification of the gene for Noonan syndrome by Gelb and colleagues (Tartaglia et al., 2001, 2002; Gelb and Tartaglia, 2006). In this syndromic disorder, PS, ASD, and hypertrophic cardiomyopathy all commonly occur, either in isolation or in some combination (Noonan and O'Connor, 1996; Marino et al., 1999). The first gene identified, that known as PTPN11, which encodes the protein tyrosine phosphatase Shp2 on chromosome 12q24, is found in patients with the predominant cardiac finding of PS. The mechanism responsible for this is not yet clear.

In addition to mutations in Jagged-1 (Krantz et al., 1999; Eldadah et al., 2001) and TBX1 (Zweier et al., 2007), tetralogy of Fallot has been found to occur due to mutations in FOG2 and Nkx2.5. FOG-2 (Friend of GATA) encodes a multi-zinc finger protein that interacts specifically with GATA proteins to modulate their transcriptional activity mainly by repressing it (Pizzuti et al., 2003). In addition, mutations in GATA4 have recently identified as well (Nemer et al., 2006).

4.7. Ebstein's anomaly

Ebstein's anomaly consists of downward displacement of the tricuspid valve so that the septal and posterior leaflets are adherent to the right ventricular wall, thus to a greater or lesser extent "atrializing" the inflow tract of the RV (Ebstein, 1968; Attenhofer Jost et al., 2005). The tricuspid valve is deformed and may assume a cribriform appearance, and it is invariably associated with tricuspid regurgitation. Some evidence has linked this anomaly to maternal use of lithium, but this has been disputed (Zalstein et al., 1990).

The physical findings show a characteristic "quadruple cadence" consisting of S_1, one or multiple clicks ("sail sound") produced by the upward motion of the billowing anterior leaflet, and wide splitting of S_2 due to right bundle branch block.

In mild cases of Ebstein's anomaly, there is little or no functional disturbance, and the condition is compatible with a normal life expectancy. When the condition is severe, the entire inflow tract is functionally atrialized, and there is little pump function of the RV. There is an associated interatrial communication, with a patent foramen ovale or an atrial septal defect in 50% of patients. If right-to-left shunting occurs, cyanosis and fatigability ensue. The large "atrium" also predisposes to atrial fibrillation.

The diagnosis is made by echocardiography to delineate the tricuspid valve, assess right ventricular function, and look for right-to-left shunting (Gussenhoven et al., 1980; Attenhofer Jost et al., 2005).

The ECG may show one of two patterns: Wolff–Parkinson–White pattern (Kastor et al., 1975) or an unusual right bundle branch block with a "splintered" QRS complex in V_1 or V_2 (Figs. 8–16). These patients may have an accessory AV pathway that predisposes to supraventricular tachyarrhythmias.

No treatment is needed when the condition is mild (Giuliani et al., 1979). Severe cases of Ebstein's anomaly, however, do not do well over the long term.

4.7.1. Genetics of Ebstein's anomaly

A variety of chromosomal deletions and duplications have been described in patients with Ebstein's anomaly (Nakagawa et al., 1999; Yang et al., 2004; Miller et al., 2005). The only gene defects described to date have been described by Ikeda et al. (2002) and Benson et al. (1999) who found mutations in Nkx2.5. Mouse models have predicted Alk3 to be responsible for cases of Ebstein's anomaly (Gaussin et al., 2005) but no human mutations in that gene have been described to date.

5. OBSTRUCTION OF THE GREAT VESSELS

5.1. Coarctation of the aorta

By far, the most common congenital aortic obstruction is coarctation of the aorta, which is typically at or just distal to the ligamentum arteriosum and the take-off of the left subclavian artery. It is twice as common in men and is sometimes seen in patients with Turner's syndrome (Bondy, 2005; McBride et al., 2005a; Volkl et al., 2005). Sometimes there is a discrepancy between the blood pressure in the two arms. This may occur because the coarctation is proximal to the left subclavian artery stenosis or because there is an anomalous origin of the right subclavian artery distal to the coarctation or because the origin of the left subclavian artery is involved in the coarctation and is stenotic.

Physiologically, there is hypertension of the arterial system proximal to the coarctation and normotension distal to it. Outside of infancy, the decreased or absent pulses in the distal arteries are not due to low flow but due to a narrow pulse pressure.

Mild coarctation may not be associated with significant hypertension, initially. In general, though, the severity of obstruction is progressive and the natural history of unoperated coarctation in the presurgical era was ominous with only 10% of patients surviving to age 50 years. Coarctation of the aorta is the most common congenital cardiovascular disease in Turner's syndrome (Lacro et al., 1988; Bondy, 2005).

5.1.1. Diagnosis of coarctation

Coarctation of the aorta presents in one of the two ways. In neonates, these patients become severely ill once the patent ductus arteriosus (PDA) closes. They are poorly perfused and commonly have findings of heart failure. The diagnosis is made clinically by the faint or absent pulses in the femoral arteries in the presence or absence of hypertension in the arms. In some instances, the femoral artery pulsations are good, especially when the child is diagnosed prior to PDA closure. However, simultaneous pressure measurements in the arms and legs may reveal the systolic pressure in the lower extremities to be substantially lower than in the arms. A bicuspid aortic valve is present in more than half the cases (Reisenstein et al., 1947; Liberthson et al., 1979). In the older children and adult, the presence of upper extremity hypertension, reduced lower extremity pulses and blood pressure, and cardiac murmurs may be found. Sometimes, intercostal collaterals occur and are palpable, and murmurs are audible. The chest x-ray study often shows a notch at the site of the coarctation just distal to the aortic arch, the "3" sign caused by dilatation of the proximal aorta because of the hypertension and poststenotic dilatation distal to the coarctation. Mild cases of coarctation are sometimes found incidentally in the evaluation of a bicuspid aortic valve (Lewin and Otto, 2005).

Echocardiography is useful to assess the anatomy and function of the aortic valve and the degree of left ventricular hypertrophy and to follow the diameter of the ascending aorta, especially after surgery. In older patients, transesophageal echocardiography may or may not visualize the coarctated site well. Magnetic resonance imaging (MRI) has emerged as the most valuable tool to completely outline the anatomy of the aortic arch in coarctation (Ho et al., 2003). MRI can provide important information that is helpful to direct surgical or interventional catheterization procedures (Lima and Desai, 2004).

Treatment includes surgical or catheter intervention techniques including aortic stenting and balloon angioplasty. Balloon dilatation is often performed in the pediatric age group, but recurrence is common in native coarctation and has limited enthusiasm (Tynan et al., 1990; Pedra et al., 2005). In adults, the most common form of coarctation is a discrete constriction, which is best resected with end-to-end anastomosis (Cohen et al., 1989; de Bono and Freeman, 2005). Sometimes, significant restenosis occurs, necessitating reintervention. Long tubular coarctations are rare in adults. Crossclamping of the aorta during surgery does not usually cause ischemia of the kidneys or spinal cord, because most of the distal aortic flow is via collaterals. Overall, most patients have been treated with surgery over the years. There appears to be a place for balloon angioplasty with stent placement as an alternative to surgery in some patients. After successful correction, hypertension often persists but is more readily controlled medically.

5.1.2. Genetics of coarctation
Studies in zebrafish have identified abnormalities in the gene called gridlock to cause a coarctation-like disorder (Towbin and McQuinn, 1995; Weinstein et al., 1995). This mutation results in disruption in the Hey2 gene (Gessler et al., 2002). Mutations in humans have been elusive. In humans, some patients with neurofibromatosis develop coarctation (Lin et al., 2000). Mutations in this tumor-suppressor gene (NF1) cause an increase in the cardiac cushion tissue which will form the valve and in particular the aortic valve.

6. ABNORMAL COMMUNICATION BETWEEN CHAMBERS OR GREAT ARTERIES

Abnormal communications between cardiac chambers may be interatrial, interventricular, atrioventricular, or aortopulmonary. They may also occur between the aorta and one of the cardiac chambers. Such communications result in shunting of blood, the direction of flow being determined by the pressure gradient, and/or the difference in resistance between pulmonic and systemic circulation. Therefore, in these conditions, the shunt may be entirely left to right, entirely right to left with resulting cyanosis, or right to left, occurring only under certain conditions, such as exercise (i.e., tardive cyanosis).

A unique physiologic situation exists if the defect results in formation of a common mixing chamber. For example, a subset of complete AV canal consists of complete absence of the atrial septum (common atrium) and a cleft in the anterior leaflet of the mitral valve, and sometimes in the septal leaflet of the tricuspid valve. In total anomalous pulmonary venous connection, the RA serves as the common mixing chamber. A functionally single or common ventricle in which the other ventricle is rudimentary would constitute a mixing chamber. Finally, the septum between the aorta and the pulmonary artery may be absent (i.e., truncus arteriosus or aortopulmonary window). In all four of these situations, there always is some degree of arterial desaturation, regardless of the PVR. Infrequently, one or more pulmonary veins may connect anomalously in the presence of an intact atrial septum and give the clinical picture of an atrial septal defect.

6.1. Atrial septal defect

Atrial level communications may be any of the following (Bedford, 1960; Vick, 1998; Qu, 2004): an ostium secundum which represents non-closure of the foramen ovale; an ostium primum that is a subset of common AV canal and is an endocardial cushion defect; a sinus venosus defect that results from failure of the sinus venosus in the proximal portion of the superior vena cava to be incorporated into the RA; and a coronary sinus defect, in which there is a communication between the coronary sinus defect, in which there is a communication between the coronary sinus and the LA, resulting in flow from the LA to the RA via the coronary sinus (Shirodaria et al., 2005). In practice, the overwhelming majority of atrial septal defects are of the secundum type, representing about 70% of all interatrial communications, with ostium primum being second. The sinus venosus defect is uncommon, and coronary sinus defect is extremely rare. Patients with uncomplicated interatrial communications frequently arrive at adulthood undiagnosed.

6.1.1. Atrial septal defect, ostium secundum type

Atrial septal defect, ostium secundum type, is the extremely common in childhood and is the most common newly diagnosed CHD in the adult, possibly matched only by the bicuspid aortic valve (Bedford, 1960; Craig and Selzer, 1968; Mattila et al., 1979; Sutton et al., 1982; Steele et al., 1987; Vick, 1998; Qu, 2004; Shirodaria et al., 2005; Engelfriet et al., 2006). It is helpful here to review the circulatory physiology in the fetus. During fetal life, the pressures in the pulmonary artery and the aorta are equal; the fetal RV, being adapted for pressure, has the same compliance as that of the LV. Blood returning from the placenta (oxygenated blood) flows via the umbilical vein through the ductus venosus and preferentially shunts across the foramen ovale. Vestigial preferential right-to-left streaming from the inferior vena cava can be demonstrated in adults with atrial septal defects by echocardiography or by indicator dilution techniques;

the clinical counterpart is paradoxical embolization. At birth, with the infant's first breath, there is an immediate drop in PVR, which gradually decreases to normal in the first few months of life. At the same time, the RV undergoes regression of myocardial hypertrophy, gradually changing from its cylindrical configuration and thick walls to that characteristic of the adult, in which the cavity is more crescentic and the wall thinner than that of the LV. Therefore, in the fetus, there is virtually no left-to-right shunting across the defect. If the defect persists, as the PVR falls and the compliance of the RV increases, left-to-right shunting results and pulmonary flow may be two to five times the systemic flow. With time, both the RA and the RV enlarge.

Significant pulmonary hypertension seldom occurs before the third or fourth decade (Mattila et al., 1979; Steele et al., 1987). The mechanism by which pulmonary hypertension develops is not well understood. It may rarely start in childhood (Sutton et al., 1982). Although high flow is implicated, it takes many decades for pulmonary hypertension to develop, and not all patients do, in fact, develop pulmonary hypertension (Sutton et al., 1982). Progressive right ventricular enlargement and hypertrophy may lead to decreased compliance of the RV compared with that of the LV, and therefore, the exclusive left-to-right shunt also yields some right-to-left shunting. Right-to-left shunting is not a direct effect of the relative pressures of the pulmonary artery and of the aorta, but rather of the compliance of the two ventricles. If left ventricular compliance also decreases, the atrial pressures rise, and classic signs of heart failure may be present. The pulmonary arterial pressure in atrial septal defects with shunt reversal is almost always significantly lower than the systemic pressure. This contrasts with ventricular septal and aortopulmonary defects, in which shunt reversal and pressure equalization go hand in hand.

Approximately 10% of patients with atrial septal defects have one or more anomalously connected pulmonary veins. Mitral insufficiency may coexist in 10–20% (Leachman et al., 1976). The insufficiency may be due to prolapse of the posterior leaflet of the mitral valve associated with secundum atrial septal defect and significant right ventricular enlargement (Leachman et al., 1976). An interesting syndrome has been described (Lutembacher, 1916) in which an atrial septal defect coexists with mitral stenosis; patients remain relatively asymptomatic until pulmonary hypertension develops because the atrial septal defect decompresses the LA, and the left atrial pressure is not elevated despite significant mitral valve obstruction (Lutenbacher's syndrome). Patients with ostium secundum-type interatrial septal defects are seldom symptomatic until they begin to experience pulmonary hypertension, usually after the fourth decade, if it occurs. It may not occur, however. Unlike the other interatrial communications, there is a 2:1 female to male preponderance and is sometimes familial (Caputo et al., 2005).

The echocardiographic picture of ostium secundum atrial septal defects shows an enlarged RA and RV. The interventricular septum moves paradoxically. The interatrial septum can be seen as a "dropout," especially by transesophageal echocardiography. Shunting is demonstrable by color-flow Doppler, and the pulmonary artery pressure can be estimated if there is tricuspid insufficiency.

6.1.2. Treatment of atrial septal defect

Because the mortality risk of surgical closure of an uncomplicated secundum atrial septal defect is ~1% or less and the adverse consequences (i.e., pulmonary hypertension, paradoxical embolism, and shortened life expectancy) have a higher risk, early closure is recommended, even when patients are asymptomatic. However, when there is severe pulmonary hypertension, especially with right-to-left shunting, closure may be contraindicated and lung transplantation may become necessary. Small atrial septal defects found in childhood may not need closure, usually before 5-years of age. There is some evidence favoring waiting because about half of such patients have spontaneous closure of the defect (age 8.4 years) (Brassard et al., 1999).

Open-heart surgery for the closure of defects in the atrial septum is currently the "gold standard" for treatment of such patients (Horvath et al., 1992; Hopkins et al., 2004). The mortality rate for this procedure is close to 0% in most contemporary reports. However, open-heart surgery is a major procedure, with its attendant morbidity, need for intensive care, and significant hospitalization. The complication rates after surgical closure of atrial septal defects in adult patients can be as high as 13%.

The pioneering work of King and Mills (1974, 1976) resulted in the development of a double-umbrella device and established the feasibility of occluding atrial septal defects with percutaneous devices. Many other devices were subsequently developed, including the Rashkind single-disc device, Locke USCI "clamshell" device, "buttoned" device, ASDOS device, Manodisk device, Das Angel Wings, Amplatzer device, and the Cardioseal device (Rashkind, 1985; Lock et al., 1987; Babic et al., 1990; Rome et al., 1990; Sideris et al., 1990; Sievert et al., 1990; Rao et al., 1991, 1992; Ruiz et al., 1992; Boutin et al., 1993; Das et al., 1993; O'Laughlin et al., 1993; Pavcnik et al., 1993; Perry et al., 1993; Hausdorf et al., 1995; Zamora et al., 1995; Sharafudin et al., 1996; Sievert et al., 1997; Pedra et al., 2003; Schrader, 2003; Butera et al., 2004; Kay et al., 2004; Purcell et al., 2004; Celiker et al., 2005). These are now standard therapies.

6.1.3. Genetics of secundum atrial septal defects

Atrial septal defects (ASDs) are seen in isolation, as well as associated with other CHD, noncardiac syndromal abnormalities, and conduction system block. Patients with Holt–Oram syndrome (HOS) (Huang, 2002), a disorder in which ASD is associated with skeletal abnormalities usually affecting the thumbs or entire arm, have been shown to occur due to mutations in the T-box transcription factor gene TBX5 (Basson et al., 1997; Li et al., 1997b; Vaughan and Basson, 2000). TBX5 mutations have also been seen in subjects with non-HOS malformed hearts (Reamon-Buettner and Borlak, 2004). Similarly, patients with sporadic ASD or ASD associated with conduction disease (atrioventricular block) have mutations in Nkx2.5, a transcription-factor gene that interacts with TBX5 (Benson et al., 1999; Ikeda et al., 2002). Further, mutations in GATA-4 have been seen in cases of ASD, familial and sporadic (Svensson et al., 2000; Hirayama-Yamada et al., 2005). In addition, mutations in CRELD1 have also been shown to cause ASD, particularly AV septal-type seen in AV canal defects (Robinson et al., 2003). Recently, mutations in the myosin heavy chain-6 gene was identified in patients with ASDs as well (Ching et al., 2005).

6.1.4. Ostium primum defect

Ostium primum defect, is an endocardial cushion defect in which the septum primum (i.e., the lower portion of the atrial septum) fails to develop, as does the membranous portion of the ventricular septum. In this situation, the anterior leaflet of the mitral valve is attached to the ventricular septum in a somewhat lower position than normal and is therefore at the same level as the tricuspid valve. The anterior leaflet is cleft and, because of its lower position, its characteristic angiographic (Girod et al., 1965) and echocardiographic (Nagvi, 2004) features, the mitral valve cleft is associated with various degrees of mitral regurgitation. The considerations governing interatrial shunting and those involved in the development of pulmonary hypertension are similar to those of the secundum type of atrial septal defect. However, if the degree of mitral regurgitation is more than mild, it may cause left ventricular dilation and both left atrial and right atrial enlargement.

The clinical diagnosis of an ostium primum ASD has many of the features of those associated with a secundum-type atrial septal defect (i.e., hyperactive RV; wide, fixed splitting of the second heart sound; and pulmonary flow murmur). However, there is usually an additional murmur of mitral insufficiency associated with a cleft in the anterior leaflet. This does not necessarily radiate to the axilla, because the regurgitant jet is directed more medially than toward the free wall of the LA. If the mitral regurgitation is severe or if pulmonary hypertension develops, decreased exercise tolerance and exertional dyspnea can be expected. Until and unless pulmonary hypertension develops, the interatrial shunting is left to right, and there is no cyanosis. When the mitral insufficiency is severe, there may not be a significant elevation of atrial pressures, because the RV is adapted for volume overload and accepts the increased left-to-right shunt. However, with a decrease in compliance of a failing or hypertrophied RV because of pulmonary hypertension, atrial pressures rises, and the pulmonary venous pressure can be estimated from inspection of the neck veins.

The natural history of ostium primum defects differs significantly from that of a secundum-type atrial septal defect. Ostium primum defects are susceptible to an infective endocarditis of the cleft mitral valve. If the mitral regurgitation is severe, patients may have dyspnea and left ventricular dysfunction, and if the right ventricular compliance is compromised, there are signs of both right ventricular and left ventricular failure.

Patients with complete AV canal seldom survive to late adulthood without cardiac surgery. The ECG shows left axis, right ventricular hypertrophy, large P waves, and various degrees of AV block.

6.1.5. Genetics of primum atrial septal defect

As noted previously, mutations in CRELD1 have been identified in patients with atrioventricular septal defects in children with Down syndrome (Robinson et al., 2003).

6.2. Ventricular septal defects

Isolated VSDs are common in childhood but infrequently seen in adults (Patrianakos et al., 2005). Fifty percent or more of VSDs close spontaneously in early childhood, even as late as adolescence (Onat et al., 1998; Mehta et al., 2000). If the VSD is small, these defects are associated with little or no hemodynamic disturbance of the LV and result in only a small left-to-right shunt and no pulmonary hypertension. Large defects are associated with equalization of pressure in the two ventricles and therefore in the pulmonary artery. The direction and the degree of shunting are determined by the relative resistances of the pulmonary and systemic circuits. The right-to-left and the left-to-right shuntings are more or less balanced if the resistances in the systemic and pulmonary circulations are equal. This is the classic Eisenmenger's complex. Large defects with a low to mildly elevated PVR have large left-to-right shunts and severe volume overload of the LV. These are almost invariably discovered by a pediatric cardiologist, and the defect is closed (Mehta et al., 2000).

The most common form of congenital isolated VSD is of the so-called perimembranous type, which is posterior and inferior to the crista supraventricularis, involving what would be the membranous septum and some of the adjacent muscular septum (van Praagh et al., 1989). This is situated just under the septal leaflet of the tricuspid valve and is subtended by the aortic valve. The bundle of His courses along the posterior rim of this defect and is therefore not affected, but it is vulnerable during surgical closure of the defect. Single or multiple muscular septal defects may also occur as isolated congenital lesions (van Praagh et al., 1989). In early infancy, up to

50% of isolated VSDs are in the trabecular septum. Later, muscular VSDs are present in about 10% of the cases of VSD. They are generally multiple and small, so that even in the presence of large shunts, there is seldom significant elevation of the right ventricular or pulmonary artery pressure. Although many of these defects close spontaneously, they may persist and are difficult to close completely at the time of surgery because of heavy trabeculation on the right ventricular aspect of the interventricular septum. The seemingly logical left ventricular approach to closure of the lesion would seriously compromise the contractility of the LV.

Surgical closure of VSDs is associated with conduction abnormalities in as many as 15% of cases, consisting of right bundle branch block and left-axis deviation (Godman et al., 1974; Roos-Hesselink et al., 2004). There are no extensive data on progression to third-degree heart block, but it appears to be uncommon, although occasional instances of sudden death have been attributed to it (Godman et al., 1974). These conduction abnormalities are related to the close proximity of the bundle of His to the posterior wall of the VSD. Nonsurgical closure in the catheterization laboratory is gaining popularity with the development of different devices (Knauth et al., 2004; Bacha et al., 2005; Moodie, 2005; Pawelec-Wojtalik et al., 2005; Chessa et al., 2006).

6.2.1. Genetics of ventricular septal defects

Mutations in a large number of genes in animal models have been associated with VSDs, usually in association with other cardiac or extracardiac abnormalities. These include knockout models of Nkx2.5, FOG2, and Hey2 genes, as well as heterozygous transgenic Tbx5 mice (Svensson et al., 2000; Vaughan and Basson, 2000; Clark et al., 2006; Olson, 2006; Srivastava, 2006; Hutson and Kirby, 2007; Niessen and Karsan, 2007). Some human syndromes and sporadic cases in which VSDs have occurred include heterozygous mutations in Nkx2.5 (Benson et al., 1999; Ikeda et al., 2002; Reamon-Buettner et al., 2004) TBX5, and GATA-4 (Reamon-Buettner and Borlak, 2006; Schluterman et al., 2007). Recently, mutations in COG7 were described in patients with a complex phenotype including VSD, microcephaly, adducted thumbs, growth retardation, and episodes of hyperthermia (Morava et al., 2007). The mutations in this gene disrupt glycosylation pathways and perhaps provides new pathway-focused insight into the development of normal ventricular septal morphogenesis.

7. PATENT DUCTUS ARTERIOSUS

The ductus arteriosus is the main path through which oxygenated cord blood perfuses distal to the aortic arch of the fetus. With delivery and the baby's first breath, there is an immediate drop in PVR, resulting in a reversal of shunt through the ductus arteriosus. Usually, the ductus arterious constricts and is functionally closed by 18 h after birth. Patency may persist for several weeks without any consequences. The pathophysiologic consequences of persistent patency of the ductus arteriosus depend on the size of the ductus and, to a lesser extent, on the length (Campbell, 1968). There is a continuous left-to-right shunt during systole and diastole through the ductus as long as the PVR is lower than the systemic resistance. Predisposing factors for a patent ductus are material rubella in the first trimester of pregnancy, prematurity, and high altitude.

Most patients with restrictive PDA are asymptomatic because the left-to-right shunt is generally mild to moderate. The pathophysiology is similar to that of aortic regurgitation, and there is a rapid runoff with some degree of left ventricular volume overload. The classical physical finding is the machinery murmur of Gibson; Gibson (1898) heard best in the left subclavicular region and continuous throughout the whole cardiac cycle. This is not to be confused with long murmurs in systole and diastole, such as may occur in combined aortic stenosis and regurgitation, in which the directional shift in flow is marked by a short hiatus between the systolic and the diastolic components. The murmur is identical to that heard in an AV fistula, which in fact resembles physiologically. The chest x-ray study in patients with a small ductus shows a normal heart size with normal pulmonary vasculature. There is a tendency, however, for the aortic arch to be somewhat wider than usual for the patient's age. With a moderate left-to-right shunt, the heart size may increase somewhat over time, and there may be a suggestion of increased pulmonary flow.

The ECG shows evidence of flow into the left pulmonary artery because the ductus is usually between the distal aortic arch and the proximal left pulmonary artery (Liao et al., 1988) shortly after bifurcation. During cardiac catheterization, it is usually possible to manipulate the venous catheter from the proximal left pulmonary artery through the ductus into the descending aorta. The pulmonary artery pressure should be measured simultaneously with the systemic arterial pressure.

7.1. Genetics of patent ductus arteriosus

No genes have been identified to cause simple PDA but a gene causing a syndrome in which PDA is a prominent feature was described. The human syndrome, known as Char syndrome, is an autosomal dominant trait characterized by PDA, facial dysmorphism, and hand anomalies (Sweeney et al., 2000). The transcription factor encoding gene TFAP2B was identified as the causative gene by Satoda et al. (2000) and Zhao et al. (2001) and later confirmed by Mani et al. (2005).

8. ENDOCARDIAL CUSHION DEFECTS

This is a complex group of anomalies that arise from failure of proper development of the endocardial cushion involving some or all of the following: the lower part of the atrial septum, the upper part of the ventricular septum, and the adjacent leaflets of the two AV valves. The partial AV canal is the ostium primum, in which the lower part of the atrial septum, the upper part of the ventricular septum, and the anterior leaflet of the mitral valve are involved, resulting in an interatrial septal defect and a cleft in the anterior leaflet of the mitral valve. The transitional type is associated with a single atrium with no atrial septum at all but with a cleft leaflet of the mitral valve, resulting in a mixing lesion, as discussed previously. The complete AV canal has all our chambers in communication, there being no fusion between the atrial septum, and the ventricular septum and the two AV valves straddle this defect. This common AV valve has four or five leaflets, with an inferior and a superior leaflet straddling the canal and the lateral leaflets attached only to the ipsilateral ventricle. There may be a small fifth leaflet, which is anterior and in the RV.

Endocardial cushion defects are the most common kind of CHD associated with Down syndrome, and patients with Down syndrome have a high incidence (about 40%) of CHD.

8.1. Genetics of endocardial cushion defects

Endocardial cushion defects, also known as AV canal defects, are most commonly seen in patients with trisomy 21 (Kallen et al., 1996). The specific causative gene has remained elusive (Pierpont et al., 2000). Maslen and colleagues have shown that mutations in CRELD1, localized to chromosome 3p25, causes some cases of nonsyndromal AV canal (Robinson et al., 2003). In mice, the involvement of tyrosine kinase receptors ErbB2 and B3, which binds neureglin in the formation of the endocardial cushion, has been proposed. No human mutations have been identified thus far.

9. COMPLEX CHD

Complex CHD is a group of diseases in which more than one cardiac abnormality is present, which may or may not be associated with cyanosis.

9.1. Without cyanosis

CHDs without cyanosis have been alluded to in separate categories. Their clinical presentation is a composite of their individual pathophysiology. Such combinations include interatrial septal defects with one or more anomalous pulmonary veins (Stewart et al., 1983; McGaughey et al., 1986; Ward and Mullins, 1998; Zagol et al., 2006). Shone's syndrome (Brown et al., 2005), Williams syndrome (supravalvar aortic stenosis) and peripheral pulmonary artery stenoses (with characteristic facies), aortic insufficiency and mitral insufficiency with Marfan syndrome (Gleason, 2005), coarctation of the aorta and/or VSD with Turner syndrome (Bondy, 2005), and various degrees of AV block with endocardial cushion defects from first- to second- to third-degree AV block.

9.2. With cyanosis

These conditions involve transposition of the great arteries. Three categories of great vessel transposition can be seen in the adult D-transposition, L-transposition, and double-outlet RV.

9.2.1. D-transposition (complete transposition) of the great arteries (D-TGA)

This transposition involves virtually a direct switch between the two great arteries so that the aorta arises anteriorly from the RV and the pulmonary artery arises posteriorly from the LV (Wernovsky, 2001). It is evident that without intercommunication, these would be two closed circuits that are incompatible with life. Therefore, there must be a shunt at the atrial level (most common) or at the ventricular level, or else a large patent ductus (Anderson et al., 1991; Rigby and Chan, 1991). In patients with atrial or VSDs, pulmonary stenosis may be present, which "protects" the pulmonary vasculature. Uncorrected D-transposition is compatible with survival into adulthood, but patients are invariably cyanotic. Because the disability is obvious in childhood, these patients now undergo surgery before they reach adulthood. Echocardiography (Rigby and Chan, 1991) is the diagnostic modality of choice. Surgical treatment for this condition has undergone considerable evolution, from "atrial switching" (actually rerouting of venous blood to the appropriate ventricle) to present-day methods, including arterial switching (Wernovsky, 2001; Hornung et al., 2002).

9.2.2. Genetics of D-transposition

No specific genes have been identified in humans but D-TGA has been found in patients with 22q11 microdeletion (Goldmuntz et al., 1998). In mice, Pitx2 null animals have been found to cause TGA (Franco et al., 2000; Muncke et al., 2005). Inactivation of genes encoding the type II activin receptor, the Cited2 coactivator, and the transduction cytoplasmic protein Dishevelled2 associated with wnt signaling, all lead to TGA, in addition to other cardiac and non-cardiac phenotypes, in animal models (Hamblet et al., 2002; Bamforth et al., 2004). Mutations in humans in the gene PROSIT240, a novel TRAP240-like gene (Muncke et al., 2003), have been described to cause D-TGA, as have mutations in CFC1 (Goldmuntz et al., 2002).

9.2.3. Tetralogy of fallot

Tetralogy of Fallot consists of a large VSD in the usual position, together with pulmonary stenosis, either valvar or infundibular, or both (Fallot, 1888). The large VSD results in equal pressure in the two ventricles. The two other features constituting the tetralogy are right ventricular hypertrophy and overriding of the interventricular septum by the dilated aorta. This malformation is compatible with surviving into adulthood (Bertranou et al., 1978; Abraham et al., 1979). The more severe the pulmonary stenosis, the greater the degree of right-to-left shunting at the ventricular level. The more complete the arterial unsaturation, the greater the disability. Such patients not only are prone to the usual complications of cyanotic heart disease (e.g., endocarditis, brain abscess, and complications of marked erythrocytosis) but may also experience sudden death, which has been attributed to spasm of the infundibulum (McGrath et al., 1991). Lesser degrees of pulmonary stenosis result in a lesser degree of right-to-left shunting, with a modestly decreased exercise capacity but with the ability to perform most normal activities. In some patients, the degree of pulmonary stenosis is such that the patient is not visibly cyanotic at rest but only with exercise; such patients are said to have tardive cyanosis (Baffes et al., 1953).

The usual clinical findings are cyanosis and clubbing and a loud, harsh pulmonary ejection murmur.

The chest x-ray study shows a somewhat globular heart of normal size, and on the lateral view, a prominent RV can be seen "hugging" the sternum. There may be a right-sided aortic arch, which deviates the esophagus and the sternum slightly to the left. If the pulmonary stenosis is purely valvar and tricuspid, as occurs in about 20% of these cases, poststenotic dilation of the pulmonary trunk and left pulmonary artery occurs. With the much more common infundibular stenosis, the left heart border is concave because the pulmonary trunk is relatively small. This stenosis usually creates a "third chamber" just subjacent to the pulmonary valve (Baffes et al., 1953).

9.2.4. Genetics of tetralogy of fallot

As previously noted, the most common associated genetic defects associated with abnormalities of the pulmonary valve and pulmonary artery is the 22q11 microdeletion syndrome associated with DiGeorge syndrome and VCFS (Goldmuntz, 2005; Driscoll, 2006). In this group of disorders, mutations in multiple genes have been identified but TBX1 is the only gene thus far associated with the cardiovascular component of these syndromes (Zweier et al., 2007). Mouse models in which Tbx1 is disrupted and cardiovascular disease occurs was initially described by Baldini and colleagues and subsequently confirmed by others (Vitelli et al., 2002; Baldini, 2005; Paylor and Lindsay, 2006). In Alagille's syndrome, mutations in the gene Jagged-1 have been identified (Spinner et al., 2001). This same gene has also been found to cause nonsyndromic pulmonic stenosis, as well as tetralogy of Fallot (Krantz et al., 1999; Eldadah et al., 2001; Raas-Rothschild et al., 2002). Jagged-1 is a ligand to the Notch receptor, which has been implicated in different aspects of organogenesis in Drosophila and mammals (Li et al., 1997a). Mutations in FOG2 and Nkx2.5 have been found in isolated tetralogy, (Pizzuti et al., 2003) as have GATA-4 mutations (Nemer et al., 2006).

9.2.5. Pulmonary atresia with VSD (pseudotruncus arteriosus)

Pulmonary atresia may be associated with a large VSD and is sometimes considered the extreme version of the tetralogy of Fallot (Waldman and Wernly, 1999). A more uncommon situation exists when the pulmonary atresia is seen with an intact ventricular septum and a small RV that is drained by large coronary sinusoids. The latter is almost never seen by cardiologists treating adults (Fenton et al., 2004).

Pulmonary atresia associated with a VSD is more usefully thought of clinically as an entity apart from the tetralogy of Fallot (Waldman and Wernly, 1999). The clinical manifestations and the surgical considerations are quite different. The pulmonary circulation is derived entirely from large anomalous arteries that arise from the descending aorta, often called *bronchial arteries* (Jefferson et al., 1972). At the junction of these arteries with the pulmonary arteries, it is common to observe significant stenosis, so that the distal pulmonary arteries may be of relatively normal pressure. Pulmonary atresia can range from mere atresia of the pulmonary valve, which is uncommon, to atresia of the pulmonary trunk and often atresia of the proximal left and right arteries. These patients are markedly cyanotic. A characteristic feature on clinical examination is the presence of a continuous murmur throughout the chest, arising from the stenoses of the bronchopulmonary anastomoses.

9.2.6. Truncus arteriosus

Truncus arteriosus is an incomplete septation of the ascending aorta and the pulmonary trunk (Collett and Edwards, 1949; Mair et al., 2001). The semilunar valve is a single truncus valve, which in most cases consists of three cusps but may be either quadricuspid or bicuspid. It is commonly insufficient, sometimes markedly so, and straddles a VSD (Becker et al., 1971; Reddy and Hanley, 1998). Truncus defects are divided into three types. Type I has a common trunk, but this gives rise distally to recognizably separate ascending aorta and pulmonary trunk. In type II, the truncus extends up to the right and left pulmonary artery bifurcations, there being no separate pulmonary trunk. In type III, the left and right pulmonary arteries come off either side of the truncus. In adults, the most common type is a type I truncus. Such patients have a regurgitant truncus valve, a common mixing chamber, and Eisenmenger's physiology. In the very young, the PVR may be somewhat less than the SVR, so that surgical correction can be attempted. By adulthood, the PVR is markedly elevated, and few if any of these patients are candidates for corrective surgery. The remaining possible treatment is heart–lung transplantation. When patients are operated on in childhood, the results are promising for survival to adulthood (Reddy and Hanley, 1998).

The clinical examination shows the patient to be cyanotic and the digits clubbed. There is usually no systolic murmur, but occasionally, one may be heard. The early diastolic blowing murmur of truncus valve insufficiency is characteristic. Depending on the degree of truncus valve insufficiency, the heart may or may not be enlarged. The chest x-ray study shows the cardiac shadow to have a wide waist, which is the markedly dilated truncus arteriosus. The pulmonary vasculature is similar to that in Eisenmenger's complex (i.e., large central pulmonary arteries and no evidence of increased pulmonary flow).

Aorticopulmonary window is a separate entity and is not related to truncus arteriosus (Newfeld et al., 1962). In this condition, there is a window in the septum between the aorta and the pulmonary trunk. The aorta and the pulmonary trunk have separate semilunar valves, and the window does not go down to and involve the semilunar valves. If the window is large, the adult patient presents with Eisenmenger's syndrome. Smaller windows may protect the pulmonary vasculature and allow closure of the defect.

9.2.7. Genetics of truncus arteriosus

As previously noted, the most common associated genetic defects associated with abnormalities of the pulmonary valve and pulmonary artery are the 22q11 microdeletion syndrome associated with DiGeorge syndrome and VCFS (Goldmuntz, 2005; Driscoll, 2006). In this group of disorders, mutations in multiple genes have been identified but TBX1 is the only gene thus far associated with the cardiovascular component of these syndromes (Zweier et al., 2007). Mouse models in which Tbx1 is disrupted and cardiovascular disease occurs was initially described by Baldini and colleagues and subsequently confirmed by others (Vitelli et al., 2002; Baldini, 2005; Paylor and Lindsay, 2006). The finding of truncus arteriosus has also been notable in mouse models in which inactivation of the genes encoding BMP2, Pax3, and the vasoconstrictor hormone Edn1 have been engineered (Delot et al., 2003; Kaartinen et al., 2004; Restivo et al., 2006; Wang et al., 2006).

10. SUMMARY

The genetic basis of human CHD is one of the final frontiers in the era of cardiovascular genetic identification and mutational analysis. Although several genes have been identified, the vast majority of disease-causing genes are not yet known. Over the next several years, these genes are likely to be identified. It is hoped that these findings will directly connect with the understanding gained to date in animal models.

REFERENCES

Abraham, K.A., Cherian, G., Rao, V.D., Sukumar, I.P., Krishnaswami, S., John, S. 1979. Tetralogy of Fallot in adults: A report on 147 patients. Am. J. Med. 65, 811–816.

Almeda, F.Q., Kavinsky, C.J., Pophal, S.G., Klein, L.W. 2003. Pulmonic valvular stenosis in adults: Diagnosis and treatment. Catheter. Cardiovasc. Interv. 69, 546–557.

Anderson, R.H., Henry, G.W., Becker, A.E. 1991. Morphologic aspects of complete transposition. Cardiol. Young 1, 41.

Attenhofer Jost, C.H., Connolly, H.M., Edwards, W.D., Hayes, D., Warnes, C.A., Danielson, G.K. 2005. Ebstein's anomaly-Review of a mulficaceted congenital cardiac condition. Swiss Med. Wkly. 135, 269–281.

Babic, U.U., Grujicic, S., Djurisic, A., Vucinic, M. 1990. Transcatheter closure of atrial septal defects. Lancet 336, 566–567.

Bacha, E.A., Cao, Q.L., Galantowicz, M.E., Cheatham, J.P., Fleishman, C.E., Weinstein, S.W., Becker, P.A., Hill, S.L., Koenig, P., Alboliras, E., Abdulla, R., Starr, J.P., et al. 2005. Multicenter experience with perventricular device closure of muscular ventricular septal defects. Pediatr. Cardiol. 26, 169–175.

Baffes, T.G., Johnson, F.R., Potts, W.J., Gibson, S. 1953. Anatomic variations in tetralogy of Fallot. Am. Heart J. 46, 647.

Baldini, A. 2005. Dissecting contiguous gene defects: TBX1. Curr. Opin. Genet. Dev. 15, 279–284.

Bamforth, S.D., Brajanca, J., Farthing, C.R., Schneider, J.E., Broadbent, C., Mitchell, A.C., Clarke, K., Neubauer, S., Norris, D., Brown, N.A., Anderson, R.H., Bhattacharya, S. 2004. Cited2 controls left-right patterning and heart development through a Nodal-Pitx2c pathway. Nat. Genet. 36, 1189–1196.

Basson, C.T., Bachinsky, D.R., Lin, R.C., Levi, T., Elkins, J.A., Soults, J., Grayzel, D., Kroumpouzou, E., Traill, T.A., Leblanc-Straceski, J., Renault, B., Kucherlapati, R., et al. 1997. Mutations in human TBX5 cause limb and cardiac malformations in Holt-Oram syndrome. Nat. Genet. 15, 30–35.

Becker, A.E., Becker, M.J., Edward, J. 1971. Pathology of the semilunar valve and persistent truncus arteriosus. J. Thorac. Cardiovasc. Surg. 62, 16.

Bedford, D.E. 1960. The anatomical types of atrial septal defect: Their incidence and clinical diagnosis. Am. J. Cardiol. 6, 568.

Belmont, J.W., Mohapatra, B., Towbin, J.A., Ware, S.M. 2004. Molecular genetics of heterotaxy syndrome. Curr. Opin. in Cardiol. 19, 216–220.

Benson, D.W., Silberbach, G.M., Kavanaugh-McHugh, A., Cottrill, C., Zhang, Y., Riggs, S., Smalls, O., Johnson, M.C., Watson, M.S., Seidman, J.G., Seidman, C.E., Plowden, J., et al. 1999. Mutations in the cardiac transcription factor NKX2.5 affect diverse cardiac developmental pathways. J. Clin. Invest. 104, 1567–1573.

Bertranou, E.G., Blackstone, E.H., Hazelrig, J.B., Turner, M.E., Kirklin, J.W. 1978. Life expectance without surgery in tetralogy of Fallot. Am. J. Cardiol. 42, 458.

Block, P.C., Bonhoeffer, P. 2005. Percutaneous approaches to valvular heart disease. Curr. Cardiol. Rep. 7, 108–113.

Bondy, C.A. 2005. New issues in the diagnosis and management of Turner syndrome. Rev. Endocr. Metab. Disord. 6, 269–280.

Boutin, C., Musewe, N.N., Smallhorn, J.F., Dyck, J.D., Kobayashi, T., Benson, L.N. 1993. Echocardiographic follow-up of atrial septal defect after catheter closure by double-umbrella device. Circulation 88, 621–627.

Brassard, M., Fouron, J.C., van Doesburg, N.H., Mercier, L.A., De Guise, P. 1999. Outcome of children with atrial septal defect considered too small for surgical closure. Am. J. Cardiol. 83, 1552.

Braverman, A.C., Guven, H., Beardslee, M.A., Makan, M., Kates, A.M., Moon, M.R. 2005. The bicuspid aortic valve. Curr. Probl. Cardiol. 30, 470–522.

Brooke, B.S., Bayes-Genis, A., Li, D.Y. 2003. New insights into elastin and vascular disease. Trends Cardiovasc. Med. 13, 176–181.

Brown, J.W., Ruzmetov, M., Vijay, P., Hoyer, M.H., Girod, D., Rodefeld, M.D., Turrentine, M.W. 2005. Operative results and outcomes in children with Shone's anomaly. Ann. Thorac. Surg. 79, 1358–1365.

Butera, G., Carminati, M., Chessa, M., Delogu, A., Drago, M., Piazza, L., Giamberti, A., Firgiola, A. 2004. CardioSEAL/STARflex versus Amplatzer devices for percutaneous closure of small to moderate (up to 18mm) atrial septal defects. Am. Heart J. 148, 507–510.

Campbell, M. 1968. Natural history of persistent ductus arteriosus. Br. Heart J. 30, 4.

Caputo, S., Capozzi, G., Russo, M.G., Esposito, T., Martina, L., Cardaropoli, D., Ricci, C., Argiento, P., Pacileo, G., Calabro, R. 2005. Familial recurrence of congenital heart disease in patients with ostium secundum atrial septal defect. Eur. Heart J. 26, 2179–2184.

Celiker, A., Ozkutlu, S., Karagoz, T., Ayabakan, C., Bilgic, A. 2005. Transcatheter closure of interatrial communications with Amplatzer device: Results, unfulfilled attempts and special considerations in children and adolescents. Anadolu Kardiyol. Derg. 5, 159–164.

Chessa, M., Carrozza, M., Butera, G., Negura, D., Piazza, L., Giamberti, A., Feslova, V., Bossone, E., Vigna, C., Carminati, M. 2006. The impact of interventional cardiology for the management of adults with congenital heart defects. Catheter. Cardiovasc. Interv. 67, 258–264.

Ching, Y.H., Ghosh, T.K., Cross, S.J., Packham, E.A., Honeyman, L., Loughnar, S., Robinson, T.E., Dearlove, A.M., Ribas, G., Bonder, A.J., Thomas, N.R., Scotter, A.J., et al. 2005. Mutations in myosin heavy chain 6 cause atrial septal defects. Nat. Genet. 37, 423–428.

Clark, K.L., Yutzey, K.E., Benson, D.W. 2006. Transcription factors and congenital heart disease. Annu. Rev. Physiol. 68, 97–121.

Clinical path conference. 1968. Coarctation of the aorta as part of Shone syndrome. Minn. Med. 51, 1617–1627.

Cohen, M., Fuster, V., Steele, P.M., Driscoll, D., McGoon, D.C. 1989. Coarctation of the aorta: Long-term follow-up and prediction of outcome after surgical correction. Circulation 80, 840.

Collett, R.W., Edwards, J.E. 1949. Persistent truncus arteriosus: Classification according to anatomic types. Surg. Clin. North. Am. 29, 1245.

Craig, R.J., Selzer, A. 1968. Natural history and prognosis of atrial septal defect. Circulation 37, 805.

Cripe, L., Andelfinger, G., Martin, L.J., Shooner, K., Benson, D.W. 2004. Bicuspid aortic valve is heritable. J. Am. Coll. Cardiol. 44, 138–143.

Das, G.S., Voss, G., Jarvis, G., Wyche, K., Gunther, R., Wilson, R.F. 1993. Experimental atrial septal defect closure with a new, transcatheter, self-centering device. Circulation 88, 1754–1764.

de Bono, J., Freeman, L.J. 2005. Aortic coarctation repair—lost and found: The role of local long term specialized care. Int. J. Cardiol. 104, 176–183.

Delot, E.C., Bahanoude, M.E., Zhao, M., Lyons, K.M. 2003. BMP signaling is required for septation of the outflow tract of the mammalian heart. Development 130, 209–220.

Disse, S., Abergel, E., Berrebi, A., Houot, A.M., Le Heuzey, J.Y., Diebold, B., Guize, L., Carpentier, A., Corvol, P., Jeunemaitre, X. 1999. Mapping of a first locus for autosomal dominant myxomatous mitral valve prolapse to chromosome 16p11.2-p12.1. Am. J. Hum. Genet. 65, 1242–1251.

Driscoll, D.A. 2006. Molecular and genetic aspects of DiGeorge and velocardiofacial syndromes. Methods Mol. Med. 126, 43–55.

Ebstein, W. 1968. A rare case of insufficiency of the tricuspid valve caused by a severe malformation of the same {translated by Schiebler GS, et al.}. Am. J. Cardiol. 22, 867.

Eldadah, Z.A., Hamosh, A., Biery, N.J., Montgomery, R.A., Duke, M., Elkins, R., Dietz, H.C. 2001. Familial tetralogy of Fallot caused by mutations in the jagged1 gene. Hum. Mol. Genet. 10, 163–169.

Engelfriet, P., Tijssen, J., Kaemmerer, H., Gatzoulis, M.A., Boersma, E., Oechslin, E., Thaulow, E., Popelove, J., Moons, P., Miejboom, F., Daliento, L., Hirsch, R., et al. 2006. Adherence to guidelines in the clinical care for adults with congenital heart disease: The Euro heart survey on adult congenital heart disease. Eur. Heart J. 27(6), 737–745.

Epstein, M.L. 2001. Tricuspid atresia. In: *Moss and Adams' Heart Disease in Infants, Children, and Adolescents* (H.D. Allen, H.P. Gutgesell, E.B. Clark, D.J. Druscoll, Eds.), 6th Edn., Philadelphia, PA: Lippincott, Williams Publishers, Chapter 37, pp. 799–809.

Fallot, A. 1888. Contribution a l'anatomie pathologique de la maladie bleue (cyanose cardiaque). Mars. Med. 25, 418.

Fawzy, M.E., Mercer, E.N., Dunn, B. 1988. Late results of pulmonary balloon valvuloplasty using double balloon technique. Int. J. Cardiol. 1, 35.

Fenoglio, J.J., McAlister, H.A., Jr., deCastro, M.C., et al. 1977. Congenital bicuspid aortic valve after age 20. Am. J. Cardiol. 39, 164.

Fenton, K.N., Pigula, F.A., Gandhi, S.K., Russo, L., Duncan, K.F. 2004. Interim mortality in pulmonary atresia with intact ventricular septum. Ann. Thorac. Surg. 78, 1994–1998.

Firulli, B.A., Hadzic, D.B., McDaid, J.R., Firulli, A.B. 2000. The basic helix-loop-helix transcription factors dHAND and eHAND exhibit dimerization characteristics that suggest complex regulation of function. J. Biol. Chem. 275, 33567–33573.

Fischer, A., Klattiq, J., Kreitz, B., Diez, H., Maier, M., Holtmann, B., Englert, C., Gessler, M. 2005. Hey basic helix-loop-helix transcription factors are repressors of GATA4 and GATA6 can restrict expression of the GATA target gene ANF in fetal hearts. Mol. Cell Biol. 25, 860–870.

Franco, D., Campione, M., Kelly, R., Zammit, P.S., Buckingham, M., Lamers, W.H., Moorman, A.F. 2000. Multiple transcriptional domains, with distinct left and right components in the atrial chambers of the developing heart. Circ. Res. 87, 984–991.

Freed, L.A., Benjamin, E.J., Levy, D., Larson, M.G., Evans, J.C., Fuller, D.L., Lehman, B., Levine, R.A. 2002. Mitral valve prolapse in the general population: The benign nature of echocardiographic features in the Framingham Heart Study. J. Am. Coll. Cardiol. 40, 1298–1304.

Freed, L.A., Acierno, J.S., Jr., Dai, D., Leyne, M., Marshall, J.E., Nesta, F., Levine, R.A., Slaugenhaupt, S.A. 2003. A locus for autosomal dominant mitral valve prolapse on chromosome 11p15.4. Am. J. Hum. Genet. 72, 1551–1559.

Garg, V. 2006. Molecular genetics of aortic valve disease. Curr. Opin. Cardiol. 21, 180–184.

Garg, V., Muth, A.N., Ransom, J.F., Schluterman, M.K., Barnes, R., King, I.N., Grossfeld, P.D., Srivastava, D. 2005. Mutations in NOTCH1 cause aortic valve disease. Nature 437, 270–274.

Garty, Y., Veldtman, G., Lee, K., Benson, L. 2005. Late outcomes after pulmonary valve balloon dilatation in neonates, infants and children. J. Invasive Cardiol. 17, 323–325.

Gaussin, V., Morley, G.E., Cox, L., Zurijsen, A., Vance, K.M., Emile, L., Tian, Y., Liu, J., Hong, C., Myers, D., Conway, S.J., Depre, C., et al. 2005. Alk3/Bmpr1a receptor is required for development of the atrioventricular canal into valves and annulus fibrosus. Circ. Res. 95, 219–226.

Gelb, B.D., Tartaglia, M. 2006. Noonan Syndrome and related disorders: Dysregulated RAS-mitogen activated protein kinase signal transduction. Hum. Mol. Genet. 15(Spec No2), 220–226.

Geron, M., Hirsch, M., Borman, J., Appelbaum, A. 1972. Isolated tricuspid valvular stenosis: The pathology and merits of surgical treatment. J. Thorac. Cardiovasc. Surg. 63, 760.

Gessler, M., Knobeloch, K.P., Helisch, A., Amann, K., Schumacher, N., Rohde, E., Fischer, A., Leimeister, C. 2002. Mouse gridlock: No aortic coarctation or deficiency, but fatal cardiac defects in Hey2−/− mice. Curr. Biol. 12, 1601–1604.

Gibbs, J.L. 2000. Interventional catheterization. Opening up I: The ventricular outflow tracts and great arteries. Heart 83, 111–115.

Gibson, G.A. 1898. *Diseases of the Heart and Aorta,* London: Young J Pentland.

Girod, D., Raghib, G., Wang, Y., Amplatz, K. 1965. Angiographic characteristics of persistent atrioventricular canal. Radiology 85, 442.

Giuliani, E.R., Futer, V., Brandenburg, R.O., Mair, D.D. 1979. Ebstein's anomaly: The clinical features and natural history of the tricuspid valve. Mayo Clin. Proc. 54, 163.

Gleason, T.G. 2005. Heritable disorders predisposing to aortic dissection. Semin. Thorac. Cardiovasc. Surg. 17, 274–281.

Godman, M.J., Roberts, N.K., Izukawa, T. 1974. Late post operative conduction disturbances after repair of ventricular septal defect in tetralogy of Fallot. Circulation 49, 214.

Goldmuntz, E., Clark, B.J., Mitchell, L.E., Jawad, A.F., Cuneo, B.F., Reed, L., McDonald-McGinn, D., Chien, P., Feuer, J., Zackai, E.H., Emanuel, B.S., Driscoll, D.A. 1998. Frequency of 22q11 deletions in patients with conotruncal defects. J. Am. Coll. Cardiol. 32, 492–498.

Goldmuntz, E., Bamford, R., Karkera, J.D., dela Cruz, J., Roessler, E., Muencke, M. 2002. CFC1 mutations in patients with transposition of the great arteries and double outlet right ventricle. Am. J. Hum. Genet. 70, 776–780.

Goldmuntz, E. 2005. DiGeorge syndrome-new insights. Clin. Perinatol. 32, 463–478.

Gussenhoven, W.J., Spitaels, S.E.C., Bom, N., Becker, A.E. 1980. Echocardiographic criteria of Ebstein's anomaly of tricuspid valve. Br. Heart J. 43, 31.

Hager, A., Zrenner, B., Brodherr-Heberlein, S., Steinbauer-Rosenthal, I., Schreieck, J., Hess, J. 2005. Congenital and surgically acquired Wolff-Parkinson-White syndrome in patients with tricuspid atresia. J. Thorac. Cardiovasc. Surg. 130, 48–53.

Hamblet, N.S., Lijam, N., Ruiz-Lozano, P., Wang, J., Luo, Z., Mei, L., Chien, K.R., Sussman, D.J., Wynshaw-Boris, A. 2002. Dishevelled2 is essential for cardiac outflow tract development, somite segmentation and neural tube closure. Development 129, 5827–5838.

Hausdorf, G., Schneider, M., Franzbach, B., et al. 1995. Transcatheter closure of large atrial septal defects with the Babic system. Cathet. Cardiovasc. Diagn. 36, 232–240.

Hepnar, A.D., Ahmadi-Kashari, M., Movahed, M.R. 2007. The prevalence of mitral valve prolapse in patients undergoing echocardiography for clinical reason. Int. J. Cardiol. [Epub ahead of print].

Hirayama-Yamada, K., Kamisago, M., Akimoto, K., Aotsuka, H., Nakamura, Y., Tomita, H., Furutani, M., Imamura, S., Takao, A., Nakazawa, M., Matsuoka, R. 2005. Phenotypes with GATA4 or NKX2.5 mutations in familial atrioventricular defects. Am. J. Med. Genet. A 135, 47–52.

Ho, V.B., Corse, W.R., Hood, M.N., Rowedder, A.M. 2003. MRA of the thoracic vessels. Semin. Ultrasound CT MR 24, 192–216.

Hopkins, R.A., Bert, A.A., Buchholz, B., Guarino, K., Meyers, M. 2004. Surgical patch closure of atrial septal defects. Ann. Thorac. Surg. 77, 2144–2149.

Hornung, T.S., Derrick, G.P., Deanfield, J.E., Redington, A.N. 2002. Transposition complexes in the adult. A changing perspective. Cardiol. Clin. 20, 405–420.

Horvath, K.A., Burke, R.P., Collins, J.J., Jr., Cohn, L.H. 1992. Surgical treatment of adult atrial septal defect: Early and long-term results. J. Am. Coll. Cardiol. 20, 1156–1159.

Huang, T. 2002. Current advances in Holt-Oram syndrome. Curr. Opin. Pediatr. 14, 691–695.

Hutson, M.R., Kirby, M.L. 2007. Model systems for the study of heart development and disease. Cardiac neural crest and conotruncal malformation. Semin. Cell. Dev. Biol. 18, 101–110.

Ikeda, Y., Hiroi, Y., Hosoda, T., Utsunomiya, T., Matsuo, S., Ito, T., Inoue, J., Sumiyosh, T., Takano, H., Nagai, R., Komuro, I. 2002. Novel point mutation in the cardiac transcription factor CSX/NKX2.5 associated with congenital heart disease. Circ. J. 66, 561–563.

Ing, F.F., Grifka, R.G., Nihill, M.R., Mullins, C.E. 1995. Repeat dilation of intravascular stents in congenital heart defects. Circulation 92, 893–897.

Jefferson, K., Simon, R., Somerville, J. 1972. Systemic arterial supply to the lungs and pulmonary atresia and its relation to pulmonary artery development. Br. Heart J. 34, 418.

Johnson, A.M. 1971. Aortic stenosis, sudden death and left ventricular baroreceptors [editorial]. Br. Heart J. 33, 1.

Jonkaitiene, R., Benetis, R., Ablonskyte-Dudoniene, R., Jurkevicius, R. 2005. Mitral valve prolapse: Diagnosis, treatment and natural course. Medicina 41, 325–334.

Kaartinen, V., Dudas, M., Nagy, A., Sridurongrit, S., Lu, M.M., Epstein, J.A. 2004. Cardiac outflow tract defects in mice lacking ALK2 in neural crest cells. Development 131, 3481–3490.

Kallen, B., Mastroiacovo, P., Robert, E. 1996. Major congenital malformations in down syndrome. Am. J. Med. Genet. 65, 160–166.

Kamath, B.M., Spinner, N.B., Emerick, K.M., Chudley, A.E., Booth, C., Piccoli, D.A., Krantz, I. D. 2004. Vascular anomalies in Alagille syndrome: A significant cause of morbidity and mortality. Circulation 109, 1354–1358.

Kan, J.S., White, R.I., Mitchell, S.E., Gardner, T.J. 1982. Percutaneous balloon valvuloplasty: A new method for pulmonary valve stenosis. N. Engl. J. Med. 307, 540.

Kastor, J.A., Goldreier, B.N., Josephine, M.E., et al. 1975. Electrophysiologic characteristics of Ebstein's anomaly of the tricuspid valve. Circulation 52, 987.

Kay, J.D., O'Laughlin, M.P., Ito, K., Wang, A., Bashore, T.M., Harrison, J.K. 2004. Five-year clinical echocardiographic evaluation of the Das Angel Wings atrial septal occluder. Am. Heart J. 147, 361–368.

Keane, J.F., Driscoll, D.J., Gersony, W.M., et al. 1993. Second natural history study of congenital heart defects: Results of treatment of patients with aortic valvar stenosis. Circulation 87(Suppl. II), 1–16.

Keefe, J.F., Wolk, M.J., Levine, H.J. 1970. Isolated tricuspid valvular stenosis. Am. J. Cardiol. 25, 252.

Kern, M.J., Bach, R.G. 1998. Hemodynamic rounds series II: Pulmonic balloon valvuloplasty. Cathet. Cardiovasc. Diagn. 44, 227–234.

Kim, Y.M., Yoo, S.J., Choi, J.Y., Kim, S.H., Bae, E.J., Lee, Y.T. 1999. Natural course of supravalvar aortic stenosis and peripheral pulmonary arterial stenosis in Williams' syndrome. Cardiol. Young 9, 37–41.

King, T.D., Mills, N.L. 1974. Nonoperative closure of atrial septal defects. Surgery 3, 383–388.

King, T.D., Mills, N.L. 1976. Secundum atrial septal defects: Nonoperative closure during cardiac catheterization. JAMA 235, 2506–2509.

Knauth, A.L., Lock, J.E., Perry, S.B., McElhinney, D.B., Gauvreau, K., Landzberg, M.J., Rome, J.J., Hellenbrand, W.E., Ruiz, C.E., Jenkins, K.J. 2004. Transcatheter device closure of congenital and postoperative residual ventricular septal defects. Circulation 110, 501–507.

Krabill, K.A., Wang, Y., Einzig, S., Moller, J.H. 1985. Rest and exercise hemodynamics in pulmonary stenosis: Comparison of children and adults. Am. J. Cardiol. 56, 360.

Krantz, I.D., Smith, R., Colliton, R.P., Tinkel, H., Zackai, E.H., Piccoli, D.A., Goldmuntz, E., Spinner, N.B. 1999. Jagged1 mutations in patients ascertained with isolated congenital heart defects. Am. J. Med. Genet. 84, 56–60.

Kyndt, F., Gueffet, J.P., Probst, V., Jaafar, P., Legendre, A., Le Bouffant, F., Toquet, C., Roy, E., McGregor, L., Lynch, S.A., Newbury-Ecob, R., Tran, V., et al. 2007. Mutations in the gene encoding filamin A as a cause for familial cardiac valvular dystrophy. Circulation 115, 40–49.

Lacro, R.V., Lyons-Jones, K., Benirschki, K. 1988. Coarctation of the aorta in Turner's syndrome: A pathologic study of fetuses with nuchal cystic hygromas, hydros fetalis and female genitalia. Pediatrics 81, 445.

Latson, L.A., Prioto, L.R. 2001. Pulmonary stenosis. In: *Moss and Adams' Heart Disease in Infants, Children, and Adolescents* (H. D. Allen, H. P. Gutgesell, E. B. Clark, D. J. Druscoll, Eds.), 6th Edn., Philadelphia, PA: Lippincott, Williams Publishers, Chapter 39, pp. 820–844.

Leachman, R.D., Cokkinos, D.V., Cooley, D.A. 1976. Associatoin of ostium secundum atrial septal defects with mitral valve prolapse. Am. J. Cardiol. 38, 167.

Levine, R.A., Slaugenhaupt, S.A. 2007. Molecular genetics of mitral valve prolapse. Curr. Opin. Cardiol. 22, 171–175.

Lewin, M.B., Otto, C.M. 2005. The bicuspid aortic valve: Adverse outcomes from infancy to old age. Circulation 111, 832–834.

Lewin, M.B., McBride, K.L., Pignatelli, R., Fernbach, S., Combes, A., Menessas, A., Lamt, W., Bezold, L.I., Kaplan, N., Towbin, J.A., Belmont, J.W. 2004. Echocardiographic evaluation of asymptomatic parental and sibling cardiovascular anomalies with congenital left ventricular outflow tract lesions. Pediatrics 114, 691–696.

Li, L., Krantz, I.D., Deng, Y., Genin, A., Banta, A.B., Collins, C.C., Qi, M., Trask, B.J., Kuo, W. L., Cochran, J., Costa, T., Pierpont, M.E., et al. 1997a. Alagille syndrome is caused by mutations in the human Jagged1 gene, which encodes a ligand for Notch1. Nat. Genet. 16, 243–251.

Li, Q.Y., Newbury-Ecob, R.A., Terrett, J.A., Wilson, D.I., Curtis, A.R., Yi, C.H., Gebuhr, T., Bullen, P.J., Robson, S.C., Strachan, T., Bonnet, D., Lyonnet, S., et al. 1997b. Holt-Oram syndrome is caused by mutations in TBX5, a member of the Brachyury(T) gene family. Nat. Genet. 15, 21–29.

Liao, P.K., Su, W.J., Hung, J.S. 1988. Doppler echocardiographic flow characteristics of isolated patent ductus arteriosus: Better delineation by Doppler color-flow mapping. J. Am. Coll. Cardiol. 12, 1285.

Liberthson, R.R., Pennington, D.G., Jacobs, M.L., Daggett, W.M. 1979. Coarctation of aorta: Review of 234 patients and clarification of management problems. Am. J. Cardiol. 43, 835.

Lillehei, C.W., Cohen, M., Warden, H.E., et al. 1955. The results of direct vision closure of ventricular septal defects in 8 patients by means of controlled cross circulation. Surg. Gynecol. Obstet. 101, 446.

Lima, J.A., Desai, M.Y. 2004. Cardiovascular magnetic resonance imaging: Current and emerging applications. J. Am. Coll. Cardiol. 44, 1164–1171.

Lin, A.E., Birch, P.H., Korf, B.R., Tenconi, R., Nimura, M., Poyhonen, M., Aemfield Uhas, K., Sigorini, M., Virdis, R., Romano, C., Bonioli, E., Wolkenstein, P., et al. 2000. Cardiovascular malformations and other cardiovascular abnormalities in neurofibromatosis 1. Am. J. Med. Genet. 9, 108–117.

Lock, J.E., Cockerham, J.T., Keane, J.F., et al. 1987. Transcatheter umbrella closure of congenital heart defects. Circulation 75, 593–599.

Lutembacher, R. 1916. De la stenose mitrale avec communication interauriculaire. Arch. Mal. Coeur 9, 237.

Mahoney, L.T. 1993. Acyanotic congenital heart disease. A trial and ventricular septal defects, atrioventricular canal, patent ductus arteriosus, pulmonic stenosis. Cardiol. Clin. 11, 603–616.

Mair, D.D., Edwards, W.D., Julsrud, P.R., Seward, J.B., Danielson, G.K. 2001. Truncas arteriosus. In: Moss and Adams' Heart Disease in Infants, Children, and Adolescents. (H. D. Allen, H. P. Gutgesell, E. B. Clark, D. J. Druscoll, Eds.), 6th Edn., Philadelphia, PA: Lippincott, Williams Publishers, Chapter 44, pp. 910–923.

Mani, A., Radhajrishnan, C., Farhi, A., Carew, K.S., Warnes, C.A., Nelson-Wiliams, C., Day, R.W., Prober, B., State, M.W., Lifton, R.P. 2005. Syndromic patent ductus arteriosus: Evidence for haploinsufficient TFAP2B mutations and identification of a linked sleep disorder. Proc. Natl. Acad. Sci. USA 102, 2975–2979.

Marino, B., Digilio, M.C., Toscano, A., Giannotti, A., Dallapiccola, B. 1999. Congenital heart disease in children with Noonan syndrome: An expanded cardiac spectrum with high prevalence of atrioventricular canal. J. Pediatr. 135, 703–706.

Mattila, S., Merikallio, E., Talia, T. 1979. ASD in patients over 40 years of age. Scand. J. Thorac. Cardiovasc. Surg. 13, 21.

McBride, K.L., Marengo, L., Canfield, M., Langlois, P., Fixler, D., Belmont, J.W. 2005a. Epidemiology of noncomplex left ventricular outflow tract obstruction malformations (aortic valve stenosis, coarctation of the aorta, hypoplastic left heart syndrome) in Texas, 1999–2001. Birth Defects Res. A Clin. Mol. Teratol. 73, 555–561.

McBride, K.L., Pignatelli, R., Lewin, M.B., Ho, T., Fernbach, S., Menessas, A., Lam, W., Leal, S.M., Kaplan, N., Schliekelman, P., Towbin, J.A., Belmont, J.W. 2005b. Inheritance analysis of congenital left ventricular outflow tract obstructive malformations: Segregation, multiplex relative risk, and heritability. Am. J. Med. Genet. A. 134, 180–186.

McGaughey, M.D., Trail, T.A., Brinker, J.A. 1986. Partial left anomalous pulmonary venous return: A diagnostic dilemma. Cathet. Cardiovasc. Diagn. 12, 110–115.

McGrath, L.B., Chen, C., Gu, J., Bianchi, J., Levett, J.M. 1991. Determination of infundibular innervation in end amine receptor content in cyanotic and acyanotic myocardium: Relation to clinical events in tetralogy of Fallot. Pediatr. Cardiol. 12, 155.

Mehta, A.V., Goenka, S., Chidambaram, B., Hamati, F. 2000. Natural history of isolated ventricular septal defect in the first five years of life. Tenn. Med. 93, 136–138.

Miller, M.S., Rao, P.N., Dudovitz, R.N., Falk, R.E. 2005. Ebstein's anomaly and duplication of the distal arm of chromosome 15: Report of two patients. Am. J. Med. Genet. A 139, 141–145.

Mohamed, S.A., Aherrahrou, Z., Liptau, H., Erasmi, A.W., Hagemann, C., Wrobel, S., Borzym, K., Schunkert, H., Sievers, H.H., Erdmann, J. 2006. Novel missense mutations (p.T596M and p.P1797H) in NOTCH1 in patients with bicuspid aortic valve. Biochem. Biophys. Res. Commun. 345, 1460–1465.

Moodie, D.S. 2005. VSD closure device in the setting of adult congenital heart disease. Catheter. Cardiovasc. Interv. 64, 213.

Morava, E., Zeevaert, R., Korsch, E., Huijben, K., Wopereis, S., Matthijs, G., Keymolen, K., Lefeber, D.J., De Meirleir, L., Wevers, R.A. 2007. A common mutation in the COG7 gene with a consistent phenotype including microcephaly, adducted thumbs, growth retardation, VSD and episodes of hyperthermia. Eur. J. Hum. Genet. [Epub ahead of print].

Muncke, N., Jung, C., Rudiger, H., Ulmer, H., Roeth, R., Hubert, A., Goldmuntz, E., Driscoll, D., Goodship, J., Schon, K., Rappold, G. 2003. Missense mutations and gene interruptions in PROSIT240, a novel TRAP240-like gene, in patients with congenital heart defects (TGA). Circulation 108, 2843–2850.

Muncke, N., Niesler, B., Roeth, R., Schon, K., Rudiger, H.J., Goldmuntz, E., Goodship, J., Rappold, G. 2005. Mutational analysis of the PITX2 coding region revealed no common cause for transposition of the great arteries (DTGA). BMC Med. Genet. 6, 20.

Murphy, J.G., Gersh, B.J., Mair, D.D., Fuster, V., McGoon, M.D., Ilstrup, D.M., McGoon, D.C., Kirklin, J.W., Danielson, G.K. 1993. Long-term outcome in patients undergoing surgical repair of tetralogy of Fallot. N. Engl. J. Med. 329, 593–599.

Nagvi, T.Z. 2004. Recent advances in echocardiography. Expert Rev. Cardiovasc. Ther. 2, 89–96.

Nakagawa, M., Kato, H., Aotani, H., Kondo, M. 1999. Ebstein's anomaly associated with trisomy 9p. Clin. Genet. 55, 383–385.

Neches, W.H., Park, S.C., Ettedgui, J.A. 1998. Tetralogy of falot and tetralogy of Fallot with pulmonary atresia. In: *The Science and Practice of Pediatric Cardiology* (A. J. Garson, J. T. Bricker, D. J. Fisher, S. R. Neish, Eds.), 2nd Edn., Philadelphia, PA: Williams & Wilkins Publishers, Chapter 60, pp. 1383–1412.

Nemer, G., Fadlalah, F., Usta, J., Nemer, M. 2006. A novel mutation in the GATA4 gene in patients with tetralogy of Fallot. Hum. Mutat. 27, 293–294.

Nesta, F., Leyne, M., Yosefy, C., Simpson, C., Dai, D., Marshall, J.E., Hung, J., Slaugenhaupt, S.A., Levine, R.A. 2005. A new locus for autosomal dominant mitral valve prolapse on chromosome 13: Clinical insights from genetic studies. Circulation 112, 2022–2030.

Newfeld, H.N., Lester, R.G., Adams, P., Jr., et al. 1962. Aorticopulmonary septal defect. Am. J. Cardiol. 9, 12.

Niessen, K., Karsan, A. 2007. Notch signaling in the developing cardiovascular system. Am. J. Physiol. Cell. Physiol 293, C1–C11.

Noonan, J., O'Connor, W. 1996. Noonan syndrome: A clinical description emphasizing the cardiac findings. Acta Paediatr. Jpn. 38, 76–83.

Nora, J.J., Nora, A.H. 1978. *Genetics and Counseling in Cardiovascular Diseases,* Springfield, IL: Charles C. Thomas.

O'Laughlin, M.P., Bricker, J.T., Mullins, C.E., et al. 1993. Transcatheter closure of residual atrial septal defect following cardiac transplantation. Cathet. Cardiovasc. Diagn. 28, 162–163.

Olson, E.N. 2006. Gene regulatory networks in the evolution and development of the heart. Science 313, 1922–1927.

Onat, T., Ahunbay, G., Batmaz, G., Celebi, A. 1998. The natural course of isolated ventricular septal defect during adolescence. Pediatr. Cardiol. 19, 230–234.

Patel, M.M., Overy, B.C., Kazonis, N.C., Hadley-Folkes, L.L. 1987. Long-term survival in tricuspid atresia. J. Am. Coll. Cardiol. 9, 338.

Patrianakos, A.P., Parthenakis, F.I., Chrysostomakis, S.I., Vardas, P.E. 2005. Ventricular special defect in the elderly: An uncommon clinical entity. Hellenic J. Cardiol. 46, 158–160.

Pavcnik, D., Wright, K.C., Wallace, S. 1993. Monodisk: Device for percutaneous transcatheter closure of cardiac septal defects. Cardiovasc. Intervent. Radiol. 16, 308–312.

Pawelec-Wojtalik, M., Wojtalik, M., Mrowczynski, W., Surmacz, R. 2005. Closure of perimembranous ventricular septal defect using transcatheter technique versus surgical repair. Kardiol. Pol. 63, 595–602.

Paylor, R., Lindsay, E.A. 2006. Mouse models of 22q11 deletion syndrome. Biol. Psychiatry 59, 1172–1179.

Peckham, G.B., Keith, J.D., Evans, J.R. 1964. Congenital aortic stenosis: Some observations in the natural history and clinical assessment. Can. Med. Assoc. J. 91, 639.

Pedra, C.A., Pedra, S.F., Esteves, C.A., Chamie, F., Ramos, S., Pontes, S.C., Jr., Tress, J.C., Braga, S.L., Latson, L.A., Fontes, V.F. 2003. Initial experience in Brazil with the helex septal occluder for percutaneous occlusion of atrial septal defects. Arq. Bras. Cardiol. 81, 435–452.

Pedra, C.A., Fontes, V.F., Esteves, C.A., Pilla, C.B., Braga, S.L., Pedra, S.R., Santana, M.V., Silva, M.A., Almeida, T., Sousa, J.E. 2005. Stenting vs. balloon angioplasty for discrete unoperated coarctation of the aorta in adolescents and adults. Catheter. Cardiovasc. Interv. 64, 495–506.

Pellikka, P.A., Sarano, M.E., Nishimura, R.A., Malouf, J.F., Bailey, K.R., Scott, C.G., Barnes, M.E., Tajik, A.J. 2005. Outcome of 622 adults with asymptomatic, hemodynami-cally significant aortic stenosis during prolonged follow-up. Circulation 111, 3290–3295.

Peppine, C.J., Gessner, I.H., Feldman, R.L. 1982. Percutaneous balloon valvuloplasty for a pulmonic valve stenosis in the adult. Am. J. Cardiol. 50, 1442.

Perry, S.B., van der Velde, M.E., Bridges, N.D., et al. 1993. Transcatheter closure of atrial and ventricular septal defects. Herz 18, 135–142.

Pierpont, M.E., Markwald, R.R., Lin, A.E. 2000. Genetic aspects of atrioventricular septal defects. Am. J. Med. Genet. 97, 289–296.

Pizzuti, A., Sarkozy, A., Newton, A.L., Conti, E., Flex, E., Digilio, M.C., Amati, F., Gianni, D., Tandoi, C., Marino, B., Crossley, A., Dallapiccola, B. 2003. Mutations in ZFPM2/FOG2 gene in sporadic cases of tetralogy of Fallot. Hum. Mut. 22, 372–377.

Pokorski, R.J. 2000. Long-term survival after repair of tetralogy of Fallot. J. Insur. Med. 32, 89–92.

Purcell, I.F., Brecker, S.J., Ward, D.E. 2004. Closure of defects of the atrial septum in adults using the amplatzer device: 100 consecutive patients in a single center. Clin. Cardiol. 27, 509–513.

Qu, J.Z. 2004. Congenital heart diseases with right-to-left shunts. Int. Anesthesiol. Clin. 42, 59–72.

Raas-Rothschild, A., Shteyer, E., Lerer, I., Nir, A., Granot, E., Rein, A.J. 2002. Jagged1 gene mutation in abdominal coarctation of the aorta in Alagille syndrome. Am. J. Med. Genet. 112, 75–78.

Ransom, J., Srivastava, D. 2007. The genetics of cardiac birth defects. Semin. Cell Dev. Biol. 18, 132–139.

Rao, P.S., Sideris, E.B., Chopra, P.S. 1991. Catheter closure of atrial septal defects: Successful use in a 3.6kg infant. Am. Heart J. 121, 1826–1829.

Rao, P.S., Wilson, A.D., Chopra, P.S. 1992. Transcatheter closure of atrial septal defect by "buttoned" devices Am. J. Cardiol. 69, 1056–1061.

Rashkind, W.J. 1985. Interventional cardiac catheterization in congenital heart disease. Int. J. Cardiol. 7, 1–11.

Reamon-Buettner, S.M., Borlak, J. 2004. TBX5 mutations in non-Holt-Oram (HOS) mal-formed hearts. Hum. Mutat. 24, 104.

Reamon-Buettner, S.M., Hecker, H., Spanel-Borowski, K., Craatz, S., Kuenzel, E., Borlak, J. 2004. Novel NKX2-5 mutations in diseased heart tissues of patients with cardiac malformations. Am. J. Pathol. 164, 2117–2125.

Reamon-Buettner, S.M., Borlak, J. 2006. HEY2 mutations in malformed hearts. Hum. Mutat. 27, 118.

Reddy, V.M., Hanley, F. 1998. Late results of repair of truncus arteriosus. Semin. Thorac. Cardiovasc. Surg. Pediatr. Card. Surg. Snnu. 1, 139–146.

Reisenstein, G., Levin, S., Gross, R. 1947. Coarctation of the aorta: Review of 104 autopsy cases of the "adult type" two years of age or older. Am. Heart J. 33, 146.

Restivo, A., Piacentini, G., Placidi, S., Saffirio, C., Marino, B. 2006. Cardiac outflow tract: A review of some embryogenetic aspects of the conotruncal region of the heart. Anat. Rec. A Discov. Mol. Cell. Evol. Biol. 288, 936–943.

Rigby, M.L., Chan, K.Y. 1991. The diagnostic evaluation of patients with complete transposi-tion. Cardiol. Young 1, 26.

Roberts, W.C. 1970. The congenitally bicuspid aortic valve: A study of 85 autopsy cases. Am. J. Cardiol. 26, 72.

Roberts, W.C. 1973. Vavular subvalvular and supravalvular aortic stenosis: Morphologic features. Cardiovasc. Clin. 5, 97.

Robinson, S.W., Morris, C.D., Goldmuntz, E., Reller, M.D., Jones, M.A., Steiner, R.D., Maslen, C.L. 2003. Missense mutations in CRELD1 are associated with cardiac atrioventricular septal defects. Am. J. Hum. Genet. 72, 1047–1052.

Rome, J.J., Keane, J.F., Perry, S.B., Spevak, P.J., Lock, J.E. 1990. Double umbrella closure of atrial septal defects: Initial clinical applications. Circulation 82, 751–758.

Roos-Hesselink, J.W., Meijboom, F.J., Spitaels, S.E., Van Domburg, R., Van Rijen, E.H., Utens, E.M., Bogers, A.J., Simoons, M.L. 2004. Outcome of patients after surgical closure of ventricular septal defect at young age: Longitudinal follow-up of 22–34 years. Eur. Heart J. 25, 1057–1062.

Roos-Hesselink, J.W., Meijboom, F.J., Spitaels, S.E., Vandomburg, R.T., Vanrijen, E.H., Utens, E.M., Bogers, A.J., Simoons, M.L. 2006. Long-term outcome after surgery for pulmonary stenosis (a longitudinal study of 22–33 years). Eur. Heart J. 27, 482–488.

Ruiz, C.E., Gamra, H., Mahrer, P., et al. 1992. Percutaneous closure of a secundum atrial septal defect and double balloon valvotomies of a severe mitral and aortic valve stenosis in a patient with Lutembacher's syndrome and severe pulmonary hypertension. Cathet. Cardiovasc. Diag. 25, 309–312.

Sandler, T.L., Klinkner, D.B., Tomita-Mitchell, A., Mitchell, M.E. 2006. Molecular and cellular basis of congenital heart disease. Pediatr. Clin. NA 53, 989–1009.

Sarkozy, A., Conti, E., D'Agostino, R., Digilio, M.C., Formigari, R., Picchio, F., Marino, B., Pizzuti, A., Dallapiccola, B. 2005. ZFPM2/FOG2 and HEY2 gene analysis in nonsyndromic tricuspid atresia. Am. J. Med. Genet. A 133, 68–70.

Satoda, M., Zhao, F., Diaz, G.A., Burn, J., Goodship, J., Davidson, H.R., Pierpont, M.E., Gelb, B.D. 2000. Mutations in TFAP2B cause Char syndrome, a familial form of patent ductus arteriosus. Nat. Genet. 25, 42–46.

Schluterman, M.K., Krysiak, A.E., Kathiriya, I.S., Abate, N., Chandalia, M., Srivastava, D., Garg, V. 2007. Screening and biochemical analysis of GATA4 sequence variations identified in patients with congenital heart disease. Am. J. Med. Genet. A 143, 817–823.

Schrader, R. 2003. Catheter closure of secundum ASD using "other" devices J. Interv. Cardiol. 16, 409–412.

Sharafudin, M.J., Gu, X., Titus, J.L., Amplatz, K. 1996. Secundum-ASD closure with a new self-expanding prosthesis in swine. Circulaiton 94, 1–57.

Shirodaria, C.C., Gwilt, D.J., Gatzoulis, M.A. 2005. Joint outpatient clinics for the adult with congenital heart disease at the distinct general hospital: An alternative model of care. Int. J. Cardiol. 103, 47–50.

Sideris, E.B., Sideris, S.E., Thampoulos, B.D., et al. 1990. Transvenous atrial septal defect occlusion by the buttoned device. Am. J. Cardiol. 66, 1524–1526.

Sievert, H., Babic, U.U., Ensslen, R., et al. 1990. Transcatheter closure of atrial septal defects. Lancet 336, 566–567.

Sievert, H., Dirks, J., Rux, S., et al. 1997. ASD and PFO closure in adults with the second generation ASDOS device. J. Am. Coll. Cardiol. 29, 143A.

Sommer, R.J., Rhodes, J.F., Parness, I.A. 2000. Physiology of critical pulmonary valve obstruction in the neonate. Catheter. Cardiovasc. Interv. 50, 473–479.

Spinner, N.B., Colliton, R.P., Crosnier, C., Krantz, I.D., Hadchouel, M., Meunier-Rotival, M. 2001. Jagged1 mutations in alagille syndrome. Hum. Mut. 17, 18–33.

Srivastava, D. 2006. Making or breaking the heart: From lineage determination to morphogenesis. Cell 16, 1037–1048.

Stamm, C., Friehs, I., Ho, S.Y., Moran, A.M., Jonas, R.A., del Nido, P.J. 2001. Congenital supravalvar aortic stenosis: A simple lesion? Eur. J. Cardiothorac. Surg. 19, 195–202.

Steele, P.M., Foster, V., Cohen, M., et al. 1987. Isolated atrioseptal defect with pulmonary vascular obstructive disease: Long-term follow-up and prediction of outcome after surgical correction. Circulation 76, 1037.

Stewart, J.R., Schaff, H.V., Fortunin, N.J., Brawley, R.K. 1983. Partial anomalous pulmonary venous return with intact atrial septum: Report of four cases. Thorax 38, 859–862.

Sutton, M.G.S., TAjaik, A., McGoon, D.C. 1982. Atrial septal defects in patients ages 60 years or older. Circulation 64, 402.

Svensson, E.C., Huggins, G.S., Lin, H., Clendenin, C., Jiang, F., Tufts, R., Dardik, F.B., Leiden, J.M. 2000. A syndrome of tricuspid atresia in mice with a targeted mutation of the gene encoding Fog-2. Nat. Genet. 25, 353–356.

Sweeney, E., Fryer, A., Walters, M. 2000. Char syndrome: A new family and review of the literature emphasizing the presence of symphalangism and the variable phenotype. Clin. Dysmorphol. 9, 177–182.

Tartaglia, M., Mehler, E.L., Goldberg, R., Zampino, G., Brunner, H.G., Kremer, H., van der Burgt, I., Crosby, A.H., Ion, A., Jeffery, S., Kalidas, K., Patton, M.A., et al. 2001. Mutations in PTPN11, encoding the protein tyrosine phosphatase SHP-2, causes Noonan syndrome. Nat. Genet. 29, 465–468.

Tartaglia, M., Kalidas, K., Shaw, A., Song, X., Musat, D.L., van der Burgt, I., Crosby, A.H., Ion, A., Kerler, R.S., Jeffery, S., Patton, M.A., Gelb, B.D. 2002. PTPN11 mutations in Noonan syndrome: Molecular spectrum, genotype-phenotype correlations, and phenotypic heterogeneity. Am. J. Hum. Genet. 70, 1555–1563.

Tassabehji, M. 2003. Williams-Beuren syndrome: A challenge for genotype-phenotype correlation. Hum. Mol. Genet. 12(Spec No2), R229–R237.

Towbin, J.A. 1999. Towards an understanding of mitral valve prolapse. Am. J. Hum. Genet. 65, 1238–1241.

Towbin, J.A., McQuinn, T.C. 1995. Gridlock: A model for coarctation of the aorta? Nat. Med. 1, 1141–1142.

Tynan, M., Finley, J.T., Fontes, V., et al. 1990. Balloon angioplasty for the treatment of native coarctation results of valvuloplasty and angioplasty of congenital anomalies registry. Am. J. Cardiol. 65, 790.

van Praagh, R., Geva, T., Kreutzer J. 1989. Ventricular septal defects: How should we describe, name and classify them? J. Am. Coll. Cardiol. 14, 1298.

Vaughan, C.J., Basson, C.T. 2000. Molecular determinants of atrial septal and ventricular septal defects and patent ductus arteriosus. Am. J. Med. Genet. 97, 304–309.

Vick, G.W., III. 1998. Defects of the atrial septum including atrioventricular septal defects. In: The Science and Practice of Pediatric Cardiology (A. Garson, Jr., J. T. Bricker, D. J. Fisher, S. R. Neish, Eds.), 2nd Edn., Philadelphia, PA: Williams & Wilkins Publishers, Chapter 51, pp. 1141–1180.

Vitelli, F., Morishima, M., Taddei, I., Lindsay, E.A., Baldini, A. 2002. Tbx1 mutation causes multiple cardiovascular defects and disrupts neural crest and cranial nerve migratory pathways. Hum. Mol. Genet. 11, 915–922.

Volkl, T.M., Degenhardt, K., Koch, A., Simm, D., Dorr, H.G., Singer, H. 2005. Cardiovascular anomalies in children and young adults with Ullrich-Turner syndrome the Erlangen experience. Clin. Cardiol. 28, 88–92.

Waldman, J.D., Wernly, J.A. 1999. Cyanotic congenital heart disease with decreased pulmonary blood flown in children. Pediatr. Clin. North. Am. 46, 385–404.

Waller, B.F., Howard, J., Fess, S. 1995. Pathology of pulmonic valve stenosis and pure regurgitation. Clin. Cardiol. 18, 45–50.

Wang, J., Nagy, A., Larsson, J., Dudas, M., Sucov, H.M., Kaartinen, V. 2006. Defective ALK5 signaling in the neural crest leads to increased postmigratory neural crest cell apoptosis and severe outflow tract defects. BMC Dev. Biol. 6, 51.

Ward, K.E., Mullins, C.E. 1998. Anomalous pulmonary venous connections, pulmonary vein stenosis, and atresia of the common pulmonary vein. In: *The Science and Practice of Pediatric Cardiology* (A. Garson, Jr., J. T. Bricker, D. J. Fisher, S. R. Neish, Eds.), Philadelphia, PA: Williams & Wilkins Publishers, Chapter 63, pp. 1431–1462.

Weinstein, B.M., Semple, D.L., Driever, W., Fishman, M.C. 1995. Gridlock, a localized heritable vascular patterning defect in zebrafish. Nat. Med. 1, 1143–1147.

Wernovsky, G. 2001. Transposition of the great arteries. In: *Moss and Adams' Heart Disease in Infants, Children, and Adolescents* (H. D. Allen, H. P. Gutgesell, E. B. Clark, D. J. Driscoll, Eds.), 6th Edn., Philadelphia, PA: Lippincott, Williams & Wilkins Publishers, Chapter 51, pp. 1027–1084.

Yang, H., Lee, C.L., Young, D.C., Shortfiffe, M., Yu, W., Wright, J.R. 2004. A rare case of interstitial del(1)(p34.4p36.11) diagnosed prenatally. Fetal Pediatr. Pathol. 23, 251–255.

Zagol, B., Book, S., Krasuski, R.A. 2006. Late "adult form" scimitar syndrome presenting with "infant form" complications J. Invasive. Cardiol. 18, E82–E85.

Zalstein, E., Koran, G., Einerson, T., Freedom, R.N. 1990. A case-controlled study on the association of first trimester exposure to lithium in Ebstein's anomaly. Am. J. Cardiol. 65, 817.

Zamora, R., Lax, D., Donnerstein, R.L., Lloyd, T.R. 1995. Transcatheter closure of residual atrial septal defect following implantation of buttoned device. Cathet. Cardiovasc. Diagn. 36, 242–246.

Zhao, F., Weismann, C.G., Satoda, M., Pierpont, M.E., Sweeney, E., Thompson, E.M., Gelb, B. D. 2001. Novel TFAP2B mutations that cause Char syndrome provide a genotype-phenotype correlation. Am. J. Hum. Genet. 69, 695–703.

Zweier, C., Sticht, H., Aydin-Yaylagul, I., Campbell, C.E., Rauch, A. 2007. Human TBX1 missense mutations cause gain-of-function resulting in the same phenotype as 22q11 deletions. Am. J. Hum. Genet. 80, 510–517.

INDEX

A

abdominal A (abdA), 10
abdominal B (abdB), 10
Aberrant ventral morphogenesis, 369
Ace mutants, 193
Actin, 161
Actin cytoskeleton, reorganization of, 346
Action potential durations (APDs), 250
Action potential, repolarization of, 53
Activin, 127, 129, 130, 139, 192
Activin receptor, type II, 402
Activin/TGFβ signaling, in *Xenopus*, 134
ActRIA, 135
Acute myeloid leukemia, 271
    AML-1, 271, 282, 284
    AML-1-ETO, 282
Acute promyelocytic leukemia (APL), 285
"Adventitial layer", 302
Aka DiGeorge syndrome, 370
AKT, 285, 347
   signaling, 313
Alagille's syndrome, 390, 404
ALK2, 106, 135, 344
ALK3, 135, 159
ALK4, 127, 129
ALK5, 37
AML. *See* Acute myeloid leukemia
Amplatzer device, for occlusion of ASD, 396
ANF. *See* Atrial natriuretic factor
Angioblasts, 305, 307, 318
Angiogenesis, 272
    ETS factors in, 308
    excessive, in trunk vasculature, 324
Angiopoietin-2 (ANG-2) expression, 352
Annulus fibrosus. *See* Cardiac skeleton
Anterior heart field, 154
Anterior visceral endoderm (AVE), 123
Antisense oligonucleotides,
    O-methyl-modified, 173
Antp (Antennapedia), 10, 276, 278, 284
Antp-gal4, 276
Aorta, 3, 10
    coarctation of, 384–385, 391–393

Aorta-gonad-mesonephros (AGM), 260, 266
Aortic stenosis, 384–385
Armadillo/β-catenin, 16
Arrhythmias, 48, 52, 85. *See also* Ventricular
    arrhythmias
    cardiac, hereditary, 232
    cardiac repolarization causing, 19
    increased risk for, 214
    increasing, 20
    inducible/lethal, 250
Arterial-venous fate determination,
    molecular pathway for, 310
Arterial-venous patterning, 322
Arteriogenesis, defective, 326
ASDOS device, for occlusion of ASD, 397
ATF-2, 136, 138
Atresia, 384
Atrial bradycardia, 251
Atrial fibrillation, 58, 214, 391
Atrial natriuretic factor, 97–98, 137, 235, 243
Atrial septal defects (ASDs), 77, 368, 383,
    391, 394, 396
    ostium secundum type, 394–396
    percutaneous devices for occlusion of, 396
Atrial septation, Tbx5 in, 77
Atrial tachyarrhythmia, 249
Atrial tachycardia, 214
Atrial ventricular septal defect (AVSDs), 368
Atrioventricular (AV) block, 58,
    241, 245, 251
Atrioventricular (AV) boundary, 97–101
Atrioventricular (AV) cushions,
    225, 231, 239
Atrioventricular (AV) groove, 221
Atrioventricular (AV) junction, 209, 221,
    231–232
    chick Msx2 mRNA in, 247
    maturation, 225
    Tbx3 expression in, 244
Atrioventricular (AV) myocardium,
    embryonic, 225
Atrioventricular (AV) rings, 224
    chick Msx2 mRNA in, 247
    Tbx5 expression in, 243

Printed and bound by CPI Group (UK) Ltd, Croydon, CR0 4YY

08/05/2025

01864966-0004